AR

ANNUAL REVIEW OF ENERGY AND THE ENVIRONMENT

ANNUAL REVIEW OF ENERGY AND THE ENVIRONMENT

VOLUME 20, 1995

ROBERT H. SOCOLOW, *Editor*
Princeton University

DENNIS ANDERSON, *Associate Editor*
The World Bank

JOHN HARTE, *Associate Editor*
University of California at Berkeley

ANNUAL REVIEWS INC. 4139 EL CAMINO WAY P.O. BOX 10139 PALO ALTO, CALIFORNIA 94303-0139

ANNUAL REVIEWS INC.
Palo Alto, California, USA

International Standard Serial Number: 1056–3466
International Standard Book Number: 0–8243–2320-3

♾ The paper used in this publication meets the minimum requirements of American National Standard for Information Sciences—Permanence of Paper for Printed Library Materials, ANSI Z39.48-1984.

Typesetting by Kachina Typesetting Inc., Tempe, Arizona; John Olson, President; Toni Starr, Typesetting Coordinator; and by the Annual Reviews Inc. Editorial Staff

PRINTED AND BOUND IN THE UNITED STATES OF AMERICA

PREFACE

Do natural scientists have foresight? Most of us concede that doctors do. We refrain from certain behaviors and participate in others, accepting the validity of the doctors' science and their prescriptions. We know the doctors' foresight is imperfect, but we have no interest in shooting the messengers when they warn us of the risk factors in ordinary behavior, or when they recommend powerful medications.

As the Washington scene today makes clear, environmental scientists do not yet have the public's trust to the same degree. Like the doctors, environmental scientists are less than sure of much of their science much of the time. But, unlike the doctors, they are not given the benefit of the doubt.

There are two reasons why the message of the environmental scientists is hard to convey. First, the message is unwelcome, and many would rather wish it away. It is like the Copernian message 500 years ago. We would rather be at the center of the universe. We would rather our planet were so large that we could never foul our nest, no matter what we did on the planet—no matter how exuberantly we consumed natural resources, for example.

Second, the message is misheard, because the audience confuses an argument from necessity with an argument from preference. Arguments from necessity apply to all people; examples are the statement that wetlands reduce downstream flooding, and the statement that wetlands are beneficial to waterfowl. Arguments from preference apply only to those who share common values, such as the argument that an agricultural region is more congenial when farmland is interspersed with unmanaged areas, or the argument that a watercourse is more desirable when plentiful waterfowl are available for hunting. The arguments of an environmental scientist are arguments from necessity. The arguments of an environmentalist are arguments from preference. When the environmental message is presented by environmentalists, it is perceived as special pleading, and the messengers are perceived as just another interest group at the trough. Many environmental scientists are environmentalists, which compounds the problem of communicating to a disbelieving public.

The democratic political system is adept at limiting the incursion into mainstream life of a new set of preferences expressed by particular citizens. Similarly, a public policy process dominated by economics easily accommodates gradual shifts in aggregate preferences. But both the democratic political system and the field of economics are strained by new arguments from necessity. Economics insists that a new idea be widely accepted before, for

(*continued*)

example, the idea is allowed to drive the allocation of public funds. A democracy whose public policy is driven by economics will typically regard those who claim to have foresight as zealots whose influence should be circumscribed, rather than as thoughtful people whose proposals should be tested.

A race is on to earn the public's trust in environmental science, and in the prescriptions proposed for modifying technology and behavior in response to environmental science. Time is short. A public willing to heed neither the message nor the prescriptions will produce large and unprecedented disturbances of the Earth's natural systems.

Annual Review of Energy and the Environment is dedicated to helping win that race. We seek to review the best work, to highlight discrepancies that must be resolved and uncertainties that must be reduced, and to give an early glimpse of ideas that seem to be growing in importance. We pay particular attention to new compilations of primary data, which are usually the result of heroic efforts; several articles in this volume are examples. And we emphasize peer review. A large number of the articles in this volume were substantially overhauled, made clearer, more complete, and less tendentious, in response to the careful readings of anonymous reviewers and members of our Editorial Committee.

The autobiographies that open this volume depict two long careers dedicated to environmental science and its implications. Thomas Malone is one of the pioneers of climate research and an architect of many of the institutions that have made climatology an effective international enterprise. Chauncey Starr, an early and effective advocate of efficiency in the use of energy, has continually added coherence and breadth of vision to the amorphous professional community that designs and manages the electricity system.

Three of the fourteen review articles in this volume address the climate problem specifically. Jozef Pacyna and Thomas Graedel discuss the development of emissions inventories, which are crucial inputs to climate models, proximate targets of all regulatory regimes, and grist for the designers of new technologies. Donald Wuebbles reviews the complex task of simplifying the structure of environmental science in order to produce environmental indices that permit more subtle environmental policy; in particular, through "ozone depleting potentials" and "global warming potentials," the impacts of individual gases can be compared and traded off. Daniel Bodansky describes the new processes and institutions that are coming into being in the wake of the 1992 United Nations Conference on Environment and Development in Rio de Janeiro.

Four articles present aspects of environmental policy analysis, all in an international context. William Baumol explores innovation in environmental technology from the viewpoints of trade theory and technical progress; he argues that environmental industries are good candidates for public support, when they have high start-up costs, scale economies, and opportunities for learning-by-doing, and when their production costs fall rapidly with increased volume. Both the scoping article by Iddo Wernick and Jesse Ausubel on materials flows and the article by Valerie Thomas on lead in gasoline introduce the reader to industrial ecology, and, in particular, to its agenda of documenting and interpreting the flows of materials in the world economy. Albert Mödl and Ferdinand Hermann review the many national programs already established that assign environmental labels to consumer products, with the aim of guiding purchases toward goods with above-average environmental performance (as measured, for example, by the quietness of a lawnmower or the recyclability of packaging).

Four articles on primary energy supply, two articles on energy demand, and an article and a Forum on electricity policy constitute the rest of this volume. Each of the three major primary energy supply routes—the fossil fuels, renewable energy, and nuclear energy—gets attention: Joyce Dargay and Dermot Gately, presenting extensive econometric analysis of the ups and downs of oil demand in response to oil price, document a sticky system, where reductions in demand for oil in response to a rise in price are largely not undone by a subsequent fall in price. Bent Sørensen reviews the extraordinary progress in wind technology and assesses the environmental impacts of deployment. The travails of the nuclear industry are explored both in the article by Jack Barkenbus and Charles Forsberg, which reviews international institutions designed to improve reactor safety, and in the article by James Flynn and Paul Slovic on the futile struggle to gain public acceptance for the disposal of high-level radioactive waste at Yucca mountain in Nevada.

The first of the pair of articles discussing energy demand is by Toufiq Siddiqi, who disaggregates (by country, fuel, and end use) the demand for energy in Asia—the part of the world where energy demand is growing most rapidly at this time. The second is by Lee Schipper, who describes the recent history of energy demand for transportation, with a focus on the industrialized countries. In another sense, Schipper's article complements the article by Dargay and Gately: Their conclusions about the strength of the demand for oil for transportation are supported by Schipper's specific documentation of the steady increase in the use of vehicles for all purposes.

Recent changes in the structure of institutions to manage electricity are assessed by Robert Bacon, with an emphasis on recent changes in Europe in the direction of less government ownership. Bacon's article is a foil to the first-ever Forum of this series, "Energy Efficiency and the Economists." In this Forum, organized by Associate Editor Dennis Anderson, several policy analysts (Charles Cicchetti; Ralph Cavanagh; Paul Joskow; Mark Levine, Jonathan Koomey, James McMahon, Alan Sanstad, and Eric Hirst; and Albert Nichols) join Anderson in debating the merits and pitfalls of electricity industry subsidies for efficient end-use devices and systems (such as motors, lighting, and thermal insulation). Their contributions were refined in a second round, after receiving each other's papers. Anderson adds a concluding essay to the Forum as well.

I judge the Forum to have been a success and would like to try others in future volumes. As always, the views of our readers about what we are trying to do and how well we are doing it would be most welcome.

ROBERT H. SOCOLOW, EDITOR
PRINCETON, NJ, JULY 1995

Annual Review of Energy and the Environment
Volume 20, 1995

CONTENTS

(*continued*)

SOME RELATED ARTICLES IN OTHER *ANNUAL REVIEWS*

From the *Annual Review of Anthropology,* Volume 24 (1995)

Consumption and Commodities, D. Miller

From the *Annual Review of Earth and Planetary Sciences,* Volume 23 (1995)

Tectonic, Environmental, and Human Aspects of Weathering and Erosion: A Global Review from a Steady-State Perspective, R. F. Stallard

From the *Annual Review of Ecology and Systematics,* Volume 26 (1995)

Antarctic Terrestrial Ecosystem Response to Global Change, A. D. Kennedy

Can We Sustain the Biological Basis of Agriculture?, C. Hoffman and R. Carroll

Ecological Basis for Sustainable Development in Tropical Forests, G. S. Hartshorn

Environmental Growth vs Sustainable Societies: Reflections on the Players in a Deadly Contest, J. G. Clark

Environmental Sustainability: Universal and Rigorous, R. Goodland

Human Ecology and Resource Sustainability: The Role of Institutional Diversity, C. D. Becker and E. Ostrom

Sustainability of Soil Use, S. Buol

Sustainability, Efficiency, and God: Economic Values and the Sustainability Debate, R. Nelson

Sustainable Exploitation of Renewable Resources, R. Hilborn, C. J. Walters, D. Ludwig

The Role of Nitrogen in the Response of Forest Net Primary Production to Elevated Atmospheric Carbon Dioxide, A. D. McGuire, J. M. Mellilo, and L. A. Joyce

From the *Annual Review of Fluid Mechanics,* Volume 27 (1995)

Thermohaline Ocean Processes and Models, J. A. Whitehead

Biological-Physical Interactions in the Upper Ocean: The Role of Vertical and Small-Scale Transport Processes, K. L. Denman and A. E. Gargett

From the *Annual Review of Public Health,* Volume 16 (1995)

The Importance of Human Exposure Information: A Need For Exposure-Related Data Bases to Protect and Promote Public Health, D. K. Wagener, S. G. Selevan, and K. Sexton

ANNUAL REVIEWS INC. is a nonprofit scientific publisher established to promote the advancement of the sciences. Beginning in 1932 with the *Annual Review of Biochemistry,* the Company has pursued as its principal function the publication of high-quality, reasonably priced *Annual Review* volumes. The volumes are organized by Editors and Editorial Committees who invite qualified authors to contribute critical articles reviewing significant developments within each major discipline. The Editor-in-Chief invites those interested in serving as future Editorial Committee members to communicate directly with him. Annual Reviews Inc. is administered by a Board of Directors, whose members serve without compensation.

Thomas F. Malone

Annu. Rev. Energy Environ. 1995. 20:1–29

REFLECTIONS ON THE HUMAN PROSPECT

Thomas F. Malone

The Sigma Xi Center, Research Triangle Park, North Carolina, 27709, and North Carolina State University, Raleigh, North Carolina, 27695-8208

KEY WORDS: environment, energy, technology, cascade of knowledge, sustainable human development

---

CONTENTS

ABSTRACT

World population and the global economy are expanding in a manner that is propelling civilization along a path that is unsustainable, inequitable, and unstable. A concerted, global effort to discover, integrate, disseminate, and apply knowledge about the natural world and human behavior would change this trajectory to a path of sustainable human development. This path would point toward the vision of a society in which the basic human needs and an equitable share of life's amenities could be met by successive generations while maintaining in perpetuity a healthy, physically attractive, and biologi-

1056-3466/95/1022-0001$05.00

cally productive environment. The scholarly community is urged to provide impetus for the pursuit of this vision. An unprecedented degree of collaboration among the disciplines will be necessary. New modes of communication and cooperation among the major sectors of society will have to be fashioned. Knowledge will become an organizing principle for society in the twenty-first century.

SEMINAL INFLUENCES

Roots

My thoughts on the human prospect are deeply rooted in youthful experiences on the windswept plains of Haakon County, in western South Dakota, during the 1920s and 1930s. In the 1920s, Haakon County was in the throes of dramatic change driven by energy-powered technology, the pursuit of economic development, and the social aspirations of a restless group of pioneering homesteaders. These forces were all being brought to bear on a fragile ecosystem.

Great tracts of prairie, covered with wheat grass, were being plowed. Crops of wheat were planted and regularly rotated with corn. Memories persist of the spectacular, wind-driven "amber waves of grain" that replaced equally magnificent waves of tall prairie grass. Cattle ranching was giving way to farming. Horse-drawn, single-bottom plows were being replaced by multiple-bottom plows pulled by kerosene-powered, Titan tractors. These steel-wheeled, two-cylinder behemoths had a proclivity for mechanical failure because of primitive metallurgy.

McCormick-Deering grain binders, drawn by four horses, were used in the 1920s to harvest the wheat. The stalks holding the grain were cut down, tied in bundles, and mechanically ejected from the machine every few hundred yards. The bundles were then gathered into shocks, collected in horse-drawn hayracks, and hauled to threshing machines where the grain was separated from the straw and loaded into wagons. Other horse teams pulled the wagons to the granary, a very slow method of transport. At the granary, the grain was shoveled by hand into bins to await ferrying, again by horses, to railroad boxcars a day's journey away.

As the 1920s came to a close with the great crash of 1929, combines, powered by Continental engines and pulled by Farmall tractors, short-circuited this cumbersome, backbreaking process. The combines cut the wheat, separated the grain from the straw, and deposited a golden torrent of grain into a tank on the combine, from which the grain was emptied periodically into Six-Speed-Special International Harvester trucks. The trucks delivered their loads to the farm, or directly to a railroad town 40 miles distant, for shipping to flour mills in Minneapolis.

As a teenage combine operator in the 1930s, I observed firsthand the dramatic leap in economic productivity created by energy-powered technology. During those halcyon days, the population of Haakon County was growing at an annual rate of one and a half to two percent. A half-century elapsed before I realized that this economic and demographic development was the result of global forces that are still central to the human prospect as the world crosses the threshold into the third millennium.

Nature's lessons are sometimes traumatic. An epidemic of anthrax ravaged cattle herds in the early 1930s. Each carcass had to be burned to prevent spread of the fatal disease within the herd and to humans. (Not until many years later did I connect this incident to the looming global threat of virulent infection as a potential factor in the human prospect.) In the middle 1930s, the rains stopped coming and the grasshoppers arrived in force. They consumed the stunted crops struggling to survive the drought. Howling winds lifted topsoil and created great drifts of dust which buried the weed-clogged, woven-wire fences separating the barren fields. The impact of the depression was also severe. The price of wheat plunged from $1.20 a bushel in 1929 to $.20 in 1932.

What I saw of nature's revenge for our tampering with a fragile ecosystem, such as the profound impact of erratic weather patterns on human existence, strongly motivated me to learn about the link between the environment and the human condition. The particulars of my circumstances led me to focus on the vagaries of weather and climate.

In Pursuit of Knowledge

My father was the eldest son of Tom Malone, a revered patriarch of the Irish community in Sioux City, Iowa, and the proprietor of one of the city's two "fancy" grocery stores. Dad had a passion for ranching and farming and was one of the South Dakota homesteaders in the early 1900s. He had two avocations: telephones and weather. He established the first barbed-wire telephone system in our community, and he had an uncanny ability to anticipate weather changes with little more than an aneroid barometer and an experienced eye (no radio or TV weather channel in those days!). My brother chose a career as an engineer in the telephone industry. I elected meteorology.

In the mind of my wise and determined mother, an education was indispensable to acquiring the knowledge that links humans to their environment and that drives human progress. Unfortunately, my high school education was interrupted from 1932 to 1934 by the hard times endemic to farm life. During those years, I spent summers working ten-hour days riding a Farmall tractor, which cultivated four rows of corn at a time. This high-technology activity was followed in the fall by backbreaking hand labor: husking corn, stalk by stalk, throwing each ear into a horse-drawn wagon. The important role of technology in easing the conversion of natural resources into goods and services was strongly impressed on me.

I resumed high school in Philip (the county seat and nearest railroad town) in 1934. Debating the issues of federal aid to education and socialized medicine under the auspices of the National Forensic League sparked my interest in the national affairs increasingly impacting Haakon County. This early experience in public speaking and thinking on my feet proved very useful in later years, both in the classroom and on the speaker's platform.

As high school graduation approached in 1936, I journeyed to Rapid City in the nearby Black Hills for career guidance from the meteorologist in charge of the local weather bureau, Harley N. Johnson. He gave succinct and sage advice: "First, get a good grounding in math and physics at the South Dakota State School of Mines here in Rapid City. Then pursue meteorological studies."

I followed his directive. Four years later, I was a graduate student with a fellowship in meteorology at MIT. However, my studies were interrupted again when the Air Force and Navy called on me to help train hundreds of weather forecasters for World War II. Then, in early 1945, restless for a more direct role in the war effort, I enthusiastically accepted an offer to serve as a civilian consultant forecasting the weather for the impending invasion of Japan.

The countless hours of studying weather maps for eastern Asia in preparation for this assignment turned out to be a waste of time. Upon my arrival in April at Air Weather Service Headquarters in Asheville, North Carolina, I learned that the need for an air supply route across northern Africa called for the creation of an upper-air forecasting center at John Payne Field in Cairo. I was directed to establish that center in collaboration with Major Guy Gosewich of the 19th Weather Squadron. Before the task was complete, the bomb was dropped, the war ended, and I returned to MIT, where I completed the doctoral program in 1946. I then remained at the Institute as an assistant professor of meteorology to teach and do research.

An opportunity for the equivalent of postdoctoral reflection and maturation came with an invitation in 1949 to edit the *Compendium of Meteorology* (1). This 1300-page book collected the views of more than a hundred authors from around the world on the state of scientific knowledge about weather and climate. Promising avenues for future research were outlined. The second half of the century was destined to be an era of explosive growth in our understanding of the physics, chemistry, and biology of the atmosphere. No better postdoc experience could be imagined. These years deepened my interest in the application of meteorology to the management of human affairs, particularly the impact of weather on business and industry.

After publication of the *Compendium,* I pursued an opportunity to deal quantitatively with the isobaric patterns on weather maps. During World War

II, Professor George Wadsworth and Dr. Joe Bryan of MIT used Tschebyshev polynomials to convert these patterns into numbers. This approach made it possible to relate the isobaric pattern on any given day to subsequent patterns and weather elements, such as temperature and rainfall (2).

The predictions involved inverting thirtieth-order matrices. Bob Miller and I used Marchant desk calculators for this procedure until we discovered the incredible computing power of MIT's ground-breaking Whirlwind computer, which was being developed in the former Polaroid Building, literally in MIT's backyard. This computer reduced the time needed to invert these matrices from weeks to minutes! The improvement in information handling had implications for economic productivity just as dramatic as those of technology cum energy years earlier in South Dakota.

New Horizons

In 1954, I was invited to establish a research center and weather service at The Travelers Insurance Company in Hartford, Connecticut. The activities of the company (then known as the Tiffany of the insurance industry) reached into every facet of business and industry. Hurricanes, floods, earthquakes, and the stability of agriculture were of particular concern to those who handled the company's property and casualty lines of insurance.

At this time, I was making speeches about the potential mutual benefits offered by a more intimate interaction between the meteorological community and the private sector. The challenge of participating in such an interaction, rather than talking about it, prompted me to leave a tenured appointment at MIT. My place at the Institute was taken by an esteemed colleague and close friend, Dr. Edward Lorenz, whom I had met when he was an Air Force cadet in one of my classes. Ed's place in history is assured because of his pioneering work in chaos.

Bob Miller and Don Friedman, my associates at MIT, soon joined me in Hartford to form the core of our research staff. We assembled a superb group of professional meteorologists skilled in radio and television communications. In invited talks around the state, I described the economic advantages of the use in decision making of probabilities from weather forecasts, because of intrinsic instabilities and uncertainties in weather processes. The classic example was a savings of 25% over a period of 182 days earned by protecting newly poured concrete from rainfall. The probability of rain was used to decide whether to protect at a known cost or to risk suffering loss because of no protection (3). The reaction to these presentations was so enthusiastic that in 1956 we introduced probability statements into our radio, television, and newspaper weather predictions. The public response was positive and the practice became widespread.

Our research group joined forces with the research department of United

Aircraft in 1958 to work on a governmental initiative to explore modernization of the system for weather observations and forecasting. I was able to attract Dr. Robert White from MIT to lead this effort. We created the independent, nonprofit, Travelers Research Center (TRC) in 1961 to pursue this task and other studies. I became chairman of the board of directors, and White was named president. Several decades later, he was elected president of the National Academy of Engineering.

The TRC group performed the weather research and forecasting started at The Travelers and expanded its research with their continuing support. When Bob White left the TRC to become Chief of the US Weather Bureau in 1963, Dr. Douglas L Brooks, another former student of mine, became president (4). In 1970, in a prescient move, he transformed the TRC into the Center for Environment and Man (CEM). (The era of inclusive language had not yet fully dawned.) Doug's initiative put humans right at the center of environmental issues, where we belong, and none too soon. I continued as chairman of the board.

Studies of how to modernize weather observation and forecasting tailed off, awaiting more sophisticated advances in understanding of atmospheric processes, and observation and communication technologies. These developments came by the 1980s, following a major expansion in research and development in the 1960s and 1970s, triggered by the landmark report of the Committee on Meteorology of the National Academy of Sciences and the initiation of the Global Atmospheric Research Program. The lesson learned was the critical importance of an adequate body of knowledge, broadly construed and properly framed.

With the untimely death of The Travelers' young president Sterling Tooker in 1969, interest in a broadly based research program waned, and CEM was transferred to the University of Connecticut (UCONN) in Storrs in 1970. I had left a senior vice presidency and a seat on the seven-man Operations and Policy Group to become Dean of the Graduate School at UCONN.

In retrospect, I see the 15 years I spent at The Travelers as a period of personal growth and transformation. In 1956, the company created a full-fledged research department and named me Director. I became involved in economic forecasting under the tutelage of Dr. Eli Shapiro, Associate Dean of MIT's Sloan School. Tooker and I brought Shapiro in to assist in the executive development stimulated by our research department. I predicted premium growth for the company from my economic forecasts and moderated the Annual Business Outlook for the Greater Hartford business community. Market research, insurance risk for nuclear power plants, and operations research came under my purview. I persuaded the company to introduce long-range planning and participated actively in that endeavor. In concert with Hartford's famed Institute of Living, we developed a prototype of a computerized information system for hospitals, although the project was abandoned after Tooker's death. I was asked to reorganize the Insurance Institute for Highway Safety and, in 1967, was appointed by President Johnson to chair the National Motor Vehicle Safety Advisory Council. In that position, I met United Auto Workers President Walter Reuther and motor-industry critic Ralph Nader.

Professional and scientific interests also occupied my time during this period. These opportunities included presidency of the American Meteorological Society (1960–1961) and the American Geophysical Union (1961–1964), chairmanship of the US National Commission for UNESCO (1965–1967), and membership on the Advisory Panel on Science and Technology for the Committee on Science and Technology, US House of Representatives (1960–1970). I also sat on the Committee on Meteorology of the National Academy of Sciences (1956–1960) and chaired its successor, the Committee on Atmospheric Sciences (1962–1968).

The American Geophysical Union represents the US scientific community in the International Union of Geodesy and Geophysics (IUGG). IUGG is one of the 20 scientific organizations in the International Council of Scientific Unions (ICSU). ICSU is an umbrella organization for, and is supported by, national nongovernmental scientific bodies, such as national academies of science, royal societies, and national research councils. In a sense, ICSU represents the nongovernmental, world scientific community. Its counterpart in the intergovernmental arena is the array of UN scientific bodies (e.g. UNESCO, the World Health Organization, the Food and Agriculture Organization, and the World Meteorological Organization).

Through my work with the American Geophysical Union, I was introduced to the ICSU and later became a vice president of this international scientific infrastructure (1970–71) and its treasurer (1978–84). I soon came to appreciate the potential of combining the creative and innovative character of the nongovernmental scientific community with the stability and resources of intergovernmental organizations.

For several decades following my first acquaintance with the ICSU, I was involved in international scientific affairs as Director of the Holcomb Research Institute at Butler University in Indianapolis, Indiana, (1973–1983), and through appointments during retirement (1983–1990) and postretirement at North Carolina State University and The Sigma Xi Center in Research Triangle Park (1990–1995).

EARLY MILESTONES ON AN ODYSSEY

National Center for Atmospheric Research

In the middle 1950s, Dr. Francis Reichelderfer, Chief of the US Weather Bureau, suggested to Detlev Bronk, President of the National Academy of

Sciences, that the Academy compare progress in meteorology with the advancement of related fields. The bureau wanted an assessment of the nature and scope of unsolved problems and the research and education needed for advances in the field.

Bronk had lost a valuable boat in a hurricane and responded in April 1956 by convening a Committee on Meteorology, chaired by the legendary Lloyd Berkner, President of the Associated Universities (the entity that managed the Brookhaven National Laboratory for the federal government). Geophysicist Berkner left his footprints in nearly every branch of geophysics. He was the moving spirit behind the International Geophysical Year (IGY) (1957–1958), which was initiated under the auspices of the ICSU when he was president of that organization. Carl Rossby, one of the world's most respected meteorologists, was named vice chairman of the committee. Another distinguished member was the brilliant John von Neumann, whose many contributions included a proposal for the operating architecture that dominated the design of high-speed computers for many years.

In the late 1940s and early 1950s, von Neumann, Rossby, Jule Charney, and other collaborators were engaged in predicting the weather by applying finite-difference methods to the integration of the nonlinear partial differential equations governing atmospheric motion. Rossby and von Neumann realized this approach was within reach owing to the advent of high-speed computers, and von Neumann first demonstrated its feasibility on the ENIAC. This was Lewis Richardson's 30- year-old dream realized. In a landmark paper for which he received the Sir Napier Shaw Prize of the Royal Meteorological Society in 1956, Norman Phillips, one of von Neumann's collaborators, showed that this approach could produce realistic patterns of world climate when energy from the sun was applied to a global atmosphere at rest (5).

The deaths of both Rossby and von Neumann led to the appointment of Charney as von Neumann's replacement. Rossby's long-time associate, committee member Horace Byers, was named vice chairman. Other notable members were physicists Edward Teller and Paul Klopsteg, as well as the sage and saintly hydrodynamicist Hugh Dryden, Home Secretary of the National Academy of Sciences and, later, Associate Director of NASA. Although not yet a member of the National Academy, I was appointed to the committee because of my experience with the *Compendium of Meteorology*. Berkner asked me to take the lead in drafting the committee's report. Formally entitled *Research and Education in Meteorology*, this report appeared in 1958 (6) and was known thereafter as *The Berkner Report*, probably the most significant policy statement on meteorology published in recent times.

Summarizing six in-depth meetings, we recommended a 50–100% increase in funding for basic meteorological research at universities and kindred institutions. The need to mount a broad interdisciplinary effort led to an additional recommendation for " ...a center of intellectual activity that would bring together scientists from meteorology and the related physical sciences...[with]...research facilities on a scale required to cope with the global nature of the meteorological problem." We proposed that the center be managed by a consortium of universities with financial support from the National Science Foundation (NSF). We suggested a capital investment of $50 million, spread over five years, with a subsequent annual budget of approximately $15 million.

In an era of governmental support for basic research that reflected the philosophy of "Science—The Endless Frontier" (7), this was a feasible initiative. The proposal for a center, however, was debated at length. Would it drain support from the universities or strengthen their programs? Within the committee, this issue was brought to a head at a meeting in Washington, DC in November 1957. With typical decisiveness, Berkner instructed Charney and me to reflect on this matter overnight and return with an unequivocal recommendation the next morning.

Jule and I debated the question half the night at the subsequently demolished Roger Smith hotel, where we were staying. We finally reached the conclusion—that the center should be proposed—that we reported to Berkner the next morning. He then charged us with visiting a dozen universities to consult on the matter. We did, and found a generally favorable reaction. The stage was set for a discussion with Academy President Bronk.

Our committee met over dinner with Dr. Bronk at his residence on the campus of Rockefeller University, where he was president. He responded warmly but, with characteristic wisdom, asked for a formal statement of support from the university community. The American Meteorological Society was to meet in New York, January 28–30, 1958. Berkner asked me, as secretary of the society, to ascertain the views of several university representatives and, if favorable, obtain a statement. We met at the executive offices of the Associated Universities in Manhattan. I was joined at the meeting by committee member Edward Teller, whose towering reputation and formidable demeanor provided great support. A statement of strong endorsement emerged from that meeting (8).

The universities created a University Committee on Atmospheric Research under the leadership of Henry Houghton, Chairman of the Department of Meteorology at MIT. I was invited to prepare a report with specific plans for what would become the National Center for Atmospheric Research (NCAR) in Boulder, Colorado. With Roscoe Braham of the University of Chicago and William von Arx of the Woods Hole Oceanographic Institution, I convened 17 two-day workshops at The Travelers Cliff House on Avon Mountain, just outside of Hartford, Connecticut. The report (9, 10) was submitted to the NSF in February 1959.

A lively debate about this ambitious proposal at a meeting of the National Science Board in Jackson Hole, Wyoming, later in 1959 culminated in a favorable decision. An eloquent statement by the NSF's associate director, Paul Klopsteg, was a deciding factor. Paul was familiar with the case from his participation in discussions with the Committee on Meteorology.

A final hurdle was getting approval from the congressional committee on appropriations of a line item on NCAR in the NSF budget. As a result of an earlier chance encounter on Capitol Hill with Dale Leipper of A&M College in Texas, I knew Texas Congressman Olin "Tiger" Teague. Teague was close to Congressman Albert Thomas, also from Texas and chair of the relevant appropriations committee. Fortuitously, I learned the day before the appropriations committee was to act on the budget item for NCAR that a rumor had reached Thomas that southern states were not to be considered as sites for the center because of their segregation policies. In addition, there was some feeling in lower levels of government that the center should be established within the government, not in the university community. Half a night of telephoning assured me that there was absolutely no basis for this rumor. I reached Tiger Teague with this information 20 minutes before the appropriations committee met. Teague reached Thomas in time to set the record straight and the NCAR budget item was approved.

The original roster of 14 universities in the consortium responsible for NCAR has grown to 61. NCAR has developed into a world-class institution with a scientific and support staff of nearly 1000 individuals engaged in a wide-ranging program of research and facility support. A detailed review for the National Science Board in 1992 confirmed NCAR's substantial contribution to what *The Berkner Report* described as " ...one of the most difficult, most important, most challenging—and yet relatively one of the most neglected—scientific problems of our times."

NCAR illustrates how institutional innovation can be brought about by cooperation between the federal government and universities on an inherently international and intrinsically interdisciplinary problem. This organization required new ways of thinking and doing. I am persuaded that NCAR suggests a mode of academic-government interaction that could be adapted to develop the "cascade of knowledge" essential for an attractive human prospect. The cascade of knowledge is the dynamic and nonlinear continuum linking the discovery, integration, dissemination, and application of knowledge concerning the nature and interaction of matter, energy, living organisms, information, and human behavior.

The Global Atmospheric Research Program

The first faint signal to me of what ultimately became a major international research program was a telephone call from Jule Charney in February 1961.

MIT physicist Bruno Rossi, who was assisting Presidential Science Advisor Jerome Wiesner, had asked Jule for suggestions for an initiative in international cooperation in weather research. President Kennedy was interested in proposals he could present to Mr. Kruschev during their summer meeting in Vienna. Jule proposed that we meet with Rossi and Richard Goody of Harvard.

Conversation at the meeting was desultory until we turned to the possibility of conscious or inadvertent human intervention in weather and climate. Bruno immediately detected a rationale for a joint endeavor with other nations. We agreed to explore the matter further. I dined with the Rossi family that evening and began to lay plans for a small meeting on February 21.

We met at the Bullfinch House on Beacon Hill in Boston at the offices of the American Meteorological Society, of which I was president. Charney, Goody, and I were joined by Henry Houghton from MIT, Sverre Petterssen from the University of Chicago, David Johnson and Harry Wexler from the US Weather Bureau, and Morris Tepper from NASA. We proposed "... a concerted international program aimed at the study of global weather processes with the intent of developing the scientific basis for weather prediction, thereby establishing a rational point of departure for investigating the feasibility of large-scale modification of weather and climate"... to be initiated jointly by the USA and USSR (11). Our report was communicated to Rossi.

The Vienna Conference was a minor diplomatic setback for President Kennedy. The proposal was not even presented. Instead, it was set aside as intrinsically sound, timely, and imaginative, meriting a more auspicious venue. That more favorable setting turned out to be the opening session of the Sixteenth General Assembly of the United Nations in September, at which President Kennedy was scheduled to make an address. I was called to a meeting at the Old Executive Office Building on August 31 to brief several members of the President's Scientific Advisory Committee. In addition to Jerry Wiesner, participants included Det Bronk, chair of the President's Scientific Advisory Committee's (PSAC) International Panel, Richard Gardner from the State Department, Pete Scoville from the CIA, and Arthur Schlesinger from the President's office.

The response was positive and the proposal was included in the President's address. His words at the UN were few, but far-ranging in consequence. He urged international cooperation "... in weather prediction and eventually in weather control...." These words led to UN Resolution 1721, inviting "... members and WMO [World Meteorological Organization] to study measures to advance the state of the atmospheric sciences and technology in order to improve existing weather forecasting capabilities and to further the study of the basic physical processes that affect climate."

My experience with NCAR in fostering a partnership between government and academia prompted me to urge direct involvement of the ICSU to assure a strong scientific foundation for the project. In 1962, I was serving as a volunteer aide to Assistant Secretary of Commerce J Herbert Hollomon, whose responsibilities included the US Weather Bureau. I proposed to him that a new UN Resolution in 1962 specifically invite the ICSU to participate in planning the research recommended in the 1961 UN Resolution.

In July 1962, Herb Hollomon called me in Hartford and asked me to fly to Washington immediately in The Travelers' corporate plane and meet with him and Richard Gardner from the State Department to discuss my proposal. When I arrived in Washington two hours later, the meeting had already been held. Gardner had agreed to invite ICSU. Herb described the meeting as "some Indian wrestling." Gardner and his chief, Assistant Secretary of State for International Organizations, Harland Cleveland, were towers of strength in this initiative, both within the Department of State and at the UN.

The 1962 UN Resolution 1802 recommended that WMO "develop in greater detail its plan for an expanded programme to strengthen meteorological services and research" and invited the ICSU through its unions and national academies "to develop an expanded programme of atmospheric science research which will complement the programmes fostered by the World Meteorological Organization."

The remaining task was to identify the ICSU body that could most effectively plan this program. The International Association of Meteorology and Atmospheric Physics (IAMAP), a subsidiary body of the IUGG, was anxious to undertake this task. I preferred ICSU's interdisciplinary Committee on Space Research (COSPAR). An important catalyst for this effort was the great potential offered by surveillance of the global atmosphere from space. However, exploratory discussions at COSPAR meetings in Vienna, Florence, and Mar del Plata (Argentina) were not fruitful. Even a fine dinner for the Director of the UK Meteorological Service Sir Graham Sutton, and Lady Sutton, at the Three Hussars restaurant in Vienna did not generate support for a COSPAR initiative. Sir Graham preferred to leave the matter in WMO hands.

The promise in a bilateral meeting between the United States and the USSR in Rome in February 1963 to discuss cooperation in applying space science and technology to meteorology was never fulfilled. Cochaired by Hugh Dryden (US) and Academician Blagonravov (USSR), the meeting was more important for fostering dialogue than for its substance. I participated as chair of the Academy's Committee on Atmospheric Sciences.

The possibility of a Special Committee within the ICSU (similar to the arrangement for the IGY) was attractive. However, a trip to Toronto in The Travelers' plane with Horace Byers (President of IAMAP) to enlist the support of Warren Godson, Secretary General of IAMAP, ended in disappointment. We were graciously hosted for lunch at the Godson residence, but Warren was understandably adamant that IAMAP undertake the task.

The impasse was finally resolved at the Triennial General Assembly of the IUGG in Berkeley, hosted by the American Geophysical Union (AGU), in September, 1963. As AGU president, my responsibilities were to chair the organizing committee for the Assembly and lead the US delegation. Once again, a vigorous debate focused on the issue of IAMAP vs a Special Committee. A fortuitous interpretation of an assembly resolution by IUGG Secretary General GR Laclavere led to the establishment of a Committee on Atmospheric Sciences under the joint auspices of ICSU and IUGG. The committee's charge was "to develop an expanded programme of atmospheric science research."

Our delegation at Berkeley was successful in securing the election of the US candidate for the presidency of IUGG, Professor Joseph Kaplan from UCLA. Kaplan had been a prime mover in the United States' participation in the IGY. He appointed Professor Bert Bolin of Sweden as chair of the new committee and asked me to serve as its secretary general.

IAMAP was represented by its new president, Professor AM Oboukhov (USSR) and by Godson on the ICSU/IUUG committee. Other relevant ICSU bodies were also included on the committee, which cooperated closely with a special working group in COSPAR, headed by Morris Tepper from NASA.We met three times at the WMO Secretariat in Geneva and established close connections to the Advisory Committee to the WMO initiative already under way. US Weather Bureau Chief Francis Reichelderfer ensured that agency's close cooperation as well. His successor, my colleague Robert White, keenly appreciated the value of the nongovernmental scientific community and was influential in subsequent national and international developments.

As chair of the Academy's Committee on Atmospheric Sciences (1962–1968), I arranged a set of panel studies. I presented my own views (12) in a symposium earlier in 1963 at UCLA on progress in the geophysical sciences as a result of IGY. The paper built upon and brought up to date ideas presented in my address at an inaugural dinner for NCAR in Boulder, Colorado, in December 1961 (13). The UCLA symposium was held just before the Berkeley meeting of IUGG, and served as a sounding board for my ideas. The enthusiastic endorsement I received from the influential oceanographer Roger Revelle gave me courage to proceed.

The first meeting of the ICSU/IUGG committee in Geneva, February 8–11, was notable for a brilliant lecture by Jule Charney that established the scientific rationale for a global program of meteorological observations and research. Jule pointed out that "advances in data-processing technology and physical understanding have so extended the scope and complexity of the numerical models that can be treated that it is becoming possible to deal with the circulation of the atmosphere as a whole and to attack directly the problem of long-range prediction, but such investigations are in danger of becoming mere academic exercises for lack of observations to supply the initial conditions and to check the calculations" (14).

Charney was a chronic late-riser and, after a late and long dinner the previous night at the "Silos," he had to be summoned from his sleep by my assistant from the National Academy of Sciences, Dr. John Sievers. WMO Secretary General DA Davies left his office on the second floor of the WMO building, came to the balcony above the conference hall, and listened intently to Jule's talk. Davies' positive reaction to Charney's lecture did much to smooth the way to subsequent cooperation with WMO.

A generous planning grant of $75,000 from the Ford Foundation was arranged though the good offices of foundation executive Carl Borglund who had turned to WMO's Davies for advice. This happy circumstance opened the way for our committee to obtain essential staff support and to proceed with the extended study conference we felt was needed to plan a major international program. A dinner meeting in The Travelers Suite at The Pierre in New York in January 1967 with Davies and Dr. Rolando Garcia (Argentina) led to Garcia's appointment as Executive Secretary to our committee, with joint support by the WMO and the ICSU.

The study conference was held in Stockholm, June 28–July 11, 1967, cosponsored by our committee, the WMO, and COSPAR. More than 50 participating scientists produced a 144-page report (15) for the Global Atmospheric Research Program (GARP). Garcia went to an IAMAP meeting in Lucerne, Switzerland, in September 1967, in a little Fiat weighted down with 100 copies of the report, arriving just as the meeting convened. Protocol required that the report be approved by IAMAP, IUGG, and the executive committee of ICSU, in that order, for subsequent consideration by the Executive Council of the WMO.

Bert Bolin, who had chaired the study conference, made a splendid presentation to IAMAP. Dr. John Mason, Director of the UK Meteorological Service was quite critical. The day was saved, however, when the UK's respected RC Sutcliffe, who had served on the WMO Advisory Committee, stood up and quietly remarked that whatever its shortcomings, the plan appeared to have considerable merit and he thought it should be approved. As chair of the session, I called for a vote. The approval was virtually unanimous. With this critical test passed, subsequent approvals were routine. (Mason later became a strong supporter of GARP. He spent several years chairing the Joint Scientific Committee for the Climate Research Program that followed GARP.)

My shuttle diplomacy between K Chandrasekharan, the ICSU's Secretary General in Zurich, and DA Davies, at the WMO Headquarters in Geneva, resulted in a formal agreement for cooperation between the two organizations. This agreement is still in effect, with modifications. A combined meeting of governments and scientists in Brussels in 1968 formally launched GARP under the guidance of a Joint Organizing Committee sponsored by WMO and ICSU and chaired for the first several years by Bert Bolin.

The first phase of GARP was a large-scale program of observations and research in 1974 on the critical interaction of the deep cumulus convective systems in the tropics with the mesoscale circulation patterns in the rain areas and with motions at synoptic scale. The GARP Atlantic Tropical Experiment (GATE) took place during the oil shortage. A measure of the international cooperation it engendered was the offer by Soviet participants to provide oil to ensure continued operation of US research vessels. The full-scale program known as the First GARP Global Experiment (FGGE) (called figgie by those in the know) was conducted in 1979.

The scientific results of GARP greatly improved the accuracy of weather forecasts and the extension of their time range. GARP laid the foundation for the World Climate Research Program (WCRP) into which it evolved. WCRP, in turn, is critical for the international assessment of global warming currently under way (16).

GARP illustrated the potential for international collaboration on a global issue when science, technology, and societal needs simultaneously reach a stage at which significant advances are within reach. An important element in GARP's success was the cooperation that allowed the imagination and insight of the scientific community (ICSU) to be combined with the stability and resources of national governments (WMO). GARP was a precursor of the kind of partnerships that will be increasingly important in the years ahead. A notable feature of GARP was the unrestricted sharing of data—a splendid tradition currently in some jeopardy.

ALTERING PERCEPTIONS OF ENVIRONMENTAL CHANGE

Weather Control

The Berkner Report recognized the implications of early experiments attempting to modify supercooled clouds with dry ice or silver iodide (17) and induce precipitation artificially (18). During the 1950s and 1960s, the subject of weather control generated lively discussion. Lawrence Spivak interviewed Reichelderfer and me on this subject on the TV program *Meet the Press.* A study mandated by an Act of Congress in 1953 suggested cautious optimism about the efficacy of cloud seeding (19). The Advisory Committee responsible for the study recommended that the "government sponsor meteorological research more vigorously..." (19) and urged continuity and stability for longer-term projects. The committee made the point that "the development of weather modification must rest on a foundation of fundamental knowledge that can be obtained only through scientific research into all the physical and chemical processes in the atmosphere" (19). Henry Houghton, Chairman of the Department of Meteorology at MIT and an eminent cloud physicist, was if anything more restrained (20). Although he said, "It would be unthinkable to embark on such a vast experiment before we are able to predict with some certainty what the effects would be," he also remarked, "I shudder to think of the consequences of a prior Russian discovery of a feasible method of weather control" (20).

The Academy Committee viewed the results as a significant, but not compelling, argument for deepening our understanding of weather and climate. The *Berkner Report* stated that "It would be an inexcusable distortion of presently available information...to seize upon the single issue of weather control to argue for an expansion of meteorological research...however, the question of weather control should not be excluded from the evidence...." As previously noted, President Kennedy's UN address included the words "weather control." The ICSU/IUGG Committee's first report for GARP cited this program as a major international research and development program " ... directed at observing, understanding, and predicting the general circulation of the troposphere...a prerequisite for the scientific exploration of large-scale climate modification" (21).

Greenhouse Warming

Weather control issues shifted during the next two decades from deliberate control to inadvertent climate change resulting from continuing economic and demographic growth on planet Earth. The increasing concentration of carbon dioxide in the atmosphere over Mauna Loa from measurements started by Charles Keeling during the IGY provided an important clue to the potential for this increase to influence climate by perturbing radiation processes in the atmosphere (22). In a keynote address at a conference on "Technological Changes and the Human Environment" at the California Institute of Technology on October 17, 1970, I called for "intensive study" of the "greenhouse effect" generated by the burning of fossil fuels (22a). This issue was highlighted in the first of a 20-report series issued in 1977 by the Geophysics Study Committee of the Geophysics Research Board in the National Research Council, which I established during my chairmanship of the board (1969–1975). The study committee for this report, *Energy and Climate,* was chaired by Roger Revelle of Scripps Institution of Oceanography.

Energy and Climate began with these words: "Worldwide industrial civilization may face a major decision over the next few decades—whether to continue reliance on fossil fuels as principal sources of energy or to invest the research and engineering effort, and the capital, that will make it possible to substitute other energy sources for fossil fuels within the next 50 years " (23).

In a foreword, my study committee cochair, Phil Abelson, and I wrote: "To reduce uncertainties and to assess the seriousness of the matter, a well-coordinated program of research that is profoundly interdisciplinary, and strongly international in scope will be required... [it]... should extend beyond scientific and technical considerations to include the complex factors and institutional innovations that will enable the nations of the world to act with wisdom and in concert before irreversible changes in climate are initiated" (23).

Years later, as chairman of the Board on Atmospheric Sciences and Climate of the National Research Council, I gave testimony on February 23, 1984, before a subcommittee of the House Committee on Science and Technology regarding a 500-page report by our Board on this matter (24). I remarked in my testimony that "The issue, and research directed at its illumination, will be with us for a long time... A successful response to widespread environmental change will be facilitated by the existence of an international network of scientists conversant with the issues and of a broad international consensus on facts and their reliability." Our report was brought to public attention when David Hartman interviewed me on ABC's *Good Morning America.* (The concept of a network of scientists connected to educators, decision makers, and the public did not emerge until advances in communication technology brought it within reach in the 1990s.)

A starting point for such a network had already been identified in 1968. Roger Revelle and I had successfully argued at the ICSU General Assembly in Paris that year for the establishment of a Scientific Committee on Problems of the Environment (SCOPE) to prepare a series of assessments of the increasing number of environmental problems confronting society. SCOPE was created in 1970. I served as Secretary General from 1970–1976. Over the past quarter century, SCOPE has published (Wiley & Sons) more than two dozen authoritative analyses of environmental issues (25).

In 1985, one of these reports (26) alerted the international community to the potential societal impact of the "greenhouse effect." Today, discussion of measures to stabilize world climate by controlling the emission of greenhouse gases takes place internationally.

Environmental Hazards of Nuclear War

The insidious threat to the climate of global warming differs sharply from the catastrophic impact of a major nuclear war. This threat was addressed by SCOPE in a pair of 1986 reports.

As Foreign Secretary of the National Academy of Sciences from 1978–1982, I became convinced that a scientific dialogue on nuclear weapons between members of our Academy and members of the Soviet Academy would complement official channels of communication between the superpowers. I had a comfortable relationship with academician George Skryabin, Chief Scientific

Secretary of the USSR Academy of Sciences, as we were both members of the Executive Board of the ICSU.

On February 12, 1980, National Academy President Phil Handler and I, in my role as Foreign Secretary, convened a meeting of colleagues that led to the creation of the Academy's Committee on International Security and Arms Control. Meanwhile, the USSR Academy had established a Scientific Council for Problems of Peace and Disarmament.

I wanted to begin a dialogue on nuclear weapons between the two academies, but little progress was made until the World Climate Conference in Geneva in 1979. Academician Yevgeny Federov, head of the USSR Hydrometeorological Service (and a member of the Soviet Academy's Peace Council) was stonewalling the Conference by insisting that a resolution on world peace should precede discussions of cooperation on climate issues. When told by US delegate and conference chair Robert White that the Soviet Academy was not responding to my overtures for joint discussions, Federov withdrew his objections and the Conference proceeded smoothly and effectively. When I described this incident to George Skryabin at an ICSU Board meeting shortly thereafter, he laughed and said he expected an imminent visit from Federov. Skryabin was right, and within a short time discussions between the two academies were on track (27).

The Soviet Academy was particularly cooperative in the SCOPE study on the environmental effects of a major nuclear war. One personal link between the US and USSR scientific communities was Vladimir Aleksandrov, a frequent visitor to the US. His 1984 disappearance in Spain remains a mystery (28).

During Thanksgiving week 1984, Notre Dame's president, Father Ted Hesburgh, and I convened a meeting of 30 USSR and US scientists and religious leaders to discuss the environmental consequences of a nuclear war. Scientists included Carl Sagan and Nobel Laureate Charles Townes from the United States, and USSR Academy Scientific Secretary George Skryabin and the head of the USSR Space Institute, Roald Sagdeev. Religious leaders included Bishop (now Cardinal) Roger Mahoney from the United States and Archbishop Kirill from Leningrad (29).

In 1983, Yevgeny Velihkov (Vice President of the Soviet Academy) and I moderated a satellite-based teleconference between a group in Washington and one in Moscow, triggered by some recent studies by Crutzen & Birks (30) and by Sagan and colleagues (31). I thought the international character of this potential threat required a critical and objective assessment by the world scientific community. SCOPE was the logical instrumentality.

I enlisted the distinguished UK engineer, Sir Frederick Warner, whom I met when he was treasurer of SCOPE, to lead the assessment. SCOPE's two-volume report made a powerful case that the global climatic impact of a large-scale nuclear war would be catastrophic (32). Apart from the immediate devastation of an atomic bomb (33), food production would be grievously impaired in noncombatant countries. Civilization would be in serious jeopardy. These predictions were supported by a subsequent UN study in which I participated. I would like to believe these assessments contributed to the reluctance of the superpowers to start a nuclear war.

International Geosphere-Biosphere Program

My experience with the environmental impact of a nuclear war deepened my conviction that closer interaction between physical and biological scientists was needed. Canadian George Garland had advanced this view in his lecture commemorating the twenty-fifth anniversary of the International Geophysical Year at the 1982 General Assembly of ICSU in Cambridge, England. He noted that many scientific mysteries unresolved by the IGY involve the interaction between physical processes and living organisms, including humans. In a 1982 NASA report intended to stimulate international interest on global habitability, Richard Goody and others made the same observation: "This is a unique time, when one species, humanity, has developed the ability to alter its environment on the largest (i.e., global) scale and to do so within the lifetime of a single species member " (34).

Herbert Friedman, chair of the National Research Council's Commission on Physical Sciences, Mathematics, and Resources urged a bold, holistic, interdisciplinary venture in global research to deepen understanding of global change in the terrestrial environment and its living systems. At a February 1983 meeting of the ICSU's Executive Board in Stockholm, I proposed that the ICSU mount an effort of this kind. The response was an invitation to Juan Roederer and me to convene a symposium at the General Assembly of the ICSU in Ottawa in September 1984 to consider this proposal. The symposium (35) led to the establishment in 1986 of the ICSU's International Geosphere-Biosphere Program (IGBP): A Study of Global Change (36). IGBP's goal is to deepen our understanding of the physical, chemical, and biological systems that regulate Earth's favorable environment for life and the role of human activity in changing that environment. Six core projects have now been initiated and more than 40 countries have created national committees. A companion program on the Human Dimensions of Global Change has also been mounted by the International Social Science Council (37).

In 1990, I was invited to chair an IGBP workshop at the Rockefeller Foundation's Bellagio Conference Center to revisit a concept developed at a conference in Aspen, Colorado, in the summer of 1971. The Aspen conference was convened by the International Institute for Environmental Affairs (IIEA) and the Aspen Institute for Humanistic Studies. I participated both as Chairman of the Academy's Committee on International Environmental Programs and as a member of the IIEA Board of Directors. The workshop's purpose was to recommend international organizational arrangements for the UN Conference on the Human Environment, which was held in Stockholm in 1972.

The workshop led to the conclusion that "the most compelling organizational requirement is to forge an intimate working relationship in international environmental affairs between the intergovernmental community and the world of science and technology...[through] a network of decentralized but cooperating regional centers of excellence in environmental affairs" (38). The centers were to be linked by a central body in the international scientific community that would interact with a similar body in the intergovernmental community. At one time, I entertained the thought that the privately endowed Holcomb Research Institute at Butler University might be the central body in the scientific community. In retrospect, I think the network concept is much to be preferred.

The result of the Bellagio workshop is START, which stands for a global system of regional networks for analysis, research, and training. This program is intended to mobilize scientific talent to address local and regional environmental issues in a global context. START fosters cooperation in interdisciplinary research and provides a mechanism for the dissemination and application of research results (39). Fourteen regions have been identified and six regional networks are now being established.

IGBP represents a significant step toward a growing perception of global change that transcends climatic change. Earlier steps in this direction were Osborne's popular 1948 book, *Our Plundered Planet* (40) and the scholarly 1956 work, *Man's Role in Changing the Face of the Earth*, edited by WL Thomas Jr. and published under the auspices of the Wenner-Gren Foundation (41). In 1962, Rachel Carson's *Silent Spring* (42) ignited public interest that led to a 1965 study of environmental quality (43) by the President's Science Advisory Committee, led by John Tukey of Princeton University. More recently, *The Earth as Transformed by Human Action* has deepened our perceptions of the impact of human activity on the environment (44).

Interest has been growing in the interaction between environmental quality and economic development, particularly for nations in the early stages of economic development. In 1974, SCOPE President Victor Kovda (USSR), Mohammed Kassas (Egypt), and I convened a scientific conference on the intertwined issues of environment and development in Nairobi, Kenya (45). A decade later the World Commission on Environment and Development released its landmark report, *Our Common Future* (46), which powerfully described the inextricable link between environmental quality and economic development. This document led to the June, 1992, UN Conference on Environment and Development in Rio de Janeiro. More than 100 heads of state gathered at what came to be called the Earth Summit (47). I attended as a special guest of the Summit's Secretary General, Maurice Strong.

A new concern is emerging: the uncontrolled spread of infectious airborne diseases (48). In 1994, Laurie Garrett remarked that "while the human race battles itself, fighting over more crowded turf and scarcer resources, the advantage moves to the microbe's court. They are our predators and they will be victorious if we...do not learn to live in a rational global village that affords the microbes few opportunities "(49).

Global Change and the Human Prospect

I passed other milestones in my personal odyssey in the first half of this decade during three meetings at which the deliberations of the preceding three decades converged. These meetings crystallized my ideas on the driving forces of global change (demographic and economic growth), stimulated my thinking on effective responses, and left me cautiously hopeful about the human prospect.

The first of the meetings was an international forum in Washington, DC, convened in 1991 by Sigma Xi, The Scientific Research Society. Five kindred organizations in the natural sciences, social sciences, and engineering participated in the forum entitled *Global Change and the Human Prospect: Issues in Population, Science, Technology and Equity*. In order to support and deepen discussions at the Earth Summit in Rio the following year, participants in the forum addressed three basic questions: What kind of a world do we have? What kind of a world do we want? What must we do to get there? I served as chair of the steering committee. Warmly received by the Secretary General for the Earth Summit, the proceedings stressed the importance of knowledge and of institutional renewal and innovation. The forum probed deeply into the forces underlying global change. The essential findings were captured in three sentences in my Foreword to the Proceedings: "Apocalypse is *not* impending. An attractive human prospect *is* within reach. However, in several respects the world is embarked on a trajectory that *could* lead to severe problems. The task for our generation is to change the forces determining that trajectory so that its terminus is that attractive human prospect" (50).

The second meeting was a 1993 international workshop given under the auspices of The Sigma Xi Center. The participants articulated a vision of the kind of world we want and focused on the role of knowledge networks in pursuing that vision. Two possible scenarios were prepared for the year 2050. In the first, economic and demographic growth continue at the rates prevailing during the 1980s and 1990s, respectively. The other explores the consequences of changes in those growth rates should society consciously choose to pursue a vision of the world we want. A Global Array of Nested Networks (GANN) was recommended to develop the cascade of knowledge that drives human progress (51).

The third meeting was the annual gathering of the National Association of State Universities and Land-Grant Colleges (NASULGC) held in Chicago in November 1994. I presented a White Paper entitled "Sustainable Human Development: A Paradigm for the 21st Century," commissioned by NASULGC. I challenged institutions of higher education to forge a new compact between academia and society to succeed the one stimulated 50 years ago by *Science—The Endless Frontier.* This compact would center on sustainable human development undergirded by conscious efforts to foster growth in the cascade of knowledge. The goal would be a new vision of society attainable in the twenty-first century (52). That vision is a society in which the basic human needs and an equitable share of life's amenities can be met by successive generations, while maintaining in perpetuity a healthy, physically attractive, and biologically productive environment.

The thoughts that emerged during these three meetings, in the context of the milestone experiences I've touched on here, left me with certain views on the course ahead, some alternative courses, the elements of a response, and a few convictions about leadership.

DOWN THE ROAD TO THE YEAR 2050

Two scenarios constructed by The Sigma Xi Center for 2050 from data compiled by the United Nations Development Programme (UNDP) show what the future may hold (52). In each scenario, the countries of the world are divided into three groups: (*a*) the 46 industrial countries, (*b*) the 65 countries at an intermediate stage of development, and (*c*) 62 less-developed countries.

In the first scenario, current population growth rates in the three groups of countries continue. World population increases threefold to 15 billion people. More people are added in the less-developed and impoverished countries than exist in the world today (i.e. five billion).

Present rates of growth in the individual capacity to convert natural resources into goods and services (i.e. economic productivity) in the three groups also continues in the first scenario. With associated population growth, the global production and consumption of goods and services increases eightfold, attaining a value of more than $200 trillion (Purchasing Power Parity dollars). The increase in the economies in the industrial countries (>$70 trillion) is two and a half times the total world economy today (i.e. $27 trillion). The 10% of the world's people that are in the industrial countries are producing and consuming 40% of the world's economic output. The 50% of the people that are in the less-developed countries are producing and consuming only 10% of that output.

Stresses on the carrying capacity of our planet increase. The demographic and economic gaps between affluent and impoverished countries widen. Social stresses are exacerbated worldwide. Poverty still exists in all countries, but is especially acute in the less-developed countries, where one billion people already live in poverty. The most disturbing aspect of this scenario is that an increasingly interdependent world travels on a path leading to a civilization that is physically and biologically unsustainable, ethically and morally inequitable, and socially and politically unstable.

The alternative scenario considers the consequences of pursuing consensus demographic and economic goals for 2050. The demographic goal is a nearly doubled, rather than tripled, world population by 2050. This growth reduction is accomplished by incentives to halve population growth rates in all countries. The world population reaches nine billion in 2050. Instead of exceeding the world's current population, the population increase in the less-developed countries is less than half the total population today.

The economic goal for the less-developed countries is to raise the production and consumption of goods and services by the average individual by 2050 up to the 1990 level in the industrial countries. This increase requires annual gains in individual economic productivity of four and a half percent. With concurrent population growth, the economies of the less-developed countries grow thirtyfold.

The economic goal for the industrial countries is to double the production and consumption of goods and services by the average individual by 2050, which requires annual economic productivity gains of one and a half percent. The economies of these countries grow threefold. The increase in their economies is 1.3 times greater than the total world economy today.

In this alternative scenario, the world economy still grows eightfold. Vigorous measures would be necessary to minimize the impact of production and consumption on the environment. The economies of the 65 countries at an intermediate stage of development (including China) grow the most, increasing to nearly four times the current world economy. With nearly 50% of the world's people, this intermediate group produces and consumes more than 40% of the global economic output.

With nearly 20% of the population, the industrial group produces just over 20% of the world's goods and services. The less-developed countries, with just over 40% of the population, produce nearly 30% of the global economy, a marked improvement over the first scenario.

OUTLINE OF A PLAN OF ACTION

Five needs must be recognized to move in the direction of the alternative scenario: (*a*) to strengthen understanding of the interacting physical, chemical, biological, and socioeconomic systems that regulate the total human environment; (*b*) to transform an energy-, technology-, and consumer-driven socioeconomic system into one that is environmentally benign; (*c*) to stabilize world population; (*d*) to reduce poverty wherever it exists; and (*e*) to reexamine societal values and goals and accord high priority to quality of life and sustainable human development.

Reduction of poverty will require economic growth, which contributes to the stabilization of population at a sustainable level. Low rates of demographic growth are usually found in affluent societies, high rates in impoverished societies. The premium placed on economic growth underscores the imperative to transform the socioeconomic system into one that is environmentally benign. The National Science and Technology Council has proposed an attractive production initiative (53). This initiative, combined with a complementary consumption initiative and a broad-based, properly framed body of knowledge, would provide a solid base for addressing the fundamental issue of sustainable human development.

Sustainable human development (*a*) places humans in the center of sustainable development, (*b*) empowers individuals through education and meaningful employment to expand their opportunities and options, (*c*) emphasizes the quality rather than quantity of economic growth, (*d*) seeks regeneration and enrichment rather than degradation and impoverishment of the total human environment, and (*e*) provides political and religious freedom and personal security.

The overarching challenge to the pursuit of sustainable human development is to recognize that human progress is driven by the cascade of knowledge. Again, the cascade of knowledge is the dynamic and nonlinear continuum linking the discovery, integration, dissemination, and application of knowledge concerning the nature and interaction of matter, energy, living organisms, information, and human behavior.

Strengthening this cascade is a prerequisite for the pursuit of sustainable human development. Major institutional renewal and innovation will be necessary. A dynamic and creative interaction must be generated among physical scientists, biologists, mathematicians, physicians, social scientists, engineers, and scholars in the humanities. Present arrangements in academia and in scholarly and professional organizations often inhibit, rather than encourage, the interaction of individuals in these disciplines; an interaction essential for the integration of knowledge. Some of the most exciting discoveries are taking place at the interface among disciplines, such as advances in modern biology (e.g. biochemistry, biophysics).

New modes of communication and cooperation need to be forged among business and industry, the several levels of government, academia, and private organizations. Individuals need to become involved through the growing number of nongovernmental grass roots organizations (NGOs). New partnerships

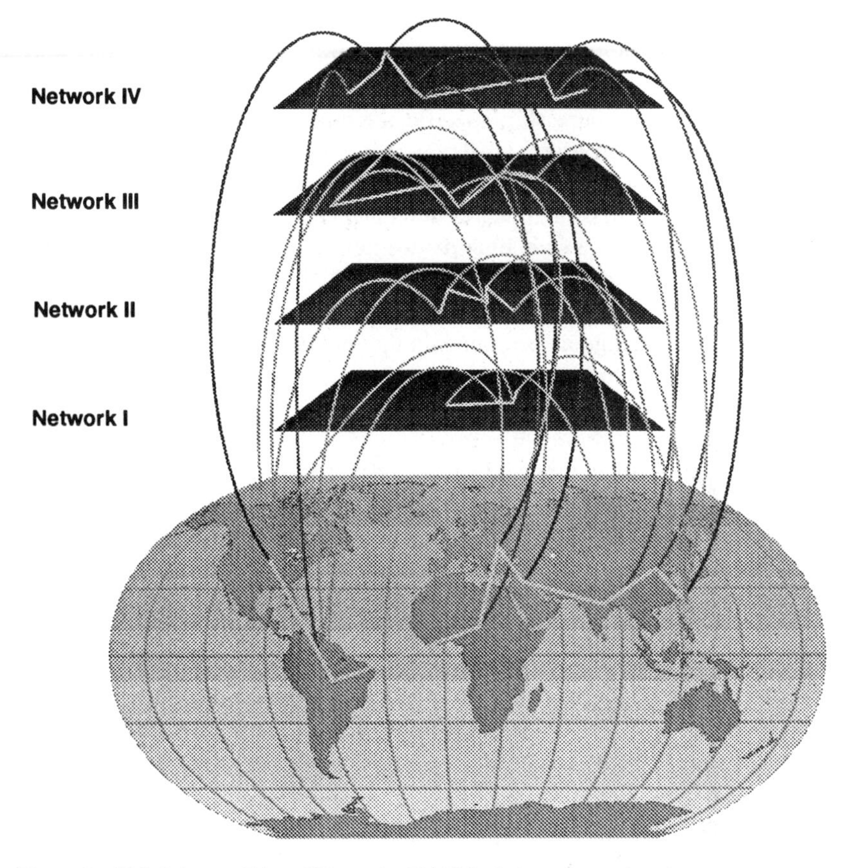

Figure 1 Global Array of Nested Networks (GANN). See text for explanation.

are needed between industrial and developing nations and among individual countries in those two groups of sovereign nations.

The global nature of the task and the overriding importance of the cascade of knowledge to the human prospect demand that a global knowledge strategy be fashioned. Rapidly emerging technologies in audio/visual information networking will soon bring within reach implementation of GANN, which will enhance the effectiveness of individuals and of national and international institutional arrangements (Figure 1).

Network I would reach individuals at grass roots levels in both developing and industrialized countries to inform, raise consciousness, provide input to the other networks, and build the public will for action. Network II would link educational institutions worldwide with each other and with the other networks.

The primary mission of Network II would be to embrace the discovery and dissemination of knowledge, with a focus on both formal and informal education. Network III would increase the effectiveness of several ongoing and planned networks in addressing specific regional and global issues, with a focus on the integration of knowledge across disciplines. Network IV would be largely responsible for integrating and interpreting the outputs of Networks I, II, and III and identifying policy options that emerge from their work. This network would serve as a switching device between decision-making mechanisms in the public and private sectors and GANN.

The societal discontinuity created by innovations in information technologies presents a challenge as profound as the one created by the invention of the printing press centuries ago. The world is still trying to solve tomorrow's problems with yesterday's economic measures, transfer of conventional technologies, and military power. Electronic transfer of currency has brought the world to the brink of a financial crisis (54). It is time to construct a new framework.

The currency of international exchange in that new framework will not be dollars, nor yen, nor marks. It will be knowledge, construed holistically and developed within the cascade of knowledge. Modest reordering of priorities would provide adequate monetary resources without new funding. A reallocation of 3% from (*a*) the $60 billion spent annually on Official Development Assistance, (*b*) the $120 billion spent annually on the military in developing countries, and (*c*) the $670 billion spent annually on the military in industrial countries, would provide $25 billion each year to underwrite a Global Knowledge Strategy (estimated from data in Reference 55).

LEADERSHIP

Imaginative leadership will be required to convert these concepts into specific initiatives that will galvanize individuals and institutions to launch a unified effort focusing on the human prospect. From my experience, this leadership is most likely to come from the scholarly community: institutions of higher education, and nongovernmental organizations that bring together wise and innovative individuals. Since the issues are global, stronger links should be established with kindred institutions in other parts of the world.

The opportunities before the scholarly community come at the end of an era of sustained growth and expansion. The compact between society and the scientific research community that grew out of *Science—The Endless Frontier* is expiring. The imperative of financial integrity, nationally and globally, is introducing painful measures of austerity. Consequent adjustments and reordering of priorities are the order of the day. Even as these sacrifices are made, expenditures for education and research must be recognized as investments for the future.

A special responsibility and opportunity faces the land-grant colleges. These institutions played a major role in transforming the United States over the past century. Their importance at the interface between natural and human systems has been noted (56). Land-grant colleges interact with the several levels of government and with the private sector. Their traditional mission of teaching, research, and extension is a point of departure for developing the cascade of knowledge.

In the end, the creativity and dynamism of individuals working within a variety of institutional instrumentalities will determine the human prospect. The cogent comment of the editors of *Foreign Policy* is germane: "It is our firm belief that the United States has yet to develop a foreign policy relevant to this post-Cold War world. Indeed, much of the foreign policy debate in the United States seems trapped in a time warp" (57).

Foreign policy and national policy are becoming inextricably intertwined. At this moment in history, the full panoply of knowledge must be brought to bear on the formulation and execution of policies that simultaneously serve national and international interests. The involvement of the scholarly community is imperative. A new and broadly based compact is needed between the community and society. The last one, forged 50 years ago, was initiated within the government. Perhaps this time the scholarly community could take the initiative.

Literature Cited

1. Malone TF, ed. 1951. *Compendium of Meteorology*. Boston: Am. Meteorol. Soc.
2. Malone TF. 1955. Application of statistical methods in weather prediction. *Proc. Natl. Acad. Sci. USA* 41(11):806–15
3. Thompson JC. 1950. A numerical method for forecasting rainfall in the Los Angeles area. *Mon. Weather Rev.* 78(7):113–24
4. Colborn R, ed. 1966. Interview with Douglas Brooks and Thomas Malone. In *The Ways of the Scientist*, pp. 360–66. New York: Simon & Schuster
5. Phillips NA. 1966. The general circulation of the atmosphere: a numerical experiment. *Quart. J. R. Meteorol. Soc.* 82:123–64
6. National Academy of Sciences, Committee on Meteorology. 1958. *Research and Education in Meteorology (The Berkner Report)* Washington DC: Natl. Acad. Sci. 38 pp.
7. Bush V. 1945. *Science—The Endless Frontier. Pub. No. VR-2, 1990.* Washington, DC: Natl. Sci. Found.
8. University Committee on Atmospheric Research. 1959. Appendix B. See Ref. 9, p. 1
9. University Committee on Atmospheric Research. 1959. *Preliminary Plans for*

the National Institute for Atmospheric Research. Boulder, CO: Natl. Cen. Atmosph. Res. 269 pp.
10. Malone TF. 1959. A national institute for atmospheric research. *Trans. Am. Geophys. Union* 40(2):95-111
11. Malone TF, convenor. 1961. *Proposal for International Collaboration in Meteorology Report.* Presented at *PSAC Panel for Possible USA-USSR Collaboration in Space,* 21 Feb., Boston
12. Malone TF. 1964. International cooperation in meteorology and the atmospheric sciences. In *Research in Geophysics,* ed. H Odishaw, 1:533–40. Cambridge, MA: MIT Press 574 pp.
13. Malone TF. 1961. Progress, purpose and potential in the atmospheric sciences. *Bull. Am. Meteorol. Soc.* 43(6):229–33
14. Charney JG. 1961. Scientific requirements for a global observation system. In *First Report of the ICSU/IUGG Committee on Atmospheric Sciences, 20 March 1967,* pp. 1–29. Paris: Int. Counc. Sci. Unions
15. Bolin B, conf. dir. 1967. The Global Atmospheric Research Programme (GARP). *Report of a Study Conference, 28 June–11 July, 1967.* Stockholm: Int. Counc. Sci. Union, Paris. 144 pp.
16. Intergovernmental Panel on Climate Change. 1991. *The IPCC Response Strategies.* Covelo, CA: Island
17. Schaefer VJ. 1946. The production of ice crystals in a cloud of supercooled water droplets. *Science* 104:457–59
18. Langmuir I. 1950. The control of precipitation from cumulus clouds by various seeding techniques. *Science* 112: 35–41
19. Orville HT, chair. 1957. Final report of the advisory committee on weather control. *Bull. Am. Meteorol. Soc.* 39(11): 583–86
20. Houghton HG. 1957. Present position and future possibilities of weather control. *Bull. Am. Meteorol. Soc.* 39(11): 567–70
21. Int. Counc. Sci. Unions/Int. Union Geod. and Geophys. Comm. Atmosph. Sci. 1961. *First Report, Summary of Meeting 8–11 February, 1967.* Paris: Int. Counc. Sci. Unions 37 pp.
22. Keeling CD, Bacastow RB, Bainbridge AE, Ekdahl CA, Guenther PR, et al. 1976. Carbon dioxide variations at Mauna Loa Observatory. *Tellus* 28: 538
22a. Los Angeles Herald-Examiner, *Warns of Peril in Fossil Fuel Burning,* October 17, 1970, p. A-13
23. National Research Council, Geophysics Study Committee. 1977. *Energy and Climate.* Washington, DC: Natl. Acad. Sci. 158 pp.
24. National Research Council, Board on Atmospheric Sciences and Climate. 1984. *Changing Climate.* Washington, DC: Natl. Acad. Sci. 500 pp.
25. White G. 1987. SCOPE: The first 16 years. *Environ. Conserv.* 14(1):7–13
26. Bolin B, Doos BR, Jager J, Warrick RA, eds. 1986. *The Greenhouse Effect, Climate Change, and Ecosystems. SCOPE 29.* Chichester, UK: Wiley. 541 pp.
27. Committee on International Security and Arms Control. 1985. *Nuclear Arms Control: Background and Issues.* Washington, DC: Natl. Acad. Sci. 378 pp.
28. Revkin AC. 1986. Missing: the curious case of Vladimir Alexandrov. *Sci. Dig.* 94(7):32–43
29. Hesburgh TM. 1984. Nuclear war: its consequences and prevention. *Origins* 14:415
30. Crutzen PJ, Birks J. 1987. The atmosphere after a nuclear war: twilight at noon. *AMBIO* 11:114–25
31. Turco RP, Toon OB, Ackerman JP, Pollack JB, Sagan C. 1983. Nuclear winter: the consequences of multiple nuclear explosions. *Science* 222:1283–92
32. Pittock AB, Ackerman JP, Crutzen PJ, MacCracken MC, Shapiro CS, eds. 1986. *SCOPE, Physical and Atmospheric Effects,* Vol. 1, *Environmental Consequences of Nuclear War.* Chicester, UK: Wiley. 359 pp.; Harwell M, Hutchinson TC, eds. 1986. *SCOPE, Physical and Atmospheric Effects,* Vol. 2, *Ecological and Agricultural Effects.* Chicester, UK: Wiley. 523 pp.
33. The Committee for the Compilation of Materials on Damage Caused by the Atomic Bombs in Hiroshima and Nagasaki. 1981. *Hiroshima and Nagasaki: The Physical, Medical, and Social Effects of the Atomic Bombing.* New York: Basic Books. 706 pp.
34. Goody RM, chair. 1982. *Global Change: Impacts on Habitability.* Pasadena, CA: Calif. Inst. Technol. (JPL D-95)
35. Malone TF, Roederer JG, eds. 1985. *Global Change. Proc. ICSU Symp. Ottawa, Canada, 25 September.* Cambridge: Cambridge Univ. Press. 510 pp.
36. Malone TF, Corell R. 1989. Mission to planet earth revisited. *Environment* 31(3):6–11, 31–35
37. Kessler E. 1994. Integrating earth system science: climate, biogeochemistry, social science, policy. *AMBIO* 23:1–103
38. International Institute for Environmental

Affairs. 1971. *The Human Environment: Science and International Decision-Making*. New York: Aspen Inst. Hum. Stud.
39. Eddy JA, Malone TF, McCarthy JJ, Rosswall T, eds. 1991. *Global Change System for Analysis, Research and Training (START). Global Change Rep. No. 15.* Stockholm: IGBP Secretariat, R. Swed. Acad. Sci.
40. Osborne F. 1948. *Our Plundered Planet*. Boston, MA: Little, Brown
41. Thomas WL, ed. 1956. *Man's Role in Changing the Face of the Earth*. Chicago: Univ. Chicago Press
42. Carson R. 1962. *Silent Spring*. Boston, MA: Houghton Mifflin
43. Environmental Pollution Panel, President's Science Advisory Committee. 1965. *Restoring the Quality of Our Environment*. Washington, DC: The White House. 291 pp.
44. Turner BL, Clark WC, Kates RW, Richards JF, Mathews JT, Meyer WB, eds. 1990. *The Earth as Transformed by Human Action*. Cambridge: Cambridge Univ. Press. 713 pp.
45. Dworkin D, ed. 1974. *Environmental Sciences in Developing Countries, Proc. SCOPE/UNEP Symp. Environmental Sciences in Developing Countries, Nairobi, Feb 11–23.* SCOPE 4, Indianapolis: Holcomb Res. Inst. 418 pp.
46. World Commission on Environment and Development, Brundtland GH, chair. 1987. *Our Common Future*. Oxford: Oxford Univ. Press
47. United Nations. 1994. EARTH SUMMIT CD-ROM. Complete source for all materials pertaining to the 1992 Rio de Janeiro Earth Summit. New York: UN publications, Dept. 136A
48. Murphy F. 1994. New emerging and reemerging infectious diseases. *Adv. Virus Res.* 43:1–52
49. Garrett L. 1994. *The Coming Plague: Newly Emerging Diseases in a World Out of Balance*. New York: Farrar, Straus & Giroux. 750 pp.
50. Malone TF, chair, Steering Committee. 1992. *Proc. Sigma Xi Forum on Global Change and the Human Prospect: Issues in Population, Science, Technology and Equity. Nov. 16–18, 1991*. Research Triangle Park, NC: Sigma Xi. 293 pp.
51. Malone TF, convenor. 1993. *International Networks for Addressing Issues of Global Change: Summary Report of a Workshop*. Aug. 24–27. Research Triangle Park, NC: Sigma Xi. 44 pp.
52. Malone TF. 1994. *Sustainable human development: a paradigm for the 21st century*. White Paper presented at Annu. Mtg. Natl. Assoc. State Univ. & Land Grant Coll. Chicago
53. National Science and Technology Council. 1994. *Technology for a Sustainable Future*. Washington, DC: Office of Science and Technology Policy, The White House. 154 pp.
54. Hot Money. 1995. *Business Week*. March 20, pp. 46–50
55. United Nations Development Programme. 1994. *Human Development Report 1994*. New York/Oxford: Oxford Univ. Press. 226 pp.
56. Dahlberg KA. 1979. *Beyond the Green Revolution*. New York: Plenum. 215 pp.
57. The editors. 1995. The new look of foreign policy. *For. Pol.* 98:3–5

Chauncey Starr

Annu. Rev. Energy Environ. 1995. 20:31–44

A PERSONAL HISTORY: Technology to Energy Strategy

Chauncey Starr
Electric Power Research Institute, Palo Alto, California 94304

KEY WORDS: nuclear energy, space power, energy research, risk analysis

CONTENTS

Abstract

This personal history spans a half century of participation in the frontiers of applied science and engineering ranging from the nuclear weapons project of World War II, through the development of nuclear power, engineering education, and risk analysis, to today's energy research and development. In each of these areas, this account describes some of the exciting opportunities for technology to contribute to society's welfare, as well as the difficulties and constraints imposed by our society's institutional and political systems. The recounting of these experiences in energy research and development illustrates the importance of embracing social values, cultures, and environmental views into the technologic design of energy options. The global importance of energy in a rapidly changing and unpredictable world suggests a strategy for the future based on these experiences which emphasizes the value of applied research and development on a full spectrum of potential options.

1056-3466/95/1022-00031$05.00

Introduction

In the past half century, I have participated in the early development of three major contributors to our energy systems: nuclear power, risk analysis, and the Electric Power Research Institute. My involvement in the energy field was the result of a career shift engendered by my World War II experiences. Prior to the war, I was interested in basic research in academia. I received an electrical engineering degree and PhD in physics from the Rensselaer Polytechnic Institute, and then did postdoctoral research at Harvard in solid-state properties at high pressures. Following this work I did cryogenic studies in high magnetic fields as a research associate at the Massachusetts Institute of Technology. Solid-state physics was my field of interest.

When World War II appeared imminent, I was asked by the Navy Department's Bureau of Ships to establish an underwater electronic research group at the David Taylor Model Basin at Carderock, Maryland. In 1942, E. O. Lawrence invited me to join his group in developing the electromagnetic separation of uranium isotopes in his radiation laboratory at the University of California at Berkeley. This technique was being applied to the massive separation plants being built at the new town of Oak Ridge, Tennessee. When schedule pressures made the time between laboratory results and engineering decisions at the site extremely important, Lawrence asked me to move from Berkeley to Oak Ridge to act as liaison between the laboratory and the engineering design, and to direct the development work at the pilot plant being built at Oak Ridge.

This was a seminal and maturing experience. I had moved from basic research to managing the transfer of applied science to the engineering of a novel large-scale system whose performance was critical to the war effort. For the first time I was directing the work of others. The pilot-plant technical staff, assigned by the Corps of Engineers, operated daily around the clock. The success of the effort was quickly evident from the increasing production of the separation plants. The personal reward came both from the sense of accomplishment and from the confidence that I could manage the development of an applied technology.

When the war ended, it was clear that the electromagnetic process would not continue as a source of separated uranium, and that the exciting postwar development would be in nuclear power—then called atomic power. I transferred to the Clinton Laboratory, then guided by Eugene Wigner and Al Weinberg (1), to be immersed in reactor theory and concepts. In those early days all imaginable combinations of materials, neutron physics, and engineering were freely proposed and considered. Some of these were eventually tried, with a few in commercial use today—notably the water-cooled reactors.

Nuclear Power Development for the Air Force

At the end of 1946, when most professional careers faced an uncertain future, colleagues at North American Aviation in Los Angeles asked me to lead a group to study the possible application of nuclear power to the propulsion of large intercontinental rockets and ram jets: an Air Force–sponsored exploration of a new energy source applied to a new, postwar military objective. I accepted this opportunity to return to California as a group leader in the Aerophysics Laboratory of North American Aviation (now Rockwell International). The task called for imaginative engineering in the then-embryonic art of large-rocket propulsion. We were searching for a nuclear reactor core that both was extremely compact and had an extremely high power output for several minutes. The feasibility of this concept depended on the interaction of engineering design options with test results on novel uranium fuel combinations at high gas temperatures and flow rates—obviously an exciting applied science challenge.

In 1947 my group issued a classified report entitled *Feasibility of Nuclear Powered Rockets and Ram Jets* (2). It was a product of a team of about two dozen engineers, physicists, and chemists, drawn from the resources of the Aerophysics Laboratory. In 444 pages, it described and analyzed every aspect of the project—design, development, and performance. Its conclusion stated, "Specifically, analysis indicates that a 10,000 mile rocket-missile, nuclear powered and hydrogen propelled, can be designed and constructed with a gross initial weight of about 100,000 pounds and a useful payload of 8,000 pounds. With slight modification, and without the payload, the rocket can escape from the gravitational field of the earth. The analysis further shows that a nuclear powered ramjet with an 8,000 pound payload has a gross weight of 17,600 pounds and can fly almost indefinitely at a speed of about 2,000 miles per hour."

In parallel with this study, the Aerophysics Laboratory (subsequently the Rocketdyne Division) had been developing chemically fueled rockets for the same mission. In comparing the outcomes, it was evident that chemical propulsion was a much better choice than nuclear propulsion for intercontinental missions in several important aspects. I therefore recommended to our company management that we advise the Air Force that the nuclear option not be pursued for this mission. This was considered an unorthodox step, as it would be today, to voluntarily turn off government funding. Nevertheless, after some discussion with the Air Force, the work was halted. Some decades later, both concepts were activated and field tested by the Division of Military Applications of the Atomic Energy Commission—Project Pluto on the ram jet, and Project Nerva as a rocket for outer space. The report NA 47-15 (2) was a seed source for both.

The company management, at my suggestion, agreed to continue support for my group while we examined the opportunity to apply its nuclear power expertise to electricity generation. This support was not needed for long. The Reactor Division of the US Atomic Energy Commission (AEC) had been aware of the analytical and experimental work embodied in the Air Force project, and volunteered to support our new efforts. At that time the AEC had already designated the Argonne National Laboratory, under Wally Zinn, as the center for nuclear power development, but private discussion with Dr. Zinn resulted in an independent contract to our group. This was the birth of Atomics International, a division of North American Aviation.

Atomics International and Electric Power

In the late 1940s and early 1950s the sense of mission in the AEC was very strong. The responsibility for harnessing the extraordinary power from nuclear fission, both for the country and mankind, pervaded all its reactor programs. The vision was more than a cliché—the target was visible because we knew that reactors worked. For the technologists engaged in this effort the principle issue was which of the many engineering paths to take. During this period there were numerous professional meetings, in the United States and Europe, to debate the potential values and problems associated with alternative combinations of fissionable fuels, moderators, and their many configurations. In those early days low-cost uranium ore seemed a limited resource, and enrichment of U_{235} was considered expensive. Only the military application in naval reactors could justify the use of highly enriched fuel. The obvious solutions for the long term were either the plutonium-based fast breeder, which could utilize the abundant fertile U_{238}, or the thorium thermal breeder. The many scientific uncertainties about the safety-related dynamic behavior of the reactor cores of both types of breeders, however, meant a requirement for meticulous exploratory research. So the near-term choices were the more certain thermal reactors based on natural or slightly enriched uranium. The AEC program included all of the above approaches, with ample support to lead to small-scale pilot demonstrations of the more promising.

The public enthusiasm for this program was unwavering for almost two decades. The Joint Committee on Atomic Energy, representing both branches of Congress, was an active leader and taskmaster, supporting the AEC and urging rapid accomplishments in the national interest. With this ambiance, many of the best scientists and engineers were eager to participate in the challenge of development. It continued the professional momentum that existed in the earlier World War II weapons program. Both the national laboratories and private engineering companies were deeply involved. It was in such circumstances that Atomics International (AI) emerged.

In our initial efforts we engaged in many studies of alternative reactor concepts, including heavy water–natural uranium arrangements. As a result, AI was asked to configure such a combination as a neutron source to make tritium for the AEC weapons program. This preliminary design became the basis for the Savannah River military complex. Although the Canadian program (Candu) pursued this concept for power, AI did not. The AI choice for power focused on reactors that operated with low-pressure coolants, based on the belief that the stored energy in high-pressure coolant systems represented a hazard that should be avoided. Two coolants were chosen for study, liquid sodium and an organic liquid, terphenyl, which was radiation resistant. The development of both was undertaken simultaneously. Their history illustrates some of the engineering pitfalls encountered in reactor development.

The Sodium-Graphite Reactor

Liquid metals are attractive reactor fuel coolants because of their high thermal conductivity. Of these, sodium has the most desirable characteristics—a very low viscosity that reduces pumping power, a boiling point higher than the top temperatures needed for the steam cycle to generate electricity, a low-neutron-capture cross-section, and most importantly, no chemical corrosion of steel. Sodium has two undesirable properties—a neutron-induced radioactivity that requires several days to decay, and intense chemical reaction with water. This latter hazard means that very careful shielding of a sodium coolant system from a steam power cycle is required. When exposed to dry air, sodium quickly develops an oxide film that inhibits burning, so that fire danger is not an important issue. On balance sodium has been considered a desirable choice as a reactor coolant.

The sodium-graphite combination was chosen to provide the comfort of a thermal neutron spectrum with the safety advantages of the liquid metal coolant. The graphite moderator was encased in hexagonal zirconium cans with a central fuel channel to contain the sodium. The designs were started in 1949 and resulted in an operating unit in 1957 at Santa Susana, California. The Sodium Reactor Experiment (SRE) produced 20 MWt, and later provided 6 MWe to the Southern California Edison Co.

The success of the SRE led to the AEC sponsoring a 75-MWe Sodium Graphite Reactor (SGR) plant at Hallam, Nebraska, which went critical in 1962. Because of the increased size of the Hallam SGR, it incorporated many modifications of the earlier SRE. The most significant was the use of stainless steel instead of zirconium to encase the graphite. After about a year of operation the stainless steel cans developed leaks, allowing the sodium to penetrate the graphite, poisoning the reactor. It was found that repeated thermal cycling had caused stress failures at the end seals of the cans. At that time the AEC was faced with serious budget constraints, and decided that it should concentrate on the already commercial light water reactor. So Hallam was abandoned, even though an engineering fix for the stress concentrations was proposed. Noteworthy, however, was the engineering success of the liquid sodium system, which was then the largest ever built for nuclear power. It provided a basis for the eventual engineering of a liquid metal cooling system suitable for a fast breeder. Subsequent to the Hallam SGR, the fast breeder programs of the United Kingdom, France, Japan, and Russia have operated much higher-power systems.

The Organic Cooled Reactors

The early studies on the radiation resistance of liquid organics suggested that those in the terphenyl group were sufficiently stable to be considered as coolants, particularly because of their low vapor pressure at high temperature (boiling point ~700° F), and the absence of corrosion problems. In accordance with the Atomics International concept that low-pressure coolant systems would lead to safer reactors, a proposal was made to the AEC in the early 1950s for testing such a design. In 1957 the Organic Moderated Reactor Experiment (OMRE) went critical at the National Reactor Testing Station in Idaho at a power level of 16 MWt. This experiment served chiefly to establish the acceptability of terphenyl in a reactor environment, and the test reactor's operation was discontinued in 1963. The early results were encouraging enough to lead to a modest power-plant demonstration reactor sponsored by the AEC. In 1963 an OMR rated at 45 MWt and 11 MWe was placed into operation at Piqua, Ohio. The system was sufficiently stable that the small staff of this municipal utility operated the power plant without difficulty. After almost three years, the first fuel loading needed replacement, but the AEC decided that because of budget constraints it would not do so. So the Piqua OMR was shut down.

In the meanwhile, however, the Canadians had selected organic cooling as an appropriate replacement for heavy water cooling in the next phase of the Candu heavy water–moderated reactor. In 1964, Atomic Energy of Canada Limited (AECL) and the AEC formally agreed to jointly support the design and eventual construction of a 500-MWe Heavy Water Organic Cooled Reactor (HWOCR). The task was given to Atomics International and Combustion Engineering, as a joint program. The technical directors were myself and Wally Zinn, then at Combustion. To support this design, AECL established a 60-MWt organic cooled research reactor (WR-1) at its Whiteshell Nuclear Research Establishment, at Pinawa, Manitoba.

The Whiteshell research program contributed valuable information on the behavior of the fuel elements. The chemically inert organic coolant permitted the use of uranium carbide fuel, whose high thermal conductivity contributes to reactor safety. WR-1 underwent a comprehensive study, which substantiated that uranium carbide was the fuel of choice. A troubling issue was the fouling of the fuel cladding by the formation of a thin organic coating after about a year of operation, impairing heat transfer. Intensive experimentation revealed that trace chlorine left in the terphenyl from its production was acting as a catalyst in growing the fouling film. Removal of the chlorine solved the problem. WR-1 ran for many years with this coolant, until the program of AECL for exploring changes in Candu was ended about 1973. Within the temperature range chosen for normal operation, organic cooling is certainly an attractive option.

In 1967, when most alternative reactor projects had been stopped by the AEC for budgetary reasons, the AEC also withdrew unilaterally from the joint HWOCR program, to the consternation of the Canadians. The HWOCR design was ready for construction, and needed a substantial financial commitment. AECL felt unable to make this commitment alone. This was a regrettable outcome, as I am strongly of the opinion that organic cooling would have opened the door to simpler and safer reactors. The experience at Whiteshell and at Piqua showed that the organic cooled reactor was trouble-free and especially easy to operate and to maintain, as compared to other concepts. In particular, the radiation field around the primary cooling circuit was very low, permitting inspection and maintenance while the reactor was operating. I believe that the importance of simplicity in man–machine interactions for economic and safe reactor operation has been underrated—by the AEC and its successors, by the regulators, and by the industry.

Space Nuclear Power

The high power density of nuclear reactors was a powerful attractant to imaginative aerospace engineers following World War II. The Air Force's Aircraft Nuclear Propulsion program was initiated early, intensively developed in several configurations, and finally abandoned when it became obvious that the mass of the reactor and shielding to protect a crew would be too large for use in a practical airplane. But the visionary possibilities of long-duration flight have always remained.

The application to unmanned space vehicles offered more promise, and in 1953 our group at Atomics International proposed the development of an electric power reactor unit, as part of the Systems for Nuclear Auxiliary Power (SNAP) program of the AEC and the National Aeronautics and Space Administration (NASA). At that time radioisotope heat and thermoelectric converters were being developed for that purpose. The objective was to make the smallest reactor/electric system capable of producing a few kilowatts of power with at least a one-year lifetime for future satellite missions. Our program had three designs: a 0.5-kW (SNAP-10A), a 3-kW (SNAP-2), and a 30-kW (SNAP-8) reactor.

Their common core concept was a homogeneous fuel mix of fully enriched uranium and zirconium hydride as a moderator, with the core reflected by beryllium. This combination resulted in a very compact assembly. The coolant was liquid NaK (a sodium–potassium alloy), which permitted operation at high temperature with low core pressure. This was the first use of zirconium hydride as moderator, which requires special treatment to control the hydrogen content. Based on the findings from an experimental core operated in 1959, the final design was chosen. SNAP-10A and SNAP-2 were the same reactor, but -10A had a thermoelectric convertor and -2 had a high-efficiency metallic vapor turbine-generator.

This program was originally initiated to meet requirements that the Air Force anticipated for a satellite defense system. In late 1963, the Air Force stated that in the absence of an immediate requirement, it was withdrawing its budgetary support. The AEC decided to complete the SNAP-10A program with its own funds in order to demonstrate a system in space. On April 3, 1965, the SNAP-10A was placed in a synchronous orbit by an Atlas booster/Agena vehicle at the Vandenberg Missile Launch Site. Four hours after lift-off, when a stable orbit was verified, a ground command initiated nuclear start-up, and full power was reached in eight hours. The operation continued smoothly for 43 days until a voltage controller on the Agena failed and automatically shut down the reactor. The unit remains in space, waiting for some future space technician to restart it.

These projects were challenging and exciting development tasks, and their progress professionally rewarding. It was disappointing to have the Air Force unable to foresee at that time the value of such space power, and halt the research and development (R&D) momentum. In contrast, the Soviets instituted a continuing program to launch such nuclear-powered satellites, their Topaz series, utilizing thermionic convertors instead of thermoelectric. In the 1980s, when the Strategic Defense Initiative became a military space objective, interest in space nuclear power re-emerged and development was initiated again, but without the hands-on experience from the earlier projects. The space application interest continues today at a modest level. Importation by the United States of a Russian Topaz assembly is being considered.

The fascination with space nuclear power will always exist. However, the practicalities of shielding personnel from radiation may also limit its future use to unmanned space vehicles. Perhaps a future space city might be large enough to accommodate a nuclear station, although solar power appears competitive in space. As a by-product of space nuclear power research, we have learned how to make small nuclear power units in the kilowatt range. There may be niche applications—the ocean bottom, for example—where such units may be useful.

Engineering Education and Risk Analysis

In 1966, I resigned as President of Atomics International to become Dean of the School of Engineering and Applied Science at the University of California at Los Angeles (UCLA). In addition to contributing to the growth of the School, I then had the time to work with the National Academy of Engineering, of which I was Vice President. I also wanted the freedom to pursue some intellectual concepts of my own. I believed then, as I do now, that some of the most fruitful frontiers of engineering are interdisciplinary. Schools without walls between departments is an often-quoted objective; it is generally the experience, however, that interdisciplinary activities have a difficult existence in the discipline-focused, professionally accredited engineering curricula. Nevertheless, with the support of other UCLA faculty members, I initiated a degree program in environmental engineering. As a joint effort with the Dean of Medicine, an Institute of Medical Engineering was founded. Such ventures into bridging conventional departments are more common today, but their acceptance within the university framework generally remains a struggle.

As a personal study, I undertook to fathom the disparity between the public and professional perception of the safety of nuclear power, an issue that came from my years at Atomics International. With the aid of a group of graduate students, I explored the data resources on the public risks associated with most of our common activities involving technical systems, such as automobiles, air transport, farming, etc, as well as the risks from natural events. This permitted me to establish a rough quantitative relationship among the benefit/risk/cost elements of public choice, and to suggest the relevant factors. This was described in a paper, "Social benefits versus technological risks," published in September 1969 in the journal *Science* (3). Apparently, this was the first time such holistic quantitative research had been done, and the paper became the seminal document for the new interdisciplinary field of risk analysis.

The study disclosed a large quantitative difference in the public acceptance of risk between voluntary and involuntary exposures. It also showed the importance of comparative risk among alternatives for achieving an end objective. It left unresolved the mysteries of the public perception of risks, as compared with quantitative risk comparisons.

There is now a Society for Risk Analysis with a professional journal and a broad interdisciplinary membership including social scientists, political scientists, physicists, chemists, engineers, biologists, physicians, public health administrators, regulators, etc. Many official regulatory processes, both state and federal, now routinely include risk analysis as part of their information resources. The subject is sufficiently complex that much research remains to be done to provide a solid basis for strategic decision making. Public attitudes remain a mystery and a social science challenge.

The Electric Power Research Institute (EPRI)

In 1972 I was anticipating a sabbatical year to write a book on risk analysis. The book never got written. That year the electrical utilities decided to form EPRI as the research center for the industry, and in December asked me to undertake the formation of this institute. They accepted my concept that EPRI provided an opportunity to enhance society's welfare by improving the technology of electrification in all its aspects. I had long believed that electricity was a powerful force for social change and betterment. The opportunity and challenge were so intriguing that I accepted. In January 1973 I became President of EPRI, and in the following six months phased into that position while phasing out as Dean at UCLA.

The precursor to EPRI was the industry's Electric Research Council (ERC), which represented an effort in the late 1960s of the utility industry, both public and investor-owned, to undertake joint research in technical areas of common interest. The ERC was the result of pressure from Joseph Swidler, then chairman of the Federal Power Commission, to fill the gap in technical development that was too long-range for the traditional equipment vendors to undertake. The ERC initiated a small number of modestly funded R&D projects, but its major contribution was the "Green Book" issued in 1971, a thoughtful and imaginative compendium of R&D opportunities projected to reach annual funding of a billion dollars (in 1971 dollars) by the year 2000. This book recommended the formation of an Electric Power Research Institute (EPRI) to carry out this program.

The triggering event was the threat by the Senate Commerce Committee to establish a federal agency to undertake utility R&D, as a response to the very costly consequences of the famous Northeast blackout in 1965 that resulted from massive electric system failures. The EPRI concept developed by the ERC was proposed as an alternative, and in 1972 EPRI was established as a nonprofit institute, to be supported by voluntary contributions from both public and investor-owned utilities. With enthusiastic support from the National Association of Regulatory Utility Commissioners, the initial financial support was generous. When I undertook the responsibility for EPRI in January 1973, the task ahead was to spend these funds effectively by filling the R&D pipeline with relevant and promising projects. In the first few months, I operated from a two-room office near UCLA, with a secretary and an ex-graduate student. In addition to funding a few obviously good projects, I engaged in a recruiting campaign for technical and administrative talent. By the late summer, a building in Palo Alto had been leased and the core staff had been hired.

The management philosophy of EPRI was based on utilizing—rather than reproducing—the best R&D facilities and talent worldwide, and particularly those in the United States. The EPRI staff was responsible for a holistic view of electrical systems, utility operations, their technological components, the choice of significant opportunities and appropriate R&D strategies, the selection of contractors to do the physical R&D, and the analysis and coordination of individual tasks. Although EPRI subsequently funded some specialized laboratories, these are contractor operated, and none is at EPRI. The R&D contractors include universities, commercial engineering and research companies, equipment vendors, and consultants. EPRI collaborates in joint projects with government agencies and laboratories, both US and foreign.

These management concepts have been successfully maintained for the past two decades, with annual expenditure levels of several hundred million dollars. The output from the R&D pipeline has been so large and valuable to the industry that EPRI has had to create a substantial technology-transfer operation to speed up the application of results—a need not anticipated originally. Another unanticipated activity was the growth of engineering effort devoted to solving immediate and pressing technical difficulties. Because of its broad matrix of scientific and engineering specialists, EPRI has become the resource that is called on to handle nonroutine technical difficulties. EPRI's Nuclear Division has been especially loaded with such tasks, because of the many novel problems that first-generation nuclear technology has developed over time.

A major part of EPRI's long-range contribution has been its pioneering R&D in advanced energy sources and end uses. It has led the way in such technologies as fuel cells, fluidized-bed combustion, integrated gasification combined cycles, wind machines, novel solar power devices, and energy storage with compressed gas and batteries. It developed solid-state devices for power transmission, which radically improve system flexibility and performance. In nuclear power development, EPRI has been the major US contributor to improved performance and technical progress. It has led the US utilities' preparation for the next generation of nuclear plants, working cooperatively with the Nuclear Regulatory Commission in obtaining approval of the advanced designs for these plants. EPRI has assumed that all feasible technical options should be kept viable for future evaluation and choice. Because the parameters that determine the optimal choice among alternatives are so subject to site-specific priorities and changing benefit/cost factors over time, it is not realistic to foreclose any workable option for future use.

Consistent with EPRI's original concept of a holistic approach to electric energy systems, a major part of EPRI's program has been focused on environmental effects and "beyond the meter" end uses by consumers, sometimes called "demand-side management." EPRI conducted many of the earliest studies on the distribution of "acid rain" and its effects on lakes. Many techniques developed by EPRI for reducing the effluents from fossil-fuel power plants have been applied by the operating utilities. Conservation programs have involved both the functional organization of consumer subsystems and the efficiency of their equipment. For example, factory dehydration of liquid mixtures by freeze-drying rather than heating was developed as an energy saver, and in the case of milk as also a flavor saver. A significant improvement in heat pumps now permits very efficient winter heating and summer cooling with one piece of equipment. Microwave drying has been developed both for industry and the home laundry. EPRI has nurtured the electric automobile from its infancy. Many such opportunities have been studied, and much remains to be done. Although, on balance, such programs have only marginal influence on the total market for electricity, they are consistent with the long-range objective of EPRI to make electricity a tool for improving society's quality of life.

The Energy Future

By the middle of the next century, global energy demand, driven by population and economic growth, will be in the range of 2–4 times the present level, depending on the uncertain use of energy efficiency and conservation globally (4). Most of this growth will occur in the presently less developed countries, where the shift from primitive agriculture to modern industrialization and modern agriculture is slowly occurring, as for example in China and India. The cumulative effect of energy use in this coming half century may strain the world's low-cost resources, particularly oil. Although coal reserves appear ample to meet even the high demand for several centuries, the environmental effects of its use may place a premium value on alternative, nonfossil energy sources, such as hydro, nuclear, biomass, solar, wind, and geothermal. Annual carbon dioxide emissions will undoubtedly increase for the foreseeable future, even with intensive global conservation and efficiency efforts.

Because electrification is the key to modern productivity, in a half century more than half of primary fuel usage will be for electricity production. Realistically, even with a maximum conservation effort, electricity demand will be more than four times present global usage. A massive expansion of nonfossil sources would be needed to slow the future annual increase in carbon dioxide emissions to the atmosphere. Most optimistically, the contribution of the environmental trinity—solar, wind, and biomass—is not likely to exceed a third of the electricity supply due to practical limitations. That fraction of the future supply would amount to more than all the electricity used globally today. The expansion of hydro and nuclear sources, uncertain as that may appear now, is needed to bring about a more significant reduction in fossil fuel use.

The Role of Technological Change

In the historical sweep of mankind's energy use, it seems that a few discrete technological developments for converting primary fuels to usable work forms have been the chief stimuli for massive social changes and large increases in energy use. The prehistoric manipulation of fire has been, of course, the foundation for all use of biomass and fossil fuels. But the conversion to useful work took a very big jump when the steam engine was developed. It quickly overcame the limitations of the waterwheel, and opened the gates to widespread industrialization. In the past century, the development of the internal combustion engine as the basis for many transportation modes has indelibly altered the infrastructure of all modern societies. Also in this century, the distribution of electricity as an intermediate energy form usable by anyone was obviously a revolutionary contributor to changing people's life-styles. It directly made possible our many information and communication systems. In the coming century, the pending commercialization of the electric vehicle may result in another substantial restructuring of energy systems if it reduces the demand for oil significantly. More significant may be an eventual acceptance of nuclear power as the next millennium's successor to fossil fuel. It is the only man-made energy source not dependent on chemical combustion, like the natural energy of the sun that gives us solar power. At issue today is whether we can learn to manage and accommodate its undesirable side-effects in a few decades, as we have done with the undesirable aspects of conventional combustion over several centuries.

In the modern world of science and technology, new frontiers and opportunities are being opened at a seemingly accelerated pace. In comparison, energy systems embedded in society seem to change very slowly. Nevertheless, some unforeseen technology of the future may radically alter today's perceptions. I believe that the flexibility of electricity as an energy intermediary will provide the common structure for both the present and future energy options. Expanding electrification globally while maintaining its flexibility is the exciting challenge.

Energy Strategy

For an unforeseeable range of social conditions, we should develop a range of energy options that might fit a variegated world. There are, however, some general criteria that will shape our future energy systems. When Archimedes recognized the power of the lever, he stated "give me where to stand and I will move the earth." With equal boldness, a technologist might state "give me enough energy and I can reshape the world." I entered the past half-century with such confidence, but today it is more evident that, once survival needs are met, an advanced society now wants such technological power to be judiciously exercised to protect the ambiance of the natural world. The growing global demand for energy, although economics and population driven, must be met with this environmental intangible in mind. This is particularly relevant to current exploratory research strategies.

A recent shift in energy planning has followed a revitalized concern with potential climate changes that might result from the carbon dioxide additions to the atmosphere produced by the use of fossil fuels. This century old suggestion (Arrhenius, 1896) was not considered seriously until a few decades ago, on the belief that any effect would be marginal. It has now become a major issue in climate research and a contentious aspect in energy debates. Although atmospheric carbon dioxide has increased about 25% in the past century, there is no clear evidence of a substantial climate effect, but models project major changes in the coming century. Unfortunately, there is no practical bulk substitute for fossil fuels, and it is almost inevitable that in the coming half century fossil fuel use will increase globally.

However, by assiduous application of energy efficiency techniques, and expanding the use of the known nonfossil sources—solar, wind, hydro, geothermal, and nuclear—it should be possible to slow the increase in carbon dioxide emissions. This would buy us time, but time for what? First, the slow adaptation of populations to climate shifts takes place constantly and given enough time accommodation to future climates could occur without societal strains. Second, it also provides more time for the creation of a spectrum of new or very improved energy systems and devices from which society can choose. For example, nuclear fission power plants are today only first stage designs, and nuclear fission is now only an embryonic concept. Unpredictable as this is, I have great faith in the potential capability of research and development to meet the needs of a future climate if it is given the opportunity and support.

Let me end therefore with a strong plea for sustained societal investment in long-range R&D. This is particularly important in the energy supply field because the physical equipment is generally large, costly, and requires decades to construct and develop for commercial use. Political attention spans are too short in comparison. As the above personal account illustrates, programs spasmodically stopped because of short-term budget needs cut off the opportunities to open important new options for society's use. Obviously, R&D is a gamble but its history shows that sustained and broadly based support has proven to be a profitable long term social investment. We need to strengthen our cultural commitment to such investment in science and technology.

Literature Cited

1. Weinberg AM. 1994. From technological fixer to think-tanker. *Annu. Rev. Energy Environ.* 19:15–36
2. Starr C. 1947. *Feasibility of Nuclear Powered Rockets & Ram Jets.* North American Aviation (now Rockwell International) Report NA 47–15 (declassified 1969)
3. Starr C. 1969. Social benefits versus technological risks. *Science* 165:1232–38
4. Starr C. 1993. Global energy and electricity futures. *Energy* 18(3):225–37

Annu. Rev. Energy Environ. 1995. 20:45–70

WEIGHING FUNCTIONS FOR OZONE DEPLETION AND GREENHOUSE GAS EFFECTS ON CLIMATE

Donald J. Wuebbles
Department of Atmospheric Sciences, University of Illinois, Urbana, Illinois 61801

KEY WORDS: ozone depletion potentials, global warming potentials, global environment, ozone policy, climate policy

---

CONTENTS

ABSTRACT

Weighing functions are extremely useful tools for evaluating the relative effects of gases affecting global atmospheric ozone and climate. In particular, the concepts of Ozone Depletion Potentials (ODPs) and Global Warming

1056-3466/95/1022-0045$05.00

Potentials (GWPs) are extensively used in policy consideration and scientific studies of the ozone and climate issues. ODPs provide a relative measure of the expected cumulative impact on stratospheric ozone from trace gas emissions, and are being used to examine the relative effects on ozone from CFCs, halons and other halocarbons currently being used and to evaluate the potential effects of possible replacement compounds. GWPs provide a means for comparing the relative effects on climate expected from various greenhouse gases. This chapter examines these weighing functions and other indices being used in evaluating concerns about global ozone and climate, discusses the science underlying these indices, and presents the current state-of-the-art for numerical indices and their uncertainties.

INTRODUCTION

Atmospheric measurements show that natural amounts of ozone in the global atmosphere have decreased globally by more than 10% since 1970 (1–3). The losses have occurred in the stratosphere, which contains 90% of the ozone, and are greatest in winter and spring at mid- to high latitudes in both hemispheres. The weight of the evidence indicates that decay products from human activity–related emissions of various halocarbons, particularly chlorofluorocarbons (CFCs), such as $CFCl_3$ and CF_2Cl_2, and halons, such as CF_3Br, are responsible for this ozone decrease (1, 2). Chlorine and bromine contained in these molecules react catalytically in the stratosphere to destroy ozone.

Atmospheric measurements also show that concentrations of several radiatively important gases, termed greenhouse gases, are changing, and in many cases, growing. Budget analyses of the sources and sinks of these gases indicate that these changes primarily result from human activities. Concern about the effects of these changes on climate has centered around carbon dioxide (CO_2), because it is an important greenhouse gas, and because its atmospheric concentration is rapidly increasing. However, other greenhouse gases have contributed about half the overall increase in the radiative forcing effect on climate (5). The most important of these greenhouse gases are methane (CH_4), nitrous oxide (N_2O), and a variety of halocarbons, including chlorofluorocarbons and halons.

In addition to the direct radiative effects of these gases, research studies show that chemical interactions in the atmosphere can lead to indirect effects on climate as well. For example, changes in the distribution of stratospheric ozone can affect climate. Numerical models of the climate system suggest that surface temperatures may increase by 3 K (the best estimated range is 2–5 K) over the next century (4–6), owing to projected changes in the concentrations of greenhouse gases.

The ozone decline has resulted in national and international actions to protect the ozone layer. Under the Clean Air Act, the United States has essentially eliminated the production and import of CFCs and halons and has put extensive controls on several other chlorine-containing halocarbons. This policy is in compliance with recommendations of the international Copenhagen Agreement modification of the Montreal Protocol on Substances That Deplete the Ozone Layer (7).

In policy considerations and in scientific studies of the ozone and climate issues, weighing functions and other indices have been extremely useful for evaluating the relative effects of the gases of interest and their emissions. This chapter examines these weighing functions and other indices being used in evaluating concerns about global ozone and climate, discusses the science underlying these indices, and presents the current state-of-the-art for numerical indices and their uncertainties.

WHY DEVELOP INDICES FOR ENVIRONMENTAL CONCERNS

Analyses of current and possible future changes in ozone are largely based on complex, numerical models of the chemical and physical processes controlling the atmosphere. Similarly, analyses of possible future changes in climate largely depend on complex, numerical, three-dimensional models of the global climate system (atmosphere, oceans, and land surface) that attempt to represent the many processes controlling climate (see Figure 1). These models are computationally expensive and, given current computer capabilities, severely limited in the number of calculations that can be and have been examined. The complexity of these models prohibits their extensive use in policy-related analyses of greenhouse gases and their effects on climate.

Simplified measures of human-related effects on ozone and climate can provide important insights for use in scientific and policy analyses. For example, the concept of Ozone Depletion Potentials (ODPs) (1, 2, 8–11) is used extensively by policymakers in rulings related to protection of the stratospheric ozone layer. ODPs provide a relative measure of the effect on stratospheric ozone expected per unit mass emission of a gas compared with the same mass emission of CFC-11, one of the halogenated gases with the strongest impact on ozone. ODPs are used to examine the relative potential effects on ozone of CFCs, halons, and other halogenated compounds, particularly those containing chlorine or bromine. The ODP concept is used to evaluate both the potential effects of new compounds and the relative effects of compounds already in commercial use.

The Intergovernmental Panel on Climate Change (IPCC) has been instrumental in examining relative indices for comparing the radiative influences of greenhouse gases on climate. The concept of Global Warming Potentials

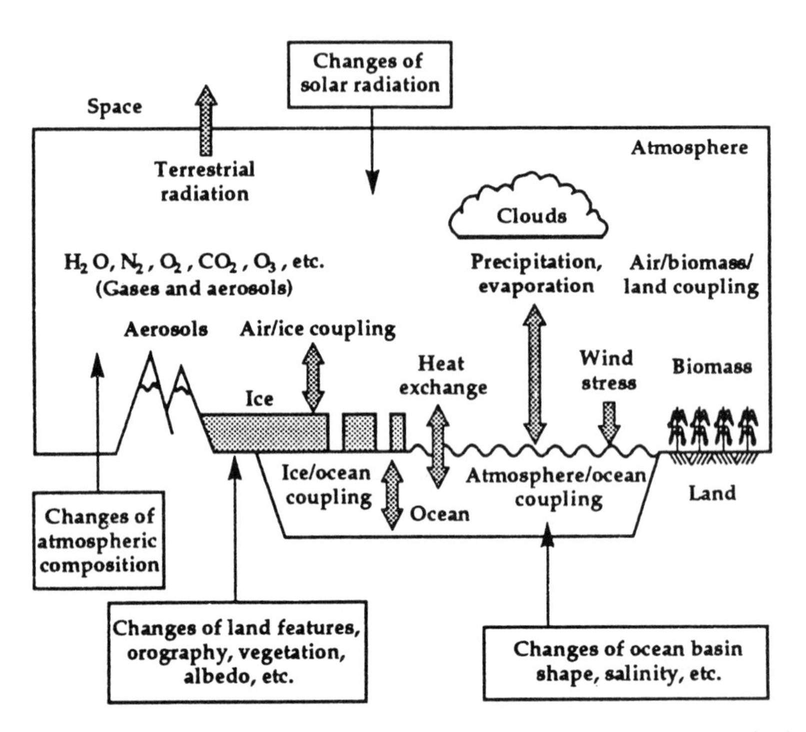

Figure 1 Schematic illustration of major components of the coupled atmosphere-ocean-ice-land climatic system (7a).

(GWPs), which is akin to the ODP concept, was developed as a means to provide a simple representation of the relative effects on climate of a unit mass emission of a greenhouse gas (4–6). Alternative measures and variations on the definition of GWPs have also been considered and reported.

A weighing function or index that places gases on an equivalent scale based on their effects on ozone or climate has considerable practical value. Such an index can be used in various analyses. Applications include:

1. Developing approaches to minimize the impact of human activities on ozone or the climate system;
2. Assessing the relative contributions of the many human activities contributing to emissions of gases that affect ozone or climate;
3. Comparing (and ranking) ozone or climate effects from competing technologies, energy uses, and energy policies;
4. Ranking the emissions from various countries;
5. Establishing a basis for comparing reductions in the ozone or climate effects in various countries; and

6. Functioning as a signal to policymakers to encourage some activities and discourage others.

Industries and governments can also use such indices to determine the best approaches for meeting their commitments to help reduce the radiative forcing on climate caused by increasing emissions and concentrations of greenhouse gases.

ATMOSPHERIC CONCENTRATIONS AND LIFETIMES

As mentioned above, the major CFCs and halons, including $CFCl_3$ (also referred to as CFC-11), CF_2Cl_2 (CFC-12), $C_2F_3Cl_3$ (CFC-113), CF_2ClBr (or H-1211), and CF_3Br (H-1301), are being banned because of their effects on stratospheric ozone. Several compounds are being evaluated as replacements for CFCs and halons, and some are already in use. Compounds currently under consideration as replacements include hydrofluorocarbons, such as C_3HF_7 (HFC-227ea), $C_2H_2F_4$ (HFC-134a), and CHF_3 (HFC-23); hydrochlorofluorocarbons, such as CHF_2Cl (HCFC-22), $C_2HF_3Cl_2$ (HCFC-123), and C_2HF_4Cl (HCFC-124); perfluorocarbons, such as C_2F_6 (also referred to as fluorocarbon 116, or FC-116), C_4F_{10} (FC-31-10), and C_6F_{14} (FC-61-14); and other compounds, such as CF_3I and SF_6. All of these compounds are also greenhouse gases, which implies that they can affect radiative forcing on climate because they absorb and emit infrared (IR) radiation. In some applications, mixtures of such compounds are being considered. Only the individual compounds are examined here, although the approximate effects of a mixture can be evaluated using the proper ratios of the individual effects.

This chapter focuses on weighing functions and hence considers only a fraction of the many compounds that may have a negative effect on ozone and climate. Table 1 lists many of the compounds of interest, their atmospheric lifetimes, and current tropospheric concentrations.

The effects of emissions of any given compound on ozone or climate depend on its atmospheric concentration and its atmospheric lifetime (the time scale for removal of a gas), along with other factors. The tropospheric concentration of a CFC, halon, or other constituent emitted at the Earth's surface depends on the rate of emission into the atmosphere and its atmospheric lifetime. The atmospheric lifetime of such compounds is an important factor in the determination of their potential effects on ozone and climate, including the calculation of ODPs and GWPs.

The atmospheric lifetime of a gas is generally defined as the ratio of total atmospheric burden to integrated global loss rate after the gas is emitted into the atmosphere (this calculation can be complicated by biogeochemical cycle exchanges with the oceans and biosphere, as discussed later). The lifetime is defined as the time the global amount of the gas takes to decay to $1/e$ or 36.8% of its original concentration (after initial emission into the atmosphere). The atmospheric lifetime integrates over spatial and temporal variations in the local atmospheric chemical loss frequencies for the compound.

The lifetime must take into account all of the processes that contribute to removal of a gas from the atmosphere, including photochemical losses within the atmosphere [typically caused by photodissociation or reaction with hydroxyl (OH)], heterogeneous removal processes (e.g. loss into clouds or raindrops), and permanent uptake by the land or ocean. The atmospheric lifetimes of several gases have been determined based on current knowledge of these loss processes. These lifetimes were recently updated for the IPCC (4) and WMO (1) assessments.

As shown in Table 1, atmospheric lifetimes for gases of interest range from

Table 1 Atmospheric lifetimes and current atmospheric concentrations for gases of concern in ozone and climate studies[a]

Species	Chemical formula	Atmospheric lifetime (years)	Current concentration (ppbv)
Carbon dioxide	CO_2	Not measurable[b]	354,000
Methane	CH_4	8–10	1,714
Nitrous oxide	N_2O	120	311
CFC-11	$CFCl_3$	50	0.268
CFC-12	CF_2Cl_2	102	0.503
CFC-113	$C_2F_3Cl_3$	85	0.082
Methyl chloroform	CH_3CCl_3	5.4	0.160
Methyl bromide	CH_3Br	1.3	0.012
H-1211	CF_2ClBr	20	0.0025
H-1301	CF_3Br	65	0.012
HCFC-22	CF_2HCl	13.3	0.105
HCFC-123	$C_2F_3HCl_2$	1.4	
HCFC-124	C_2F_4HCl	5.9	
HCFC-141b	$C_2FH_3Cl_2$	9.4	
HFC-23	CHF_3	250	
HFC-32	CH_2F_2	6.0	
HFC-125	C_2HF_5	36	
HFC-134a	CH_2FCF_3	14	
HFC-143a	$C_2F_3H_3$	55	
HFC-152a	$C_2F_2H_4$	1.5	
Perfluoromethane	CF_4	50,000	
FC-116	C_2F_6	10,000	
Sulfur hexafluoride	SF_6	3,200	0.0025
Trifluoroiodo-methane	CF_3I	<0.005	

[a] Taken from References 1, 4.
[b] CO_2 does not have simple atmospheric lifetime. Decay of CO_2 is complex function of the carbon cycle.

a few days (e.g. CF_3I) to thousands of years (e.g. SF_6 and several perfluorocarbons). These values are based on the recent international assessments (1, 4) and studies referenced therein, except for the values for perfluorocarbons, which are based on Ravishankara et al (12). The lifetime for C_3F_8 is not available, but judging by other perfluorocarbons, it is likely to be on the order of 5000 years.

Lifetimes for the CFCs are quite long, ranging from about 50 to several hundreds of years. These chemicals are essentially unreactive in the troposphere but are dissociated in the stratosphere; there, their chlorine can react with ozone. The atmospheric lifetimes of the perfluorocarbons and SF_6 are extremely long, so emissions of these gases will likely remain in the atmosphere for well over a thousand years. Lifetimes of HCFCs and HFCs are much shorter because of their reactivity with OH. The atmospheric lifetime of CF_3I is extremely short (on the order of a few days), because of its photolysis at near-ultraviolet wavelengths.

Complex exchange processes with the oceans and biosphere preclude a simple representation of an atmospheric lifetime for CO_2 (4–6). CO_2 added to the atmosphere is exchanged between reservoirs that have various turnover times, with time scales ranging from several years to several thousand years. Thus, the turnover time of 5–7 years for atmospheric CO_2, which was deduced from the removal rate of $^{14}CO_2$ produced during atmospheric nuclear bomb tests in the 1950s and 1960s (4), is relevant to the initial response of the carbon system but is not representative of the much slower, long-term response of atmospheric CO_2. The *e*-folding lifetime for atmospheric CO_2, as determined from current carbon-cycle models (4, 18–19) that attempt to represent the effects of the oceans and the biosphere, is about 70–100 years.

THE CONCEPT OF OZONE DEPLETION POTENTIALS

Laboratory studies, atmospheric measurements, and numerical models of the atmosphere provide important evidence of the effect chlorine and bromine have had on stratospheric ozone in the past few decades (1, 2). The chlorine and bromine catalytic mechanisms are particularly efficient at destroying ozone. These catalytic cycles can occur thousands of times before the catalyst is converted to a less reactive form such as HCl or HBr. Because of this cycling, relatively small concentrations of reactive chlorine or bromine can have a significant impact on the amount and distribution of ozone in the stratosphere. Atmospheric and laboratory measurements indicate that heterogeneous chemistry (reactions between gases and particles) enhances the effect of chlorine and bromine on ozone in the lower stratosphere. This heterogeneous chemistry helps convert less reactive species of bromine and chlorine to the forms that can react catalytically.

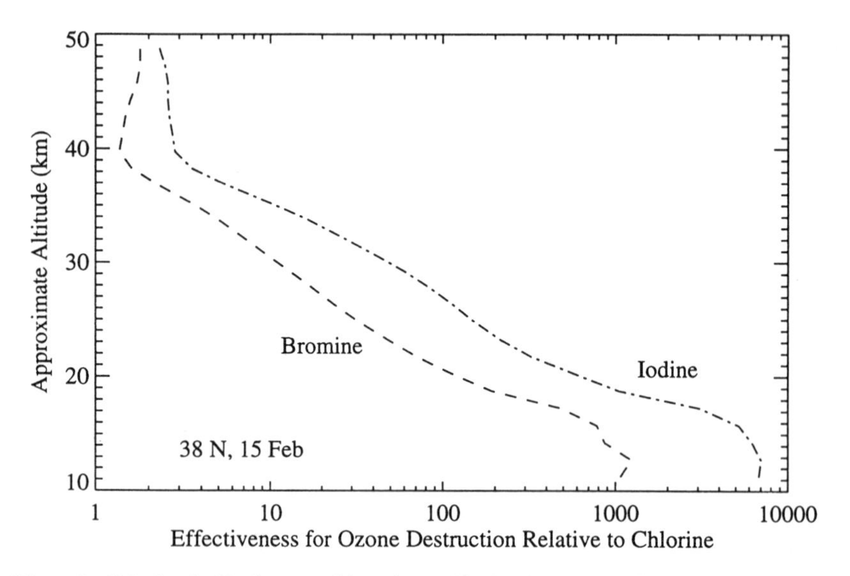

Figure 2 **Calculated effectiveness of bromine and iodine in destroying stratospheric ozone at mid-latitudes relative to chlorine. Based on LLNL zonally averaged chemical-radiative-transport model; similar results also found by Solomon et al (14).**

Chlorine catalytic cycles are extremely efficient and account for much of the existing ozone depletion. On a molecular basis, however, bromine is much more effective than chlorine at destroying ozone in the lower stratosphere. Figure 2 shows the effectiveness of bromine and iodine, relative to chlorine (shown on a per-molecule basis), in destroying ozone at mid-latitudes. The figure is based on calculations from the zonally averaged, chemical-radiative-transport model of the global atmosphere developed at Lawrence Livermore National Laboratory (LLNL) (13). The bromine catalytic cycles are more than 100 times more efficient than the chlorine catalytic mechanisms at destroying ozone in the lower stratosphere, especially below 20 km. However, emissions of brominated compounds are much smaller than those of the chlorinated compounds. As a result, the current impact of bromine on the atmosphere, while not negligible, is smaller than the effect of increasing amounts of chlorine.

CF_3I and other compounds containing iodine might serve as replacements for halons. Figure 2 shows that iodine in the stratosphere is more effective than bromine, and over a thousand times more effective than chlorine, at destroying ozone in the lower stratosphere. Solomon et al (14) calculated similar results from their atmospheric model. In another study, Solomon et al (15) suggest that natural sources of iodine, despite very small concentrations, may have contributed significantly, through extremely fast reactions with bromine and chlorine, to the recent lowering of stratospheric ozone depletion.

Other suggested halon replacements are composed primarily of carbon, fluorine, and hydrogen. None of these compounds, in the anticipated production and emission amounts is likely to affect ozone significantly. CF_3 radicals, produced by the dissociation of some of these compounds, such as HFC-134a, may affect ozone. However, recent studies indicate that these radicals are unlikely to have any significant effects on ozone (1).

Increased understanding of the depletion of stratospheric ozone has led to the need for simple measures for comparing the impact of different compounds on ozone that can be used as scientific guides to public policy. The Ozone Depletion Potential concept has proven to be a useful index of the effects on ozone from CFCs, halons, and their replacements.

Necessary Criteria and Definitions

The concept of ODPs provides a relative cumulative measure of the expected effects on ozone of the emissions of a gas (1, 2, 8–11), relative to one of the gases of most concern to ozone change, namely CFC 11. The ODP of a gas is defined as the change in total ozone per unit mass emission of the gas, relative to the change in total ozone per unit mass emission of CFC-11. This definition provides a single-valued, relative measure of the maximum calculated effect on ozone of a given compound compared to the effect calculated for CFC 11. As a relative measure, ODPs are subject to fewer uncertainties than estimates of the absolute percentage of ozone depletion caused by different gases.

ODPs are an integral part of national and international considerations on ozone-protection policy, including the Montreal Protocol and its Amendments (7) and the US Clean Air Act. ODPs provide an important means for analyzing the potential for a new chemical to affect ozone relative to CFCs, halons, and other replacement compounds.

Since ODPs are defined in terms of the steady-state ozone change, they are not representative of the relative, transient effects expected for shorter-lived compounds during the early years of emission. Time-dependent ODPs can also be defined that provide information on the shorter time scale effects of a compound on ozone. However, the steady-state values generally are preferred and used in regulatory considerations.

Model Calculated and Semiempirical Approaches

ODPs are currently determined by two different means: calculations from two-dimensional models of the global atmosphere (1-2), and the semi-empirical approach developed by Solomon et al (10). The two approaches give similar results.

The numerical models attempt to account for all of the known chemical and physical processes affecting chemical species in the troposphere and stratosphere. Each compound is assumed to enter the atmosphere at ground level and then be transported in the atmosphere by dynamical processes. These compounds can also react in the models by a variety of pathways, depending on their molecular structure. The compounds may undergo photolytic breakdown by ultraviolet or near-ultraviolet light, react with hydroxyl in the troposphere and stratosphere, or react with excited, atomic oxygen. Products resulting from these reactions, such as atomic chlorine and bromine, can react in the modeled atmosphere, which in turn may affect the calculated distribution of ozone. Although various models have been used to derive ODPs worldwide, in general the results differ only within the expected range of uncertainty for the mechanisms that affect ozone.

A major uncertainty in the models is the amount of tropospheric hydroxyl, OH. Because few reliable measurements of OH are available, the global distribution has not been directly measured. However, measurements of methane, methyl chloroform, and other gases whose destruction primarily results from reaction with tropospheric OH provide a separate calibration of the tropospheric OH amount. The observation-based tropospheric lifetime for methyl chloroform is often used to calibrate the atmospheric lifetimes of HCFCs and HFCs in various models to compare derived ODPs.

The semiempirical approach for determining ODPs is based on direct measurements of select halocarbons and other trace species in the stratosphere. The observed fractional dissociation is used to determine the amount of chlorine and bromine released and is then compared with the observationally derived ozone-loss distribution, with the assumption that the ozone loss results only from halogen chemistry. The correlation between different compounds is determined on the basis of their relative reactivity in the troposphere and stratosphere. This semiempirical method avoids some of the demanding requirements of accurate numerical simulation of source gas distributions (i.e. of the CFCs, HCFCs, and other compounds) and of the resulting ozone destruction. This approach also may be more realistic than current models that still underrepresent ozone destruction at lower stratospheric altitudes, particularly near the poles.

Calculated Ozone Depletion Potentials

Table 2 presents steady-state ODPs for the CFCs, halons and possible replacement compounds, based on the recent international ozone assessment (1). By definition, the ODP for CFC-11 is 1.0. The calculated ODPs for other CFCs being banned are all greater than 0.4. The Clean Air Act currently calls for policy actions on compounds whose ODPs are greater than 0.2. The ODPs for halons are all extremely large, much greater than 1.0, reflecting the reactivity of bromine with ozone.

Table 2 Steady-state Ozone Depletion Potentials (ODPs) for several of the compounds of interest (1)[a]

Species	Chemical formula	Ozone depletion potential
CFC-11	$CFCl_3$	1.0
CFC-12	CF_2Cl_2	0.9
CFC-113	$C_2F_3Cl_3$	0.9
Methyl chloroform	CH_3CCl_3	0.12
Methyl bromide	CH_3Br	0.6
H-1211	CF_2ClBr	5.1
H-1301	CF_3Br	13
HCFC-22	CF_2HCl	0.04
HCFC-123	$C_2F_3HCl_2$	0.014
HCFC-124	C_2F_4HCl	0.03
HCFC-141b	$C_2FH_3Cl_2$	0.10
Trifluoroiodo-methane	CF_3I	< 0.008

[a] HFCs and perfluorocarbons have ODPs near zero.

The ODPs in Table 2 for the HCFCs being considered as CFC or halon replacements are all small, with values of 0.02–0.1. The effect on ozone from a unit mass emission of one of these HCFCs would be less than one tenth that of the CFC or halon they would replace. The reduced effect of these HCFCs on ozone results from their short lifetimes. ODPs for all of the HFCs, PFCs, and for sulfur hexafluoride are near zero, owing to the low reactivity of their dissociation products with ozone.

Although iodine is extremely reactive with ozone, the ODP for surface emissions of CF_3I is less than 0.01 (14). Because of its reactivity in the troposphere, very little iodine would be expected to reach the stratosphere from surface emissions of CF_3I. However, this ODP value is subject to significant uncertainty, owing to limited understanding of iodine chemistry and the physical processes affecting the iodine distribution in the troposphere and stratosphere.

Time-Dependent Effects

Time-dependent ODPs are useful because they provide insight into short-term, postemission effects on stratospheric ozone, whereas steady-state ODPs reflect integrated effects over longer time scales. Figure 3 presents the calculated, time-dependent ODPs for several compounds. As discussed in the recent international assessments (1, 2), the ODPs for the HCFCs (and several other compounds, such as CH_3Br and CF_3I) are much larger on short time scales (a few years) than they over longer periods. The short atmospheric lifetimes of these compounds imply that they also release chlorine (or bromine or iodine)

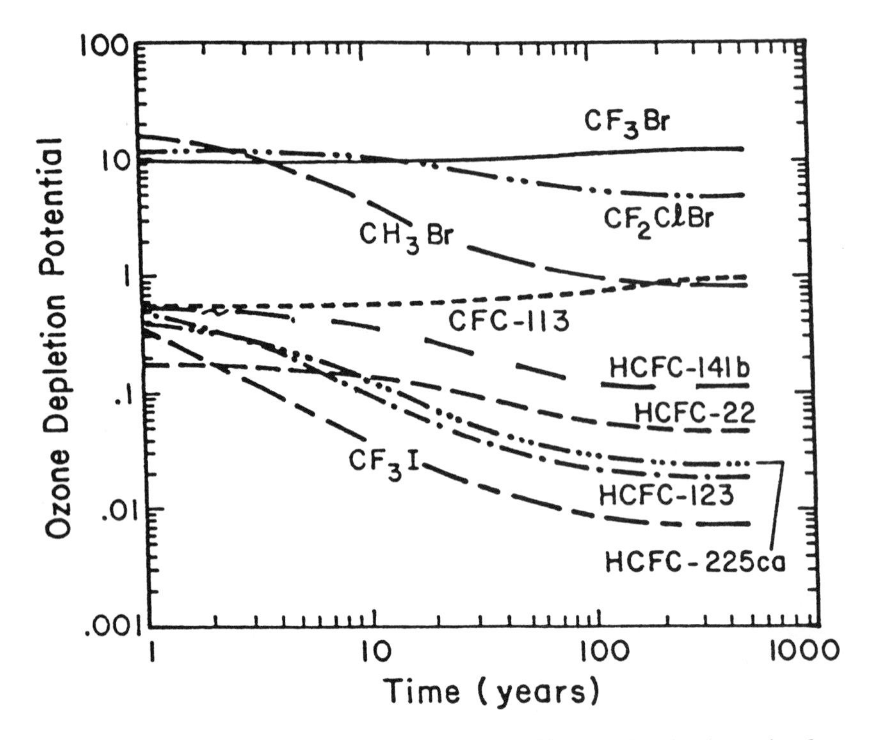

Figure 3 Time-dependent ODPs for several compounds of interest. Note that the *x*-axis refers to the time since reaching the stratosphere, not the total time, which is about 3–5 years longer (1).

in the lower stratosphere more quickly than CFC-11 and can result in a more immediate, (but generally small in absolute) effect on ozone as compared to CFC-11.

GREENHOUSE GAS INDICES

Uncertainties about key climatic processes and feedbacks can significantly influence the accuracy of detailed quantitative predictions of future climate changes. These uncertainties have strong implications for any models used to predict the magnitude of future regional and global climate changes caused by greenhouse gases. (This is also the reason for much of the uncertainty about temperature range in the next century.) Despite all these difficulties, policy-makers still need ways to compare the effects of different greenhouse gases on climate. The concept of radiative forcing on climate allows us to circumvent many of the uncertainties about the magnitude of climate change.

Necessary Criteria and Definitions

Any indexing approach must be scientifically well grounded, should be subject to a minimum of uncertainties, and should be applicable to the policy concerns of interest. Scientifically, an indexing approach must consider several factors including the radiative properties of the gases, their atmospheric lifetime, differences in the resulting effects on climate, and any indirect effects on other radiatively important constituents.

RADIATIVE FORCING Chemical species in the atmosphere differ markedly in their ability to absorb infrared radiation. The greenhouse gases absorb the Earth-emitted infrared radiation in specific energy (or wavelength) bands determined by the quantum-mechanical properties of the specific molecule. Water vapor is the single most important greenhouse gas in the Earth's atmosphere because of its large tropospheric concentrations and its many infrared-absorption features. Carbon dioxide is the second most effective greenhouse gas; its 15-μm band dominates the infrared absorption from 12 to 18 μm. Other important greenhouse gases in the current atmosphere include ozone (O_3), methane (CH_4), nitrous oxide (N_2O), and the CFCs, (particularly $CFCl_3$ and CF_2Cl_2).

The radiative forcing of the surface-troposphere system (due, for example, to a change in greenhouse-gas concentration) is defined (4–6) as the change in net irradiance (in Wm^{-2}) at the tropopause after allowing for stratospheric temperatures to readjust to stratospheric equilibrium. It is appropriate for radiative forcing to be calculated at the tropopause because it is considered in a global and annual mean sense that the surface and troposphere are closely coupled; the net irradiance at the tropopause is thus indicative of the change in radiation affecting the surface–troposphere system.

A key factor affecting radiative forcing for a gas is the location of the wavelengths for its absorption of infrared radiation. The spectral region from about 8 to 12 m is referred to as the window region because of the relative transparency to radiation over these wavelengths. Most of the non-CO_2 greenhouse gases with the potential to affect climate, including O_3, CH_4, N_2O, and the CFCs, have strong absorption bands in the atmospheric window region. Relatively small changes in the concentrations of these gases can produce a significant increase in radiative forcing.

As the concentration of a greenhouse gas becomes high, it can absorb most of the radiation in its energy bands. Once any of its absorption wavelengths near saturation, it is less able to absorb more energy at that specific wavelength, and further increases in its concentration have a diminishing effect on climate. This is called the band-saturation effect. For example, the radiative forcing from further increases in carbon dioxide concentrations in the current atmosphere will increase as the natural logarithm of its concentration because of this effect. Methane and nitrous oxide also exist at sufficient quantities that significant band saturation is occurring; it is found that their forcing is approximately proportional to the square root of their concentration. Also, at the wavelengths where water vapor and carbon dioxide strongly absorb IR radiation, the greenhouse effect of other gases will be minimal. On the other hand, absorption by other gases such as the chlorofluorocarbons or other halocarbons at wavelengths that are not saturated varies linearly with concentration.

Another important consideration in radiative absorption is the band-overlap effect. If a gas absorbs at wavelengths that are also absorbed by other gases, then the effect on radiative forcing of increasing its concentration can be diminished. There is significant overlap between some of the absorption bands of methane and nitrous oxide that need to be carefully considered in determining their radiative forcing on climate.

EFFECTS ON CLIMATE For a number of years, calculations made with climate models have shown, at least for well-mixed greenhouse gases, that there is approximately a linear relationship between the global mean radiative forcing at the tropopause and the resulting change in global mean surface temperature. Climate model calculations have shown that for a variety of forcing mechanisms the relationship is relatively unaffected by the nature of the forcing, whether, for example, it be due to a change in greenhouse gas concentration or a change in the solar flux.

Recent studies (16, 17) suggest that the relationship between global mean radiative forcing and global mean surface-temperature change is not a simple one for gases and aerosols having strong variations spatially. The effects of these findings are likely to be most important for species with short atmospheric lifetimes (such as sulfuric aerosols) or for species like ozone that have strong variations in concentration changes occurring as a function of latitude and altitude.

ATMOSPHERIC LIFETIMES As with effects on ozone, the atmospheric lifetime of a constituent is also important to determining its resulting effects on climatic radiative forcing. As mentioned earlier, the atmospheric lifetime for carbon dioxide is not a single number like most of the other greenhouse gases. Concentrations of carbon dioxide emitted into the atmosphere decay in a highly complex fashion initially showing a very fast removal in the first few decades, a more gradual decay over roughly the next 100 years, and an extremely slow decline for thousands of years. The response decay curve of a pulse of CO_2 into the atmosphere also depends on the background content of CO_2 in the atmosphere-ocean-biosphere system and possibly on the state of the climate system, and will thus be scenario-dependent (18). This response primarily

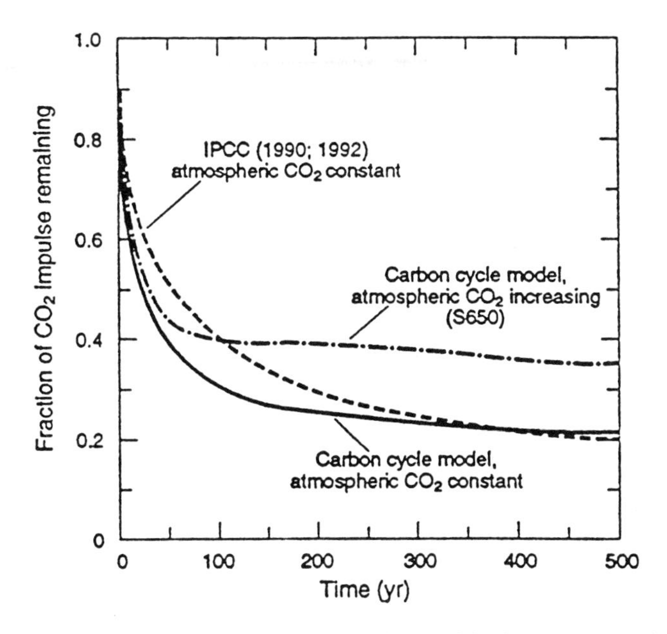

Figure 4 Response of atmospheric CO_2 following an impulse injection, based on results for IPCC (4) from the Bern carbon-cycle model (19).

reflects processes transferring CO_2 between the atmosphere, the biosphere, the ocean mixed layer, and the deep ocean. Figure 4, based on results for IPCC (4) from the Bern carbon-cycle model (19), shows response curves for CO_2 decay in the atmosphere following a pulse emission that demonstrate these effects.

Several recent studies (20, 21) indicate that the atmospheric response–time for methane, following a pulse emission into the current atmosphere, is significantly larger (corresponding to an *e*-folding time as large as 17 or 18 years) than its nominal atmospheric lifetime of about 10 years. This longer response time results primarily from the nonlinear coupling between CH_4, CO, and OH in the troposphere; increased methane leads to additional CO and reduced OH, the major sink for CH_4.

INDIRECT EFFECTS In addition to the direct forcing effect from emission of a gas into the atmosphere, the net radiative forcing can also be modified through indirect effects relating to chemical interactions on other radiatively important constituents. For example, emissions of CFCs result in stratospheric chlorine that can destroy stratospheric ozone, which is also a greenhouse gas (22–25). Emissions of methane result in several indirect effects. Oxidation of methane is thought to result in decreased concentrations of OH (hydroxyl largely determines the oxidation capacity of the atmosphere), increased tropospheric and lower stratospheric ozone, increased stratospheric water vapor, and increased carbon dioxide. Such indirect effects need to be considered when evaluating potential climate-effect indices.

FORMULATIONS OF GREENHOUSE GAS INDICES

No single indexing approach or tool is likely to meet all of the needs of policymakers. Policy analyses often need a variety of tools including the complex as well as simpler climate modeling studies, along with careful consideration of the state of the science. There really is no universally accepted methodology for combining all of the relevant factors into a single approach. Several different indices have been used as measures of the strength of the radiative forcing from different greenhouse gases. Some of the more important indices are described below.

Relative Radiative Forcing per Molecule or Mass

This measure compares the radiative forcing on a molecule-per-molecule basis or kilogram-per-kilogram basis of the different greenhouse gases. It is generally given relative to CO_2. A radiative transfer model of the atmosphere is used to determine the radiative forcing for small perturbations of these gases relative to present-day conditions. Small perturbations are necessary to prevent the marked nonlinear absorption of carbon dioxide, methane, and nitrous oxide from affecting the radiative forcing for these gases. Table 3 shows radiative

Table 3 Radiative forcing relative to CO_2 on a per-molecule basis for a few greenhouse gases[a]

Gas		Relative radiative forcing
Carbon dioxide	CO_2	1
Methane	CH_4	21
Nitrous oxide	N_2O	206
CFC-11	$CFCl_3$	12,400
CFC-12	$CF2Cl_2$	15,800
HCFC-22	CF_2HCl	10,700
HCFC-123	$C_2F_3HCl_2$	9,920
HCFC-124	C_2F_4HCl	10,790
HFC-134a	$C_2F_4H_2$	9,550
Perfluoromethane	CF_4	5,460
Sulfur hexafluoride	SF6	38,400

[a] Perturbation is relative to a background atmosphere for 1990 concentrations (4–6).

forcing on a per–molecule basis, relative to CO_2, for a few of the important greenhouse gases.

Hammond et al (26) and Handel (27) have used such instantaneous radiative forcing in developing approaches for examining the relative importance of greenhouse gas emissions from various countries. Others (e.g. 28) have criticized this approach because it does not adequately consider the time horizon over which emissions of a gas can affect climate.

Past, Present, and Future Changes in Greenhouse Gas Concentrations

It is also useful to examine how the radiative forcing for atmospheric concentrations of greenhouse gases have changed or might change over various time periods. Published analyses (4–6, 29–31) have shown significant changes have occurred in the radiative forcing since preindustrial times and depending on the scenario, significant changes are quite likely over the next century. Total radiative forcing from greenhouse gases has increased since the 18th century by about 2.45 Wm^{-2} (4). CO_2 accounts for 1.56 Wm^{-2}, with 0.47 Wm^{-2} for CH_4, 0.14 Wm^{-2} for N_2O and 0.28 Wm^{-2} for CFCs and other halogens. Changes in ozone could add another 0.2 to 0.6 Wm^{-2}, depending on actual changes in tropospheric ozone (4). Changes in amounts of tropospheric aerosols, particularly sulfates, over this time, would tend to counteract this warming, with a cooling tendency of 0.3 to 3.0 Wm^{-2} subject to large uncertainties. In contrast to these large human-induced effects, natural changes due to solar variability are thought to have increased radiative forcing since 1850 by no more than 0.5 Wm^{-2} (4).

Global Warming Potentials

The indexing approach for greenhouse gases that has gained the widest acceptance is the Global Warming Potential concept originally developed for the IPCC (5). As discussed below, this concept has continued to be modified and reevaluated in more recent assessments. Similar approaches to the IPCC definition were also developed by Rodhe (32) and Lashof & Ahuja (33).

Global Warming Potentials are expressed as the time-integrated radiative forcing from the instantaneous release of a kilogram (i.e. a small mass emission) of a trace gas expressed relative to that of a kilogram of the reference gas, CO_2 (5, 6):

$$\mathrm{GWP}(x) = \frac{\int_0^n a_x \cdot [x(t)]dt}{\int_0^n a_{\mathrm{CO_2}} \cdot [\mathrm{CO_2}(t)]dt} \qquad 1.$$

where n is the time horizon over which the calculation is considered, a_x is the climate-related, radiative-forcing response due to a unit increase in atmospheric concentration of the gas in question, and $[x(t)]$ is the time-decaying concentration of that gas and the corresponding quantities for the reference gas are in the denominator. The radiative forcing responses, a, are derived from radiative transfer models. The trace gas concentrations, $[x(t)]$ remaining after time t are based upon the atmospheric lifetimes of the gas in question. The reference gas has been taken generally to be CO_2, since this allows a comparison of the radiative forcing role of the emission of the gas in question to that of the dominant greenhouse gas that is emitted as a result of human activities.

The definition above assumes a linear relationship between the incremental concentration change and the resulting change in radiative forcing. This is appropriate for the small mass emission implied for the IPCC definition of GWPs. A more general definition of GWPs would be:

$$\mathrm{GWP}(x) = \frac{\int_0^n F([x(t)])dt}{\int_0^n F([\mathrm{CO_2}(t)])dt} \qquad 2.$$

where $F[x(t)]$ is the radiative forcing in response to the changing concentration of species x after the pulse emission at time $t = 0$.

Unlike the other greenhouse gas indices discussed above, GWPs are determined relative to the emissions of greenhouse gases (rather than concentration). Changes in emissions are particularly relevant to evaluating or formulating policies for reductions in greenhouse gases.

The choice of the reference molecule and the choice of integration time horizons require special discussion.

REFERENCE MOLECULE As indicated above, CO_2 has been used as the reference gas in determining GWPs. The atmospheric residence time of CO_2 has the largest scientific uncertainties of the major greenhouse gases. As a result, GWPs are quite sensitive to uncertainties in the understanding and treatment of the carbon cycle and its effects on the amount of atmospheric CO_2. Several papers (18, 34) have noted the importance of uncertainties in the carbon cycle for calculations of GWPs when CO_2 is used as the reference. It is important to note that the budget of CO_2 must be carefully balanced in some way, with detailed accounting of trends, sources, and sinks. While recognizing these issues, Caldeira & Kasting (35) discuss feedback mechanisms that tend to offset some of these uncertainties for GWP calculations.

Alternative reference molecules, such as CFC-11 or a CO_2-like gas, with simplified assumptions about the atmospheric lifetime, could be used. These have the advantage that the denominator would not be subject to major uncertainties and separate GWP values for CO_2 could be determined. However, such approaches also have major disadvantages for policy analyses that are often oriented at comparisons and trade-offs relating to CO_2.

INTEGRATION PERIOD The best choice of integration time horizon in evaluating GWPs has been the subject of much discussion and controversy (4–6). Unlike ODPs, the complexities of treating CO_2 and the carbon cycle prevent integration of GWPs to steady state. There is, however, no given value of integration time for determining GWPs that is ideal over the range of uses of this concept. GWPs are generally calculated over three time horizons; 20, 100, and 500 years. It is believed that these three time horizons provide a practical range for policy applications (6). GWPs determined for the longer integration period provide a measure of the cumulative chronic effects on climate, whereas the integration over the shorter period is representative of near-term effects. GWPs evaluated over the 100-year period appear generally to provide a balanced representation of the various time scales for climatic response (6).

The best choice of time horizon will depend on the specific analysis being considered. One needs to balance the effects of near-term responses in comparing greenhouse gases with consideration of the long-term persistence of any environmental effects from long-lived gases.

Absolute Global Warming Potentials

A variant formulation (4, 36) is to give simply the integrated radiative forcing of the gas in question:

$$\mathrm{AGWP}(x) = \int_0^n a_x \cdot [x(t)]dt \qquad \mathrm{W} \cdot yr/m^2 \qquad 3.$$

The advantage of this formulation of Absolute Global Warming Potentials (AGWPs) is that the index is specific only to the gas in question; changes in the integrated radiative forcing for CO_2 will not affect the GWPs for all other gases. The disadvantage with AGWPs is that it is no longer a relative measure.

Other GWP Formulations

Fisher et al (37) and WMO (11) presented GWP calculations for various halocarbons integrated over the entire lifetimes of the gases, using CFC-11 as the reference gas. These analyses were primarily interested in the relative radiative forcing for CFCs and potential substitutes.

Various studies (33, 38–39a, 46) have suggested the application of discount rates–future radiative forcing, reflecting the increased uncertainty with time (e.g. to account for the possibility of new technologies). As discounting essentially acts to reduce the effective atmospheric lifetime, it reduces consideration of long-term effects from greenhouse gas emissions. Although discounting can be quite important to policy analyses the type of discounting used will be dependent on the specific policy study.

Other authors, such as Reilly & Richards (40), have argued for the consideration of damage factors, which could offset future discounts in using GWPs. Harvey (41) proposes an alternative GWP index that accounts for the duration of capital investments in the energy sector. The possibility of coupling such factors into a revised GWP definition requires detailed study of the economics and policy implications and is beyond the scope of this report.

EVALUATION OF GWPs

Direct and indirect GWPs have recently been reevaluated for the new international climate and ozone assessments (4, 1). The values for GWPs represent clear improvements in the science over earlier assessments. In prior assessments (5, 6), the decay response curve for CO_2 used in evaluation of the GWPs was based on results from an unbalanced carbon-cycle model that did not correctly calculate the current CO_2 concentration based on a realistic emission history. The new assessments base the CO_2 response curve on results from more sophisticated carbon-cycle models with improved treatments of ocean and biospheric effects that are "balanced," giving a consistent determination of past and present concentrations of CO_2.

GWPs are quite sensitive to the assumed background concentration for CO_2. The constant-concentration background assumed in prior GWPs, and also used in the evaluation presented in Table 4, is clearly unrealistic. This assumption implies that the emission occurs into a system in equilibrium, whereas, the carbon-cycle system is currently far from equilibrium as a result of past emissions and CO_2 concentrations are likely to increase into the foreseeable future.

However, the choice of which future scenario to assume is also uncertain. The definition for the background atmosphere is chosen to provide consistency with evaluations in earlier assessments at the same time, as will be presented below, information is provided that allow determination of GWPs for other assumptions of future CO_2 background concentrations.

Direct GWPs

Table 4 presents a summary of the results for GWPs included in the recent assessments (1, 4). Because their radiative forcing is of similar magnitude, the CFC and halon replacements with the longer atmospheric lifetimes have the largest GWPs. The shorter-lived compounds, such as HCFC-123, HCFC-124, and HFC-32, have appreciably smaller GWPs, particularly at the 100-year and 500-year integration periods. The GWPs for CF_3I are extremely small (< 5

Table 4 Direct Global Warming Potentials, referenced to the AGWP for the Bern carbon-cycle model CO_2 decay response and future CO_2 atmospheric concentrations held constant at current levels (1, 4)

Species	Chemical formula	Global warming potential (time horizon in years) 20	100	500
CO	CO_2	1	1	1
CFC-11	$CFCl_3$	5,000	4,000	1,400
CFC-12	CF_2Cl_2	7,900	8,500	4,200
CFC-113	$C_2F_3Cl_3$	5,000	5,000	2,300
HCFC-22	CF_2HCl	4,300	1,700	520,000
HCFC-123	$C_2F_3HCl_2$	300	93	29
HCFC-124	C_2F_4HCl	1,500	480	150
HCFC-141b	$C_2FH_2Cl_3$	1,800	630	200
Methyl Chloroform	CH_3CCl_3	360	110	35
H-1301	CF_3Br	6,200	5,600	2,200
HFC-32	CH_2F_2	1,800	580	180
HFC-125	C_2HF_5	4,800	3,200	1,100
HFC-134a	CH_2FCF_3	3,300	1,300	420
HFC-152a	$C_2H_4F_2$	460	140	44
HFC-143a	$C_2H_3F_3$	5,200	4,400	1,600
Chloroform	$CHCl_3$	15	5	1
Methylene chloride	CH_2Cl_2	28	9	3
Sulfur hexafluoride	SF_6	16,500	24,900	36,500
Perfluoromethane	CF_4	4,100	6,300	9,800
Perfluoroethane	C_2F_6	8,200	12,500	19,100
Methane[a]	CH_4	62 ± 20	24.5 ± 7.5	7.5 ± 2.5
Nitrous oxide	N_2O	290	320	180
Trifluoroiodomethane	CF_3I	<5	$<<1$	$<<<1$

[a] Includes direct and indirect components

even for the short integration time) because of its few–day atmospheric lifetime (14).

The GWPs for the perfluorocarbons and for SF_6 are all much larger than any of the values for CFCs or halons. The very long atmospheric lifetimes of these gases lead to extremely large GWPs. These large GWPs imply potentially large effects on climate over long timescales with the actual effect on climatic radiative forcing dependent on the magnitude of emissions into the atmosphere.

GWP values for methane in Table 4 include the indirect effects of methane on ozone, water vapor, OH, and CO_2 (see below). The methane GWPs show a range representing uncertainties in these indirect effects.

The direct GWPs given in Table 4 can be readily converted to AGWPs or to GWPs that depend on a changing background atmosphere (e.g. the scenario

Table 5 Absolute GWPs (AGWPs) in units W m^{-2} yr $ppmv^{-1}$ (4)[a]

Case	Time horizon (in years)		
	20	100	500
CO_2, Bern Carbon-Cycle Model, fixed CO_2 (354 ppmv); used in Table 4	0.235	0.768	2.459
CO_2, Bern Carbon-Cycle Model, S650 scenario stabilized at 650 ppmv)[b]	0.225	0.702	2.179
CO_2, Wigley Carbon-Cycle Model, S650 scenario	0.248	0.722	1.957
CO_2, LLNL Carbon-Cycle Model, S450 scenario (stabilized at 450 ppmv)[b]	0.247	0.821	2.823
CO_2, LLNL Carbon-Cycle Model, S650 scenario (stabilized at 650 ppmv)[b]	0.246	0.790	2.477
CO_2, LLNL Carbon-Cycle Model, S750 scenario (stabilized at 750 ppmv)[b]	0.247	0.784	2.472
CO_2-like gas, IPCC decay function,[c] fixed CO_2 (354 ppmv)	0.267	0.964	2.848

[a] Multiply these numbers by 1.291×10^{-3} to convert from per ppmv to per kg.
[b] See Reference 4.
[c] See Reference 5.

labeled S650, developed for the carbon-cycle chapter of IPCC, 1994, which stabilizes CO_2 at 650 ppmv). Table 5 shows the relevant factors necessary to converting to these approaches. To convert to AGWP units, multiply the numbers in Table 4 by the AGWP values for fixed CO_2 from the Bern carbon-cycle model (line 1 in Table 5). By comparing the appropriate AGWPs, one can also determine the effects from using other background scenarios for CO_2.

Also shown in Table 5 is the effect of using different carbon-cycle models (rather than the Bern model used by IPCC); this shows that the assumptions about background CO_2 concentrations and the carbon-cycle model used can affect the GWPs by up to 20% or so.

Indirect GWPs

A large number of possibly important indirect effects resulting from chemical interactions can be identified. Table 6 summarizes some of the more important indirect effects that have been recognized. The effects arising from such processes are difficult to quantify in detail, largely because evaluating their effects depends on uncertainties associated with treating the full interactive chemistry in current models of the atmosphere. GWPs for interactions involving emissions from short-lived gases such as NO_x and nonmethane hydrocarbons are particularly difficult to evaluate because of the complexities and uncertainties associated with tropospheric chemical processes, and cannot be reliably determined at this time. Like NO_x, the inhomogeneous spatial distri-

Table 6 Several of the important indirect effects on GWPs

Species	Indirect effect	Sign of effect on GWP
CH_4	Changes in response times due to changes in tropospheric OH	+
	Production of tropospheric O_3	+
	Production of stratospheric H_2O	+
	Production of CO_2	+
CFCs, HCFCs,	Depletion of stratospheric O_3 bromo-carbons	−
CO	Production of tropospheric O_3	+
	Production of tropospheric CO_2	+
NO_x	Production of tropospheric O_3	+
NMHCs	Production of tropospheric O_3	+
	Production of tropospheric CO_2	+

butions of aerosols make it extremely difficult to determine meaningful GWPs for these atmospheric constituents. Better known are the expected sign of the indirect effects relative to the direct GWPs; the signs of these effects are shown in Table 6.

Analyses of the indirect effects resulting from emissions of methane are considered in IPCC (1994) based on modeling results from several research groups (e.g. 21, 43, 44). Indirect effects from methane include the effects on its response time due to interactions with hydroxyl, effects on tropospheric and stratospheric ozone, effects on stratospheric water vapor, and effects on carbon dioxide produced from dissociation of methane from nonbiogenic sources. The uncertainty range in the total GWP for methane relate primarily to uncertainties in the modeling of effects on the methane response time and on the effect on ozone in the troposphere. These uncertainties largely relate to our lack of knowledge on the distribution of nitrogen oxides in the troposphere and on the simplifications of tropospheric transport and chemical processes in the models used for these evaluations. While current understanding indicates that the GWPs for methane should be within the range shown, there is no basis for selecting one value over another at this time.

Possible effects on ozone from emissions of halons or their replacements also need to be considered. Daniel et al (45; see also 4) have estimated the indirect GWPs for effects on ozone from a variety of halocarbons, including CFCs, halons, and HCFCs in an attempt to clarify the relative radiative roles of different classes of ozone-depleting compounds. Decreased ozone from CFCs and halons should primarily decrease the radiative forcing on climate. They find that the net GWPs of halocarbons depend strongly upon the effectiveness of each compound for ozone destruction. Halons are likely to have negative net GWPs while those of CFCs are likely to be positive over both 20- and 100-year time horizons (4). These analyses, however, are still subject to remaining uncertainties about the cause of ozone decreases in the lower stratosphere.

CONCLUSIONS

Weighing functions, plus other simple measures and indices, can provide powerful tools for policy analyses relating to concerns about ozone and climate. ODPs and GWPs have recently been reevaluated for IPCC (4) and WMO (1) assessments, and these values are recommended for current analyses.

The concept of ODPs provides a relative measure of the expected cumulative impact on stratospheric ozone from trace gas emissions. This concept has been particularly useful in analyses to examine the relative effects on ozone expected from CFCs, halons and other halocarbons currently being used and to evaluate the potential effects of possible replacement compounds.

The concept of GWPs, as developed by IPCC, provides a means for comparing the relative effects on climate expected from various greenhouse gases. GWPs are a relative measure of the radiative forcing on climate from emissions of greenhouse gases. The GWP concept assumes that radiative forcing is a direct indicator of expected climate change. Both direct effects on climate and indirect effects due to chemical interactions can be accounted for in the GWP concept. GWPs reflect the cumulative forcing on climate over a chosen period of time, with the choice of time horizon dependent on policy analysis considerations and on the climatic impacts of concern.

It is important to recognize the limitations and uncertainties associated with these indices. They are designed to provide a simplified measure of the relative effects on ozone and climate from emissions of important gases, but they do not provide a representation of the full complexity of effects on ozone and climate. Both ODPs and GWPs are based on global average responses and do not provide information on more localized or regional effects. The GWP concept is difficult to apply to gases or to aerosols that are unevenly distributed in the troposphere. GWPs do not include climatic or biospheric feedbacks nor do they consider resulting impacts on the environment or social systems; this is both a blessing and a curse, as it means that GWPs are not subject to the significant uncertainties associated with climatic feedbacks, but it also means that these weighing functions are not directly based on quantities some consider drivers for actual policy considerations. The GWP concept is based on the science of greenhouse gas effects on climate and does not account directly for economics and policy implications, such as the use of damage functions.

Users need to bear in mind that these weighing functions were not designed to be used in isolation without considering more fully the many complexities affecting ozone and climate change and the resulting impacts on humanity and the biosphere. For example, policy analyses also need to consider the range of possible scenarios that emissions could undergo from human (and natural) activities and the resulting effects expected on ozone and climate. Users of ODPs and GWPs also need to be aware that advancements in science will likely result in changes in the numerical indices presented here. This implies that uses of such weighing functions and other indices need to be flexible to such changes.

Literature Cited

1. World Meteorological Organization. 1995. *Scientific Assessment of Ozone Depletion, 1994, Global Ozone Research and Monitoring Project—Report* No. 37. Geneva: WMO
2. World Meteorological Organization. 1991. *Scientific Assessment of Ozone Depletion, 1991, Global Ozone Research and Monitoring Project—Report,* No. 25. Geneva: WMO
3. Reinsel GC, Tiao GC, Wuebbles DJ, Kerr JB, Miller AJ, et al. 1994. Seasonal trend analysis of published ground-based and TOMS total ozone data through 1991. *J. Geophys. Res.* 99: 5449–64
4. Intergovernmental Panel on Climate Change. 1994. *Climate Change 1994: Radiative Forcing of Climate Change.* Cambridge, UK: Cambridge Univ. Press
5. Intergovernmental Panel on Climate Change. 1990. *Climate Change, The IPCC Scientific Assessment,* ed. JT Houghton, GJ Jenkins, JJ Ephraums. Cambridge, UK: Cambridge Univ. Press
6. Intergovernmental Panel on Climate Change. 1992. *Climate Change 1992, The Supplementary Report to the IPCC Scientific Assessment,* ed. JT Houghton, BA Callander, SK Varney. Cambridge, UK: Cambridge Univ. Press
7. United Nations Environment Programme. 1992. *Rep. Meeting of the Parties to the Montreal Protocol on Substances that Deplete the Ozone Layer, 4th, Copenhagen, Nov. 23–25, 1992.* Nairobi: UNEP

7a. Wuebbles DJ. 1993. Global climate change due to radiatively active gases. In *Global Atmospheric Chemical Change,* ed. CN Hewitt, WT Sturges. New York: Elsevier Applied Science.

8. Wuebbles DJ. 1981. The relative efficiency of a number of halocarbons for destroying stratospheric ozone. *Lawrence Livermore Natl. Lab. Rep. UCID-18924.* Livermore, Calif.
9. Wuebbles DJ. 1983. Chlorocarbon emission scenarios: potential impact on stratospheric ozone. *J. Geophys. Res.* 88: 1433–43
10. Solomon S, Mills MJ, Height LE, Pollock WH, Tuck AF. 1992. On the evaluation of ozone depletion potentials. *J. Geophys. Res.* 97:825–42
11. World Meteorological Organization. 1989. *Scientific Assessment of Stratospheric Ozone: Global Ozone Research and Monitoring Proj.—Rep.* No. 20. Geneva: WMO
12. Ravishankara AR, Solomon S, Turnipseed AA, Warren RF. 1993. Atmospheric lifetimes of long-lived species. *Science* 259:194–99
13. Wuebbles DJ, Kinnison DE, Grant KE, Lean J. 1991. The effect of solar flux variations and trace gas emissions on recent trends in stratospheric ozone and temperature. *J. Geomagn. Geoelectr.* 43: 709–18
14. Solomon S, Burkholder JB, Ravishankara AR, Garcia RR. 1994. On the ozone depletion and global warming po-

tentials for CF3I. *J. Geophys. Res.* 99: 20929–35
15. Solomon S, Garcia RR, Ravishankara AR. 1994. On the role of iodine in ozone depletion. *J. Geophys. Res.* 99:20491–99
16. Wang WC, Dudek MP, Liang XZ, Kiehl JT. 1991. Inadequacy of effective CO_2 as a proxy in simulating the greenhouse effect of other radiative active gases. *Nature* 350:573–7
17. Taylor K, Penner JE. 1994. Anthropogenic aerosols and climate change. *Nature* 369:734–37
18. Wuebbles DJ, Jain AK, Patten KO, Grant KE. 1995. Sensitivity of direct global warming potentials to key uncertainties. *Climatic Change*. 29:265–97
19. Siegenthaler U, Joos F. 1992. Use of a simple model for studying oceanic tracer distributions and the global carbon cycle. *Tellus* 44B:186–207
20. Prather MJ. 1994. Lifetimes and eigenstates in atmospheric chemistry. *Geophys. Res. Lett.* 21:801–4
21. Wuebbles DJ, Grossman AS, Tamaresis JS, Patten KO, Jain A, Grant KE. 1994. Indirect global warming effects of ozone and stratospheric water vapor induced by surface methane emissions. *Lawrence Livermore Natl. Lab. Rep. UCRL-ID-118061*. Livermore, Calif.
22. Lacis AA, Wuebbles DJ, Logan JA. 1990. Radiative forcing by changes in the vertical distribution of ozone. *J. Geophys. Res.* 95:9971–81
23. Ramaswamy V, Schwarzkopf MD, Shine KP. 1992. Radiative forcing of climate from halocarbon-induced global stratospheric ozone loss. *Nature* 355: 810–22
24. Schwarzkopf MD, Ramaswamy V. 1993. Radiative forcing due to ozone in the 1980s: dependence on altitude of ozone change. *Geophys. Res. Lett.* 20: 205–8
25. Wang WC, Zhuang YC, Bojkov RD. 1993. Climate implications of observed changes in ozone vertical distributions in middle and high latitudes of the northern hemisphere. *Geophys. Res. Lett.* 20: 1567–70
26. Hammond AL, Rodenburg E, Moomaw W. 1990. Accountability in the greenhouse. *Nature* 347:705–6
27. Handel MD. 1991. Letter to Nature. *Nature* 349:468
28. Enting IG, Rodhe H. 1991. Letter to Nature. *Nature* 349:468
29. Kiehl JT, Briegleb BP. 1993. The relative roles of sulfate aerosols and greenhouse gases in climate forcing. *Science* 260:311–14
30. Ko MKW, Sze ND, Molnar G, Prather MJ. 1993. Global warming from chlorofluorocarbons and their alternatives: time scales of chemistry and climate. *Atmos. Environ.* 27:581–87
31. Shi G, Fan X. 1992. Past, present and future climatic forcing due to greenhouse gases. *Adv. Atmos. Sci.* 9:279–86
32. Rodhe H. 1990. A comparison of the contributions of various gases to the greenhouse effect. *Science* 248:1217–19
33. Lashof DA, Ahuja DR. 1990. Relative contributions of greenhouse gas emissions to global warming. *Nature* 344: 529–31
34. Wigley TML. 1995. Ozone depletion effects on halocarbon GWPs and future radiative forcing. *Geophys. Res. Lett.*
35. Caldeira K, Kasting JF. 1993. Global warming on the margin. *Nature* 366: 251–53
36. Wigley TML. 1995. The effect of carbon cycle uncertainties on global warming potentials. *Geophys. Res. Lett.*
37. Fisher DA, Hales CH, Wang W-C, Ko MKW, Sze ND. 1990. Model calculations on the relative effects of CFCs and their replacements on global warming. *Nature* 344:513–16
38. Smith KR, Ahuja DR. 1990. Toward a greenhouse equivalence index: the total exposure analogy. *Climatic Change* 17: 1–7
39. Ellington RT, Meo M, Baugh DE. 1992. The total greenhouse warming potential of technical systems: analysis for decision making. *J. Air Waste Manag. Assoc.* 42:422–28
39a. Schmalensee R. 1993. Comparing greenhouse gases for policy purposes. *Energy J.* 14:245–55
40. Reilly JM, Richards KR. 1993. Climate change damage and the trace gas index issue. *Environ. Res. Econ.* 3:41–61
41. Harvey LD. 1993. A guide to global warming potentials (GWPs). *Energy Policy* 24–34
42. Deleted in proof
43. Brühl C. 1993. The impact of the future scenarios for methane and other chemically active gases on the GWP of methane. *Chemosphere* 26:731–38
44. Clerbaux C, Colin R, Simon PC, Granier C. 1993. Infrared cross sections and global warming potentials of 10 alternative hydrohalocarbons. *J. Geophys. Res.* 98:10491–97
45. Daniel JS, Solomon S, Albritton DL. 1995. On the evaluation of halocarbon radiative forcing and global warming potentials. *J. Geophys. Res.* 100:1271–85

Annu. Rev. Energy Environ. 1995. 20:71–81

ENVIRONMENTAL INDUSTRIES WITH SUBSTANTIAL START-UP COSTS AS CONTRIBUTORS TO TRADE COMPETITIVENESS

William J. Baumol

Professor of Economics and Director, CV Starr Center for Applied Economics, New York University, New York, NY 10003 and Professor (Emeritus) of Economics and Senior Research Economist, Princeton University, Princeton, New Jersey 08540[1]

KEY WORDS: retainable industries, multiple equilibria, acquisition of industries, environmental equipment trade, government support of environmental equipment

---

CONTENTS

ABSTRACT

This article describes a second reason beside externalities that can justify government support of new environmental-protection techniques. A new analysis of industries with high start-up costs and scale economies shows that the country in which such an industry is located is not dictated by comparative

[1]Address correspondence to: Department of Economics, 269 Mercer Street, New York University, New York, NY 10003.

1056-3466/95/1022-0071$05.00

advantage. Hence, gain of a scale-economies activity need not entail loss of another and can increase income in the acquiring country. High start-up costs also impede entry by a prospective rival country, making it easier for the current producer country to retain it. Private enterprise may not find it attractive to undertake such socially-beneficial fields because of the substantial private risks that result from their high start-up costs. So, new environmental-protection techniques with high start-up costs or scale economies can contribute to a nation's prosperity even when unattractive to private enterprise in the absence of government support.

Introduction

In the relatively young and burgeoning industrial field of environmental cleanup and protection, the costs of much of the new technology have declined rapidly, presumably attributable in good part to learning by doing and economies of scale. Recent research by Ralph Gomory and myself (1; RE Gomory & WJ Baumol, submitted) indicates that such industries are what we call retainable, meaning that a country in which they are currently located is likely to be able to keep them despite efforts by other countries to enter the field.[2] In short, a retainable industry may contribute to an economy's trade competitiveness, and is likely to contribute to a country's relative national income.

The main point of this paper is that retainable environmental industries are good candidates for government encouragement because, in addition to their beneficial effects on the environment, their high start-up costs and scale economies make them relatively invulnerable in the world marketplace. I suggest, however, that government support for the purpose should be provided only for a limited time, because once launched, retainable industries should be able to prosper on their own. Governmental programs of support should therefore indicate exactly when or under what circumstances start-up help will end.

Environmental Activities Whose Costs Decline Over Time

The costs of several new techniques for protection and/or cleanup of the environment have declined sharply in the years since their introduction. Young technology is relatively primitive, almost by definition, and workers are naturally inexperienced in its use. These and other handicaps can be overcome only gradually. Still, the number of environmental-protection techniques whose costs have declined since their introduction and the magnitude of savings are impressive. Examples of such technology include substitutes for chlorofluorocarbons (CFCs), new methods for industrial pollution control, and renewable energy sources such as photovoltaic cells, solar-thermal power,

[2]Our work in this arena is closely related to a variety of important recent writings on the role of scale economies in international trade (see e.g. 2).

wind power, and fuel cells. Advanced biomass power–generation techniques, which reduce net carbon dioxide emissions, may reduce the danger of global warming at a lower cost than previously expected.

The magnitude of the decline in cost is easily illustrated. According to Anderson (3, p. 34)

> ...it is still not widely enough appreciated what remarkable developments have taken place over the past two decades. In 1970 the cost of PV [photovoltaic] units was roughly $200,000 per peak KW [kilowatt] of capacity (KWp) in 1989 prices, and applications were largely confined to aerospace programs and other specialized uses; by 1980 costs had fallen five-to-eightfold to the $25,000–50,000 range, and by 1990 another five-to-eightfold decline had occurred, with current costs for grid-tied systems being about $6000/KWp. The US Department of Energy has projected costs of $2100/KWp by year 2000 falling to $1500/KWp a decade later, with costs being prospectively $1000/KWp in the long-term. In high insolation areas [areas that receive a lot of sunshine] this would put generation costs in the range 4 to 6 US cents per KWh, and would make solar schemes both competitive with fossil fuels and less costly than nuclear power, even ignoring environmental factors.

During the 1970s and 1980s, many observers argued that some new environmental-protection techniques would always prove too costly to be practical. Although the subsequent decline in costs has reduced this concern, it also raises questions about the need for government support for the new techniques. If new techniques can become sufficiently cheap to transform them into profitable propositions, why can't private enterprise pursue them on its own? If the techniques can never repay the public sector investment needed to launch them, the justification for the initial outlay requires careful examination.

Externalities, the Gomory Analysis, and Public Encouragement of Private Activities That Benefit the Environment

I turn now to the central issue of this article: how the high start-up costs and scale economies of some environmental-protection techniques affect the case for governmental support of such measures. Environmental issues clearly continue to be dominated by the detrimental externalities of the sort that AC Pigou called to our attention in 1911 (4). However, I focus on externalities of a different sort: the benefits an industry gains from nearby ancillary activities and the social benefits of a growing industry that has overcome the handicap of substantial start-up costs. In this section, I review a new analysis [first devised by Ralph Gomory and then expanded by Gomory and myself (1; RE Gomory & WJ Baumol, submitted)] that deals directly with the implications of industry or product benefits for the welfare of the economy and with the appropriate role of the public sector as it relates to such an industry or product.

The issues I raise and the new analysis I describe apply to many industries other than environmental protection.

Ever since Pigou (4) first made the point, economists have recognized that if an economic activity generates external benefits (i.e. benefits that accrue, at least in part, to persons other than the immediate parties to the transaction), some type of governmental support, most notably a subsidy, will promote the public interest. Industrial activities that facilitate environmental protection or reduce its cost are obvious candidates for subsidies because they benefit society as a whole, and not just the firms that carry out the activities. Subsidies are appropriate for such activities because the incentive to invest in the activities that serve the public interest is reduced by the fact that the supplier does not reap all the benefit. Consequently, in the absence of a subsidy or some other such stimulus, the amount invested in environmental protection by private enterprise will fall short of the quantity that best serves the public interest.

The Gomory analysis strengthens the argument for such governmental subsidies by showing that appropriate amounts of investment in an environmental-protection activity that has substantial start-up costs and scale economies can benefit society economically as well as environmentally. This indirect benefit of such governmental support is promotion of the nation's role as a supplier of the environmental services not only to itself, but also to other nations, in other words, the nation's competitiveness as a supplier of the environmentally-beneficial products and equipment. The industries that supply environmental cleanup or protection equipment are by no means the only fields to which the analysis applies. Yet they are an important case and the Gomory analysis gives additional weight to the argument for public support of investment in the environment.

The Classical Trade Model: The Rule of Comparative Advantage

The Gomory analysis indicates that scale economies and high start-up costs make environmental-protection industries extremely attractive because a nation that enters the field first can hope to become the world's leading supplier of the products, a position from which it cannot easily be dislodged. Thus, these industries may provide an opportunity for a nation to do well by doing good. This scenario is in marked contrast to that implied by the classical model of world production and trade.

The high start-up costs for an industry that, along with scale economies, are the underpinning of the new analysis can result from the absence of experience in the operation of such an industry, from the absence of nearby ancillary industries, and from still other sources beside the need, because of scale economies, to make very large investments before one can hope that entry will succeed.

The textbook version of the classical model of international trade suggests that pursuit of primacy in world trade is tantamount to pursuit of a chimera, a mythical beast that exists only in the minds of an ill-educated majority. In the Ricardian world, industries are allocated among nations by market forces largely on the basis of comparative advantage. Thus, even if a country is an inefficient producer of every good, that nation can still prosper by producing those goods at whose manufacture it is least bad, that is, at which it has a comparative advantage even if it suffers an absolute disadvantage.

The doctrine is illustrated (somewhat inaccurately) by the allegory of a successful attorney who is better at typing than her secretary but is driven by market forces to devote her working time exclusively to the more remunerative lawyering activity, leaving the typing to her secretary. Analogously, market forces will drive each country to produce only those items for which it has a comparative advantage. Any industries potentially offering a comparative advantage will be sought out by profit-hunting entrepreneurs who will see entry and reallocation of some of the nation's labor force into this industry as an opportunity to gain, just as our lawyer gained by shedding her secretarial tasks.

A striking implication of the analysis is that for every industry a nation gains, it can expect the gain to be offset by loss of another. If one starts from an international trade equilibrium, a nation can expect a change in its industries only if technological or other related developments change the industries in which it has a comparative advantage. But by definition, when the comparative advantage in one industry increases, then in at least one other industry its comparative advantage must be reduced. If our attorney's skill and earning power in legal activity grows, her comparative advantage in an ancillary activity such as library research must have fallen, so that it may become appropriate for her to abandon this activity to a paralegal with research skills inferior to her own.

Thus, the classical model suggests that a nation should not expect to gain substantially from the acquisition of an industry, because benefits from the acquisition will be offset, at least in part, by the detrimental effects of the loss of another industry. Hence, in the classical trade model scenario, the pursuit of industries for acquisition is generally a misdirected effort.

The Gomory Analysis in a Setting of Scale Economies and Substantial Start-up Costs

The Ricardian model relies on the assumption that the world's industries do not have scale economies or significant start-up costs. In such a world, by definition, production on a small scale is cheaper than on a large scale, so production of a particular good can be expected in a substantial number of countries, with no one country producing the bulk of world output. This is, in fact, what we see in agricultural products, where scale diseconomies do indeed prevail beyond some fairly modest level of output. And in such products successful entry can occur on a small scale precisely because smaller is more efficient. The Ricardian model describes a world where pursuit of new industries has little purpose, as we have seen.

More recent work, culminating in Gomory's analysis, differs fundamentally from the Ricardian model by assuming economies of scale and high start-up costs, which prevail in many actual industries. Such circumstances tend to produce specialization. Rather than many countries producing a product, as in the case of wheat, coffee, or clothing, one or a few countries will capture the role of producer. Scale economies mean that larger-volume production is cheaper, so a country with a high output of good Z is largely immune from competition from small-scale producers of Z in other countries. Indeed, if the scale economies are not exhausted beyond some level of output, a natural monopoly will result, because the larger an economy's output, the better positioned it is to outdo smaller producer countries. In addition, high start-up costs will inhibit entry by prospective rivals.

A more startling implication of the scale economies/high start-up cost model is that any one of the many possible patterns of specialization may be selected by market forces. For example, a situation in which France alone produces good A and the Czech Republic is the sole producer of good B is no more ruled out than the reverse. The presence of scale economies and high start-up costs dictates that once a country achieves a large market share in some industry, its position in that industry becomes relatively invulnerable to attack. An enormous range of alternative outcomes in the market mechanism's allocation of industries among countries is therefore possible. Moreover, a country can hope to influence the outcome by taking steps to become a relatively large producer in several industries. High start-up costs mean that such a program is not cheap or easy, but the expenditure can eventually reward the first entrant by making entry difficult for would-be followers.

Another implication of the Gomory analysis is that a country can become the leader in many industries with scale economies, leaving diminishing-return industries such as agriculture to be shared with other nations. In the world of scale economies and substantial start-up costs, comparative advantage no longer determines industry distribution among nations because the savings achieved through large-volume production can offset comparative disadvantage. Thus, if a nation captures one industry, it is not generally condemned to lose another. A country that is fortunate or skilled in the acquisition of industries may become a leader in a substantial share of the world's industries, whereas another country may suffer the opposite fate.

Finally, in a world of scale economies and high start-up costs the acquisition of an industry generally raises the relative national income of the acquiring country.[3] For our purposes here, the main implication of the Gomory analysis is that it is worth pursuing new industries with high start-up costs and scale economies, unlike the classical world in which there is no corresponding reward for the acquisition of industries.

Why the Free Market May Be Incapable of the Task, Particularly in Environment-Benefiting Industries

One may well ask why acquisition of promising industries cannot simply be left to private enterprise. Are there not profit-seeking entrepreneurs ready to jump at new opportunities and do whatever is necessary to launch an industry and expand its output to achieve the low costs that protect it from competition?

There are several answers, as we will see, and they are particularly strong in the case of industries that benefit the environment. Success in start-up environmental industries may require the adoption of both parts of a two-pronged policy. First, the industry may not be viable without an arrangement such as an emission charge that reduces the environmentally damaging activities. Second, new industries require stimulation of the investment and, hence, the risk of large amounts of capital. Because of scale economies and start-up costs, an entrepreneur cannot succeed by embarking on such a project incrementally. Investment sufficient to permit operation on a substantial scale and to cover the start-up costs is required from the beginning. The risk this entails can prevent private enterprise from pursuing such opportunities.

These difficulties are exacerbated by the fact that investment in an industry in which one has little previous experience, and on a large scale to boot, entails risks that may be far greater than those normally accepted in business. Furthermore, the risks to the private investors may be far greater than the risks to society. Economic history provides many examples of new industries in which the initial investors lost all they put in, but others acquired the assets at distress-sale prices, using them to provide substantial returns both to their new owners and the community.

Furthermore, the success of such a new venture may depend on the availability of ancillary industries that can supply information and labor skills, as well as inputs, at reasonable cost, or purchase its products on profitable terms. Entrepreneurs hoping to launch new industries cannot be expected to take on the task of creating support industries as well.

[3]An important exception to this generalization is the situation in which industry acquisition is carried too far, even from the selfish point of view of the acquiring nation. At some point, the only industries left for a nation to acquire are those in which it has a marked comparative disadvantage. Reassignment of its labor force to these industries is similar to a decision by our attorney again to do the secretarial jobs in her office. This change reduces the net earnings of the entire enterprise, meaning a possible loss in income to her, even if her share of the incomes earned at the office rises to 100%.

In addition, when new industries are launched, much of the resulting benefit may go to those who supply or use the products of the new venture. As we have seen, when substantial portions of the benefit go to others, the level of investment in a new industry will probably be lower than that called for by the public interest. In an environmental-protection industry, much of the benefit is by its nature external, thus reducing the profit prospects for private investors and making pursuit of new ventures without government assistance unattractive. The prospect that the investment can benefit the nation in two ways, by environmental improvement and by addition to the nation's economic prosperity, in accord with the Gomory analysis, increases the likelihood that the cost of launching the industry will be justified from the viewpoint of the nation as a whole, even if it does not promise to be profitable to private investors.

These obstacles to establishment of environmental-protection industries by the unaided effort of private enterprise also apply to investment in the research and development required to design workable new technology. The economic literature has long emphasized the forces that contribute to underinvestment in innovation. As with the start-up of new industries, much of the benefit of innovation accrues to persons other than those who undertake the risk and provide the necessary effort. Such spillovers, along with the risk of large investment and the need for ancillary industries, must be overcome if private enterprise is to have the incentive to invest sufficiently in socially beneficial innovation for environmental protection.

One of the objections most frequently raised to governmental stimulation of environmental-protection industries is that it requires the government to pick future "winner" industries to which to devote its largess. A winner industry is presumably one whose outputs will prove to be in great demand in the future. But the selections by bureaucrats are likely to be distorted by political pressures, and government officials are probably less capable than private entrepreneurs at judging the prospects of industries in which the latter have direct experience.

However, experience in innovative environmental-protection techniques is often lacked by public servants and businesspersons alike, particularly experience entailing investment on a large scale. Widely held reservations about government's ability to select promising industries for public support are not groundless. Nevertheless, when an industry is launched, there may be little relevant experience anywhere in the economy, and in those cases private investors lose much of their edge.

Moreover, the new analysis suggests that retainability rather than prospect of winning is the more appropriate criterion for selection of industries that merit public support. It is much easier to determine whether an industry is retainable currently than to judge reliably whether it is likely to be a future winner.

To determine whether an industry is retainable, one must determine its start-up costs and whether it benefits from scale economies. Any industry with substantial scale economies and/or large start-up costs can be reasonably deemed retainable if these attributes seem unlikely to evaporate in the near future. For example, they will evaporate if start-up costs need be incurred only once, as in getting the bugs out of a new type of equipment, so that all future users are spared replication of the initial outlays. In contrast, if high start-up costs must be incurred by every entrant and scale economies are substantial, the industry is retainable, and a government agency should be reasonably able to determine if this is true of an industry.

What Form Should Governmental Encouragement Take?

The public interest is promoted more effectively by certain types of public support for emerging environmental-protection technology. Some of the measures advocated for the purpose are gratuitously damaging to economic welfare because they directly impede market choice, such as special quotas, and unjustifiable product safety requirements really adopted to impede competing imports (which are ultimately paid for by consumers who get little benefit in return). Various other restrictive measures often employed to protect an infant industry are far more damaging to the public interest than a simple tariff. Of course, neither quotas nor tariffs are likely to promote the public interest, but this example illustrates the sort of misunderstanding that economic analysis can help the practitioner to avoid.

Two complementary approaches seem more promising. First, as economic analysis teaches, when a product yields an external benefit, like that likely to derive from equipment that improves the environment, a direct subsidy for the item's production serves the public interest. For each unit of the item produced, the subvention should approximate the magnitude of the item's marginal social benefit. Such a subsidy restores to investors the benefit from the product that they would otherwise lose to others—it compensates them for benefits lost to spillovers. Investors are thereby provided with the incentive to invest the amount called for by the public interest. A direct subsidy of this sort, then, is a form of support for an environmental-protection product that is firmly recommended by economic analysis.

Subsidies provided by one administration can, however, all too easily be withdrawn, often rapidly and without advance notice, by another. The risk of such a change in policy obviously undermines the effectiveness of a subsidy program and limits the amount of investment private enterprise is willing to devote to a given project. Banks that provide investment for subsidized projects are said, consequently, to prefer public sector contributions to capital costs (as opposed to operating costs), because if support policies are reversed the investments in capital equipment can still be used and may even offer a profitable return.

The Pigouvian analysis also suggests that government subsidies for capital constitute a promising form of public support because of imperfections in the capital market and because investors bear all the capital cost but derive only part of the benefit of investment in projects that are promising to society. Capital cost subsidies are an appropriate remedy because they address the source of the problem directly.

Another point in favor of capital subsidies is that their incentive effects are relatively neutral. Some programs tend to dictate business decisions, as is true, for example, of subsidies favoring the installation of scrubbers in smokestacks rather than the use of low-sulfur coal.[4] Government agencies are often at their worst when they use this sort of technique-specific approach because they deny businesspersons the freedom of decision on matters in which they are often at their best. By allowing businesspersons to use capital as they see fit within reasonable limits, a government program will gain the best from both the public and private sector worlds.

Subsidies: Permanent or Temporary?

Do environmental-protection subsidies need to continue indefinitely, or should they be provided only while a new industry or type of technology is establishing itself? The answer is mixed. If the subsidy is provided to internalize beneficial externalities, it is not appropriate to offer it only temporarily because the process will indefinitely offer those who supply it less of a return than that gained by society unless the public sector makes up the difference. If that difference is not made up, then private investment will be reduced below the level that the public interest requires. This result can be expected to follow any future discontinuation of the subsidy.

On the other hand, if the subsidy is intended to facilitate the establishment of a retainable industry, then it is appropriate to discontinue the subvention once the industry attains maturity, as long as the industry is in fact retainable. However, this subsidy should be withdrawn in a manner that does not subject the socially valuable activity to unnecessary risk. Government should make clear in advance exactly when the subsidy will be discontinued: after x years, or after industry revenues reach some prespecified level, or after some other

[4]Here, a reviewer asks, quite appropriately, "What about the classic points that capital subsidies distort choice of techniques and fail to reward production?" This problem is undeniable unless public-sector support of capital is a Pigouvian subsidy corresponding to an unrecognized external benefit of investment. In any event, a global measure such as a general capital subsidy still seems preferable to narrowly targeted measures that invite debilitating bureaucratic controls and are apt to have unpredictable consequences.

readily observable criterion is satisfied. Failure to undertake such a precommitment not only risks prolongation of government funding of the industry long after any justification has evaporated, but it also increases the uncertainty with which the industry must deal, raising the costs that will ultimately be borne by the general public.

ACKNOWLEDGMENT

I am very grateful to the Alfred P. Sloan Foundation and the CV Starr Center at New York University for their support of the work reported in this paper. I also thank Dennis Anderson for suggesting this topic to me and for providing several studies that were extremely helpful for the preparation of this article. I must apologize for the number of references to my own work, but I have no other sources for the results cited.

Literature Cited

1. Gomory RE. 1994. A Ricardo model with economies of scale. *J. Econ. Theory* 62:394–419
2. Helpman, Elhanan, Krugman PR. 1985. *Market Structure and Foreign Trade.* Cambridge, MA: MIT Press
3. Anderson D. 1991. Energy and the environment. In *Special Briefing Paper No. 1.* Edinburgh: Wealth Nations Found.
4. Pigou, AC. 1911. *The Economics of Welfare,* London: MacMillan 1st ed.

Annu. Rev. Energy Environ. 1995. 20:83–118

YUCCA MOUNTAIN: A CRISIS FOR POLICY: Prospects for America's High-Level Nuclear Waste Program

James Flynn and Paul Slovic
Decision Research, 1201 Oak Street, Eugene, Oregon 97401-3575

KEY WORDS: Nuclear Waste Policy Act, public perceptions of high-level nuclear waste, opposition to high-level nuclear waste facilities

---

CONTENTS

ABSTRACT

The federal government's attempts to site a permanent high-level radioactive waste repository have been frustrating and so far unsuccessful. Many of the

1056-3466/95/1022-0083$05.00

problems were recognized more than a decade ago. In 1982, the US Congress passed the Nuclear Waste Policy Act and established a program to site and develop an underground repository. By 1987, this program was widely considered to be a failure. It was plagued by public opposition, intergovernmental conflict, poor management, scientific questions and concerns, and substantial cost overruns. In December 1987 Congress amended the act and selected Yucca Mountain, Nevada, as the only location to be studied as a potential repository. This halted further work on other potential sites in the western US, the eastern US sites for a second repository, and the monitored retrievable system facility that the US Department of Energy (DOE) wanted to locate in Tennessee. These amendments did not address the basic causes for the failures between 1983 and 1987. As a result, although the program was greatly simplified with only one site, it was beset by the same problems that created the original collapse. Now there are new and widespread calls for a congressional review and restructuring of the program from federal, state, and local officials, government agencies, industry, and public groups. This article examines the lessons from these two failed attempts and makes recommendations for devising a new policy.

BACKGROUND

Two forms of radioactive waste must be managed and disposed of in a high-level nuclear waste (HLNW) program. One is spent fuel, which is the uranium-based fuel that has been irradiated in a nuclear power reactor to the point where it no longer contributes efficiently to the nuclear chain reaction. Spent fuel is highly radioactive and thermally hot. The second form, often called defense waste, is created when highly radioactive materials are reprocessed from spent fuel, primarily in connection with nuclear weapons production. This form of HLNW is produced through a chemical process and attains a liquid form, which must be solidified for disposal.

The amount of HLNW to be managed and disposed of depends on future nuclear power generation and production of nuclear weapons. Current estimates, assuming no new nuclear power plant construction, are for about 100,000 metric tons of heavy metal (MTHM) from commercial reactors and perhaps the equivalent of 15,000 MTHM from the nuclear weapons complex (1). Commercial spent fuel is now stored at more than 70 sites, the majority located east of the Mississippi River. The lack of a HLNW program acceptable to the public has been repeatedly cited as one of the major reasons for the current moribund condition of the nuclear power industry and its poor prospects for the future. The possibility that HLNW might be stored for extended periods at or near the reactors where it was produced, in lieu of an operating waste facility, has motivated the communities and states near these reactors to call for prompt and effective actions to remove those wastes.

The federal government has assumed responsibility for final disposal of HLNW. For more than 30 years, the Atomic Energy Commission, the Energy Research and Development Agency, and the US Department of Energy (DOE) (successor agencies) have attempted to locate and develop a permanent underground repository site. In 1982, the US Congress passed the Nuclear Waste Policy Act, establishing the goals and processes for HLNW disposal. This legislation has been modified several times by subsequent congressional action, and major revisions are being considered in Congress as this chapter goes to press.

This chapter focuses on the social and political issues that directly affect relations between the federal government and Nevada's state government concerning Yucca Mountain, Nevada, as a site for the nation's first HLNW repository. Many other subjects are of great importance to the management of HLNW, and the literature on each is voluminous. Transportation is a major issue and in the case of Yucca Mountain would involve more than 40 states and hundreds of cities and smaller communities. Other important issues relate to whether a permanent underground repository is the optimal choice, the ability of science to provide adequate information and forecasts for repository performance, how the program should be regulated, institutional control of a repository over long periods of time, the process by which public policy and programs should be managed and decisions made and shared, the potential effects of future scientific and technical developments on managing HLNW, the potential social and cultural impacts on citizens, groups, and communities, and the real costs of the various options. In writing on the subject we have chosen, we do not intend to slight or deny these other essential topics. Rather, we wish to contribute to an understanding of one set of reasons for the existing impasse in assembling a viable national policy for HLNW storage. In the concluding section, we outline some guidelines that we believe will greatly improve the chances for establishing a socially and politically acceptable program.

THE FEDERAL HIGH-LEVEL NUCLEAR WASTE PROGRAM IN 1995

Current federal policies for managing HLNW in the US are failing. The existing federal program is years behind schedule and billions of dollars over budget. The site at Yucca Mountain, Nevada, presents difficult and unusual technical problems and is not guaranteed to result in a functioning facility. DOE has struggled unsuccessfully for over a decade to bring the program under control. Moreover, because the state of Nevada is adamantly opposed to hosting such a facility, the struggle between state and federal governments could raise important constitutional questions with uncertain results. Countless parties interested in the program—from federal government agencies to electric power utilities with nuclear reactors to affected state and local governments, congressional representatives, environmental and public advocacy groups, and independent policy analysts—have called for a review and restructuring of the existing program.

In the most optimistic view, the program at Yucca Mountain must be seen as a troubled and high-risk effort. The federal government is gambling that Yucca Mountain will be suitable and acceptable. Success is possible, but serious doubts remain about this site. And there is no alternative site, creating more doubts about the HLNW repository program. According to the director of the DOE Office of Civilian Radioactive Waste Management (OCRWM), no contingency plan exists in the event Yucca Mountain is found unsuitable. The director characterized this as a great deficiency because DOE must return to Congress for new instructions if it finds that Yucca Mountain will not do. For all practical purposes, returning to Congress means starting over, and choosing, studying, and developing another site could take several decades (2).

The apparent strategy of DOE and its allies in Congress is to force the state of Nevada to accept a HLNW facility at Yucca Mountain. However, even if this strategy prevails and a repository or monitored retrievable storage (MRS) facility is located at Yucca Mountain, the public policy must still be considered a failure. Such an assertion needs explanation. How could development of a repository be considered evidence of a failed public policy?

This question can be placed in perspective by other questions. How could the program be considered a success if a repository were built only at exorbitant cost, after a long and bitter intergovernmental struggle, and without a fair and equitable process and outcome according to community, state, and public opinions? How could the program be considered a success if a special set of environmental rules were devised to uniquely qualify a site that would otherwise be disqualified? In addition, wouldn't such a process cast doubt on the federal government's ability to manage technological hazards of all kinds?

The federal government claims the exclusive right to regulate radioactive waste activities. It is also the originator, developer, and promotor of modern nuclear technologies (i.e. nuclear weapons production and testing, nuclear power generation, radioactive waste management). Thus, the federal government is motivated to act as the major stakeholder in attempts to construct HLNW storage facilities. This position inevitably puts the federal government in opposition with affected states and communities because the local response has been unswerving opposition. State and local opposition is based on deeply held negative public opinions and attitudes about living near radioactive waste facilities.

The crux of the public policy problem is the opposition of the public, state governments, and local communities to federal policy on HLNW management. This opposition is based on (*a*) concern about controlling health and environmental hazards of radiation; (*b*) distrust of DOE, other federal agencies assigned regulatory oversight, the nuclear power industry, and Congress; and (*c*) a belief that project proponents are unwilling to address a range of potentially serious socioeconomic impacts, such as stigmatization of host areas.

THE NUCLEAR WASTE POLICY ACT OF 1982

The current program is a far cry from congressional intentions when the Nuclear Waste Policy Act (NWPA) was passed in 1982. The background to that legislation is discussed by a number of writers and will only be summarized here (3–6).

From the beginning, the US nuclear waste management program has had a fractious and troubled history that pitted federal efforts to locate waste storage sites against the opposition of states and communities. In several early attempts, for example at Lyons, Kansas, and Alpena, Michigan, the federal government was forced to abandon its plans for a facility to store HLNW (4).

The NWPA attempted to address some long-standing problems revealed in these early failures by establishing a set of democratic principles and ethical rules to guide the process of selecting a repository site (7–9). Several principles were designed to produce an equitable outcome. The NWPA mandated the selection of two repositories: one in the West, where DOE had already done some site characterization studies, and one in the East, where most of the US's nuclear wastes are generated. The NWPA mandated that those who benefit from the repository pay for it and required fees charged on nuclear-generated energy to go into a Nuclear Waste Fund to finance repository development. Monetary compensation was authorized for those who would live near the repository.

Other provisions of the NWPA were intended to ensure an equitable process of site selection. They required DOE to provide information about all activities associated with selecting and building the repository, including scientific data and analyses, to affected stakeholders (i.e. state governments, Indian tribes, and the public). Program managers were to employ an objective selection process based on technical criteria. In addition, the decision process used to select or eliminate candidate sites was to be open to outside scrutiny, making it difficult for DOE to select a site on arbitrary or capricious grounds. DOE was to consult and cooperate with affected states and Indian tribes before making key decisions. (However, this provision did not give stakeholders authority to control the siting process.) Monies from the Nuclear Waste Fund were provided to affected states and Indian tribes to oversee DOE studies and conduct socioeconomic assessments.

In addition, the NWPA permitted any state designated to host a repository to file a notice of disapproval—essentially a veto of the site. Although this provision exceeded previously granted state authority over the siting of a federal facility, it was still a weak power because any state veto could be overturned by Congress.

The NWPA contained important provisions to protect public health. The act instructed the US Environmental Protection Agency (EPA) to set radiation exposure standards. EPA subsequently determined that a repository could cause no more than 1000 excess cancer deaths over its 10,000-year lifetime. DOE was to demonstrate how it would meet these standards, which would be used by the US Nuclear Regulatory Commission (NRC) in deciding whether to grant a construction license. DOE was exempted from preparing an environmental impact statement or obtaining NRC permits and licenses for on-site studies and investigations (called site characterization).

The NWPA appeared to many observers to be a reasonable political compromise—a good faith effort to forge a successful siting process. The equity and public safety provisions were intended to ensure fairness in site selection and make the eventual choice acceptable to those affected directly. And, indeed, the NWPA appeared to have succeeded—at least to the extent of attracting, in 1982, the support of most congressional representatives from states then identified as potential repository host sites.

RESISTANCE, PUBLIC OPINION, AND POLITICS BETWEEN 1983 AND 1987

The NWPA, which President Reagan signed into law on January 7, 1983, authorized the creation of the Office of Civilian Radioactive Waste Management (OCRWM) within DOE. The history of OCRWM from 1983 to 1987 records a shaky and uncertain start that led to collapse. By 1987 OCRWM's inability to continue became clear (4).

Potential first-round sites, mostly in the West, were limited to those already under investigation by DOE. These sites were allowed under a grandfather clause to meet the demands of Congress' optimistic schedule, which called for opening the first repository by 1998. In May 1986, sites in Nevada, Washington, and Texas were selected for formal site characterization. Three possible sites in Tennessee were identified for an MRS facility. The second-round selection in the East involved preliminary identification of 235 sites with an eventual reduction to 20 locations in seven states (Wisconsin, Minnesota, Maine, New Hampshire, Virginia, Georgia, and North Carolina).

These three major elements of the DOE program—the first-round selection, the MRS site selection, and the second-round selection—were strongly resisted by states and the public (10). The selection of the final three sites for the first-round repository, especially the inclusion of Hanford, Washington, was criticized widely. As a result, DOE's decision process was questioned. The Board on Radioactive Waste Management of the National Academy of Sciences was asked to review DOE's multiattribute utility analysis approach. Although they endorsed the approach, the board did not review the data or critique the final choices (11). Other reviewers were critical of the final choice and suggested that DOE had established a reputable technique but then had abandoned it to make a politically motivated decision (12–14).

While the community of Oak Ridge, Tennessee, supported an MRS facility, the state of Tennessee and the governor were strongly opposed. They claimed that the state could be stigmatized and its economic development efforts undermined. Other concerns expressed in the more than 2000 pages of reports and appendices that state researchers prepared in response to the MRS proposals were about the need for and efficiency of an MRS, the ability of DOE to operate a facility, DOE's lack of experience as an NRC licensee, the opposition of the majority of Tennessee citizens, and the claim that scientific data used to evaluate sites were flawed and inadequate (15).

In early 1986, about 18,000 people attended public briefings and hearings on the second-round sites, and more than 60,000 comments were provided for DOE review (16). Congressional office holders running for reelection in 1986 from the seven second-round states felt threatened by association with the second-round selection process (17). DOE suspended further work on these sites in May 1986, with the explanation that revised and lowered projections of HLNW meant that a second repository could be deferred. The head of OCRWM stated that politics was not a part of the decision, but internal DOE documents specifically identified "immediate political relief" as an important benefit of ending the second-round site work (4).

A widespread failure of intergovernmental relations became apparent during these early years of the program. States filed numerous law suits against the program. This parade to court was joined by public advocacy, community-based, and industry groups. In 1986, the US General Accounting Office (GAO) began to include a section on litigation with subsections entitled pending cases, new litigation this quarter, and completed litigation in their quarterly reports to Congress. At one time more than 20 cases, some of which grouped plaintiffs, were active (18).[1]

Several other problem areas were apparent by 1987. DOE management and administration of the program was widely considered inadequate. According to the GAO series of quarterly reports, OCRWM reorganized in July 1984 and made further changes in September 1985, reassigning responsibilities to new

[1]See GAO quarterly reports to the Senate Committee on Energy and Natural Resources from October 1985 to September 1987.

offices and creating revised administrative lines of reporting. Further organizational changes were made in 1986 and 1987. GAO reported that program officials said they had difficulty hiring quality personnel, owing in part to the noncompetitive federal salary schedule.

Schedules began to slip shortly after the program began. For example, the Mission Plan, which was intended to guide the entire program, was delayed from August 1984 to July 1985 (19).

On April 30, 1985, the president, following a DOE recommendation, exercised an option under the NWPA to include defense HLNW with HLNW from civilian nuclear power reactors as part of the OCRWM program. Thus, the program became the US's only effort to provide for disposal of HLNW. The combination made sense in many ways. Economies of scale were achieved because DOE was responsible for both the US's nuclear weapons program and the HLNW disposal program. DOE was able to focus on a single HLNW program, rather than two separate efforts.

At the same time, this program had the unfortunate effect of linking the civilian nuclear industry's HLNW problems with the nuclear weapons complex, just when revelations of past practices at the weapons complex became sensational stories in the news media. These stories included reports of hazardous conditions at places such as Hanford, Washington, Fernald, Ohio, Rocky Flats, Colorado, and Savannah, Georgia. News accounts of accidental and deliberate releases of radioactivity from these weapons sites during the Cold War were broadcast widely. These stories damaged the reputation of DOE and its claims to provide expert management of radioactive processes and wastes (20).

The issue of quality assurance and DOE oversight of contractor work was highly visible by 1987. DOE was forced to issue stop work orders to key contractors such as the national laboratories, the US Geological Survey, and private management and consulting companies (21). NRC, as early as 1985, warned that the DOE quality assurance plan did not meet regulatory requirements, an opinion reiterated in 1986 and again in 1988 (22).

In 1987 DOE changed the opening date of the first repository from 1998 to 2003 for partial operations and 2008 for full operations. This rescheduling, though inevitable, was seen as a sign of DOE management failure. The change also highlighted the possibility that DOE might not be able to accept HLNW from the nuclear power utilities in 1998 according to the NWPA schedule and as DOE had agreed to do in contracts signed in 1985. Numerous other requirements of the NWPA, such as the mandate that DOE negotiate cooperation and consultation agreements with potential host states, were not completed. In fact, these agreements have never been negotiated.

Other problems in dealing with intergovernmental and public stakeholder issues, especially the role of states in program oversight and decision processes, were a source of conflict. A February 1987 report by GAO on institutional relations describes these issues in understated terms (23). DOE's 1986 decision to abandon the search for a second repository site in the East, with the claim that the single western facility would meet the US's needs, stands out (24). This decision was seen as a capitulation, in the midst of an election period, to the political interests of congressional representatives from populous eastern states that were being considered for sites. This decision especially angered political leaders from the three western states with sites selected for study in the first round.

Program cost estimates escalated rapidly between 1982 and 1987 and were the subject of congressional concern. In a September 1987 report, GAO pointed out that the 1981 DOE estimates of site characterization were $60 to 80 million per site. By 1987 these costs were estimated at close to $2 billion per site (25).

We do not intend to blame DOE for these failures. The DOE implementation of the program had many limitations and problems, and DOE faced tasks it was poorly equipped to handle in terms of management skills, department culture, past performance, and public reputation. However, the real failure was one of congressional policy. The purpose of this section is to make clear the problem faced by Congress in 1987 when it was forced to do something about the program. How well, or badly, Congress acted in passing the 1987 revision of the US's HLNW policy can only be evaluated by understanding what went wrong in the original formulation of the 1982 NWPA, which became clear between 1983 and 1987.

The discussion above outlines the problems introduced by the NWPA of 1982, which can be grouped into four main areas. First, the choice of DOE as the agency to implement the NWPA was unfortunate for many reasons. Any agency or office mandated to site a HLNW repository might eventually face extreme public opposition, as DOE personnel have claimed at numerous public meetings. However, few agencies would have brought to the task the full range of liabilities that DOE did. These liabilities included no past experience with meeting the environmental regulations or regulatory performance standards of other federal agencies such as the NRC and EPA; a very restricted ability to conduct relations with various other government entities (e.g. states, Indian tribes, counties, communities, multigovernmental associations); and a DOE culture that was formed during the Cold War when secrecy and suspicion surrounded nuclear weapons production. Even the O'Scannlain Panel, appointed by the Secretary of Energy, recommended against DOE management of the HLNW repository program in a 1984 report. The panel cited concerns about administration, management, fiscal ability, credibility, and technical competence (26).

Second, the NWPA mandated an impossible schedule that almost by definition precluded a scientifically credible process for selecting and developing a repository site. This process was further compromised by the general expectation that some kind of comparative evaluation process within and among various geological options would take place.

Third, public opinion was stronger and its opposition to HLNW facilities much more adamant than was recognized during the process of drafting the NWPA. What became clear between 1983 and 1987 was that a majority of citizens in states identified as a potential site of a HLNW facility felt victimized by the federal government. Opposition from the public and their state representatives was rapidly mobilized and was highly effective in frustrating federal efforts (10).

Fourth, although Congress recognized that public opposition and concerns about HLNW facilities were potentially critical limitations in developing any repository, no effective policy initiatives were developed to address these problems. What was developed was a series of statements about cooperation, consultation, public involvement, and oversight for the HLNW program. These statements stop short of providing any decision authority outside the direct control of Congress and the federal agencies—DOE, NRC, and EPA. In other words, the NWPA allowed states, Indian tribes, and other stakeholders to play roles but gave them no authority over important program decisions. A case in point is the state veto, which can be overridden by vote of Congress. This veto will come at a point in the process when the entire US program depends on overriding it. Thus, the veto will become meaningless, because after the final site recommendation is made no other options will be available. For DOE the message was certainly clear that the final arbiter was Congress.

What was poorly understood was that Congress could not mandate an end to public opposition in the same manner that it mandated DOE to undertake a HLNW program. In fact, Congress made the opposite assumption that writing the NWPA as the nation's first congressionally mandated policy on HLNW would make the public accept the program outcomes. This was not the case. A full review in 1987, which Congress struggled to avoid, should have cleared up this misunderstanding. The evidence of widespread and effective public opposition between 1983 and 1987 to the first-round and second-round selection processes and the MRS projects was voluminous.

THE NUCLEAR WASTE POLICY ACT AMENDMENTS

When the failure of DOE's HLNW program became clear to Congress in 1987, two options presented themselves (4, 5). Congress could revisit their earlier policy decision and write new legislation. This choice promised a significant delay in the program and an uncertain outcome, given the intense interest of stakeholders. However, this option had important advocates. Representative Morris Udall of Arizona recommended a moratorium on the program and a systematic policy review by Congress. The other option was to modify the

NWPA in some way and attempt to get the program back on track. This latter approach, taken by Senator J Bennett Johnston (D-La.), who is credited as the author of the 1987 amendments, prevailed (5).

The Nuclear Waste Policy Act Amendments of 1987 (NWPAA) made several changes in the US's program and restructured much of the policy perspective that had been incorporated in the NWPA (27). The NWPAA selected only Yucca Mountain, Nevada, for study as a potential repository site and directed DOE to terminate all work at Hanford, Washington, and Deaf Smith County, Texas. If Yucca Mountain proves unsuitable, DOE will report back to Congress for further guidance.

The second-round program was cancelled because of public opposition from the midwestern, northeastern, and southern states that had been identified as possible second-round repository study sites (28).

Congress cancelled DOE plans for an MRS facility in Tennessee, even though the state had lost a court case challenging the selection, and the local community of Oak Ridge supported it. The NWPAA established a commission to examine the need for an MRS and prepare a report (27). The NWPAA also specified that a future MRS could not be selected until the president approves a site for development of a repository [Section 145(b)]. This section assures that an MRS facility will not become a de facto permanent HLNW facility, and Section 145(g) states that no MRS may be constructed in Nevada, assuring that Nevada will not be forced to accept the whole burden of the nation's HLNW. The amendments also provided for an Office of Nuclear Waste Negotiator (Title IV, sections 401–411) to conduct negotiations with communities willing to discuss hosting an MRS facility or a HLNW repository.

Phil Sharp, a Democrat from Indiana, then chair of the House Subcommittee on Energy and Power, was very concerned about oversight issues and was largely responsible for establishing the independent Nuclear Waste Technical Review Board (Title V, sections 501–510) to evaluate the technical and scientific validity of activities undertaken by the Secretary of Energy.

The NWPAA provided participation rights and financial assistance to units of affected local governments similar to those originally afforded to states and Indian tribes. An affected local government was described as a unit of local government "with jurisdiction over the site of the repository or monitored retrievable storage facility" and at the discretion of the Secretary of Energy could include "units of local government that are contiguous with such unit" (27). In the case of Yucca Mountain, affected units of local government were Nye County, the host jurisdiction, and nine contiguous counties, including Clark County, where Las Vegas is located, and Inyo County in California.

The amendments authorized payment of benefits to potential repository host states, Indian tribes, and local governments in the amount of $10 million per year during site characterization and $20 million annually once waste was delivered to the facility. In exchange for these benefits, recipients would be required to surrender the right to veto selection of the site, forego impact mitigation assistance, and agree to cooperate with DOE in the siting process. Several other stipulations were included: a socioeconomic report by DOE (Section 175) plus certain requirements for transporting wastes and for studying alternative methods of waste disposal.

Passage of the NWPAA raised important policy issues about the relationship between the amendments and the original NWPA of 1982. Senator Johnston received credit for steering the amendments through a complex and often byzantine process in Congress. By attaching the amendments to the Omnibus Budget Reconciliation Act of 1987, Johnston avoided an open policy review, such as had preceded the NWPA, and confined the substantive discussions to committees where the allure of a quick fix (selecting a single state) was instrumental in overturning the selection process mandated in the NWPA. Veteran congressional observers were not surprised that Nevada, politically weaker than Washington and Texas (the other two finalists at that point), was chosen to be the single repository site (5).

Important provisions of the NWPA were overturned, including the role for potential host states based on equity and fairness provisions in the NWPA. DOE would not be required to cooperate and consult with the state of Nevada, and Congress was willing to side with DOE to develop a repository in Nevada. One of the major decisions of the NWPAA was to endorse and support DOE as the agency to continue the US's HLNW program. Ironically, while the NWPAA attempted to force Nevada to accept a repository site, Congress accepted Udall's idea of a volunteer program for the MRS and established the Nuclear Waste Negotiator's Office.

Many underlying problems that undermined the original effort went unaddressed in the process of enacting the NWPAA because the US's nuclear waste policy was not thoroughly reviewed and the lessons of 1983–1987 were ignored. The strategy seems to have been to make the technical task simple enough for DOE, by selecting a single site for study, so that a revised program could go ahead. Nothing in the NWPAA effectively addressed public opposition, lack of trust and confidence in DOE, scientific and technical problems of uncertainty and risk, DOE's paucity of management and administrative expertise, the possibility of alternative options for HLNW management, or the difficulties presented by intergovernmental conflict. To address these problems, the NWPAA appeared to rely only on the authority of a congressional mandate.

The failure to conduct a proper review and address the issues and problems that were so clearly illuminated during the first five years of the HLNW program meant that the US's HLNW program would continue to struggle. The program required constant further interventions by a Congress that had become less focused on making policy to solve a national problem and more focused on implementing a chronically troubled and deeply flawed program. In retrospect we can see that the NWPAA was not a solution to the HLNW program problems but a prolongation of an obsolete and historically failed approach.

LIVING WITH THE NUCLEAR WASTE POLICY AMENDMENTS, 1987–1995

Nevada's reaction to the NWPAA is well characterized by Harry Swainston, Nevada deputy attorney general:

> Prior to the Amendments Act, Nevada's opposition to the use of Yucca Mountain as a nuclear waste repository was emphatic but generally restrained and respectable. The court cases filed by the state sought judicial review of programmatic decisions by the DOE that Nevada considered not in conformance with the NWPA or that restricted Nevada's rights of participation.
>
> Following the Amendments Act, however, the state took the gloves off as its political leaders became almost instantly galvanized into a united opposition to the repository. Most commentators concede that the state was, at that point, fully justified. (29)

The politicians, news media, and public immediately characterized the NWPAA as the "screw Nevada bill." The outrage expressed by this term was largely the result of unmet expectations of a fair process and an equitable outcome.

Intergovernmental Conflict

Official opposition by the state of Nevada increased substantially. At its first opportunity following the NWPAA, the Nevada State Legislature passed Assembly Joint Resolution (AJR) 4 and AJR 6. AJR 4 stated the "adamant opposition to the placement of a high-level radioactive waste repository in the state of Nevada" (29). AJR 6 "prohibited the federal government from establishing a repository at Yucca Mountain without the prior consent of the Nevada Legislature of a cession of jurisdiction pursuant to Chapter 328 of the Nevada Revised Statutes, which consent and cession were refused" (29). These two joint resolutions were passed on April 6, 1989, and transmitted to the president, the US Senate, and the US House of Representatives on April 19, 1989.

Assembly Bill 222 passed the Nevada Legislature on June 28, 1989, and was signed by the governor on July 6, 1989. This law made storage of high-level radioactive waste by any person or government entity in Nevada illegal (29).

The screw Nevada strategy was the wrong approach for additional reasons: It was a repudiation of the NWPA's original equity and fairness standards and, in changing the comparative site evaluation strategy, it bet the US's entire

HLNW program on a single site. The latter was hardly a prudent choice in any case and introduced an unknown but significant bias into the project's management. In effect, DOE was given a mandate to find Yucca Mountain suitable or to declare that their effort and the congressional strategy was a failure. More important than these difficulties was the fact that the NWPAA provoked an adversarial relationship that introduced important uncertainties about the final outcome. In addition, the adversarial relationship complicated implementation for an agency (i.e. DOE) that had few skills in dealing with state and local governments.

What difficulties were introduced with this adversarial relationship? Joseph Rhodes, Jr, a public utility commissioner from Pennsylvania, noted the polls and reports on opposition from Nevada citizens and political leaders and told the 1990 meeting of the National Association of Regulatory Utility Commissioners, "I can't imagine that there will ever be a usable Yucca Mountain repository if the people of Nevada don't want it....There are just too many ways to delay the program" (30). This observation echoes an earlier statement by a former DOE official, John O'Leary, in a 1983 interview with Luther Carter, " 'When you think of all the things a determined state can do, it's no contest,' O'Leary told me, citing by way of example the regulatory authority a state has with respect to its lands, highways, employment codes, and the like. The federal courts, he [O'Leary] added, would strike down each of the state's blocking actions, but meanwhile years would roll by and in a practical sense DOE's cause would be lost" (4).

The first test of the adversarial relationship between Nevada and DOE followed quickly after the passage of the two resolutions (AJR 4 and AJR 6) and Assembly Bill 222. Five months after AJR 4 and AJR 6 were transmitted to Congress, Nevada Governor Miller asked the Nevada attorney general for an opinion about the status of environmental permits requested by DOE for the Yucca Mountain project. The governor stated his belief that the two legislative resolutions constituted a valid and effective notice of disapproval according to the terms of the NWPA [Section 116(b)] following congressional selection of Yucca Mountain as the US's only potential HLNW repository. According to the NWPA, following official notice of disapproval by the governor or legislature of a state, Congress has 90 days to override the state veto and endorse DOE selection of a repository site. The attorney general's opinion, dated November 1, 1989, advised the governor that the state's actions and the failure of Congress to override the disapproval meant that the Yucca Mountain repository had been disapproved and processing environmental permits for DOE work was "unnecessary because they were moot" (29, 31). The governor then ordered state agencies to stop work on the permit requests and return the applications to DOE.

This litigation battle was only one of many between Nevada and DOE (the first suit was filed in 1985, the most recent in 1994), but demonstrates the potential for intergovernmental legal conflict. On January 5, 1990, Nevada filed a case with the Ninth Circuit Court of Appeals (the court of original jurisdiction for Nevada litigation under terms of the NWPA) asking for a review of the DOE refusal to recognize Nevada's disapproval of the Yucca Mountain project and its failure to terminate site characterization work. In February 1990, this case was combined with *Nevada vs Watkins* (86-7308, 9th Cir.), which claimed that DOE jurisdiction and control over the site violated constitutional guarantees to states and was counter to NRC regulations.

Meanwhile, DOE filed suit against the state of Nevada in the US District court in Las Vegas on January 25, 1990, asking for a declaratory judgment that Nevada's notice of disapproval was premature and for an injunction requiring state agencies to process DOE applications for environmental permits. The state responded by gaining approval for a stay of the court proceedings, which lasted until March 20, 1991. During this interim period, the Ninth Circuit Court of Appeals ruled against the state in the case of *Nevada vs Watkins* on September 18, 1990. The Appeals Court found that the state's disapproval was premature, because the federal action was to continue site characterization and not to select a repository location. The court also found, on narrow grounds, against the state's claims that congressional actions violated constitutional provisions (29).

This case was then appealed to the US Supreme Court, but a hearing was denied, effectively upholding the Appeals Court decision against the state. At this point, the Nevada agencies informed DOE they would resume processing requests for environmental permits. The Federal District Court agreed to dismiss the DOE suit based on satisfactory progress by the state agencies. Thus, more than three years after the passage of the NWPAA, the state began to process DOE permits.

DOE claimed these actions by Nevada agencies were obstructionism that substantially delayed the site characterization studies and were evidence of a "scorched-earth battle plan." This last phrase was used on March 5, 1991, by John Bartlett (then director of the DOE's OCRWM) in testimony to the US House of Representatives, and by James Watkins, then Secretary of Energy, in his March 21, 1991, testimony in the US Senate. These two DOE officials supported provisions in the National Energy Strategy bill (Subtitle B of Title V), which would have removed state of Nevada licensing, permitting, and regulatory control over DOE site characterization activities at Yucca Mountain. This attempt to remove Nevada permitting-authority was defeated in the final days before the energy bill was passed because of several factors. The state was in the process of issuing and processing DOE requests for permits and was, therefore, not openly in defiance of Congress. In addition, which program delays resulted from Nevada state opposition and which from DOE management, funding, and administrative decisions was not clear (32). Finally, at a critical time in the legislative calendar, a Nevada senator threatened to filibuster the energy bill unless the objectionable sections pertaining to state permitting-authority were removed.

This account is necessarily brief and is intended to clarify the difficulties of intergovernmental relations regarding HLNW facilities. The state was determined to protect its areas of authority and did not simply give in and produce the DOE permits. DOE requested three permits, two of which were issued in July of 1991. The final permit was a water permit, which the state opposed in hearings before the state engineer. The state claimed that there were two reasons to reject the DOE request: No unappropriated water was available in the source applicable to the permit and the project was not in the public interest. The state argued that a HLNW facility could stigmatize the state and the area near the repository, resulting in adverse psychological, social, and economic impacts. In the end, the state withdrew their argument about potential stigmatization and the public interest, and the state engineer found available water for site characterization activities. A permit with restrictions and stipulations was issued on March 3, 1992.

These moves and counter moves must be considered skirmishes in the larger battle between Nevada and DOE over the acceptability of Yucca Mountain as a repository site. Still to be tested are basic constitutional questions about whether other states, through federal legislation, can force Nevada to host a repository against its will.

Scientific Problems

Yucca Mountain, composed of welded tuff (compressed volcanic ash with a rock-like texture), is far from any large community in a very dry desert area with a potential repository space several hundred feet above the water table. The location is entirely on federally owned lands adjacent to the Nuclear Test Site (NTS) where atomic weapons have been detonated for four decades.

The argument that the existence of the NTS should qualify Nevada as a location for the nation's HLNW repository has been strongly resisted by the state's officials. They point out the Cold War circumstances that led to locating the NTS in Nevada and the fact that the state legislature approved that facility. They argue that HLNW, by contrast, is an additional risk, and that the state legislature has definitely voted against a repository. Also, Nevada officials argue that the state's acceptance of a central role in national defense is very different from submitting to hosting the nuclear power industry's HLNW.

Yucca Mountain's apparent attractions have led more than one person to misjudge the difficulties posed by the site. In 1987, science writer Luther Carter, citing such qualifications (4, 33), recommended Yucca Mountain in two publications that apparently had great influence in Congress.[2] Some DOE people came to a similar conclusion based on the preliminary data from the environmental reviews of the first-round sites and an optimistic outlook.[3] In 1987 congressional testimony, Donald Vieth, then DOE's Nevada manager of the Yucca Mountain work, said he thought the Yucca Mountain site would exceed the applicable safety standards by a factor of 10.

The question of whether the Yucca Mountain site can safely isolate HLNW from the surrounding environment and communities is subject to considerable scientific uncertainty. DOE is expected to demonstrate that nearby populations will receive minimal radiation exposure, but determining how much radiation might escape over the lifetime of a repository at Yucca Mountain challenges the abilities of physical and social scientists (35, 36).

Concerns have been raised that catastrophic geologic events or human intervention may threaten the security of a repository at Yucca Mountain. For example, one of DOE's scientists, Jerry Szymanski, theorized that Yucca Mountain is subject to upwelling of groundwater—water forced up through the mountain by a major seismic event, such as an earthquake. He argues that geological evidence, especially from calcite-silica deposits at the location called Trench 14, shows that upwelling has occurred in the past. If upwelling were to occur with a repository in place, water would presumably flood the waste storage area, possibly causing the canisters to break and releasing radiation into the environment. A plume of contaminated water powered by the thermal heat of HLNW could have widespread and disastrous consequences (35, 37).

No scientific consensus has been reached on the plausibility of this scenario; its likelihood has been disputed by scientists in government agencies and academic institutions. A DOE-sponsored review of the theory by the National Academy of Sciences concluded that such a catastrophic event was unlikely (38). However, research by the state of Nevada may resurrect the issue (36).

The licensability of Yucca Mountain is also threatened by the possibility that strong earthquakes might occur in the region. DOE and the state of Nevada have publicly disagreed about whether the fault system at Yucca Mountain

[2]In 1987, Carter recommended a negotiated agreement with the state of Nevada as the basis for doing site characterization work, but this recommendation was ignored by Congress. Subsequently, Carter backed a more authoritarian approach and a get-tough stand with Nevada; a 1993 article called for a combination MRS and repository at Yucca Mountain (currently prohibited by the NWPAA) to be accomplished by Congress "declaring unequivocally that the NTS [Nuclear Test Site] is to become the nation's center for nuclear storage and by directing that a spent-fuel surface storage facility be built and ready to operate by 1998." In Carter's opinion the technical problems were manageable (34).

[3]For a summary of the environmental assessments of the first-round sites and a comparative analysis, see reference 11.

will give rise to earthquakes that would breach the containment system. This question was given added prominence on June 29, 1992, when a magnitude 5.6 quake occurred at Little Skull Mountain, just 12 miles from the proposed repository site. Reaction to this incident provides yet another demonstration of the gulf between proponents and opponents of the Yucca Mountain project. DOE officials argued that the event demonstrated the geologic structure was robust enough to warrant a repository, whereas Robert Loux, executive director of the Nevada Nuclear Waste Project Office (NWPO), described it as confirmation that the whole of southern Nevada is a young, active, geologic environment (39). His interpretation was reinforced by a magnitude 6.0 quake on May 17, 1993, about 100 miles west of Yucca Mountain and 35 miles southeast of Bishop, California. The Ghost Dance Fault, which cuts directly through Yucca Mountain from south to north, is of particular interest. As government geologists explained in May 1993, they do not know when this fault was last active or whether it is in a zone connected to other faults (40).

Scientists have also found evidence that two volcanoes located within 27 miles of the Yucca Mountain site erupted as recently as 5000 years ago, which suggests that volcanoes could erupt again within the 10,000 years that repository wastes will remain dangerously radioactive (41).

Another potential problem is the release of radioactive carbon-14. Scientists agree that some carbon-14 will escape over the repository's lifetime. In fact, a relatively large fraction of the carbon-14 present when the spent fuel rods are emplaced will eventually reach the environment (42) because carbon-14, which has a half-life of 5730 years, will still be present in significant quantities when the waste canisters exceed their effective life span of 300–1000 years. Once a canister is breached, the carbon-14 will readily transform into a gaseous state and escape to the surface before decaying to harmless levels.

Charles Bowman and Francesco Venneri, physicists at DOE's Alamos Laboratory, initiated a recent debate with the claim that the geologic structure at Yucca Mountain might allow the HLNW to "erupt in a nuclear explosion." This theory is disputed by other DOE scientists who say Bowman and Venneri's basic assumptions contain serious flaws. In a front page article in the *New York Times,* William Broad (43) pointed out that Bowman is also an advocate of a process to neutralize HLNW through transmutation using particle accelerators.

Finally, future human intrusion, either accidental or deliberate, has been suggested as a potential threat to the integrity of the repository. Intentional mining of HLNW could conceivably occur if future generations come to regard the wastes as valuable, either because an effective means of reprocessing is developed or because another use for them is discovered. Accidental intrusion could occur if future generations forget that the wastes are buried at Yucca Mountain. Whether human memory of the repository site can survive 10,000 years of cultural, climatic, and geologic changes that will probably result in settlement patterns, economies, and languages that bear little resemblance to those of today is certainly open to question (44, 45).

These risk scenarios suggest that a repository at Yucca Mountain might not isolate radiation to the degree required by EPA standards. In fact, DOE may not find sufficient evidence that Yucca Mountain is a suitable repository site. In 1990, the National Research Council's Board on Radioactive Waste Management identified several sources of scientific uncertainty that prevent DOE from proving various radiation release events are sufficiently unlikely. These sources include scientists' inability to collect all relevant data, the evolving nature of geohydrologic computer modeling, and the extremely long period during which risks will persist. The board said these uncertainties make precise estimates of the likelihood of many of the release events that would be covered by EPA standards impossible (46).

The role of scientific uncertainty and the DOE's strategies for addressing these questions have produced several legal actions. A 1994 Nevada suit, *Nevada vs O'Leary* (94-70148: 9th Cir.), claimed that DOE failed to adequately characterize the nature and origin of calcite-silica deposits in Trench 14 at Yucca Mountain. The import of this case is that the NWPA requires DOE to terminate all site characterization activities if at any time Yucca Mountain is found to be unsuitable, which (if the state's analysis is correct) would require DOE abandonment of the site under current EPA guidelines. DOE, however, has not established procedures for making a finding of unsuitability as a result of their site characterization studies.

Nevada argues that the decision by DOE to discontinue study of the possible disqualifying conditions put forward in the Szymanski theory, a strongly contested technical issue between the state and DOE, is a failure to carry out appropriate site characterization activities at Yucca Mountain. The state asked the court to establish the state's right to a valid scientific process (8). A court order in support of the state would require DOE to establish procedures for making a finding of unsuitability and limit the ability of the department to set aside study of contested issues such as those raised by Szymanski and Bowman.

Another suit now before the 9th Circuit Court also addresses the issues of how DOE decisions are made and what record will be maintained. This case, *Nevada vs O'Leary*, asks that the state be allowed to take depositions from 27 scientists who reviewed the studies of the Szymanski theory (8).

Congress made a partial response in the 1992 Energy Policy Act, which mandated a less stringent method of calculating radiation risks for the Yucca Mountain site than EPA had planned. This new approach, which applies only to Yucca Mountain, will make acquisition of a construction license from NRC easier for DOE. The Energy Policy Act also attempted to eliminate the problem of potential human intrusion by requiring DOE to act as perpetual custodian over the repository (47). How a legislative edict like this can, in practice, ensure 10,000 years of continuous stewardship is unclear. No theoretical or historical evidence shows that any societal or government institution can endure for such a long time. DOE, for example, was created in 1978 and succeeded the Atomic Energy Commission (which existed for less than 20 years) and the Energy Research and Development Agency (which existed for 4 years).

Although the 1992 Energy Policy Act eases some of the obstacles DOE faces in licensing Yucca Mountain, it does not ensure approval of a repository. If the scenarios suggested by Szymanski, Bowman, or others are credible, they will certainly preclude the NRC from issuing a license for Yucca Mountain, regardless of the standards EPA eventually promulgates. In the meantime, DOE and the state of Nevada continue to disagree on the site's technical suitability. DOE contends that its data analyses support the view that Yucca Mountain is an appropriate site, or at least that more information is needed before a determination can be made (48). Many important scientific and technical uncertainties remain unresolved, so much so that different groups of analysts can look at the same data and legitimately reach exactly opposite conclusions (49).

Management Problems

The NWPAA did not solve the DOE management problems. Many complaints from various critics have highlighted program costs and the slow pace of progress. An example is the project managers' recent failure to order conveyor equipment that matches the capacity of the new tunnel boring machine used to excavate the underground exploratory studies facility. The result is that tunnel work will go forward at only about one-third the rate possible with the proper equipment, and ordering and installing a proper conveyor system is expected to take more than a year. DOE does not lack the desire to move ahead. In some cases, it appears willing to take risks with scientific data collection to push the schedule. For example, in October 1994, NRC sent a letter to DOE about its concerns over the lack of an effective quality assurance program. The issue was whether DOE should proceed with its tunneling program even if it might destroy portions of the site needed to study the gas pathways. A DOE spokesman in Las Vegas said the department was concerned about the NRC warning and that "We don't want our data to be in question" (50). OCRWM head Daniel Dreyfus told the NWTRB later in October that the department intends to proceed with the exploratory tunnel despite concerns about the quality assurance program (51). A history of management problems such as these has resulted in a series of recommendations and calls for substantial changes at DOE.

These recommendations seem to increase as each failure of legislative or reorganizational fixes becomes clear. A November 1993 letter to the Secretary of Energy from the NWTRB, which was established by Congress to review the scientific and technical aspects of the repository program, suggested a major program review. An official recommendation followed this letter in February of 1994. In March of 1994, 10 US senators asked President Clinton to appoint a special commission, independent of DOE, to review the US's nuclear waste programs. These senators pointed to budget, management, schedule, and public acceptance problems with low-level radioactive wastes, temporary waste storage facilities, handling of transuranic wastes, and the HLNW repository program. They complained that current federal efforts are piecemeal, are not cost effective, and do not provide acceptable nuclear waste management at any level.

Other groups joined in the call for investigation and review of the Yucca Mountain program (52). Nevada congressional representatives asked for a review in June of 1993, which was followed by a similar request from the Nevada governor in July. The Western Governors' Association request for a review was contained in a June 1993 resolution. Several public interest groups asked President Clinton to initiate a comprehensive review in July of 1993. Congressional members sent letters to the Secretary of Energy and to the president in August and October 1993. In August of 1994, the Nuclear Waste Strategy Coalition, consisting of state regulators, utility executives, and attorney general representatives, released a 1994 draft report, *Redesigning the US High-Level Nuclear Waste Disposal Program for Effective Management* (53), which reviewed 18 federal government documents produced between 1982 and 1994 about DOE management issues and problems. The report recommended that DOE be removed from HLNW management and its duties reassigned to a federally-chartered corporation.

In 1993, the Secretary of Energy commissioned "an independent review of the financial and management" performance of DOE's work on the Yucca Mountain project. DOE appointed a private industry executive to oversee the evaluation, and the Nevada governor appointed the other overseer, a Nevada public service commissioner. These overseers asked GAO to comment on the statement of work prepared by DOE. This statement served as the basis for a contract with a consulting firm that conducted the study. The GAO report said that the review proposed by DOE "is too narrow and may result in a product that, while useful, will not address many of the major issues confronting the disposal program" (54). GAO also noted that the funding and time allowed for the review work were probably inadequate.

GAO reported in May 1993 that more than 60% of allocated funds were being spent on infrastructure activities (i.e. management, administration), and only about 22% on essential scientific and technical activities at Yucca Mountain (49). The program costs are now estimated at more than $6 billion for the site characterization alone, three times the per-site estimate in 1987 and 100 times the 1981 estimate. In a September 1994 report, GAO said that recent DOE review initiatives "are too narrow in scope and lack sufficient objectivity to provide the thoughtful and thorough evaluation of the program that is needed" (55). The report stated that an independent review is needed now more than ever, and claimed that the unresolved issues are fundamental and must be addressed objectively. GAO also noted that it first recommended a comprehensive review in 1991 (55).

The Clinton administration, which took office in January 1993, filled the key DOE offices. Daniel Dreyfus began work in February 1993, although he was not confirmed as the director of OCRWM (i.e. the Yucca Mountain project) until October 1993. In interviews with the industry publication *Nuclear Energy*, Dreyfus recounts that he found a troubled HLNW program and that within DOE there was "widespread concern that the program could not realize the expectations of its clients," which the article identifies as "the electric utilities that operate nuclear power plants and their customers" (56).

Department of Energy's Response

In response to the calls for action, DOE initiated a new study of its program which identified three major problems. First, DOE determined that the work at Yucca Mountain was not properly focused. Second, DOE found that program management needed improvement. Third, DOE needed a substantial initiative to meet the goal of accepting HLNW from the utilities by 1998 (56). This deadline is the only one of many mandated by Congress in the NWPA that is still considered by DOE to be operative; all other dates have long since been missed or rescheduled.

This recent DOE review concluded that existing program obligations were beyond the department's abilities and funding resources. In short, DOE could no longer continue as it had planned and still meet the 1998 date to begin accepting HLNW from the utilities or the 2010 date to begin operation of a repository at Yucca Mountain. This reality about the program and its schedule has been pointed out by people outside DOE many times. In May 1993, GAO estimated that the repository operations were 5 to 13 years behind the scheduled 2010 opening date (49). Both the National Academy of Sciences and NWTRB have cautioned that DOE's schedule objectives are unrealistic and are inappropriately driving the program (46, 57).

In the spring of 1994, DOE proposed a major restructuring of the HLNW program (58). The new Proposed Program Approach (PPA) seeks to provide waste acceptance by 1998 with the use of multiple purpose canisters (MPC), which DOE would design, produce, and provide to utilities with nuclear reactors. The MPC would allow HLNW to be stored on-site at the reactors, to be shipped to an interim storage facility, and, potentially, to be used for permanent storage at a repository—all in the same canister. This approach would make use of specially designed overpacks for storage, transport, and disposal.

However, as GAO pointed out in its September 1994 report, not enough is known about the Yucca Mountain site to design a waste canister (55). In the words of the GAO report, "The Nuclear Waste Technical Review Board and others have repeatedly pointed out, and DOE program managers have acknowledged, that more information about the potential repository site and the potential effects of the heat from the waste on the repository will be needed before a disposal package with a high degree of safety assurance can be developed" (55). The GAO report concluded that DOE runs the risk of having to redesign the waste canister, provide new engineered barrier systems, or "accept certain safety risks" (55). Of course, DOE would not be alone in having to accept safety risks.

In addition, the PPA would make major changes in scientific and technical work on site characterization, site suitability determination, licensing under NRC regulations, and repository construction and operation. Instead of an 8- to 10-year study period to demonstrate reasonable assurance (NRC's licensing standard) that Yucca Mountain can isolate HLNW for the required 10,000 years (the EPA standard), DOE would make an early determination of technical suitability by 1998. This determination would be made with available data and expert judgments to compensate in areas where data are not available or sufficient. DOE would seek an NRC license to construct and operate the facility on the basis of a finding of technical suitability. A final suitability determination of the site as a repository would be postponed for 50 to 100 years to a performance confirmation period. During this performance confirmation period, DOE would attempt to collect data needed to fully demonstrate that Yucca Mountain complies with NRC and EPA regulations for waste isolation and reasonable assurance of repository performance (56, 58).

This PPA is a significant departure from past DOE plans to study Yucca Mountain and obtain authorizations and licenses for construction and operation of the site as a HLNW facility. DOE maintains that its new approach does not require changes to the NWPA or to federal regulations. The state of Nevada contends the opposite—that this approach requires numerous changes. The state further contends that the PPA is a veiled attempt to build the repository and emplace HLNW at Yucca Mountain without first addressing the serious performance deficiencies the state believes to be present at the site (59). DOE, on the other hand, presents the PPA as an innovative and cost effective alternative.

The PPA attempts to address schedule, cost, and scientific uncertainty issues associated with Yucca Mountain. It does not confront key institutional, organizational, and management issues. Issues of public and intergovernmental participation, public acceptance, fairness and equity, voluntarism, and the need for improved management of the US's HLNW are not addressed in this latest program design. The new approach continues to rely on nonvoluntary siting of a facility at Yucca Mountain by relaxing standards for scientific study, analyses of data and information, and licensing requirements. PPA does not address the need for alternatives to Yucca Mountain in the event the site proves unsuitable. At best, PPA moves the final suitability determination forward as much as a century. Converting Yucca Mountain to a de facto interim storage site would address DOE's perceived need for an immediate national storage site for HLNW. However, this new program creates even more uncertainty while DOE avoids examining existing uncertainties fully (8). What precedent would the federal government set by circumventing the requirements for licensing HLNW facilities? And how would this approach influence future public acceptance of other radioactive waste facilities or even other hazardous technologies subject to government regulation? Opponents to the Yucca Mountain project can be expected to argue that this approach exhibits similarities to the culture and rationale that dominated HLNW management at the nation's nuclear weapons complex, and that such a strategy risks a similar unfortunate outcome.

PUBLIC OPINIONS AND CONCERNS ABOUT HIGH-LEVEL NUCLEAR WASTE

Resistance by state officials to DOE plans for HLNW facilities is firmly grounded in the public's strong opposition. Only a few communities—usually those historically associated with other nuclear industries such as power plants or weapons manufacturing—have shown any willingness to host a nearby repository. Elsewhere, people find HLNW materials to be the least acceptable hazardous wastes (10). State and community officials' expressions of these attitudes and opinions in response to the federal HLNW program should come as no surprise. Given our political system, strongly held public opinions will have significant influence on the formulation of public policy (60).

The public believes that HLNW management and facilities involve very high risks (61–64). This belief is in stark contrast to that of technical experts, including those assigned to the management of HLNW, who believe that the risks are controllable with current methods (4, 63). Sanguine reassurances from technical experts usually do little to counteract public skepticism; in fact, these assurances often reinforce the determination of stakeholders to gain greater access to, and control over, the decision-making and policy processes. This determination, in turn, can lead to conflict with managers of the nuclear enterprise, who often regard dealing with public concerns primarily as a complication in need of careful, even manipulative, handling. Scientists, engineers, and technically-oriented managers are often baffled by what they perceive as scientifically unsophisticated emotionalism on the part of the public. These managers generally find the public's skepticism about the technological fix approach difficult to accept and mistakenly believe that if they redouble efforts to educate the public opposition will simply melt away (65).

This approach to managing public opposition does not allow for the fact that people who are asked (or forced) to host a nuclear facility are apt to mix value-laden criteria into their assessment of risks. These criteria include concerns for the environment and property values, risks to social and cultural life, quality of life considerations, psychological factors (such as tolerance for risk-taking), and value judgments (such as the relative weight given to risks and benefits). The public's value judgments may differ markedly from the judgments of those in the nuclear industry (10, 63, 66–68).

Numerous surveys demonstrate that trust and confidence in DOE is severely lacking (69, 70). Such difficulties stem in part from the fact that DOE was given two distinct mandates that often conflicted with each other: to develop and promote nuclear technology and to ensure the public's safety from radioactive hazards. As early as 1982, the congressional Office of Technology Assessment (OTA) warned that "distrust may indeed be the single most complicating factor in the effort to develop a waste disposal system that is acceptable technically, politically, and socially" (71). Some critics believe DOE's actions have shown a distinct bias in favor of its programs to promote nuclear technology (5). Moreover, evidence of past deceptions and mismanagement has damaged DOE's reputation with the public (72, 73).

An Advisory Board Task Force commissioned by the Secretary of Energy in 1991 to address the problems of trust and confidence made its report in November 1993 (74). The focus was on the management of nuclear wastes, particularly HLNW. The report found that an atmosphere of distrust pervaded the high-level radioactive waste program, that DOE evoked little trust and confidence from any sector of the public, and that the lack of trust and confidence was a direct consequence of the public's experiences with DOE. The report also says this lack of trust is not an irrational reaction, nor can it be discounted as a manifestation of the not-in-my-backyard syndrome. The report's authors state that in their hearings senior managers from OCRWM pointed to several conditions of distrust that constrained their program. These managers also presented their opinion that congressional legislation "created an institutional context that almost seems purposely designed to stimulate distrust" (p. 56). The idea that DOE must operate in a context of distrust appears to be strong among department personnel. The Task Force report notes as typical of DOE employees and contractors the comment that the department "has no friends, just temporary allies" (p. 34).

The NWPA was intended to address some of these problems. Unfortunately, the act assigned DOE the responsibility of implementing the innovative provisions relating to state and community participation in the repository program. DOE was apparently unprepared for this role. The department seems to view proposed host states and communities not as possible allies, and certainly not as partners, but, at best, as potential service providers for solving the nuclear industry's waste problems—and, at worst, as probable adversaries.

This attitude, of course, did little to allay the concerns of those opposed to the Yucca Mountain repository. Numerous studies of public attitudes and perceptions showed extreme perceptions of risk (10, 61–64, 73). These studies had two main goals: to document attitudes of Nevada residents to the repository and to explore how opinions about a repository at Yucca Mountain might affect the willingness of people outside Nevada to vacation in the state, move there for employment or retirement, or invest there. Between 1986 and 1994, more than 25 surveys were conducted in Nevada, southern California (the major source of visitors to Nevada), and across the nation (75).

The median distance respondents preferred to live and work from a HLNW repository was 200 miles, twice the distance recorded for chemical waste landfills, and far beyond the distances listed for oil refineries (25 miles), pesticide plants (50 miles), and even nuclear power plants (60 miles).

Two-thirds of survey respondents believed that rail and highway accidents will occur in transporting HLNW to a repository. They also expected problems resulting from earthquakes or volcanic activity, contamination of underground water, and accidents in handling wastes during burial operations.

Two-thirds to three-quarters of those surveyed said a state that does not generate nuclear wastes should not be forced to host a HLNW repository. Most also believed that building a single national repository was the least fair approach, compared with building two national repositories, or storing wastes in more than one region or state or at each reactor site.

Two-thirds to three-quarters of respondents expressed serious distrust of DOE; they believed the agency would not be forthright about accidents or problems with the HLNW management program.

Surveys of Nevada residents have repeatedly demonstrated strong opposition to the repository program. A 1989 survey revealed that 70% would vote against the project, and nearly 75% believed the state should persist in fighting the repository even if this fight meant relinquishing benefits offered by the federal government (64). Follow-up surveys in 1990, 1991, 1993, and 1994 confirmed that very high levels of opposition and distrust persisted; the percentage of Nevadans who would vote against a Yucca Mountain repository has remained in the 70% range, and opposition has continued to outstrip support by margins of three or four to one. Nevadans believe strongly that the state should decide whether to accept the repository, and that state and local officials should be involved in decisions about Yucca Mountain. Finally,

Nevadans expressed concerns about the potential for stigmatization of the state as a nuclear dump to adversely affect the tourist and visitor industries, which dominate the state's economy.[4]

Several studies were done outside Nevada to assess whether the mere existence of a nuclear repository within a 100-mile radius would reduce a community's attractiveness as a place to attend a convention, take a vacation, raise a family, retire, or locate a new business. Surveys of the general population nationwide, as well as special groups, such as convention planners and real estate experts, indicated that the Yucca Mountain repository could stigmatize Nevada and particularly Las Vegas, the major metropolitan area in the state. A large majority of respondents said a repository would reduce the desirability of a community for raising a family. A majority said a repository would deter them from visiting for a vacation or a convention. The repository also might deter job seekers from Las Vegas and reduce job opportunities (7, 8, 10, 64).

At least 30% of convention planners surveyed said they would reduce their rating of Las Vegas as a meeting site if a repository was located nearby. If the facility should experience accidents or incidents that are given extensive media coverage, 75% of planners would reduce their rating further, and nearly 50% said they would no longer consider Las Vegas an acceptable convention site (7, 10).

Real estate executives believed that the existence of a repository within 100 miles of a community would detract from its suitability as a location for administrative offices, business and professional services, and businesses that serve the hospitality industry (76).

Socioeconomic surveys such as these are limited in their ability to predict actual future behavior, especially in relation to a unique facility with which no one has had any experience. Therefore, another set of studies based on the concept of environmental imagery was designed to test three propositions: (*a*) that people associate images with different environments and places, and that these images can affect their behavior with regard to those places; (*b*) that a HLNW repository evokes extremely negative images; and (*c*) that negative images associated with the Yucca Mountain repository will extend to Nevada and Las Vegas (63, 64). If true, these propositions suggest a mechanism by which the Yucca Mountain repository might generate significant social and economic stigma effects.

In these environmental imagery surveys, respondents were asked to state the first thought or image that came to mind when presented with stimulus phrases such as "Las Vegas," "Nevada," and "nuclear waste repository." The findings, which supported all three propositions, indicated that the more posi-

[4]Reports on the surveys are available from the Nevada Nuclear Waste Project Office, Carson City, Nevada, or from the authors.

tive an image a city or state elicited, the more likely it was to be preferred over other places for visiting, raising a family, retiring, or locating a business. Follow-up interviews with respondents 16 to 18 months later revealed that using their image scores improved prediction of their vacation behavior during the intervening time. Those who associated Nevada with things nuclear gave Nevada lower imagery and preference ratings than those who did not. The most common words associated with a nuclear waste repository were extremely negative, evoking images of danger and death. These verbal reactions indicated a profound aversion to nuclear wastes, comprised of feelings of dread, revulsion, and anger—the raw materials of stigmatization and political opposition (10, 63, 78–79).

THE NEED FOR NEW DIRECTIONS

The HLNW management program in the US is failing badly, beset by technical difficulties, poor management, scientific uncertainties, cost overruns, equivocal political support, state opposition, and profound public distrust and antipathy. The program or its managers' ability to overcome these obstacles is doubtful. The DOE's PPA is as much of an admission of failure as we can expect given DOE's congressional mandate and their institutional history.

The unfairness of the procedures used to limit site characterization efforts to Yucca Mountain, Nevada, and the stubborn disregard for local objections to the project virtually guarantee a continuation of the fractious, messy conflicts that have dogged the waste program from its earliest days.

A new approach to managing the program and finding a site for the repository is urgently needed. The following recommendations outline some crucial elements of such an approach (7, 8, 80–82).

Reevaluate the Commitment to Underground Geologic Disposal

Congress should place a moratorium on the current program and remove the deadline of 2010 for beginning operation of the repository. Flexible and realistic timetables would allow more time for further research to be done on technical problems associated with the repository and on comparative advantages and disadvantages of different geological structures. More effort should also be devoted to developing multiple engineered barriers to isolate HLNW from the environment. A moratorium on geologic disposal would also allow time to evaluate alternative techniques, such as seabed disposal.

A more important advantage of delay is that it would allow the federal government to make a genuine effort to gain public acceptance and political support for the program. Delay would provide the leeway needed to establish a voluntary process for selecting a repository site.

Use Interim Storage Facilities

Above ground storage in dry casks at reactor sites or a centralized MRS facility could be used to store wastes for 100 years or more. Although not without problems, an MRS facility would cost considerably less than a permanent geologic repository and would buy time for additional research and public consultation processes. The federal government should, therefore, abandon its current policy prohibiting the development of an MRS facility until a site for a permanent underground repository is found.

Evaluate More Than One Site

Program failure will be less likely if more than one site is evaluated. Further study of alternative geologic options is needed, as is further research on alternatives to a geologic repository. Every effort must be made to find several states and communities willing to be considered as the location of an interim or permanent storage facility. Keeping several options open until very late in the selection process is crucial, because the repository is a first-of-its-kind facility with a great many associated uncertainties and a well-demonstrated ability to evoke intense public and political opposition. The arbitrary selection of Yucca Mountain as the only site to be characterized closed off all other options, which, given Nevada's strong and unrelenting resistance to the project, creates the very real possibility that the US may be left without any likely repository site.

Employ a Voluntary Site-Selection Process

The arbitrary selection of Yucca Mountain for site characterization has been a major source of conflict and has evoked fierce public and political opposition, as have numerous other proposed nuclear waste sites over the past three decades. To avoid such conflicts in the future, Congress should mandate that no community will be forced to accept a repository against its will and that potential host communities should be encouraged and permitted to play a genuine and active role in the planning, design, and evaluation of the repository. Experience from other countries, such as Sweden, shows that developing an effective siting process that encourages such local participation is possible (83–85). If more than one community is willing to participate in a voluntary program, a competitive process might even result.

The approach taken by the US nuclear waste negotiator to site an MRS facility was sensible; interested communities are given planning grants that do not carry with them an obligation to host the facility. These grants enable communities to learn about the technical aspects of the storage process and to determine whether local residents are really interested in hosting the facility.

A voluntary process requires not only public participation, but also an agreed-on procedure by which communities and states would decide to accept a facility (e.g. public referendum with a two-thirds plurality). Such a voluntary process must be given every chance to work.

The apparent demise of the negotiator's office is unfortunate because this position is worth a better effort and more time than was invested. The current program has not accomplished more than the negotiator despite more than three decades of effort and an expenditure of billions of dollars.

Negotiate Agreements and Compensation Packages

A voluntary MRS or repository siting program must offer sufficient benefits to potential host communities and regions so that their residents feel their situation will improve. Adequate compensation is not only fair to the host state and communities, but also to the rate payers because it internalizes genuine costs that are otherwise pushed on to the general public or to host communities.

Acknowledge and Accept the Legitimacy of Public Concerns

The attitude that HLNW disposal is merely a technical problem to be solved by experts must be abandoned. A repository program has social and economic dimensions that will seriously affect the quality of life in neighboring communities. Most notably, such a project has the potential to stigmatize these communities, making them less attractive to residents, visitors, businesses, and in-migrants. Negotiated compensation packages should take into account the full range of potential socioeconomic effects including stigma effects that could have extremely negative long-term economic and social consequences.

Guarantee Stringent Safety Standards

Many people associate nuclear wastes with danger and death and react to the idea of a HLNW repository with feelings of fear and dread. Their trust in the ability of waste managers to protect them from danger is not enhanced by political moves like the 1992 Energy Policy Act, which set less stringent radiation-exposure standards for Yucca Mountain than had originally been contemplated, standards which only apply to evaluation of this one site.

Public acceptance of the repository program requires assurances that public safety will be a priority. The federal government must negotiate contingent agreements with any community or region that agrees to host a repository and specify what actions will be taken should there be accidents or unforeseen events, interruptions of service, changes in standards, or the emergence of new scientific information about risks or impacts.

Restore Credibility to the Waste Disposal Program

Creating a new management organization and adopting a new management approach may restore credibility to the waste disposal program. The history of the nation's HLNW management program underscores one glaringly obvious point: DOE has failed in its management role and is incapable of overseeing such an extraordinarily complex and uncertain program. Nor can DOE overcome widespread mistrust and skepticism about its competence, methods, and motives. Consequently, Congress should establish a new agency or organization to manage the civilian HLNW program, separate from the management of the nuclear weapons complex cleanup.

No matter which agency is assigned the job, however, a radical new management approach is needed—one committed to implementing the recommendations listed above and doing so in an open, consultative, and cooperative manner that does not seek to deny or avoid the serious social and economic dimensions of the HLNW disposal problem.

Given the large uncertainties inherent in the repository program, and the likelihood of surprises, the management approach must emphasize flexibility and adaptability. Only by frankly acknowledging the limitations of the available information and predictive techniques can program managers promote social trust and elicit readiness to tackle problems in a new way when necessary.

This strategy requires a cautious and deliberate approach. Repository development must proceed in stages, allowing time to discover whether the predictions and expectations associated with the project are reasonable. Assessment techniques and models must be viewed as learning tools, not as crystal balls that can foretell the future with precision. And because we have no analytical tools that can completely eliminate the element of surprise, we must develop methods for coping with it. Such methods would put into place (*a*) monitoring systems that tell us what is actually happening and help us learn as we go; (*b*) redundancy in safety systems; (*c*) technical reversibility or repairability; (*d*) multiple geologic and engineered barriers; and finally, (*e*) multiple disposal sites.

Conducting the program in a careful, step-by-step fashion will permit—in fact, encourage—everyone to stop and reflect and, if necessary, make mid-course corrections mandated by new technical developments and changing social values. All of these measures will enhance the adaptability of the management system, reduce the impact of mistakes and mishaps, and increase the chances of recovering—and learning—from mistakes, while retaining public support and confidence.

Above all, we must remember that building a permanent, underground HLNW repository is essentially an experiment and that its social and economic dimensions are uncertain, unpredictable and likely to remain so. Citizens must decide how, and perhaps even whether, to proceed with this experiment in the face of the many unknowns and potential risks it presents.

SUMMARY

The doubts and uncertainties presented by Yucca Mountain cannot be addressed unless we downgrade our standards for studying and licensing a repository. Yet these repository standards have been developed over the past decade as essential protection for human life and the environment.

The rush to build at Yucca Mountain compromises our future ability to achieve acceptable disposal of HLNW. Yucca Mountain very well could be unacceptable on any terms. There are no alternative plans, only the directions in the NWPA to return to Congress for further instructions if Yucca Mountain is unsuitable. If Yucca Mountain is found to be unsuitable in 30, 40, 50, or 100 years, because no genuine site study was conducted, what options will exist? The nuclear power plants that produce the money for the HLNW program will have long been closed, and they no longer will be a source of funding. The Nuclear Waste Fund will have been spent. Loading Yucca Mountain with HLNW as if it were a repository, after compromising site selection standards, and then finding that Yucca Mountain is unsuitable or that some other option for HLNW management is necessary, will place tremendous burdens on future generations, complicating rather than simplifying the management of HLNW.

The fact that Yucca Mountain is a failed program must be recognized. This message will prompt resistance from DOE, the nuclear industry, and their supporters because it substantiates outside claims that the 1987 decision to look only at Yucca Mountain was a mistake.

The past decade has taught us many things about managing nuclear wastes. Public support and acceptance for HLNW facilities at the community, state, and public levels is clearly needed. We have begun to learn that cooperation and agreement among communities, states, and the federal government are vital. We know the institution in charge of HLNW management must be trusted and competent.

We must face the fact that a new policy on HLNW is needed. HLNW can be safely stored for a century or more with dry-cask technology. We have time and ample reason to rethink this program and come to a better solution than Yucca Mountain with its many risks and vulnerabilities. A new policy based on what we have learned during the past decade can be a step toward solving the nation's HLNW problems.

Acknowledgments

Although no funds were allocated by the Nevada Nuclear Waste Project Office for this chapter, it is based in part on the research we have participated in since 1986 with the state of Nevada Yucca Mountain Socioeconomic Study Team. We would like to acknowledge our indebtedness to Jim Chalmers, Doug Easterling, Kay Fowler, Ronald Little, Roger Kasperson, Richard Krannich, Howard Kunreuther, Al Mushkatel, David Pijawka, and Jim Williams. People associated with the Nevada Nuclear Waste Project Office who have contributed to our past work on the subject of Yucca Mountain include Lydia Dotto, Steve Frishman, Bill Freudenburg, John Gervers, Bob Halstead, Bob Loux, Joe

Strolin, and Harry Swainston. Over the years we benefited from formal and informal discussions with members of the Technical Review Committee established by the Nevada Nuclear Waste Project Office to oversee and advise the socioeconomic research. We would like to thank Gilbert White, Michael Bronzini, William Colglazier, Bruce Dohrenwend, Kai Erikson, Reed Hansen, Allen Kneese, Todd LaPorte, Richard Moore, Edith Page, Roy Rappaport, and Clifford Russell. We would like to acknowledge the helpful reviews provided by Ellen Silbergeld and two anonymous reviewers. Our thanks are also extended to Robin Gregory and CK Mertz, our friends and colleagues at Decision Research, for sharing their work, knowledge, and insights with us, and to Kari Nelson who did everything that was needed to prepare the manuscript for publication. Funding for work on this chapter was provided by a grant to Decision Research from the Alfred P Sloan Foundation, for which we are very grateful. Finally, we would like to state that any opinions, findings, conclusions, or recommendations expressed in this article are those of the authors and do not necessarily reflect the views of the persons or organizations named here.

Literature Cited

1. Holdren JP. 1992. Radioactive-waste management in the United States: evolving policy prospects and dilemmas. *Annu. Rev. Energy Environ.* 17:235–59
2. Rogers K. 1994. Policy holds no alternative to Yucca Mountain. *Las Vegas Rev.-J.* Oct. 13:1A
3. Colglazier E. 1982. *The Politics of Nuclear Waste.* Elmsford, NY: Pergamon
4. Carter L. 1987. *Nuclear Imperatives and Public Trust: Dealing With Radioactive Waste.* Washington, DC: Resources for the Future
5. Jacob G. 1990. *Site Unseen: The Politics of Siting a Nuclear Waste Repository.* Pittsburgh: Univ. Pittsburgh Press
6. Clary B. 1992. Enactment of the nuclear waste policy act of 1982: a multiple perspectives explanation. *Policy Stud. Rev.* 10(4):90–102
7. Easterling D, Kunreuther H. 1995. *The Dilemma of Siting a High-Level Radioactive Waste Repository.* Boston: Kluwer
8. Flynn J, Chalmers J, Easterling D, Kasperson R, Kunreuther H, et al. 1995. *One Hundred Centuries of Solitude: Redirecting America's High-Level Nuclear Waste Policy.* Boulder, CO: Westview
9. *Nuclear Waste Policy Act.* 1982. 42 U.S.C. §10101-10226. PL 97–425
10. Dunlap RE, Kraft ME, Rosa EA, eds. 1993. *Public Reactions to Nuclear Waste: Citizens' Views of Repository Siting.* Durham, NC: Duke Univ. Press
11. US Dep. Energy Off. Civilian Radioactive Waste Management. 1986. *A multiattribute utility analysis of sites nominated for characterization for the first radioactive waste repository: A decision-aiding methodology.* DOE/RW-0074. Washington, DC: US GPO
12. Merkhofer M, Keeney R. 1987. A multiattribute utility analysis of alternative sites for the disposal of nuclear waste. *Risk Anal.* 7:173–94
13. Keeney R. 1987. An analysis of the portfolio of sites to characterize. *Risk Anal.* 7(2):195–218
14. Gregory R, Lichtenstein S. 1987. A review of the high-level nuclear waste repository siting analysis. *Risk Anal.* 7(2):219–24
15. McCabe A, Fitzgerald M. 1992. Pros-

pects for monitored retrievable storage of high-level nuclear waste. *Policy Stud. Rev.* 10(4):167–79
16. Kraft M, Clary B. 1993. Public testimony in nuclear waste repository hearings: a content analysis. In *Public Reactions to Nuclear Waste: Citizens' Views of Repository Siting,* ed. R Dunlap, M Kraft, E Rosa. Durham, NC: Duke Univ. Press
17. Hershey R. 1986. US suspends plan for nuclear waste dump in east or midwest. *New York Times* May 29:1, 10
18. US Gen. Account. Off. 1988. *Nuclear waste: quarterly report on DOE's nuclear waste program as of September 30, 1987.* GAO/RCED-88-56FS, Appendix I. Washington, DC: US GAO
19. US Gen. Account. Off. 1987. *Nuclear waste: status of DOE's implementation of the Nuclear Waste Policy Act.* GAO/RCED-87-17, p. 20. Washington, DC: US GAO
20. Mushkatel A, Pijawka K. 1992. *Institutional trust, information, and risk perceptions.* NWPO-SE-055-92. Carson City, NV: NWPO
21. US Gen. Account. Off. 1987. *Nuclear waste: status of DOE's nuclear waste site characterization activities.* GAO/RCED-87-103FS, pp. 37–40. Washington, DC: US GAO
22. US Gen. Account. Off. 1988. *Nuclear waste: repository work should not proceed until quality assurance is adequate.* GAO/RCED-88-159. Washington, DC: US GAO
23. US Gen. Account. Off. 1987. *Nuclear waste: institutional relations under the Nuclear Waste Policy Act of 1982.* GAO/RCED-87-14. Washington, DC: US GAO
24. US Gen. Account. Off. 1986. *Issues concerning DOE's postponement of second repository siting activities.* GAO/RCED-86-200FS. Washington, DC: US GAO
25. US Gen. Account. Off. 1987. *Nuclear waste: information on cost growth in site characterization cost estimates.* GAO/RCED-87-200FS. Washington, DC: US GAO
26. US Dep. Energy Advisory Panel on Alternative Means of Financing and Managing Radioactive Waste Facilities. 1984. *Managing nuclear waste—A better idea.* Rep. US Secretary of Energy. Portland, OR: US Dep. Energy Advisory Panel Alt. Means Financ. Manage. Radioact. Waste Facil.
27. *Nuclear Waste Policy Act Amendments of 1987. Title V, Energy and environment programs, Subtitle A, Nuclear waste amendments.* 1987. PL 100-203, 101 Stat. 1330-227
28. Kraft ME, Rosa EA, Dunlap RE. 1993. Public opinion and nuclear waste policymaking. In *Public Reactions to Nuclear Waste,* ed. RE Dunlap, ME Kraft, EA Rosa, pp. 16–17. Durham, NC: Duke Univ. Press
29. Swainston H. 1991. The characterization of Yucca Mountain: the status of the controversy. *Fed. Facil. Environ. J.* Summer:151–60
30. Rhodes J. 1990. *Nuclear power: waste disposal: new reactor technology, pyramids underground.* Annu. Meet. Natl. Assoc. Regul. Util. Comm., 102nd, Orlando, FL
31. McKay B, Swainston H. 1989. Letter to Nevada Governor Robert Miller. Carson City, NV: 21 pp.
32. US Gen. Account. Off. 1992. *Nuclear waste: DOE's repository site investigations, a long and difficult task.* GAO/cRCED-92-73. Washington, DC: US GAO
33. Carter LJ. 1987. Siting the nuclear waste repository: last stand at Yucca Mountain. *Environment* 29(8):8–13, 26–32
34. Carter L. 1993. Ending the gridlock on nuclear waste storage. *Issues Sci. Technol.* Fall:73–79
35. Shrader-Frechette K. 1993. *Burying Uncertainty: Risk and the Case Against Geological Disposal of Nuclear Waste.* Berkeley: Univ. Calif. Press
36. Shrader-Frechette K. 1994. High-level waste, low-level logic. *Bull. Atomic Sci.* Nov./Dec:40–45
37. Broad W. 1990. A mountain of trouble. *New York Times Mag.,* Nov. 18:37–39, 80–82
38. Natl. Res. Council/Natl. Acad. Sci. Board Radioactive Waste Management, April 1992. *Ground Water at Yucca Mountain: How High Can It Rise?* Washington, DC: Natl. Acad. Press
39. Rogers K. 1992. Quake rattles nuke dump's future. *Las Vegas Rev.-J.* June 30:A1
40. Rogers K. 1994. Yucca site faults deeper than thought. *Las Vegas Rev.-J.* May 26:1B, 4B
41. Shetterly C. 1988. Scientists find two volcanoes at Yucca. *Las Vegas Rev.-J.* Oct. 14:8B
42. van Konynenburg R. 1991. *Gaseous release of carbon-14: why the high-level waste regulations should be changed.* In *High-Level Radioactive Waste Management: Proceedings of the Second Annual International Conference,* pp. 313–19. La Grange, IL: Am. Nucl. Soc. Am. Soc. Civ. Eng.

43. Broad WJ. 1995. Scientists fear atomic explosion of buried wastes. *New York Times,* Mar. 5:1
44. Erikson K. 1994. Out of sight, out of our minds. *New York Times Mag.,* Mar. 6:34–41, 50, 63
45. Erikson K, Colglazier E, White G. 1994. Nuclear waste's human dimension. *Forum Appl. Res. Public Policy,* Fall:91–97
46. Natl. Res. Council/Natl. Acad. Sci. Board Radioactive Waste Management. 1990. *Rethinking high-level radioactive waste disposal: a position statement of the board on radioactive waste management.* Washington, DC: Natl. Acad. Press
47. *Energy Policy Act.* 1992. PL 102-486
48. Younker J, Andrews W, Fasano G, Herrington C, Mattson S, et al. 1992. *Report of early site suitability evaluation of the potential repository site at Yucca Mountain, Nevada.* SAIC-91/8000. Washington, DC: US Dep. Energy
49. US Gen. Account. Off. 1993. *Nuclear waste: Yucca Mountain project behind schedule and facing major scientific uncertainties.* GAO/RCED-93-124. Washington, DC: US GAO
50. Rogers K. 1994. NRC calls Yucca oversight flawed. *Las Vegas Rev.-J.* Oct. 14:B1
51. Rogers K. 1994. Case needed for fast approach to waste dump study. *Las Vegas Rev.-J.* Oct. 20:6B
52. Nevada Nuclear Waste Project Office. n.d. *Collection of calls for an independent, comprehensive review of the Department of Energy's civilian radioactive waste program.* Reprints of reports, testimony, letters, etc. Carson City, NV
53. Nuclear Waste Strategy Coalition. 1994. *Redesigning the US high-level nuclear waste disposal program for effective management.* St. Paul, MN: Minn. Dep. Public Serv. Draft
54. US Gen. Account. Off. 1994. *Comments on the draft statement of work for the financial and management review of the Yucca Mountain project.* GAO/RCED-94-258R. Washington, DC: US GAO
55. US Gen. Account. Off. 1994. *Nuclear waste: comprehensive review of the disposal program is needed.* GAO/RCED-94-299. Washington, DC: US GAO
56. Special report: Nuclear waste. 1994. *Nuclear Energy* 2nd Quart.:2, 18–35
57. Nuclear Waste Technical Review Board. 1993. *NWTRB Spec. Rep. Congr. Secr. Energy.* Arlington, VA: NWTRB
58. OCRWM. 1994. The proposed program approach from the Office of Civilian Radioactive Waste Management. *OCRWM Bull.* DOE/RW-0448(Summer/Fall)
59. Loux R. 1994. *Statement before the US House of Representatives committee on energy and commerce, subcommittee on energy and power.* Carson City, NV: NWPO
60. Kraft M. 1992. Public and state responses to high-level nuclear waste disposal: learning from policy failure. *Policy Stud. Rev.* 10(4):152–66
61. Desvousges WH, Kunreuther HS, Slovic P, Rosa EA. 1993. Perceived risk and attitudes toward nuclear wastes: national and Nevada perspectives. See Ref. 10, pp. 175–208
62. Kunreuther H, Desvousges WH, Slovic P. 1988. Nevada's predicament: public perceptions of risk from the proposed nuclear waste repository. *Environment* 30(8):16–20, 30–33
63. Slovic P. Flynn J, Layman M. 1991. Perceived risk, trust, and the politics of nuclear waste. *Science* 254:1603–7
64. Slovic P, Layman M, Kraus N, Flynn J, Chalmers J, et al. 1991. Perceived risk, stigma, and potential economic impacts of a high-level nuclear waste repository in Nevada. *Risk Anal.* 11: 683–96
65. Flynn J, Slovic P, Mertz CK. 1993. The Nevada initiative: a risk communication fiasco. *Risk Anal.* 13:643–48
66. Slovic P. 1987. Perception of risk. *Science* 236:280–85
67. Flynn J, Slovic P, Mertz CK. 1993. The Nevada initiative: a risk communication fiasco. *Risk Anal.* 13:497–502
68. Flynn J, Slovic P, Mertz CK. 1993. Decidedly different: expert and public views of risks from a radioactive waste repository. *Risk Anal.* 13(6): 647–52
69. Flynn J, Burns W, Mertz CK, Slovic P. 1992. Trust as a determinant of opposition to a high-level radioactive waste repository: analysis of a structural model. *Risk Anal.* 12:417–30
70. Pijawka K, Mushkatel A. 1992. Public opposition to the siting of the high-level nuclear waste repository: the importance of trust. *Policy Stud. Rev.* 10(4): 180–94
71. Off. Tech. Assess. 1985. *Managing the nation's commercial high-level radioactive waste.* OTA-O-171. Washington, DC: OTA
72. Mushkatel A, Pijawka K. 1994. *The 1994 Clark County, Nevada Survey: Key Findings.* Carson City, NV: NWPO
73. Slovic P. 1993. Perceived risk, trust, and

democracy: a systems perspective. *Risk Anal.* 13:675–82
74. US Secr. Energy Advisory Board Task Force Radioactive Waste Manage. 1993. *Earning Public Trust and Confidence: Requisites for Managing Radioactive Waste.* Washington, DC: US Secr. Energy Adv. Board Task Force Radioactive Waste Manage.
75. Yucca Mountain Socioeconomic Study Team. 1993. *The state of Nevada, Yucca Mountain socioeconomic studies, 1986–1992: an annotated guide.* NWPO-SE-056-93. Carson City, NV: NWPO
76. Mountain West. 1989. *Yucca Mountain socioeconomic project: an interim report.* NWPO-SE-024-89. Carson City, NV:NWPO
77. Deleted in proof
78. Slovic P, Layman M, Flynn J. 1991. Risk perception, trust, and nuclear waste: lessons from Yucca Mountain. *Environment* 33:6–11, 28–30
79. Slovic P, Layman M, Flynn J. 1993. Perceived risk, trust, and nuclear waste: lessons from Yucca Mountain. See Ref. 10 pp. 64–86
80. Flynn J, Kasperson R, Kunreuther H, Slovic P. 1992. Time to rethink nuclear waste storage. *Issues Sci. Technol.* 8(4): 42–48
81. Kunreuther H, Easterling D. 1992. Gaining acceptance for noxious facilities with economic incentives. In *The Social Response to Environmental Risk: Policy Formulation in an Age of Uncertainty,* ed. D Bromley, K Segerson. Boston: Kluwer
82. Kunreuther H, Fitzgerald K, Aarts TD. 1993. Siting noxious facilities: a test of the facility siting credo. *Risk Anal.* 13: 301–18
83. Ahlström P. 1994. Sweden takes steps to solve waste problem. *Forum Appl. Res. Public Policy,* Fall:119–21
84. Kasperson R, Kasperson J. 1987. *Nuclear Risk Analysis in Comparative Perspective: The Impacts of Large-Scale Risk Assessment in Five Countries.* Boston: Allen & Unwin
85. US Gen. Account. Off. 1994. *Nuclear waste: foreign countries' approaches to high-level waste storage and disposal.* GAO/RCED-94-172. Washington, DC: US GAO

Annu. Rev. Energy Environ. 1995. 20:119–43

PRIVATIZATION AND REFORM IN THE GLOBAL ELECTRICITY SUPPLY INDUSTRY

R. W. Bacon

Lincoln College, University of Oxford, Oxford, United Kingdom

KEY WORDS: power sector, regulation, independent power producers, restructuring, competition

---

CONTENTS

Abstract

This paper reviews the origins of the current global interest in the privatization and reform of the electricity supply industry. Particular emphasis is placed on the issues of restructuring the industry into separate firms and the difficulties of attracting private capital into an industry in which some state ownership remains. The targets to be met, the means of achieving them, and potential difficulties in implementing such changes are analyzed. An overview of actual experience is provided, along with a discussion of the problems of assessing whether these changes have been successful. Finally, some lessons are drawn for the future development of the process.

1056-3466\95\1022-00119$5.00

INTRODUCTION

During the past forty years, the global electricity supply industry (ESI) has experienced few changes in ownership or structure. Two patterns of ownership have dominated: the vertically–integrated private-sector monopolies and the national (publicly-owned) monopolies found in many European countries and many developing countries. Until recently, virtually no country has been pressured to change the structure of the industry (by separating or integrating the principal components of generation, transmission, distribution, and supply), the nature of ownership, or the method of government control (if any) over the major decision variables of price, investment, and quality. The industry has been content to let these traditional forms remain distinct and effectively unchanging.

In the 1980s, however, a series of changes led to major shifts in the industry and will almost certainly be followed by continuing change in ownership, structure, and regulation. The major influences on these developments were:

1. a need to rescue the very poorly performing publicly-owned ESI in Chile and other Latin American countries, coupled with the inability or unwillingness of such governments to finance further capital investment;
2. the ideological stance against public-sector ownership in the UK;
3. the change of the form of government in Eastern Europe and the former Soviet Union, and the desire to establish private-sector business throughout industry;
4. US dissatisfaction with the state level monopolies in power and the reliance on complex regulation to control these firms.

The experience with these initial cases has rapidly generated great enthusiasm for extending the benefits of private-sector ownership and competition to other countries, particularly developing countries. Dissatisfaction in many countries has led people to call privatization efforts a reform movement. Although much of the discussion has focused on experiences in the US and the UK, many other countries have either initiated privatization programs or are at the early stages of privatization. A review of these experiences is also useful.

This paper gives an overview of the reform process. The first section examines the background for these changes and the aims of individual countries. The second section considers the reform process as a whole: the targets to be achieved and the means to achieve them, along with the problems of implementing such a program. The third part of the paper provides an overview of the privatization and reform of the global ESI to date and a discussion of the problems of assessing success. From this analysis some lessons for the future are drawn.

RECENT PERSPECTIVES ON THE GLOBAL ESI REFORM PROCESS

The Starting Point

At the beginning of the 1980s, the global ESI appeared to be a highly stable, even traditional, industry. The traditions varied among countries with, for example, individual US states exhibiting private-sector ownership of integrated monopolies, whereas many European countries exhibited public-sector ownership. Although mixed-ownership systems existed (e.g. in the US approximately 20% of electricity sales are from public or cooperatively-owned enterprises), the patterns of ownership and structure seemed fixed and permanent. A common argument was that the industry as a whole was a natural monopoly because economies of scale would make production of the same amount of electricity more expensive with two or more firms than with one. From this argument, discussed further below, followed the idea that in order to curtail excessive pricing by the monopoly, the state had to make pricing and investment decisions either directly through its ownership of the industry (1), or indirectly through the powers of a government-appointed regulator (2). The former was seen as an appropriate choice for countries with a tradition of active intervention, the latter for countries accustomed to passive intervention.

Within approximately a decade, the industry picture has changed dramatically. Several countries have undergone complete reforms by changing both structure (splitting the previous single firm into several firms) and ownership, in part or completely. Other countries, though leaving the existing firm untouched, have encouraged entry of new private firms. Additionally, many countries are currently planning to follow these examples. The interest in exploring the potential gains from such reform is almost universal. This combination of new structure and ownership, in what is for most economies the largest industry, must be seen as one of the major global industrial changes of the postwar era.

These developments were initiated and encouraged by a series of rather different experiences. However, common to all was the thought that private-sector ownership would tend to be more efficient (in reducing costs) than public-sector ownership and that private-sector competition would reduce costs more than a private monopoly (even if regulated). Although the driving force for reform in many less-developed countries (LDCs) has been the need to stem the deterioration of the ESI's financial performance and the drain of using subsidies for electricity, the proposed solution—to involve the private sector—has been very similar.

The First Reformers

This section reviews four starting points for the privatization and reform process.

CHILE Chile was the first country to consider and then carry out power-sector privatization and reform in the recent past (3–5). Tariffs in the industry (6000 MW), which was predominantly publicly-owned, were set by the Ministry of Economy prior to 1978. The political desire to keep consumer prices low resulted in a failure to raise tariffs as fast as costs (at a time when inflation was extraordinarily rapid), so the industry suffered large losses which had to be subsidized by the government out of general finances. The reform program, which began in 1978, had the twin aims of establishing a regulatory framework that would effectively eliminate government tariff-setting and of privatizing the industry in stages, as it regained efficiency. These moves were undertaken in a framework of general macroeconomic restructuring.

In the privatization program, generation was partly separated from transmission and transmission from distribution. The generation sector was split into eleven main companies, of which one (ENDESA) accounts for about 40% of total capacity. Most of these companies have now been privatized. The transmission sector is owned by the large generator, ENDESA, and thus is in private hands. Virtually all of the distribution sector has been privatized and there are some 23 privately-owned companies, along with three small municipal utilities and seven cooperatives.

Bulk prices are negotiated for customers with demand over 2 MW, but otherwise are regulated, as are transmission and distribution prices. Dispatch is determined by a separate company acting on behalf of all the generators, so as to minimize overall system cost, based on estimated short-run marginal costs of the individual plants, rather than on competitive prices. Three features of the reform stand out:

1. Regulation was introduced to remove direct government intervention, thus permitting prices to rise in order to cover costs.
2. Virtually the whole industry was privatized, but this change took approximately a decade.
3. The industry was substantially restructured, but the new structure has been criticized because it has a single dominant generator and this generator owns the transmission network. On both grounds, the potential for effective competition from existing and new generators is weakened.

Chile is the country where reform has been longest established and therefore provides the best opportunity to assess the program's results. A recent study (6) shows a large postprivatization increase in productivity and profitability for one of the generating companies, owing mainly to increased plant utilization, but also directly attributable to the change in ownership. A distribution company showed gains in profitability owing to both reduced electricity loss from theft and improved productivity.

THE UNITED KINGDOM The reform process in the United Kingdom was extensively discussed both before (7, 8, 9, 10, 11, 12) and after (13, 14, 15, 16, 17, 18, 19) it was implemented. Several factors have been identified as important in leading the Conservative government of Lady Thatcher to undertake this radical reform.

The ESIs (in England and Wales) were run for many years as public-sector monopolies. Generation and transmission were run by the Central Electricity Generating Board (CEGB), and 12 Area Boards distributed power at a regional level. The Conservative government decided to both restructure and privatize the industry for several reasons. First, private ownership was seen as a way to improve the efficiency of the industry and lower consumer costs. Second, the receipts from selling the industry and from removing any need to subsidize the industry would improve the government's Public Sector Borrowing Requirement and permit a lowering of tax rates. Third, the ESI had provided large subsidies to the UK coal industry through what were seen as favorable long-term contracts. Given the Conservative government's wish to reduce the effective power of the coal-mining union, shifting the burden of ownership to the private sector, which would not renew coal contracts on such favorable terms, was seen as an important step in removing constraints on government behavior.

The new arrangements, which began in 1990, split the CEGB both vertically and horizontally. Generation of more than 60,000 MW was split by three companies: National Power (52%), Power Gen (33%), and Nuclear Electric (15%). The former two were privatized in stages beginning in 1991, with the final shares to be sold in 1995. The decision was made, at the last minute, not to privatize the nuclear sector, largely because of the enormous uncertainty regarding waste disposal and decommissioning costs; however, this decision is currently being reconsidered. New firms were able to enter the generation market and remarkably, despite considerable excess capacity at the time of privatization, substantial new private capacity has effectively replaced older, inefficient plants. While generation was being restructured, transmission was separated from generation, and the new transmission company is jointly owned by the 12 new privatized distribution companies (Regional Electricity Companies) created from the former Area Boards.

Power is sold from the generators into a power pool on a competitive bidding basis. To this wholesale price are added regulated transmission charges and distribution and retail charges. Retail customers with peak demands less than 100 kW must buy from their local Regional Electricity Companies (REC) until 1998, but customers with greater needs are free to use any supplier.

The scheme was designed both to introduce private-sector ownership and to introduce competition where possible, while regulating those aspects that were not considered capable of meaningful competition.

The features of the reform that have created the most interest and controversy are:

1. the complete separation of the industry stages and their subsequent privatization (with the exception of nuclear generation);
2. the attempt to create competition in the wholesale power market through competitive bidding between generators to supply power and energy and limited competition in supply (retail competition) by allowing high-level consumers to choose their supplier;
3. the forms of regulation established to control the noncompetitive industry sectors.

After nearly five years of private-sector ownership and regulatory reviews of every industry sector, several issues have been highlighted: (*a*) What have been the productivity gains resulting from the reforms? (20, 21); (*b*) Is the generation structure truly competitive? and (*c*) Has the regulatory regime proved its worth as a distinct alternative to other forms of regulation? These questions are discussed below in relation to general problems of privatization.

The England and Wales experiment has created great interest throughout the global ESI and countries considering privatization have tended to take this model as a blueprint for action, despite some of the obvious problems.

EASTERN EUROPE AND THE FORMER USSR The collapse of Communist regimes at the end of the 1980s, along with Western interest in assisting the rapid transition to market economy structures in these countries, has strongly impacted the energy sectors. Most of these countries had subsidized electricity to a large extent and had built a substantial excess of generating capacity. The need to reduce the burden on the central governments by reducing consumer subsidies and improving very inefficient economic operation of the ESIs led to a strong push for restructuring and privatization. However, apart from the former Eastern Germany (22, 23), where events have been determined by the unification with the former West Germany, the elaborate plans for privatization have not proceeded very far. Even in Poland (24, 25), which has made some advances, uncertainty remains about the number of independent generators to create. In several countries, the lack of a clear framework in which a private-sector company could operate (laws, institutions, etc) has inhibited any substantial private-sector participation to date. Nevertheless, the evident need to do something has led to considerable interest in determining optimal industry structure and ownership and designing and regulating a system for selling electricity effectively.

THE UNITED STATES The US tradition of private-sector ownership in the ESI is very well established, but until the 1980s allowing vertically-integrated monopolies to exist was considered adequate, provided a regulatory system adequately checked undesirable practices (26).

The Public Utility Regulatory Policy Act (PURPA), passed in 1978, allowed independent power producers that built generating units of certain sizes and technologies to sell to the utilities without facing discrimination. The PURPA was not motivated by a conscious decision to change the structure of the US ESI. Rather, the motive was to improve energy efficiency; restructuring was an unexpected consequence (27). In 1980, the Federal Energy Regulatory Commission (FERC) issued the rules governing the maximum prices to be paid by utilities to those independent suppliers. These changes have led to the creation of generating plants, some still under construction, that account for 5–10% of the total installed generating capacity and that belong to nonutility firms. Many firms have entered this market, including subsidiaries of gas and electric utility companies, manufacturing companies, construction companies and other types of firms. As a result of the competitive bidding to enter the market, the costs of supply (as typically expressed through long-term contracts) have been reduced below the utilities' own estimates of the cost of building and running equivalent generation capacity.

The very large scale of the total US market has generated considerable experience in the construction and operation of independent private generation, in absolute terms. This experience clearly has a major impact in the global market as such firms look overseas for new opportunities.

Despite this increased liberalization of the market, the following factors may hamper a fully efficient operation.

1. The lack of real retail competition in the market, with firms unable to bid constantly to supply and consumers not free to choose their supplier, makes the pressure to keep costs low and pass on such benefits to the consumer lower than it could be (28).
2. The regulatory system, designed to prevent excessive prices, has largely focused on tying prices to costs. This scheme is argued to give too little incentive to actually reduce costs, since few benefits of doing so accrue to the firm (29).

The US experience can be seen as a tentative step towards a more competitive structure for the ESI, but enormously important because of the size of the market. The wealth of experience in competitive bidding to build and supply electricity has shown that the entry of new firms into the generating market is technically straightforward. The real issue in many countries is whether market conditions are sufficiently attractive to encourage such entry (30).

Aims of Reform

The experiences of other countries will be brought into the discussion when appropriate, but they all point to the unifying theme of the desire for a better deal for consumers as a whole (including taxpayers) and investors. Direct goals, such as allowing prices to cover costs, and indirect goals, such as permitting competition, are desired because they are seen as steps toward improving the overall economic welfare of the society. These goals are discussed separately even though they interact strongly.

Much of the reform program is based on some version of the following argument: Market competition will drive prices towards the marginal cost structure of an industry (long-run marginal costs at peak demand, short-run marginal costs at off-peak demand) and will ensure that these costs are the lowest possible. Prices that equal marginal costs will also result in demand for the good and production of the good that is allocatively optimal. Competition is therefore required in order to attain the best economic outcome. Several important parts of this argument, along with some crucial caveats, will be discussed below.

ALLOCATIVE EFFICIENCY Central targets of reform have been to permit the industry to price to cover marginal costs and, once such pricing is feasible, to ensure that it does so. The marginal costs for a business include costs of investment as well as operating costs (1). The reason economists emphasize setting prices at marginal costs is that one of the fundamental theorems of welfare economics shows that if all industries price in such a way, and a series of other assumptions hold, then any reallocation of production between firms will lead to at least one consumer being worse off than before. However, this so-called Pareto optimum condition does not distinguish between situations in which some individuals lose and those in which some gain, and so is of somewhat limited practical relevance. For example, most attempts by a government to allow electricity prices to rise to cover costs by removing subsidies would decrease demand and increase unemployment. If unemployed workers were not reemployed, then even if industry prices conformed with the conditions required for an allocative optimum (consumers demand and pay for a good at a price just up to the total costs of producing the marginal unit of that good), the economy would not necessarily be better off.

In many developing countries, prices in the ESI are clearly far from marginal costs (31, 32, 33) and as a result, output and employment are also far from the associated target level. For instance, with an industry that prices 30% below long-run costs, and where long-run elasticity of demand is 0.5, raising prices to cost-recovery levels would decrease demand (over the long run) to 15% below where it would otherwise have been. A demand fall this large could be expected to have substantial impact on employment in an industry that was unable to carry excess workers.

Such potential unemployment is a major problem for reformers in countries where long-run unemployment is already high. Governments may create voluntary redundancy schemes to persuade workers to leave (thus bearing some of the short-run costs of the adjustment), but creating unemployment may be politically infeasible at certain moments in a government's life. The high levels of general unemployment in former Communist countries have proved a major barrier to raising prices for electricity and reducing its costs by reducing employment because governments have been reluctant to create immediate and substantial adverse effects on the population, despite the substantial long-term benefits of a more efficient ESI.

A second issue in pricing below costs is that certain sectors of the society may be receiving larger subsidies than others. Often the industrial sector pays the full cost (or even more) whereas consumers are heavily subsidized. This cross-subsidy tends to distort the use of electricity between sectors, and reform programs often target the removal of such distortions. Again, the relative importance of the groups in the political process will help to determine how easily this change in pricing can be done.

A particularly important impact of incorrect pricing strategies is exhibited when demand is high relative to supply (encouraged by low prices), but the state industry cannot invest to remove the potential bottleneck because such intervention would create further financial loss at the current subsidized prices. In such a case, power shortages and blackouts start to occur, which place extremely high costs on those consumers who had not anticipated this outcome. Firms that invested in industries requiring continuous sources of power would find they also needed suddenly to invest in expensive standby generators. Raising electricity prices to cover costs would both reduce demand, thereby reducing shortages, and make future investment more attractive.

X-INEFFICIENCY A second and equally important source of inefficiency is poor operating (or investment) techniques. If the costs of production in a state company are unnecessarily high, consumers will pay too much, or the government will provide an unnecessarily large subsidy.

Although abundant anecdotal evidence shows that state ESI companies in both developed and developing countries have unnecessarily high costs in several areas (e.g. too many employees to carry out a particular function; excessive amounts of electricity lost because of poor system design, theft, or failure to charge or collect bills), the establishment of benchmark levels of good practice is not straightforward (34). Differences in circumstances between countries can lead to differences in physical factor usage and in costs, so obtaining relatively precise ideas of efficiency targets may not be possible. However, high levels of inefficiency tend to be obvious and the state may consider cutting the costs of subsidies to increase the potential marketability of its ESI. Again, such cuts could lead to substantial increases in unemployment, depending on the nature of the inefficiency, and this possibility can reduce the government's willingness to take such action. Because the ESI is often a major employer in the industrial sector of many developing countries, the shedding of a substantial fraction of its labor force can have a very noticeable impact on the labor market.

Although some cost-cutting actions may be available that are obvious to those outside the industry, such as civil servants, an effective mechanism for constantly searching out cost reductions and implementing them is needed. Reliance on the civil service to monitor performance and direct conduct risks encouraging political interference, e.g. tariff-setting or employment policies that favor certain voters. A mechanism for establishing control over costs without exposing the industry to political manipulation must be found.

INCENTIVES FOR BETTER MANAGEMENT Several mechanisms are designed to change the industry by giving management greater independence and a greater incentive to reduce costs. These mechanisms range from the corporatization and commercialization of the state-owned utility (through the use of contracted management) to partial or full privatization in a form in which the market is truly competitive. These techniques, which are discussed below, aim to give the managers incentives to reduce costs and operate at a profit by removing state interference. In addition to giving managers freedom to act, providing a positive incentive to lower costs and hence prices is also considered desirable. This can be done through a direct performance incentive, as in a management contract, or through privatization in which the owners desire profits and choose their managers accordingly. Exposure to competition requires more aggressive profit-making by managers, although at the same time prices cannot be raised too far above costs because other firms will be attracted into the industry, increasing supply and lowering profit margins.

The desire to privatize interacts with the need to follow cost–reflective pricing strategies. Public-sector firms, in which prices are held below costs, are unlikely to be successful candidates for privatization unless prices can be set freely or are at least allowed to rise to cover costs. Tension inevitably results when prices are raised prior to privatization: The private sector is blamed for increasing the visible burden on consumers, whereas the government (which had created an indirect burden through subsidies) may avoid the unpopularity created by its former actions.

The desire to establish a competitive market also conflicts with the desire to achieve successful privatization. The more profitable the market appears to potential entrants, the more willing they will be to enter and the higher the price they will pay to enter. If several players of roughly equal size and with roughly equal costs are in the market, then profits will decrease and entrants will be less enthusiastic.

If markets cannot be made competitive, then privatization is usually proposed in tandem with the creation of a regulatory agency. This agency's principal duty is often portrayed as ensuring that prices are not set excessively high relative to costs, as in the US and the UK. However, in developing countries that are reforming the ESI, the major effort of the regulator may turn out to be ensuring that prices are regularly allowed to rise to cover costs and not kept down to satisfy political demands.

If a regulator is seen to be potentially effective in holding prices below where private industry would wish to set them, the sector's attractiveness to private capital is reduced. Conversely, if the regulator allows prices to rise to cover costs private capital will be more willing to enter the sector. The possible reduction of incentives to enter the sector is an important consideration in debates on the nature and powers of the regulatory body (13).

Although the profit motive, even in the absence of competition, can be expected to result in a search for lower costs, monopolistic firms are often argued to be less assiduous than competitive firms in seeking cost savings (especially when shielded from entry of new firms). This difference has broad implications for the amount of restructuring needed to induce competition. If the profit motive alone will ensure that the majority of potential cost savings are actualized, then restructuring for the sake of introducing competition is less necessary. If restructuring introduces costs (e.g. losses of economies of scale or of the benefits of vertical integration), the arguments for extensive restructuring will be weaker. If only competition can ensure the continued search for cost savings, then restructuring to facilitate ample competition becomes more desirable.

RESTRUCTURING OF THE ESI Following the examples of Chile and England and Wales, many countries have proposed ESI restructuring as an integral part of the reform process.

Restructuring, which can involve both vertical separation (separating transmission from generation) and horizontal separation (splitting the generation sector into several entities), has been proposed for both the public and private sectors. If the sector is still public, vertical separation creates units with a single activity, which makes it easier to relate costs to output and to avoid hidden cross-subsidization. Splitting generation and distribution into multiple units makes management decisions clearer for each part.

If generation is to be privatized and competition introduced, the argument for horizontal separation is that it creates almost instantaneous competition, as opposed to the delay of waiting for new entrants. Horizontal separation requires that the sector be split into several units, with no single unit dominant in size. Reform of a small sector (35) immediately raises issues about the optimal scale of a unit. Too great a disaggregation might lead to a substantial increase in unit costs.

When competition is introduced into generation, either through horizontal separation and privatization or entry by independent power producers (IPPs), vertical separation of transmission from generation is considered desirable. This separation ensures that one generator cannot block a rival's sales by using its ownership of transmission to discriminate overtly or covertly against the rival (13). If generation lacks competition, and is likely to do so for an extended period, vertical separation may incur higher costs than would obtain in an integrated company. Because competitive industries naturally exhibit large-scale and vertically-integrated production, proposals for very disaggregated structures should be examined carefully to avoid creating an industry that would tend to rapidly reintegrate under competitive conditions (36).

Distribution tends by nature to operate at a local monopoly level, although the optimal scale in the global context is less clear. Models range from a city to a region or province, but in all cases the pressure to provide more than one source of distribution investment (low-voltage wires etc) has been low. The argument for vertical separation from transmission is that it creates a focused management but does not risk loss of substantial economies of scale.

In some countries, notably the UK, Norway, and New Zealand (37, 38), the supply function (selling electricity to consumers) has been seen as a separate activity from distribution, and one in which competition is possible. Several reforming countries have made comparison shopping among rival suppliers possible for large customers (over 2 MW peak demand). Customers with over 100 kW in the UK can now choose a supplier and all customers will be able to do so by 1998. New Zealand has allowed all consumers to choose a supplier since 1994. This final step in vertical disaggregation is likely to introduce some retail competition, but involves increased costs, such as metering, which may be substantial in comparison to the 100 kW market. However, retail competition is scarcely a proven success at this stage and cannot be used by itself to justify horizontal separation of the ESI distribution sector or the vertical separation of distribution from transmission in small or poor economies in which such a system would be expensive to operate. In Chile (6), the privatized distribution company was more efficient, but the extent to which this efficiency resulted from private ownership or from separation is not clear.

IMPACT ON PUBLIC FINANCES ESI reform that includes the sale of state assets as well as the removal of state subsidies has substantial impacts on public finances at the time of sale. Although such a sale merely exchanges one asset (the company) for another (cash), governments dominated by short-term considerations may prefer cash, even if the company must be sold below its true value. In the UK, the emphasis on reducing the Public Sector Borrowing Requirement led to a one-off financial improvement for the government through the sale of the ESI. This sale allowed a temporary large tax reduction or a permanent smaller reduction corresponding to the smaller interest payments made possible by issuing less government debt.

In countries that are motivated to improve efficiency, rather than to raise short-term revenue for the government, partial privatization has been seen as sufficiently radical to remove state influence and introduce managerial incentives. For example, in Malaysia (39), which in 1992 privatized its integrated ESI as a single company, only 23% of the shares in the newly commercialized and corporatized entity were sold to the private sector. In other countries, where a gradual transition was desired for various reasons, the first step was to separate and privatize, or contract out, those parts of the business that can easily operate as stand-alone operations without interfering with the efficiency of the principal business. Some maintenance and repair functions, as well as construction activities, have been targeted for early separation and privatization in Eastern European countries that have chosen to reform in stages.

DIMENSIONS OF CHANGE

Corporatization and Commercialization

In those countries in which ESI was given a measure of independence and more responsibility for its own actions, an almost inevitable first step is to turn the state enterprise into a corporation and then submit this corporation to commercial discipline (38, 40).

A distinctive approach to submitting the utility to commercial discipline, while maintaining it in public ownership, has been the French system of management contracts used both in France and in some African countries (41), as well as in El Salvador and Bolivia. In France, the company and the government agree on performance targets to be met over the contract period. In some cases, an outside company contracts to manage the company and meet such targets. The company's remuneration is related to its success in meeting the targets (as specified in the contract). A case study of this approach in Bolivia (42) highlighted the potential problems of government involvement as one party to the contract. The government could either fail to honor the contract or could interfere in the running of the company, thus reducing the opportunities for independent management to improve its performance.

Privatization

If governments view corporatization and commercialization as insufficiently powerful to achieve the goals of independence and efficiency, they often see privatization as the next step in reform. The goal of privatization is not merely to transfer ownership but to do so in a way that maximally benefits the economy. The mechanism of privatization becomes important in this context.

Deciding what is to be privatized is generally described as the first step, although decisions about privatization must actually be made simultaneously. Next comes a discussion of how the privatization is to take place, including how to insure competition for the market. Once the what and how of privatization have been determined, the degree of possible competition in the market can be analyzed. Given the competitive structure, or lack of it, and the nature of the political process, the design of the associated regulatory structure can be evaluated. Finally, the impact of this process on the ESI's efficiency and profitability is evaluated.

WHAT TO PRIVATIZE The simplest approach to introducing or extending private-sector ownership in the ESI is to ensure that further investments are made by new entrants, as opposed to public or private incumbent firms. This strategy preserves the existing structure and efficiencies acquired through experience or scale of operation but aims to expand the industry at least cost. In addition, new enterprises may create competition with the existing industry, which could benefit consumers. In this model, generation is the natural target for increased private capital investments because generating plants are usually fairly large in relation to the system and have the potential to compete against other plants. Transmission, like all network investment, is seen as a natural monopoly. In this case, the least expensive form of expanding an integrated system would be for the incumbent firm to own the extra lines. Distribution is also seen as a local natural monopoly so that little cost advantage could be obtained by introducing a second distributor into a local market.

Given these considerations, the fact that countries that did not want to privatize or restructure their existing ESI have looked to investment in generation as the vehicle for private-sector participation is scarcely surprising. The dominant example of this approach is various US states in which independent power producers (IPPs) have begun to sell their electricity to the existing vertically-integrated company. China, Mexico, and India are strongly encouraging private generation because privatization of state companies is difficult or undesirable. Many other smaller countries are also encouraging IPPs to enter their markets. The cost of such an investment program will clearly be high and those governments will naturally look to the private sector for a substantial amount of the financing. However, the government may need to assume a contingent liability in return for the private investment by issuing some form of guarantee to the private firm.

In countries that have decided to privatize the existing system, a much larger short-run exercise is involved. In these cases, decisions must be made on structure and timing. Malaysia, for example, decided to privatize the whole ESI as an integrated industry (with about 6000 MW capacity) and to do so in gradual stages. Approximately 20% of the company shares was sold in the first step. Chile took nearly ten years to complete the selling process, whereas in England and Wales separation was quick and a majority of the shares was sold within a two-year period.

HOW TO SELL The state is naturally anxious to achieve large revenues by selling all or some of the ESI, whereas potential bidders wish to pay as little as possible. Competition for the market, a form of auction in which bidders will offer what they believe to be the maximum value of the company in order to secure it from rival bids, is seen as vital for achieving the maximum return possible. Several auction processes have been debated and more than one has been implemented. Some systems have proven highly unsatisfactory, such as the method in which the winning bidder pays the amount offered by the second-highest bidder (used in nonpower privatizations for certain New Zealand industries). The implementation of selling methods is an area that clearly requires care.

IPPs in the US are commonly requested to bid to supply a certain amount of power (rather than to build the generating unit). Several successful US contracts have been analyzed (28), and such contract design clearly contains many variable elements, such as capacity price, energy price, a formula for fuel costs and efficiency, and force majeure clauses. Two additional considerations are important for foreign companies considering bidding to supply from an IPP in developing countries: the need to ensure that profits can be exported and the risk associated with exchange rate fluctuations. Finding a contract that gives the prospective producer a fair return under these conditions and does not require the domestic government to absorb too much of the risk can be a major hurdle to increasing private-sector involvement in countries where the commercial return, excluding these factors, is itself only moderately attractive.

Such risks have been seen as a major impediment to foreign involvement and negotiations in these cases have on occasion been extremely lengthy. Implementation is usually thought to be more rapid under private-sector ownership, so the management of these risks is crucial if the private sector is to remain attractive (43, 44).

Preparing state companies for sale to the private sector has required much effort, particularly in Eastern Europe where many individual enterprises were designated for sale. Several steps have proven important for making the company attractive to potential buyers (45). The treatment of debt and the company's operating position are especially crucial. Companies in the state sector are often unable to set prices to recover investment costs, and if the company has large debts, the potential private investor may be unwilling to take them on, even at a low sale price. Debt write-down by the government has become standard practice in such cases (46). Countries such as Bolivia, which is in the middle of an active privatization program, have devised a capitalization scheme in which the necessary injection of financial capital into the newly privatized company is part of the privatization deal.

A similar problem arises when the government proposes that the newly privatized company should be subject to the same tax regime as other private-sector companies. The newly privatized company is likely to be suddenly faced with a large tax bill, particularly when taxes are related to profits, which may increase as a result of revaluation of assets. Such a change can impose a large temporary financial strain on the company, and governments have had to make special transition arrangements to handle the large one-off adjustments required simply by the transfer to private-sector ownership. For example, Belize allowed a seven-year adjustment period before increasing taxes to private-sector rates.

Another important parameter is the golden share, or residual control, kept by the government, as in the UK, Malaysia, and New Zealand. Such a device is politically useful for blocking opposition to foreign control but will inevitably make the company less valuable to potential bidders.

COMPETITION IN THE MARKET In a truly competitive market, firms are motivated to cut costs so they can undercut rivals' prices, thereby increasing their own market share and profits. The achievement of the competitive paradigm rests on several important assumptions, some of which do not hold in some ESIs, at least in the short run. The most important feature of a competitive market is a sufficient number of rival firms to prevent any one firm from manipulating the market to its own advantage (e.g. by withholding energy supply from some of its plants so as to increase revenue by driving up prices). In England and Wales, where there are two private generators, lack of competition has persisted, and in 1994 the UK regulator had to cap pool prices and order the two large generators to divest themselves of some capacity in order to increase the market share of rivals more rapidly. Theoretical studies (47, 48) have suggested that up to five firms would be needed to achieve effective competition.

Another requirement is that no firm can be dominant. A single firm controlling 60% of generation, coupled with four other firms controlling 10% each, could still manipulate the market to its own benefit. Such manipulation apparently occurred in Norway where the dominant generator temporarily flooded the market with cheap energy, presumably to bring other generators into line with its own pricing guidelines. A third requirement for competition is that firms' cost structures be fairly similar. For example, in a system with a run-of-river hydro scheme, this capacity will always be run (when water is available) and other firms will not be able to compete with it. In such a case the incentive for other firms to reduce costs and improve efficiency is weakened.

If the size of the steps in the merit order (the order in which plants would be run to minimize costs to meet increasingly large demand) is very large, then one firm has little real possibility to cut costs enough to improve its position in the merit order, which would have enabled it to operate for a longer period each day. In a system with a capacity shortage, because the plant will be used most of the time, the potential gain in market share as a result of competition will be rather small in the short run. Competition will be fiercer in a system with excess capacity. Finally, if entering the market is easy (through new IPPs), then the threat of new low-cost suppliers will give incumbent suppliers an incentive to seek efficiencies and to reduce costs.

These criteria for strengthening competition can be built into the design of the reformed system. For example, in Argentina, the restructured industry was designed so that no single generator would control more than 15% of total capacity. Other countries, however, have found that grouping generating plants into separate companies that would lead to a competitive structure is virtually impossible.

In countries with a very small total generating capacity, dividing the system into three or four units of roughly equal size and cost structure is impossible. In 1990, 107 countries had capacity under 1000 MW, 90 countries under 500 MW. In any of these countries, the existing system may not be suitable for introducing a competitive structure. Moreover, opening the market to new investment by IPPs may only increase the number of units very slowly because of the need to delay expansion until economies of scale in plant are achieved. For example, adding 50–100 MW every decade may be much more cost effective than adding 10 MW every one to two years.

Even if the generating-sector structure has the potential for competition, a scheme is needed for selling electricity that transfers this competitive pressure into lower prices. If all electricity is sold by generators, either to the transmission company or to the distribution companies on long-term contracts, competitive pressure will come into play only at the time the contract is made (one-time competition). Between contracts the pressures to be efficient may be reduced unless they can be maintained through contract terms, and even if the desire for profits keeps efficiency high, the benefits will not be passed on to consumers until the contract is renewed. For this reason, the England and Wales reform and the Norwegian reform (49) created wholesale power pools in which some power is sold a day before it is supplied, and some is also sold on a contract basis. The Norwegian power pool, which has many sellers, appears to have many features of true competition, but little power is actually sold through the pool.

However, establishing and operating such a system requires considerable sophistication. The dispatch system must coordinate the bids to supply at various prices with the demands on the system and the associated settlement system. The pricing system must reward the supply of energy and the supply of power in a way that generates an optimal amount of investment and permits firms to cover operating costs. Learning how to market the energy also requires sophistication. Experience so far suggests that truly competitive power pools will become ever more important, but that they are not fully developed at present and are therefore suitable only for larger and more advanced countries.

Several countries have also introduced limited competition at retail. Large consumers, usually defined as those over 2 MW or 1 MW peak demand, are permitted to buy from any distributor at unregulated prices. In the UK, the size threshold was reduced to 100 KW and there are now approximately 50,000 potential customers. In 1998, the UK will completely phase out the size limit, as New Zealand has done. Little hard evidence is available on the impacts of this competition (19) on prices paid by these larger customers, although in the UK certain industrial customers have been strongly critical of prices they believe to be unnecessarily high. However, the costs of metering such an arrangement apparently become more important as the threshold is lowered, and the supply and fitting of such meters will tend to slow the true impact of retail competition. Several reforming countries have adopted such a scheme for the larger customers and this appears to be a viable option in many systems.

REGULATION As parts or the whole of the ESI are transferred to private ownership, the state has to decide whether the industry needs to be overseen by a regulatory authority, and if so, how. In the England and Wales reform, and in Chile and Argentina, competition in generation was considered sufficient to obviate regulation of generators' prices. For transmission, distribution, and supply, areas without competition, prices were to be regulated. The concern in these countries, and in mature economies such as the US, is that an unregulated company will use its position to restrict output and force prices well above costs in order to maximize profits. The test of acceptable behavior has therefore been whether prices are excessive relative to costs. However, a subsidiary concern has been the quality of private-sector service, because monopolies can reduce quality below competitive market standards in order to increase profits.

Establishing a new regulatory system, which will be needed in many countries where privatization is under active consideration, requires that several important questions be answered. What should be the size and form of the regulatory body? Who should be appointed to it, by whom, and for how long? What should be the powers of the regulatory body, and what powers that apply to the ESI should be maintained in an electricity law? How often should the regulatory body review the tariff levels?

The US has the most experience in regulating the ESI (as well as other sectors) and much has been written on all aspects of regulation there (50). Key aspects are the use of court-style hearings with emphasis on due process and rules developed from case law. The commissions exist at both the state and federal level and have a panel of commissioners. Decisions can be appealed in state courts and the whole process can be very lengthy. The focal point of the regulation is that prices should be justified by costs, appropriately calculated.

The UK system of regulation, though created much more recently than the US system, has also generated much discussion (12, 13, 14, 15, 17). Key differences are the creation of a body with a single member (the regulator), the use of price caps to control prices, and price cap reviews at fairly lengthy intervals (typically four years). The single regulator clearly is likely to avoid compromise and permit more decisive judgements, but the system is more dependent on the quality of the individual and on the means of his/her appointment and removal. A driving force in the privatization movement is to ensure a substantial degree of independence from government interference for the ESI, so the powers and nature of the regulatory body can severely compromise the possible benefits of privatization.

The UK model is perhaps the most open to abuse in the hands of a government or minister unwilling to relinquish its powers. A single regulator, appointed by the minister, and difficult to remove before a lengthy term of office has expired, could effectively carry out the government's wishes by proxy. Solutions of the form that remit certain details to the energy ministry, such as setting retail tariffs or granting generation licences, mean that all effective power remains in the government and that, for example, prices will continue to be used as a politically-driven subsidy, leaving the industry underfunded. In light of such concerns, reforming countries, such as Colombia, have appointed a panel of regulators including more than one minister to represent various interests (e.g. the Minister of Finance to express funding concerns) and more than one nonminister.

The UK model of price control, along with the form and composition of the regulatory body, has aroused global interest (13). Unlike the US, where prices must always be related to justified costs (thus giving little incentive to lower costs), the UK sets a target price, over a period, independent of the movement of some costs. For example, the RPI–X (retail price index less X%) formula was designed to allow a maximum increase for retail electricity prices that is lower than that of the general retail price index, thus forcing the industry to improve its efficiency to cover the X%. Because certain costs are unavoidable, such as fuel costs or transmission charges paid by distributors, one part of the cost can be fully carried over into prices, whereas the remainder is price-capped. With price caps at each stage of the industry, apart from generation, the system is designed to improve efficiency at all stages. After the three- or four-year reviews, the price caps were reset and this process is expected to continue.

Two problems have emerged from this approach. The act of privatization itself requires a starting (or vesting) price on which this index formula can be based. If the starting price is too high the firms will make large profits. The second problem is determining a reasonable value for the X factor—the hoped-for annual productivity gain. When profits are high in the UK, even when they are a result of increased productivity passed on only in part through lower consumer prices, demands for higher X-factors are inevitable. The regulator, in resetting these factors, has been drawn in turn to examining whether the industry is making a reasonable return on capital or an excessive return. Attempts to check performance, particularly of distribution companies, using other companies' costs or an ideal company as a benchmark (as in Chile) complete the circle of relating prices to costs and narrow the difference between the UK and the US.

Some have argued that the US system of regulation is excessively adversarial, as well as time-consuming, and creates delay because of such features as the way in which companies have to present evidence on costs. However, the first price-cap reviews in the UK also generated many technical submissions by the companies, and the reviews were not all completed rapidly. The main difference between the systems may turn out to be the institutionalized review system in the UK, although highly excessive profits or crippling losses caused by unusual cost movements might force extra reviews. Recent suggestions by the regulator that some price caps would have to be reviewed before the scheduled date indicate that the process is sensitive to public pressure and to profit levels.

THE GAINS FROM PRIVATIZATION The principal gains sought from privatization and reform are lower costs and prices that reflect costs. Thus, in OECD countries, concern has centered on potential overpricing, as well as X inefficiencies and high costs. In LDCs, the concern has been to remove severe underpricing as well as the X inefficiencies and high costs, thus encouraging investment. Measuring such gains is particularly difficult because reform includes removing subsidies, which causes prices to rise independent of ownership or structural changes. For this reason, comparing prices before and after privatization may be inadequate to isolate the effects of the change of ownership and the introduction of competition.

Changes in costs and various productivity measures can also be difficult to interpret but have been commonly used to assess progress (6, 20, 51, 52, 53). A cost- or productivity-based measured improvement is not by itself a guarantee that consumers are better off; most or all of the benefits may have gone into profits, as may have happened in the UK. The extent to which increasing profits is necessary (in order to finance future investment etc) before passing on efficiency benefits to consumers needs to be determined before the full gains can be evaluated.

A second issue, which is important for future policy design, is to determine how much of any consumer gains was caused by the pressure of competition, which passes some or all gains on to consumers, as well as increases the gains through efficiency improvements, and how much of the gains resulted from the profit motive irrespective of competition (i.e. improvements in productivity that the regulatory system ensured were passed on in part or in total to consumers). If the profit motive, when subject to regulation, is enough to capture most of the benefits of private ownership, then the more elaborate forms of market structure used to insure competition are not essential and simpler solutions can be used.

If competition is the essential ingredient for improving consumer welfare, then countries that are too small or otherwise ill-suited for effective competition may find that the principal reason for reform is merely to remove decision-making as far as possible from the direct control of government. The limited evidence available (from those countries privatized long enough to allow a comparative analysis of pre- and post-privatization) shows large productivity gains, especially in generation. The benefits to consumers in the form of price reductions have been less marked, so that much of the impact has been on profit levels of the privatized companies.

EXPERIENCE TO DATE

Country Involvement

Although many countries have actively encouraged private-sector participation or ownership in their ESI, relatively few have actually carried out extensive changes.

Complete commercialization, corporatization, and privatization have occurred in Chile, Argentina, England and Wales, Scotland, Malaysia, and Northern Ireland. Norway, New Zealand, Spain and Portugal have experienced commercialization and corporatization.

This list is notable in that it is composed of industrialized or middle-income developing countries. Also, with the exception of Northern Ireland at 2400 MW, the systems are all in the medium-to-large range (in a global sense). The England and Wales system, which has received so much attention, is by far the largest system privatized. Colombia and Peru have also almost completed the process, and Jamaica, Costa Rica, and Bolivia have made substantial progress. These countries will be important examples of privatization at a less advanced stage of development.

The US has not yet seen the radical restructuring of the ESI as a whole that is needed to establish a truly competitive industry, although such a move is in progress. In Europe (54), only the UK and Norway have really shown enthusiasm for private-sector ownership and competition. Although Japan has seen limited privatization (55) in certain industries, it has made no move as yet in the ESI.

Globally, the introduction of IPPs is a much more widespread form of private-sector participation. The US has been a leading example both in the volume of capacity added by IPPs and in experience with designing and implementing contracts under which energy and power are sold to the integrated system. Many developing countries, from the smallest to the largest, already have some experience with this form of private-sector involvement.

The Future

The experience of the past few years shows clearly that no single, simple recipe will improve the ESI's performance through private-sector participation in every country. Several different goals may be involved, such as improving financial performance by removing government control on prices, improving operating performance, and raising revenue for the government and encouraging popular capitalism by widespread share sales (56). Individual goals, as well as important differences in local conditions, particularly the nature of the system (size, plant mix, growth), mean that different cases will need different solutions (57). Politics also plays an important role in determining what governments want (58) and what they regard as feasible. Because governments are the fundamental agents of reform, their assessments of successes and failures in other countries will be very important.

The need for new capacity and the financing to undertake it is likely to play an important role in shaping future ownership patterns in the ESI in developing countries. The widespread use of government subsidies to cover the underpricing of electricity makes new investment programs, designed to meet the inexorably increasing demand, increasingly difficult for public ESIs because of the constraints on government budgets. The need to look elsewhere for financing is not only pushing countries to encourage the entry of IPPs (generation is the easiest part of the sector to handle in this way), but is also increasing the pressure for reform and for less government involvement as the lenders (whether private firms or international agencies) insist on a market that is not biased against independent private producers. As larger shares of the market are opened to private-sector participation, the contract terms for selling power and energy will become more important to governments.

Reinforcing this trend toward partial private-sector involvement in the global

ESI is likely to be the realization that retail (supply) competition (as in New Zealand and England and Wales) and the true competition created by wholesale power pools (as found in England and Wales and in Norway) are at best options for large and highly-developed markets. Symptomatic of the complexity of such developments is the development in the England and Wales market of various financial instruments for avoiding market risks created by volatility of the pool prices. Few developing countries are likely to want to undertake a major restructuring that requires skills so unfamiliar for the sake of as yet largely unproven benefits.

At the time of writing, the lack of actual experience of the operation of fully-privatized nonintegrated ESIs means correspondingly little firm analysis is available on the costs and benefits of such a move. Improvement in industry performance and a good deal of enthusiasm for greater involvement of the private sector are needed, but experience to date must be regarded as a series of experiments from which definitive lessons may take some time to emerge. The best path appears to be country-specific and depends on the size and sophistication of the sector, the political willingness to reform, and the country's economic potential.

Literature Cited

1. Rees R. 1976. *Public Enterprise Economics.* London: Weidenfeld & Nicholson
2. Berg SV, Tschirhart J. 1988. *Natural Monopoly Regulation. Principles and Practice.* Cambridge: Cambridge Univ. Press
3. Covarrubias AJ, Maia SB. 1994. Reforms and private participation in the power sector of selected Latin American and Caribbean and industrialized countries, Vols. 1, 2. *Latin Am. Caribb. Tech. Dep., Reg. Stud. Prog., Rep. 33.* Washington, DC: World Bank
4. Bernstein S. 1988. Competition, marginal cost tariffs and spot pricing in the Chilean electric power sector. *Energy Policy* 16: 369–77
5. Hachette D, Luders R. 1993 *Privatization in Chile: An Economic Appraisal.* San Francisco: Int. Cent. Econ. Growth
6. Galal A, Jones L, Tandon P, Vogelsang I. 1994. *Welfare Consequences of Selling Public Enterprises. An Empirical Analysis,* pp. 181–295. New York: Oxford Univ. Press
7. Littlechild SC. 1991. Office of electricity regulation: the new regulatory framework for electricity. In *Regulators and the Market: An Assessment of the Growth of Regulation in the UK,* ed. C Veljanovski, pp. 107–18. London: Inst. Econ. Affairs
8. Roberts J, Elliott D, Houghton T. 1991. *Privatizing Electricity: the Politics of Power.* London: Belhaven
9. Vickers J, Yarrow G. 1988. *Privatization: an Economic Analysis,* pp. 285–316. Cambridge, MA: MIT Press
10. Vickers J, Yarrow G. 1991. The British electricity experiment. *Energy Policy* 19:187–232
11. Weyman-Jones TG. 1989. *Electricity Privatization.* Aldershot: Avebury
12. Yarrow G. 1986. Regulation and competition in the electricity supply industry. In *Privatisation and Regulation—the UK Experience,* ed. J Kay, C Mayer, D Thompson, pp. 191–209. Oxford: Clarendon
13. Armstrong A, Cowan C, Vickers J. 1994. *Regulatory Reform. Economic*

Analysis and British Experience, pp. 277–321. Cambridge, MA: MIT Press
14. Helm D. 1994. Regulating the transition to the competitive electricity market. In *Regulating Utilities: The Way Forward*, ed. ME Beesley, pp. 89–110. London: Inst. Econ. Affairs
15. Littlechild SC. 1993. New developments in electricity regulation. In *Major Issues in Regulation*, ed. ME Beesley, pp. 119–39. London: Inst. Econ. Affairs
16. Thomas S. 1994. Will the UK power pool keep the lights on? *Energy Policy* 22:643–47
17. Weyman-Jones TG. 1993. Regulating the privatized electricity utilities in the UK. See Reference 60, pp. 93–107
18. Woolf F. 1994. Retail competition in the electricity industry: lessons from the United Kingdom. *Electr. J.* June:56–63
19. Yarrow G. 1994. Privatization, restructuring, and regulatory reform in electricity supply. See Reference 62, pp. 62–88
20. Burns P, Weyman-Jones TG. 1994. *The Performance of the Electricity Distribution Business—England and Wales, 1971–1993*. London: Cent. Stud. Regul. Ind.
21. Pollitt MG. 1994. Productive efficiency in electricity transmission and distribution systems. *Univ. Oxford, Appl. Econ. Discuss. Pap. Ser. 161*
22. Bös D. 1993. Privatization in East Germany. See Ref. 22a, pp. 81–92
22a. Ramanadham VV, ed. 1993. *Constraints and Impacts of Privatization*. London: Rutledge
23. Vincentz V. 1993. Privatization in East Germany: regulatory reform and public enterprise performance. See Ref. 59, pp. 141–57
24. Jedrzejczak GT. 1993. Privatization in Poland. See Ref. 59, pp. 82–102
25. Nuti DM. 1993. Privatization of Socialist Economies: General Issues and the Polish case. See Ref. 60, pp. 373–90
26. Joskow PL, Schmalensee R. 1985. *Markets for Power: an Analysis of Electric Utility Deregulation*. Cambridge, MA: MIT Press
27. Tenenbaum B, Lock R, Barker J. 1992. Electricity privatization. *Energy Policy* 20:1134–60
28. Kahn E, Milne A, Kito S. 1993. *Lawrence Berkeley Lab. Rep. LBL-34578, UC-350*, Lawrence Berkeley Lab., Berkeley, Calif.
29. Tenenbaum B, Henderson JS. 1991. Market based pricing of wholesale electricity services. *Electr. J.* Dec:36–45
30. Munasinghe M, Sanghvi A. 1989. Recent developments in the US power sector and their relevance for the developing countries. *Industr. Energy Dep. Work. Pap., Energy Ser. 12*, Washington, DC: World Bank
31. Glen JD. 1992. Private sector electricity in developing countries: supply and demand. *Discuss. Pap. 15*, Int. Finance Corp., Washington, DC
32. Schloss M. 1993. Sub-Saharan energy financing: the need for a new game plan. *J. Energy South. Afr.* 4:4–9
33. World Bank. 1993. *The World Bank's Role in the Electric Power Sector*. Washington, DC: World Bank
34. London Economics. 1993. *Measuring the Efficiency of the Australian Electricity Supply Industry*. Sydney: Electr. Supply Assoc. Aust.
35. Bacon R. 1994. Restructuring the power sector: the case of small systems. *FPD Note 10*, World Bank, Washington, DC
36. Oliveira AD, MacKerron G. 1992. Is the world bank approach to structural reform supported by experience of electricity privatization in the UK? *Energy Policy* 20:153–62
37. McLay JK. 1993. *Electricity deregulation and the open market: the New Zealand experience; is it transportable?* Presented at Assoc. Electr. Supply Ind. East Asia West. Pacific, Maramorna
38. Spicer B, Bowman R, Emanuel D, Hunt A. 1991. *The Power to Manage. Restructuring the New Zealand Electricity Department as a State-owned Enterprise—the Electricorp Experience*. Auckland: Oxford Univ. Press
39. Jomo KS. 1993. Privatization in Malaysia: For What and for Whom? See Ref. 60, pp. 437–54
40. Bollard A, Mayes D. 1993. Corporatization and Privatization in New Zealand. See Ref. 60, pp. 308–36
41. Lentz T. 1993. The case for the "affermage" system of private management of public services: example of the power sector in Côte d'Ivoire. In *Private Financing of Public Infrastructure*, ed. C Martinand, pp. 113–21. Paris: Minist. Public Works, Trans. Tour.
42. Mallon RD. 1994. State-owned enterprise reform through performance contracts: the Bolivian experiment. *World Dev.* 22:925–34
43. World Bank/US Agency Int. Dev. 1994. Submission and evaluation of proposals for private power generation projects in developing countries. *Ind. Energy Dep., Occas. Pap.* 2. Washington, DC: World Bank
44. Besant-Jones JE, ed. 1990. Private sector participation in power through BOOT schemes. *World Bank Ind. En-*

ergy Dep. Work. Pap. 33, World Bank: Washington, DC
45. Guislan P. 1992. Divestiture of state enterprises: an overview of the legal framework. *World Bank Tech. Pap. 186.* Washington, DC: World Bank
46. Kikeri S, Nellis J, Shirley M. 1992. *Privatization: The Lessons of Experience.* Washington, DC: World Bank
47. Green RJ, Newbery DM. 1992. Competition in the British electricity spot market. *J. Polit. Econ.* 100:929–53
48. van der Fehr NM, Harbord D. 1993. Spot market competition in the UK electricity industry. *Econ. J.* 103:531–46
49. York DW. 1994. Competitive electricity markets in practice: experience from Norway. *Electr. J.* June:49–55
50. Train KE. 1991. *Optimal Regulation: The Economic Theory of Natural Monopoly.* Cambridge, MA: MIT Press. 338 pp.
51. Hutchinson G. 1992. Efficiency gains through privatization of UK industries. See Reference 61, pp. 87–107
52. Cao AD. 1993. Privatization of state-owned enterprises: a framework for impact analysis. In *Constraints and Impacts of Privatization,* ed. VV Ramanadham, pp. 313–27. London: Rutledge
53. Parker D, Hartley K. 1992. Status change and performance: economic policy and evidence. See Ref. 61, pp. 108–25
54. McGowan F. 1993. Ownership and competition in community markets. See Ref. 60, pp. 70–90
55. Yamamoto T. 1993. An analysis of the privatization of Japan national railway corporation. See Ref. 60, pp. 337–59
56. Grout P. 1994. Popular capitalism. See Ref. 62, pp. 299–312
57. World Bank. 1993. Power supply in developing countries: will reform work? *Industry and Energy Occasional Pap. 1.* Washington, DC: World Bank
58. McGowan F. 1994. What is the alternative? Ownership, regulation and the Labor Party. See Ref. 62, pp. 265–89
59. Ramanadham VV, ed. 1993. *Privatization: A Global Perspective.* London: Rutledge
60. Clarke T, Pitelis C, eds. 1993. *The Political Economy of Privatization.* London: Rutledge
61. Ott AF, Hartley K, eds. 1992. *Privatization and Economic Efficiency: A Comparative Analysis of Developed and Developing Countries.* London: Elgar
62. Bishop M, Kay J, Mayer C, eds. 1994. *Privatization and Economic Performance.* Oxford: Oxford Univ. Press

Annu. Rev. Energy Environ. 1995. 20:145–78

THE RESPONSE OF WORLD ENERGY AND OIL DEMAND TO INCOME GROWTH AND CHANGES IN OIL PRICES

Joyce Dargay
Transport Studies Unit, Oxford University, 11 Bevington Road, Oxford OX2 6NB United Kingdom; Dargay@vax.ox.ac.uk

Dermot Gately
Economics Department, New York University, 269 Mercer St., New York, NY 10003; GatelyD@fasecon.econ.nyu.edu

KEY WORDS: price reversibility, asymmetry

CONTENTS

ABSTRACT

This paper reviews the path of world oil demand over the past three decades, and the effects of both the oil price increases of the 1970s and the oil price decreases of the 1980s.

Compared with demand in the industrialized countries, demand in the Less Developed Countries (LDC) has been more responsive to income growth, less responsive to price increases, and more responsive to price decreases. The

1056-3466/95/1022-0145$05.00

LDC has also exhibited much greater heterogeneity in income growth and its effect on demand.

We expect a smaller demand response to future price increases than to those of the 1970s. The demand response to future income growth will be not substantially smaller than in the past. Finally, given the prospect of growing dependence on OPEC oil, in the event of a major disruption the lessened price-responsiveness of demand could cause dramatic price increases and serious macroeconomic effects.

INTRODUCTION

In this article we review energy and oil demand over the past three decades, with a focus on three main regions: the industrialized countries of the Organization for Economic Cooperation and Development (OECD) which includes North America, Western Europe, Japan, Australia, and New Zealand; the LDC (Less Developed Countries) Oil Exporters, OPEC and Mexico; and the Other LDC, excluding China unless otherwise noted. In addition to the historical determinants of total energy demand, we also examine the determinants of total oil demand and its two main components, transportation oil and nontransportation oil. This disaggregation of total oil is important because the two have moved differently over time, especially within the OECD.

Also for purposes of improving our understanding of demand determinants among the non-OECD countries, we examine the heterogeneity among the 61 LDC for which we have good data. This heterogeneity is most apparent in their growth rates of per capita income and in the relationship between demand growth and income growth. In particular, we examine separately three subgroups of the Other LDC, based on their average annual growth rate of per capita income from 1970–1991:

1. Growing Income—those with positive and steady growth in per capita income;
2. Declining Income—those with negative growth in per capita income;
3. Slow and Uneven Income Growth—those with relatively sluggish and variable growth.

These 3 subgroups have approximately 15 countries each, as do the LDC Oil Exporters, which are identified below.

In addition to the effects of income growth on demand in the OECD and the various LDC regions, we also examine the effects on demand of the dramatic changes in world oil prices: the two major price increases of the 1970s and the almost equivalent price decreases of the 1980s (see Figure 1). We shall see that—at least for the OECD—the demand effects of price decreases and the demand effects of price increases are not symmetric. That is,

Table 1 Data sources

Real income (GDP)
OECD main economic indicators; United Nations statistics
Real price of crude oil
British Petroleum, energy statistics
Energy and oil demand
energy statistics of OECD countries (Paris: International Energy Agency)
energy statistics of non-OECD countries (Paris: International Energy Agency)
energy balances of non-OECD countries (Paris: International Energy Agency)
DOE: US Department of Energy
BP: British Petroleum, energy statistics

the demand reductions caused by the price increases of the 1970s have not been reversed by the price cuts of the 1980s. Data sources for this article are listed in Table 1.

We also address a related issue: whether in the event of an oil price increase, either by OPEC or by increased domestic oil taxes, we can expect as great a demand reduction as we experienced after previous price increases. Although the evidence is less clear on this issue, we believe the answer to this question is also negative, at least for the OECD. That is, the demand response to future price recoveries will be smaller than the response to the price increases of the 1970s.

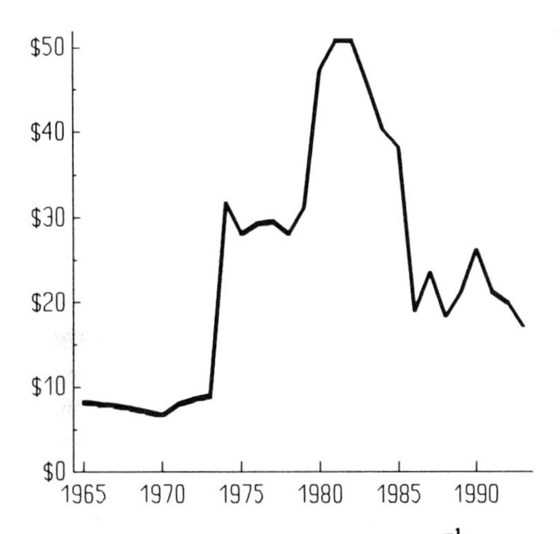

Figure 1 Real price of crude oil, 1965–1992, 1992 dollars barrel^{-1} [source: British Petroleum (BP)].

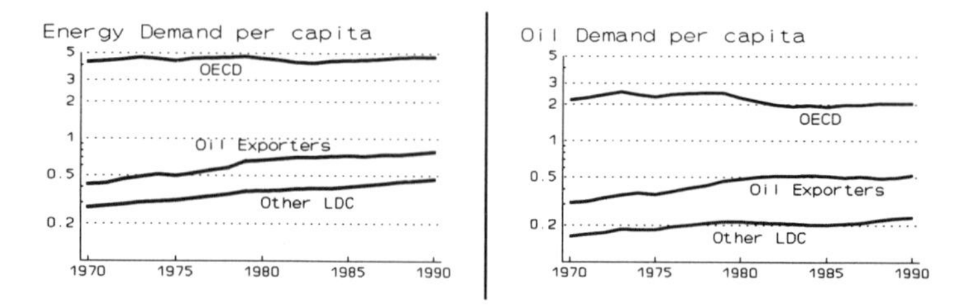

Figure 2 Per capita energy and oil demand, 1970–1990 (tons of oil equivalent per person per year).

Until recently, the majority of energy demand studies[1] were based on so-called perfectly price-reversible models, so questions such as the two posed above could not be analyzed empirically. Despite the inability of these traditional models to explain the sluggish growth in demand following the oil price collapse in 1986, some analysts, most notably Hogan (1), have been reluctant to abandon the conventional demand specification. However, as we have shown (2), the assumption of perfect price-reversibility must be rejected, especially for OECD nontransportation oil demand.

In order to provide an overview of the data, we first present some summary graphs. Shown in Figure 2 are the per capita values for energy and oil demand in each of our three main regions. Shown in Figure 3 are population, total income, and per capita income. The vertical scales in each graph are logarithmic, so that the slope of the curves indicate the percentage rate of growth: Steeper curves have higher rates of growth.

There are two major differences between the OECD on the one hand and the Oil Exporters and Other LDC on the other:

1. OECD demand per capita is several times greater than for the other two groups: four to five times greater for energy and three to six times greater for oil.
2. The rate of change in the OECD: Per capita energy demand has been relatively flat, and it has declined for oil, especially after the two price increases of the 1970s—declines that were not reversed by the price cuts of the 1980s. In contrast, per capita energy demand in the other two groups has grown during these years. Per capita oil demand has also grown, despite being roughly constant in the early 1980s, following the 1979–1980 price increase.

[1]For good examples of earlier work on energy and oil demand, see Griffin (3), Pindyck (4), and Hogan (5).

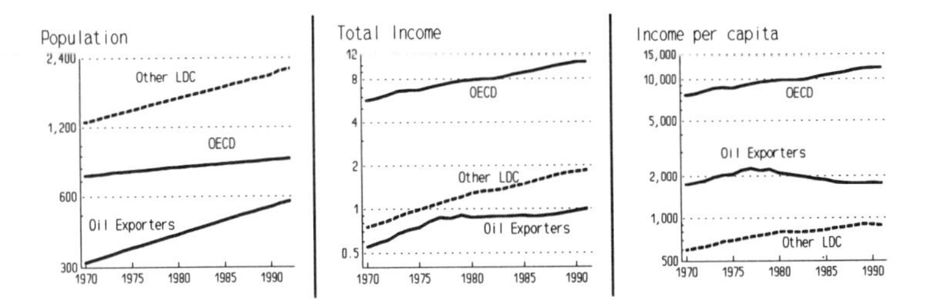

Figure 3 **Population (millions), total income (trillions, 1980 dollars), and per capita income (1980 dollars), 1970–1991.**

ENERGY AND OIL DEMAND SINCE 1965: BY REGION, BY FUEL, BY SECTOR

Energy Demand

Figure 4 summarizes the recent history of energy consumption, from 1965 to 1992, disaggregated by fuel: oil, natural gas, coal, nuclear, and hydro. Four regions are shown separately: OECD, Oil Exporters, Other LDC, along with Eastern Europe and the Former Soviet Union. Several observations can be made.

Total energy demand has been relatively flat since 1973 in the OECD but continues to grow in the other regions. Total energy demand grew similarly in all regions from 1965 to 1973, at an average annual rate greater than 2%. But during the high-price era from 1973 to 1985, demand remained flat in the OECD and to a lesser extent in Eastern Europe and the Commonwealth of Independent States (CIS) but continued to grow rapidly in the other regions. Since the 1986 oil-price collapse, energy demand has begun to grow again in the OECD, but at half the pre-1973 growth rate. In Eastern Europe and the CIS, however, energy demand flattened out and declined in the late 1980s, a time of both political and economic transition.

Total oil demand shows a similar pattern: Since 1973 demand has been relatively flat in the OECD and in Eastern Europe and the CIS. But demand has been increasing in the other regions. The renewed growth in energy demand since the mid-1980s in the OECD was not accompanied by a comparable growth in oil demand. Even in the LDC regions, oil demand grew less rapidly than total energy demand.

Oil's share of total energy is higher in the OECD than in other regions, except for the Oil Exporters. However, oil dependency within the Other LDC varies widely, although this variation is hidden in the aggregate figures. The Other LDC's low oil share reflects primarily its two largest economies, China

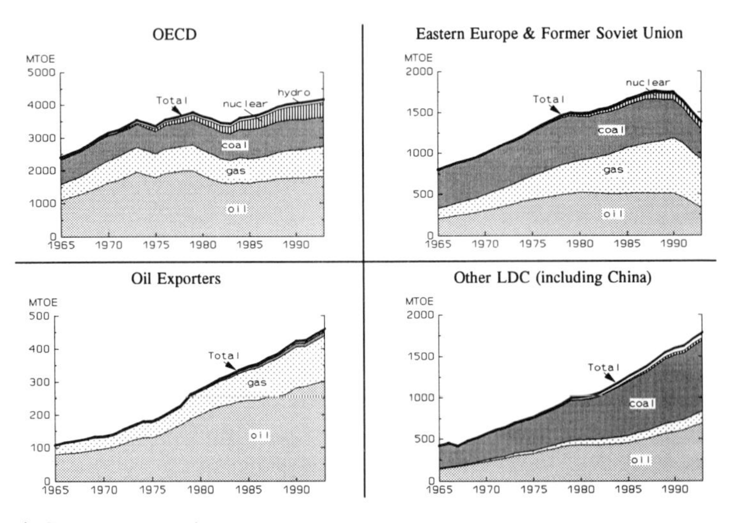

Figure 4 **Energy consumption by fuel (oil, natural gas, coal, nuclear, hydro): 1965–1992 (source: BP).**

and India, which are also large coal producers. They account for half the total energy consumption in the Other LDC region. Excluding these two countries, the oil share of total energy for the rest of the LDCs was 55% in 1993, which is significantly higher than the 43% share in the OECD. The shares for China and India are 20% and 31% respectively.

Oil's share of total energy demand has declined in most regions worldwide and from 50% in 1973 to 40% in 1993. Although the OECD is responsible for the bulk of this reduction, almost all countries have reduced their oil dependency.

Figures 5 and 6 show energy use, disaggregated by fuel, in two of the most important energy-using sectors:[2] electric power (Figure 5) and industry (Figure 6). In these two sectors, substitution for oil is easier than in most other sectors of the economy, especially transportation, in which fuel substitution away from oil is most difficult. Several points should be noted regarding electric power generation in Figure 5:

1. The demand for electric power has been growing in all regions.
2. Oil use in electric power generation declined sharply in the OECD after the

[2]The country-coverage for the Oil Exporters and Other LDC does not correspond exactly to that shown in Figure 4. Figures 5 and 6 cover about 60 of those non-OECD countries whose data are provided by the International Energy Agency (IEA); fortunately, these include the largest and most important. The IEA data are more detailed for specific countries but are not as comprehensive geographically as the BP data in Figure 4.

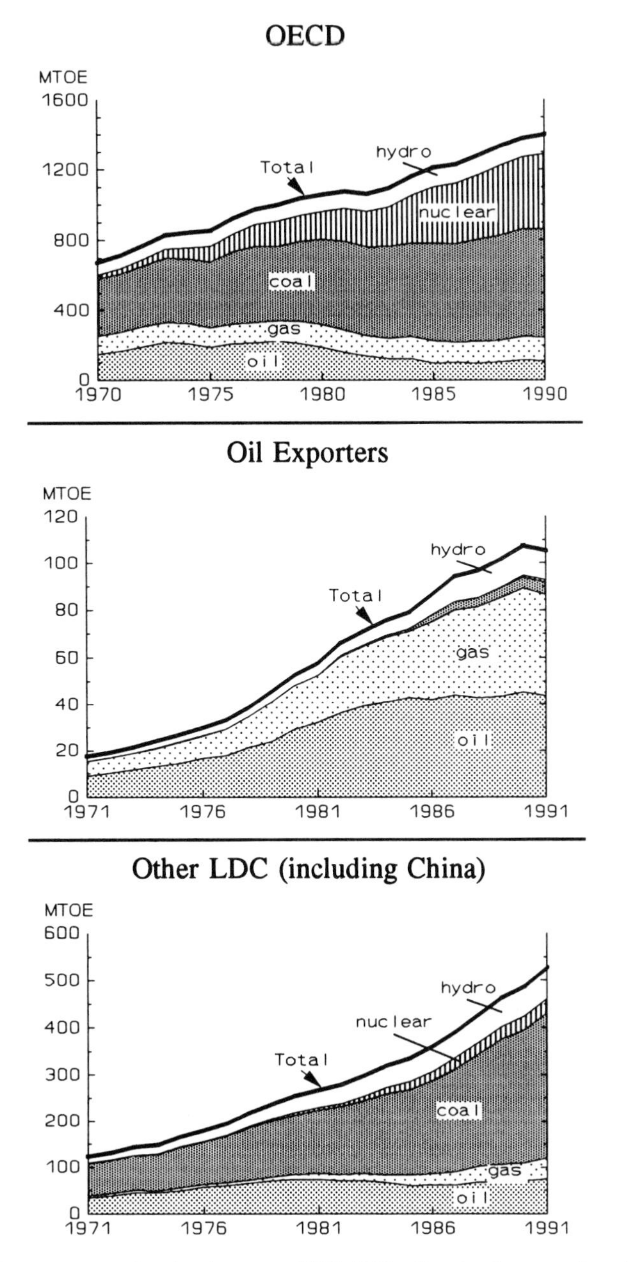

Figure 5 **Energy sources for electric power, 1971–1991: oil, natural gas, coal, nuclear, hydro (source: IEA).**

1979–1980 price increase, and this decline was not reversed by the oil price declines of the 1980s.

3. In the non-OECD regions, oil use in electric power generation has been flat since the early 1980s, and oil's fuel-share has been declining. This demand reduction has not been reversed by the price cuts of the 1980s, except in a few countries such as South Korea (to be discussed below).
4. Coal use has grown in all regions except the Oil Exporters. Similarly, nuclear use has increased in the OECD and a few LDC (Taiwan, South Korea), and natural gas use has expanded in virtually all regions.

Although similar in some ways, energy use in the industrial sector differs from energy use in electric power generation. Total energy use in the industrial sector has actually fallen in the OECD since the early 1970s but has continued to rise in the Other LDC. Growth has been most rapid in those parts of the Other LDC that have had the most growth in both income and industrial production—primarily Asia. For the Oil Exporters, industrial energy consumption grew rapidly until the mid-1980s but has declined since then.

Industrial use of oil has been relatively flat in most regions. In the OECD oil use declined substantially in response to the price increases, which were not reversed by the oil price decreases. Since the 1986 oil price collapse, only in the Pacific Rim has oil use increased substantially, primarily in South Korea.

Industrial use of oil decreased owing to overall energy conservation and some fuel-switching to natural gas. Oil's share of industrial energy use has declined since the early 1970s in all regions. In both the OECD and the Other LDC, oil's share was reduced from 40 to 30% (approximately). But oil's share of energy use varies significantly within the Other LDC. In India (and also in China, which is not shown in Figure 6), coal predominates in the industrial sector, and oil's share is only 15%. On the other hand, in the Pacific Rim, oil is still the dominant energy source in industry, accounting for nearly 55% of total consumption.

Energy use in the residential and commercial sectors (not shown) has followed patterns similar to those of the industrial sector.

Oil Demand

To understand the determinants of oil demand, we must distinguish (at least) between oil's two main uses: transportation and nontransportation. Fuel substitution for oil is far easier in most nontransportation uses, such as space heating and water heating, and for industrial processes and electric power generation. This disaggregation is shown in Figure 7.

Transportation oil use continues to grow in all regions. On the other hand, at least in the OECD, nontransportation oil demand has declined over the past two decades, especially after the 1979–1980 price increase. But in the LDC,

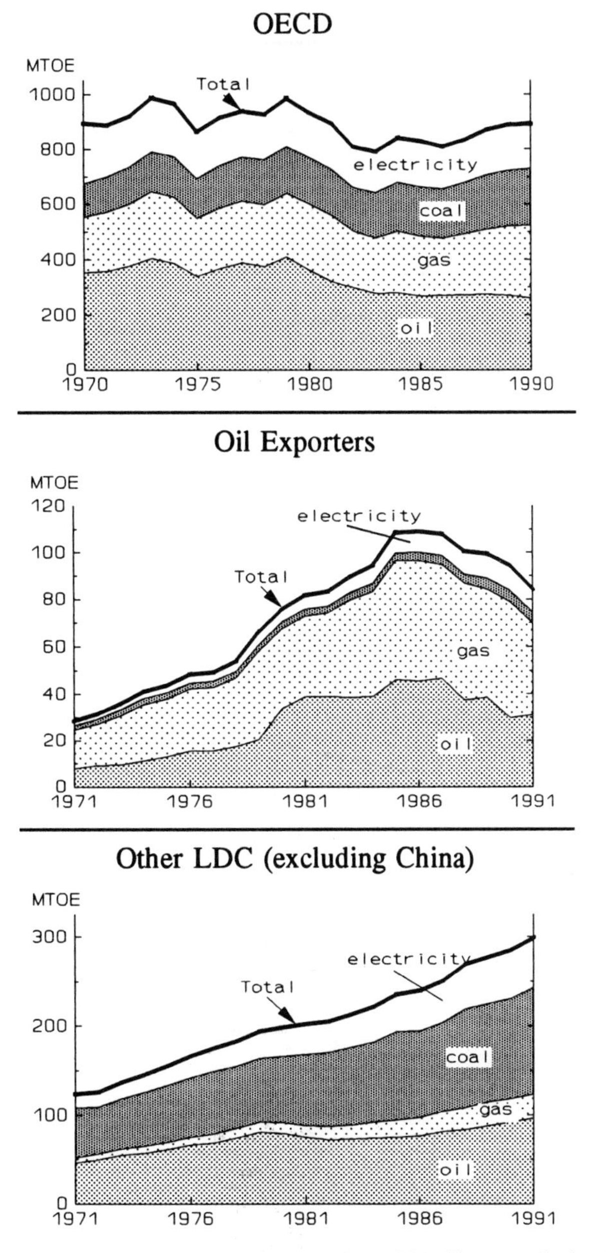

Figure 6 **Energy sources for the industrial sector, 1971–1991: oil, gas, coal, electricity (source: IEA).**

OECD

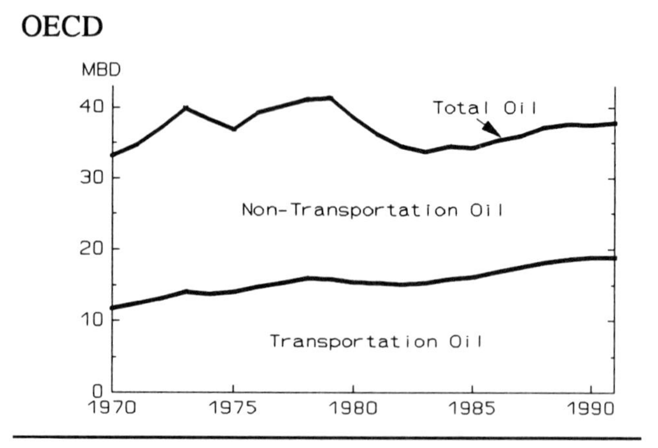

Oil Exporters

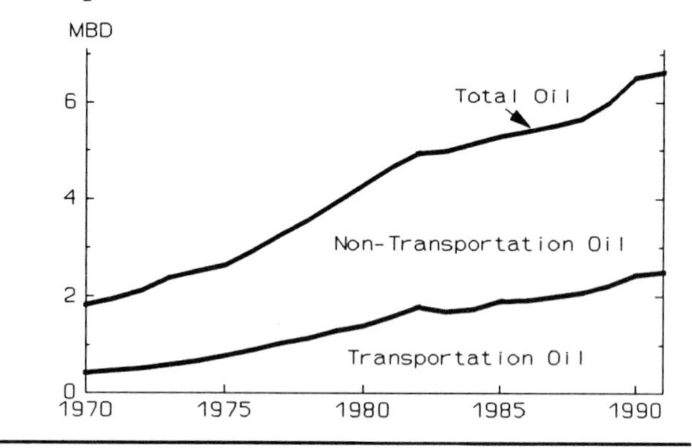

Other LDC excluding China

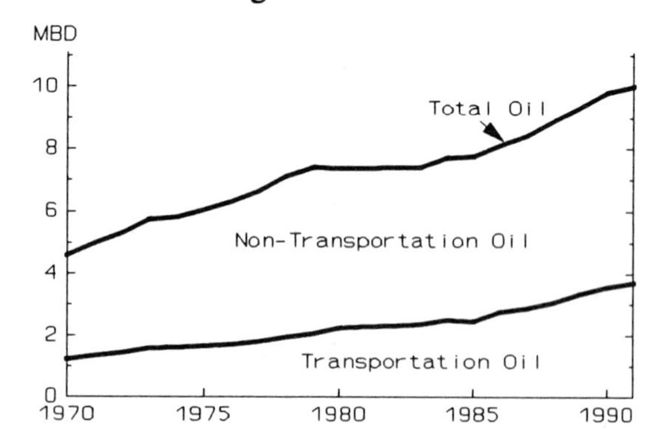

Figure 7 Oil consumption, 1970–1991: transportation oil and nontransportation oil [Source: Department of Energy (DOE)].

nontransportation oil demand continues to grow, although at a slower rate than transportation oil demand.

Transportation oil's share of total oil demand is higher in the OECD than in the other regions and now equals the share of nontransportation oil. This share has also increased in the LDC—both in the Oil Exporters and the Other LDC—from about 25% in 1970 to nearly 40% by 1991.

Nontransportation oil demand in the OECD has declined over the past two decades, especially since the 1979–1980 price increase. But in the Other LDC and the Oil Exporters, such demand has grown at about the same rate as transportation oil demand.

Thus, since 1970, total oil demand has increased significantly in all regions except the OECD, quadrupling within the Oil Exporters, and doubling in the Other LDC. Within the OECD, the decline in nontransportation oil demand has been almost matched by the continued growth of transportation oil demand, so that by 1991 total oil demand was near its 1973 level.

DEMAND AND THE EFFECTS OF INCOME GROWTH

In addition to energy and oil demand over time, both absolute and per capita, examining demand relative to real income growth is useful. A standard way of doing this is to show the ratio of demand to income, the demand/Gross Domestic Product (GDP) ratio (see Figure 8).

Within the OECD, demand/GDP ratios have been declining, except for transportation oil, which has remained roughly constant relative to GDP. Among the Oil Exporters, on the other hand, these ratios have increased. Initially, in the 1970s, the increases reflected demand growing even more rapidly than the Oil Exporters' income. Since the late 1970s, however, their income has been flat, while demand has continued to rise (although more slowly than previously), so that the demand/GDP ratio has increased.

Within the Other LDC, the demand/GDP ratios for both energy and transportation oil have increased slowly, whereas this ratio has decreased moderately for nontransportation oil and declined slightly for total oil. Figure 9 presents another view of these data. For each of the four demands (total energy, total oil, transportation oil, and nontransportation oil), respectively, each panel shows how demand has changed relative to total income from 1970 to 1991. The left column shows total demand vs total income, and the right column shows per capita demand vs per capita income. The scales are logarithmic, so that absolute distances (horizontally and vertically) measure percentage changes, not absolute changes as they normally would. Thus we see that the Other LDC experienced the greatest total income growth in percentage terms (the greatest horizontal movement).

Also shown are dashed, diagonal lines indicating equi-proportional growth.

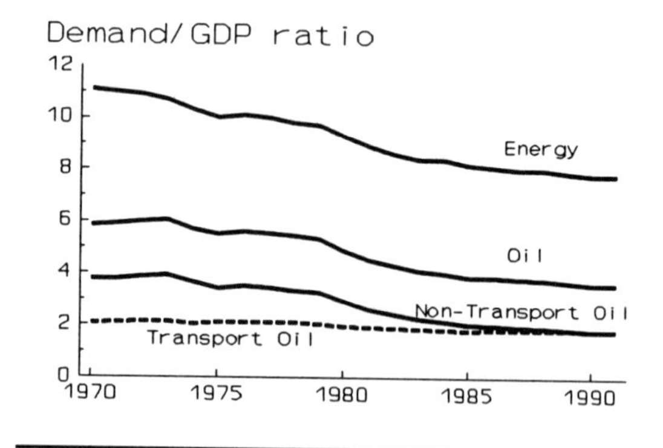

Oil Exporters

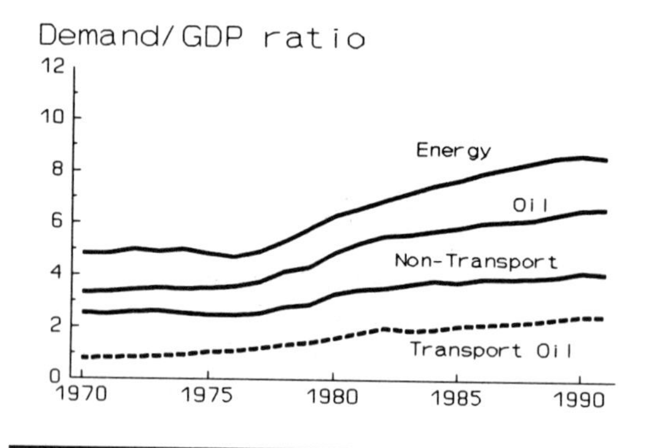

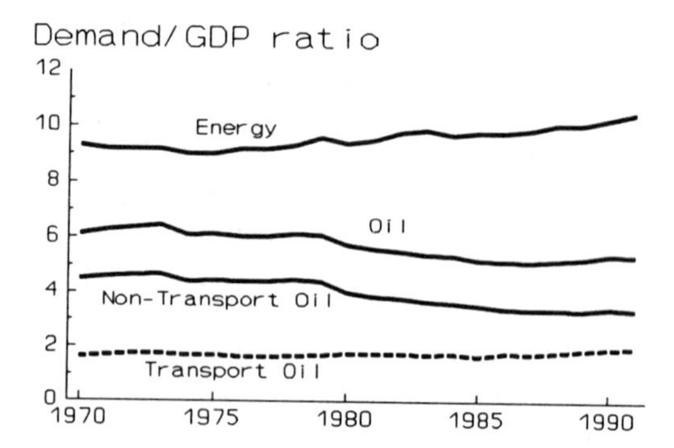

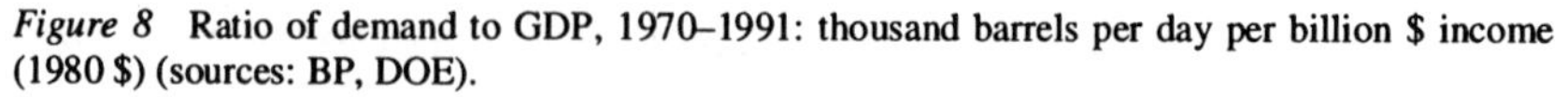

Figure 8 Ratio of demand to GDP, 1970–1991: thousand barrels per day per billion \$ income (1980 \$) (sources: BP, DOE).

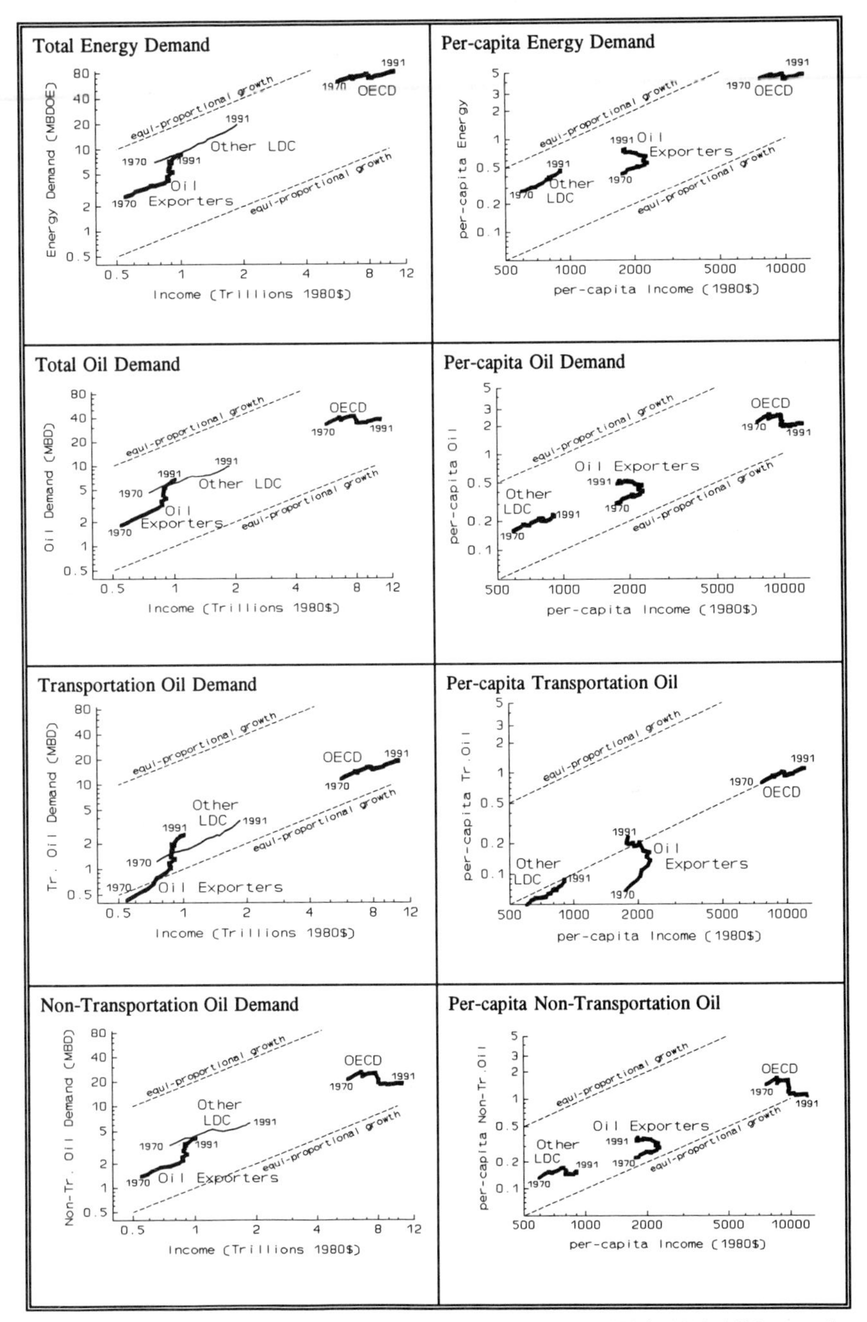

Figure 9 Energy and oil demand (MBD) vs real income (trillions of 1980 $), 1970–1991;per capita demand (tons) vs per capita income; logarithmic scales (sources: BP, DOE, UN).

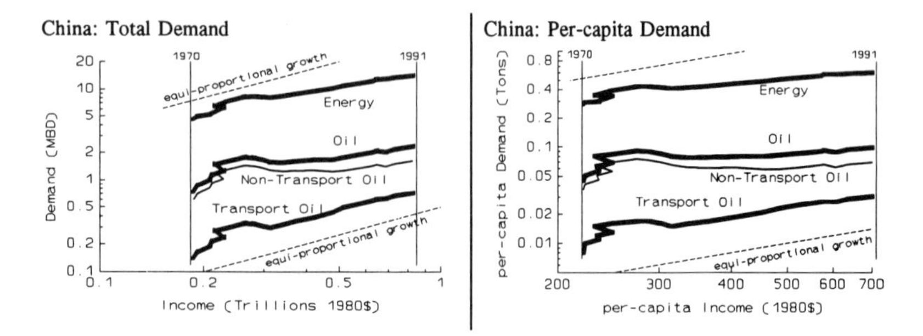

Figure 10 China: energy and oil demand (MBD) vs real income (trillions of 1980 $), 1970–1991; per capita demand (tons) vs per capita income; logarithmic scales (sources: IEA, DOE, UN).

Movement along or parallel to these lines indicates demand growing in equal proportion to income (e.g. total transportation oil in the OECD and Other LDC). Ignoring the effects on demand of any price changes, this would be equivalent to an unitary income elasticity. Steeper movement indicates demand growing faster than income (an income elasticity greater than one, ignoring price changes). Less steep, but positive movement indicates demand growing less rapidly than income (an income elasticity less than one, ignoring price changes), such as, for example, total energy demand in the OECD.

We can also observe declines in demand, as well as flat demand, as income grows, such as total OECD nontransportation oil in 1979–1983 and 1983–1991, respectively. Finally, for the Oil Exporters we observe in the left graphs total income growth (rightward movement) until the late 1970s, followed by stagnant total income but continued total demand growth (vertical movement).

The movement of per capita values is similar to that of the totals, but higher population growth rates in the Oil Exporters and Other LDC (Figure 3) result in slower increases in per capita income than those seen in the OECD. In fact, for the Oil Exporters, when total income remains roughly constant in the 1980s (vertical movement in the left graphs), we see per capita income declining (bending back in the right graphs). But despite this decline in per capita income, their demand continues to increase, especially for transportation oil.

Shown separately in Figure 10 are analogous graphs for China. They show the significant growth in both total and per capita income over this period. Although starting at a low level, China managed to sustain one of the highest rates of growth of per capita income in the world. Although transportation oil increased as rapidly as income, overall energy and oil demand did not increase nearly as rapidly. Energy demand has increased only half as rapidly as income. Nontransportation oil was relatively flat, so that total oil increased much less rapidly than income.

In general, the relationship between energy demand and income growth for any individual country is determined by several factors: the stage of economic development, the state of technology, energy endowments and energy pricing policy. All of these can and do change over time. Economic development theory suggests that economic development is initially accompanied by an increasing energy/income ratio which reaches a maximum and then declines as incomes continue to increase. In the initial stage, the mechanization of agriculture and the shift from an agrarian economy to an industrial economy will require larger inputs of energy per unit output. At the same time, larger and larger segments of society are replacing noncommercial with commercial energy supplies, especially with growing urbanization. During this phase, energy demand increases more rapidly than income. However, as the economy continues to grow, the relative importance of heavy industry diminishes, the sectoral share of services increases, and household energy consumption reaches saturation limits. The energy requirements of increasing income will diminish, so that energy consumption will increase less rapidly than GDP.

These ideas are consistent with Figures 8 and 9: Total energy consumption has not increased as rapidly as income in the OECD, but has increased more than proportionately to income in the LDC. Income elasticity declines from greater than unity in the developing countries to less than unity in the mature economies of the OECD.

DEMAND AND THE EFFECTS OF OIL PRICE INCREASES AND DECREASES

We now address the question of how energy and oil demand have been affected by the oil price increases of the 1970s and by the oil price declines of the 1980s. As we have argued in other articles, this relationship has not been symmetric (2, 6). The oil demand reductions caused by the oil price increases have not been reversed by the oil price decreases. This is most obvious in data for OECD oil demand, especially nontransportation oil demand.

Such a view of demand being imperfectly price-reversible, or of demand not responding symmetrically to price increases and price decreases, contrasts with conventional demand analysis. In particular, a conventional demand relationship between price and quantity demanded is symmetric with respect to price increases and decreases. That is, demand is perfectly price-reversible: Demand reductions caused by a price increase would be reversed by an equal price decrease.

In contrast, we argue that demand need not respond symmetrically. Demand can be imperfectly price-reversible: Demand reductions caused by a price increase need not be completely reversed by an equal price decrease. This idea is graphically summarized in Figure 11.

For evidence of this imperfect price-reversibility, especially for nontrans-

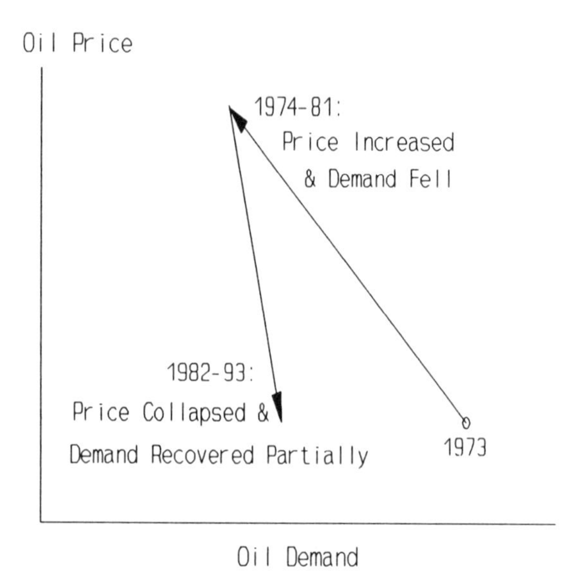

Figure 11 Imperfectly price-reversible demand.

portation oil in the OECD, see Figure 12. This figure shows the time-path, from 1970 to 1991, of the real price of crude oil vs the demand/GDP ratio in each of the three regions for each of four demands: total energy, total oil, transportation oil, and nontransportation oil. Although we could have shown the absolute level of demand, we chose to show the demand/GDP ratio in order to simplify the interpretation of the graphs, which would be complicated by natural demand growth due to income growth.

In all cases we graph the real price of crude oil, primarily because we do not have real end-use prices for oil products outside the OECD. Although the decline in real product prices in the 1980s generally has not been as large as the drop in the price of crude oil, the decline has been significant. In the OECD, real end-use product prices declined by about one-third after the 1986 collapse in crude oil prices (6a).

Figure 12 shows important regional differences in the movement of the demand/GDP ratios. In the OECD these demand/GDP ratios have all declined since the first oil price increases of 1973–1974. The greatest decline was for the ratio of nontransportation oil to GDP, which has fallen by half since 1973–1974. That reduction was not reversed by the price cuts of the 1980s, although the decline slowed after the 1986 oil price collapse. Smaller reductions have occurred in the OECD demand/GDP ratios for transportation oil,

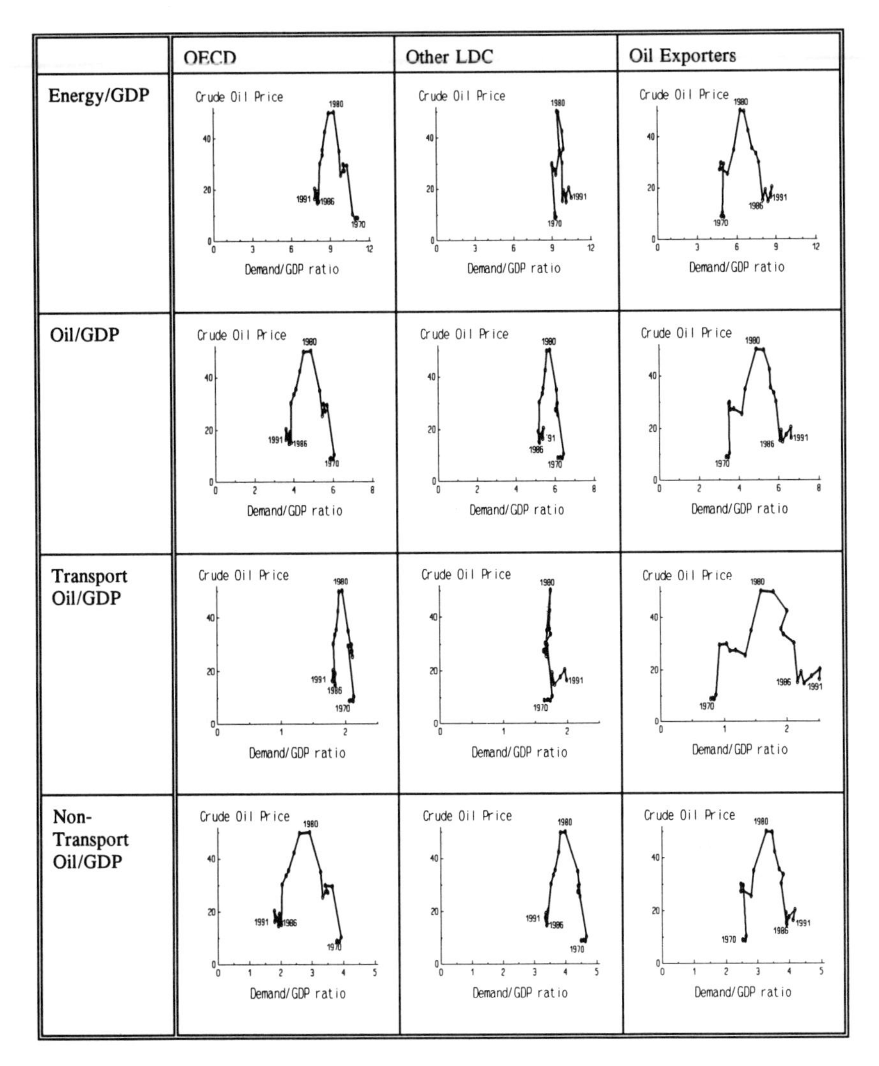

Figure 12 Crude oil price (1985 $ per barrel) vs demand/GDP ratio (Th.BD billion per $), 1970–1991 (source: DOE).

total oil, and total energy. In general, the inverted U pattern, typical of imperfect price-reversibility, is apparent in the data for the OECD.

The Other LDC has also experienced a decline in its demand/GDP ratio for nontransportation oil, even though the absolute level of nontransportation oil has risen. This decline followed the price increases of the 1970s and has not been reversed by the price cuts of the 1980s. But for transportation oil and for

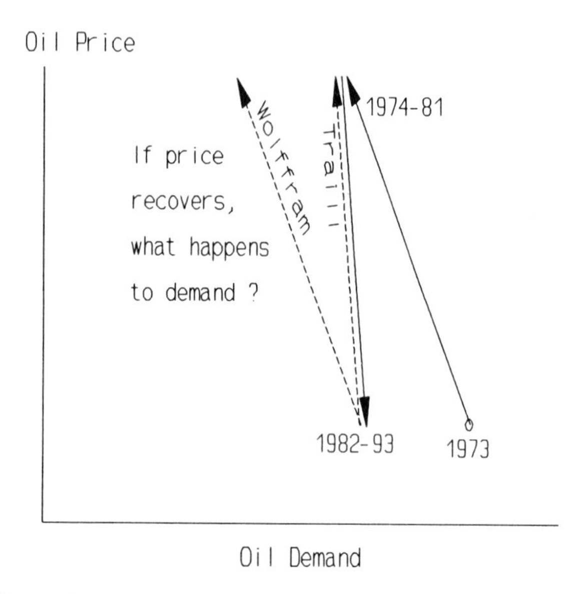

Figure 13 Effect on demand of a price recovery.

total energy, the demand/GDP ratios have increased over time: Demand increased faster than income. Either demand has not responded much to the price increases, or the response has been outweighed by an income elasticity substantially greater than unity.

For the Oil Exporters, in contrast to the OECD and the Other LDC, all these ratios have risen over time, especially after the late 1970s when their real income had flattened out but their energy and oil demand continued to increase.

ECONOMETRIC ESTIMATION OF DEMAND RESPONSES TO CHANGES IN PRICES AND INCOME

The graphic evidence suggests that, at least in the OECD, the demand response to price cuts in the 1980s has not reversed the demand reductions caused by the price increases of the 1970s. However, other important questions concern the effect on demand of future price recoveries, that is, price increases which do not exceed the historic maximum levels of the early 1980s. Will the demand response to a future price recovery be as great as it was to the price increases of the 1970s? Or will a price recovery only reverse the small demand increase caused by the price cuts? Or will it be somewhere in between?

These possibilities are depicted in Figure 13. The greatest demand reduction from a price recovery, equal to that for the price increases of the 1970s, is labeled Wolffram. The Wolffram specification assumes that all price increases—both increases in the maximum historical price and (submaximum) price recoveries—have the same effect. The smallest demand reduction from a price recovery is labeled Traill; it merely reverses the small, partial reversal of the 1980s demand increase. The Traill specification assumes the response to a price recovery is equal to the (small) response to a price cut. The third alternative would lie somewhere in between: Demand responds more strongly to price rises than to price declines but not quite as much to price recoveries as to increases in the maximum historical price.

To allow for the possibility of an asymmetric response between price increases and decreases, we need to be able to measure separately the effects upon oil demand of oil price increases and decreases. To do this, we have adapted a technique from the literature on agricultural supply (7, 8). This technique involves the decomposition of the price series P_t (price at time t) into three component series, each of which is monotonic: maximum historical price $P_{max,t}$ (positive and nondecreasing), the cumulating series of price cuts $P_{cut,t}$ (nonpositive and nonincreasing), and the cumulating series of price recoveries $P_{rec,t}$ (nonnegative and nondecreasing):

$$P_t = P_{max,t} + P_{cut,t} + P_{rec,t} \qquad \text{1a.}$$

$$P_{max,t} \equiv \max\,(P_0,\ldots,P_t) \qquad \text{1b.}$$

$$P_{cut,t} \equiv \Sigma_{i=0}^{t} \min\,\{0,(P_{max,i-1}-P_{i-1}) - (P_{max,i}-P_i)\} \qquad \text{1c.}$$

$$P_{rec,t} \equiv \Sigma_{i=0}^{t} \max\,\{0,(P_{max,i-1}-P_{i-1}) - (P_{max,i}-P_i)\}. \qquad \text{1d.}$$

Figure 14 shows the real price of crude oil, together with its three-way decomposition. We see the jump in P_{max} in 1973–1974 and 1979–1980; it is always positive and nondecreasing. The cumulating series of price cuts, P_{cut}, is negative and nonincreasing; it shows the dramatic price declines of the 1980s. Also shown is the cumulating series of price recoveries, P_{rec}, which is positive and nondecreasing; but such price increases have been relatively few and small.

We specify per capita demand in its simplest form as a log-linear function of real per capita income, the real price of crude oil, and lagged demand. Two different demand specifications are estimated econometrically, depending upon whether we assume that demand is perfectly price-reversible or whether it is estimated as being imperfectly price-reversible:

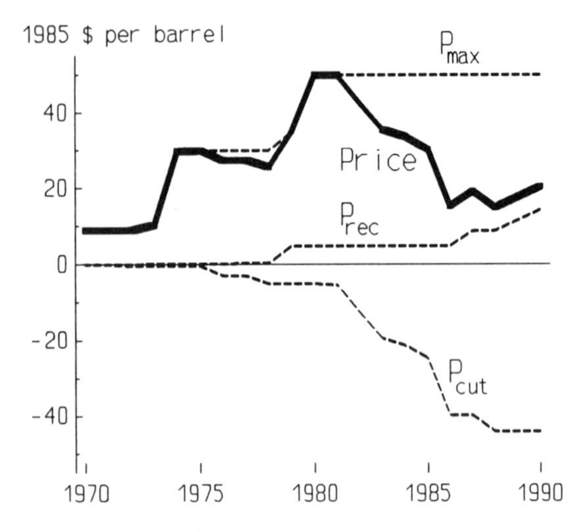

Figure 14 Decomposing price into three series.

Perfectly Price-Reversible:

$$\log \text{Demand/Population}_t = \{a\} + \{g\} \log \text{Income/Population}_t + \{b\} \log \text{Price}_t + \{f\} \log \text{Demand/Population}_{t-1} \quad \text{2a.}$$

Imperfectly Price-Reversible:

$$\log \text{Demand/Population}_t = \{a\} + \{g\} \log \text{Income/Population}_t + \{b\}_m \log P_{max,t} + \{b\}_c \log P_{cut,t} + \{b\}_r \log P_{rec,t} + \{f\} \log \text{Demand/Population}_{t-1}. \quad \text{2b.}$$

Alternatively, we could estimate total demand as a function of total income and price.

Industrialized Countries

Our econometric results for OECD oil demand were recently summarized (6); some of this work appears in Table 2. In the energy-demand equation, the real price of crude oil was used as the price variable, whereas real after-tax oil product prices were used in the respective oil product equations. Clearly, a fuel-weighted energy price would be preferable in the energy demand equation, but the non–oil price data were not readily available for all the OECD countries. Our conclusions regarding OECD demand include the following.

Demand response to the price decreases of the 1980s has been asymmetric and smaller than the response to the price increases of the 1970s. This difference holds for both transportation and nontransportation oil. Reductions in

Table 2 OECD: Elasticity estimates from imperfectly price-reversible demand specifications

		Long-run demand elasticities			
			Changes in price		
Oil product[a]	Region	Income	Increases in P_{max}	Price cuts P_{cut}	Price recoveries[b] P_{rec}
Per-capita energy demand (2b) above	OECD	0.81	−0.30	−0.20	−0.30 (Wolffram)
Total non-transportation oil (2)	OECD	1.09	−0.76	−0.03	−0.54
Total transportation oil (10)	US	0.70	−0.21	−0.04	−1.08 (Wolffram)
Gasoline per driver (11)	US	0.79	−1.08	−0.46	−1.08
Road transportation oil (12)	France	1.29	−0.80	−0.45	−0.80 (Wolffram)
	Germany	1.71	−0.44	−0.02	−0.44 (Wolffram)
	UK	1.49	−1.50	−0.10	−0.10 (Traill)

[a] The results for the oil product equations were obtained from slightly different specifications than that shown in Equation 2b. See the original papers for a description of these: (2), (10), (11), (12).

[b] If the results employed a specification in which the decomposed price coefficients were constrained, they are so indicated: Wolffram ($b_m = b_r$), or Traill ($b_c = b_r$). Otherwise, the three price coefficients were estimated separately.

gasoline demand are irreversible primarily because efforts to improve automobile fuel efficiency were not reversed by the price cuts; rather these cuts resulted from irreversible improvements in technology and the continuation of government policies such as fuel-efficiency standards. Although more fuel-efficient vehicles combined with cheaper fuel have lowered the per-mile costs of transportation, neither greater vehicle use nor a partial return to less fuel-efficient vehicles has fully reversed the demand reductions caused by the price increases. For nontransportation oil, on the other hand, the irreversibility is better explained by fuel-switching that was not reversed by the price cuts, but efficiency improvements have also played a role.

For nontransportation oil, the market responds more to price rises above the previous maximum price than it does to price recoveries. We would therefore expect the demand response to future price recoveries to be smaller than that to the price increases of the 1970s. For the OECD countries, this result is a mixed blessing. On the one hand, they may be comforted by now having both low oil prices and lessened oil-import dependence; they need not fear a full reversal of the demand reductions achieved in the past two decades. On the other hand, in the event of an oil price recovery (due either to OPEC or to domestic tax increases), they may not be able to rely upon the same demand reductions in the future as they experienced in the past. Fuel-switching or conservation measures, done in response to the price increases of the 1970s but not undone by the price declines of the 1980s, cannot be redone if price recovers in the 1990s. The possibilities for demand reduction may be smaller than in the 1970s, because the uses of oil that remain significant today are probably those less conducive to other energy sources, and for which further efficiency improvements are more costly. Nontransportation oil demand might be sharply reduced by a price recovery only if demand is still substantial and fuel-switching is relatively easy.

The demand response to future income growth will be only slightly smaller than in the past. If demand were wrongly assumed to be perfectly price-reversible, then the successive-year inclusion of post-1986 data—with moderate income growth and very low prices but little demand growth—would cause the estimate of income elasticity to be reduced, and sometimes to become negative (2, 6, 9)! With the imperfectly price-reversible specification, however, no such error occurs: the estimated income elasticity is relatively unaffected by the inclusion of the post-1986 data.

The estimates shown in Table 2 are based on a relatively simple, reduced-form model. Several modifications might be attempted. One would be a simultaneous equations approach to analyze the effects of energy price changes on technological progress, on the capital stock, and on energy demand. Another modification might relax our implicit assumption that income changes are independent of changes in energy prices. It could be argued that the recession-induced declines in demand in 1974–1975 and 1980–1982 were the indirect result of the oil price increases of 1973–1974 and 1979–1980. For example, Mory (13) showed that US income growth was negatively affected by the two oil-price shocks. Taking account of this in the model specification would have the effect of reducing the estimated income-elasticity of oil demand.[3]

Less-Developed Countries

In this section we examine the demand behavior of the LDC, both the Oil Exporters and the Other LDC; good surveys can be found in Dahl (14, 15) and Bates (16). In this section we use IEA data for about 60 non-OECD countries, rather than the more complete (but less detailed) coverage of BP or DOE statistics. Fortunately, these 60 countries include the largest and most important of the LDC.

In contrast to the steady income growth of the OECD countries, the performance of the LDC countries has varied. One-third of the 60 LDC countries have experienced negative average annual growth in per capita income, that is, their total income grew less rapidly than population.

Figure 15 shows the contrast between the heterogeneous income-growth

[3]This point has been made by Anthony Finizza and James Wallace, about the results in an earlier paper (10), not shown in Table 2. They argue that the income-elasticity of US nontransportation oil demand could have been overestimated, owing to such a specification error. Subsequent to that paper, we examined a two-stage least squares approach for US nontranportation oil demand. We estimated the determinants of US income growth, following the specification of Mory (13), and then estimated the determinants of oil demand. This procedure reduced the estimated income-elasticity by about half. This alternative specification is not reflected in Table 2.

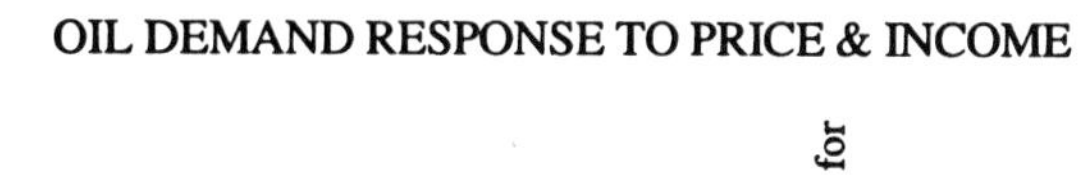

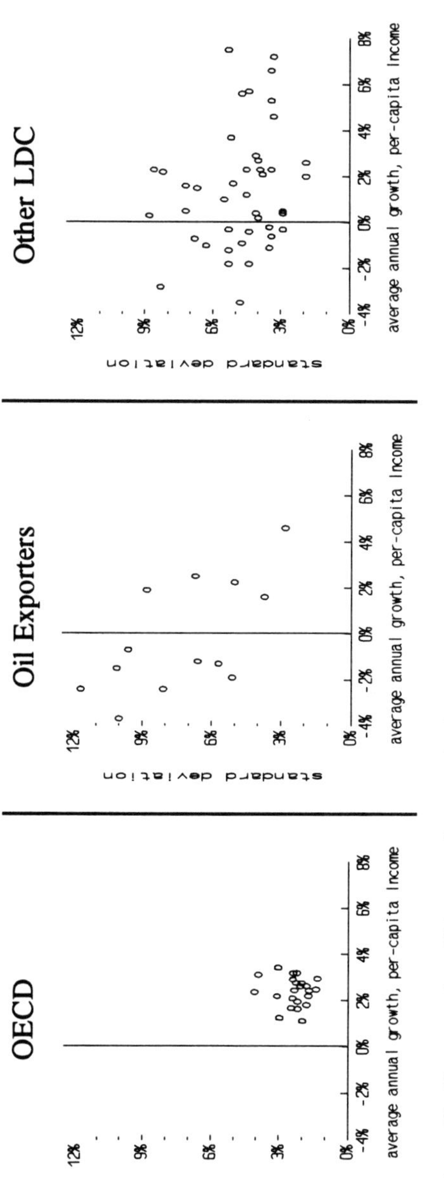

Figure 15 Heterogeneity within groups in per capita income growth rates, 1970–1991: average annual growth and standard deviation of annual growth for each country.

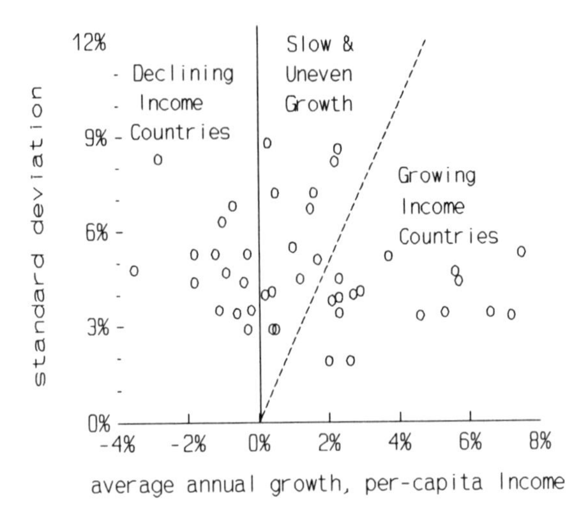

Figure 16 **Disaggregating other LDC into three subgroups.**

performance of these countries with that for the OECD countries. Each circular marker graphs a single country's average annual growth rate of per capita income from 1970–1991 and the standard deviation of its annual growth rate of per capita income.

The OECD countries are clustered together. They have each experienced steady growth in per capita income, averaging about 2–3% annually, with a standard deviation of 1–3%. In contrast, both the Oil Exporters and the Other LDC groups are widely dispersed. One-third of them have negative growth in per capita income, and the standard deviation of their annual growth rates ranges widely, from 2–10%.

Given the heterogeneous experience of these LDC with respect to per capita income growth, and given that one of our primary interests is to understand the effect of income growth on LDC energy demand, we focus on those countries that have experienced income growth most consistently.

We draw upon the work on LDC patterns of development by Chenery et al (17). They found it useful to examine the behavior of individual countries rather than aggregates. In particular, they examined common patterns of behavior within clusters of countries. This is what we do.

We examine separately each of three clusters of countries within the Other LDC, based on the idea that an understanding of the effects of income growth on demand requires that we cluster the countries according to their performance in per capita income growth. Considering all these countries as an aggregate would only cloud the analysis. The clusters shown in Figure 16 are as follows:

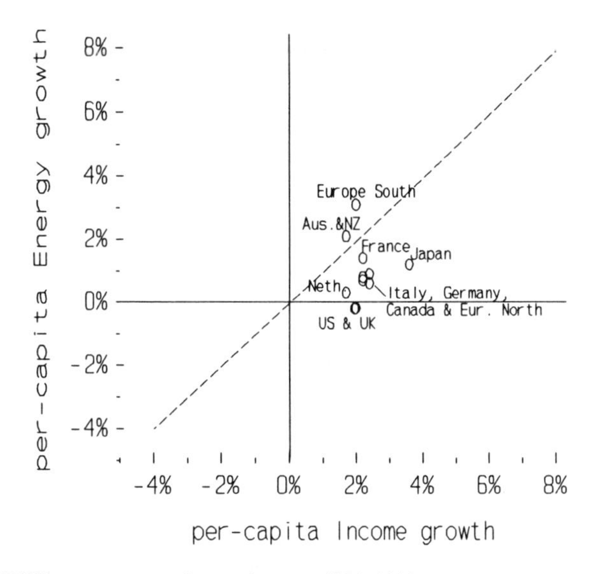

Figure 17 OECD average annual growth rates, 1971–1992: per capita energy and per capita income.

1. Other LDC, Growing Income: those with relatively high average annual growth rate in per capita income and relatively low standard deviation in the annual growth rate of per capita income (less than 2.5 times the average growth rate): Brazil, Colombia, Egypt, Hong Kong, India, Malaysia, Morocco, Pakistan, Paraguay, Singapore, South Korea, Sri Lanka, Taiwan, Thailand, Tunisia;
2. Other LDC, Declining (Per capita) Income: those with negative average annual growth rate in per capita income: Angola, Argentina, Benin, Bolivia, Ethiopia, Ghana, Ivory Coast, Jamaica, Mozambique, Peru, Senegal, South Africa, Tanzania, Zaire, Zambia, Zimbabwe;
3. Other LDC, Slow and Uneven Income Growth: those with relatively low (but positive) average annual growth rate in per capita income and relatively high standard deviation: Bangladesh, Cameroon, Chile, Congo, Guatemala, Kenya, Myanmar, Nepal, Panama, Philippines, Sudan, Trinidad, Uruguay.

Although the three subgroups within the Other LDC have roughly the same number of countries, the Growing Income cluster comprises a disproportionately large share (two-thirds) of the total population.

China, although it would be within the cluster of Growing Income countries in Figure 16, is analyzed separately because of its size and its different history.

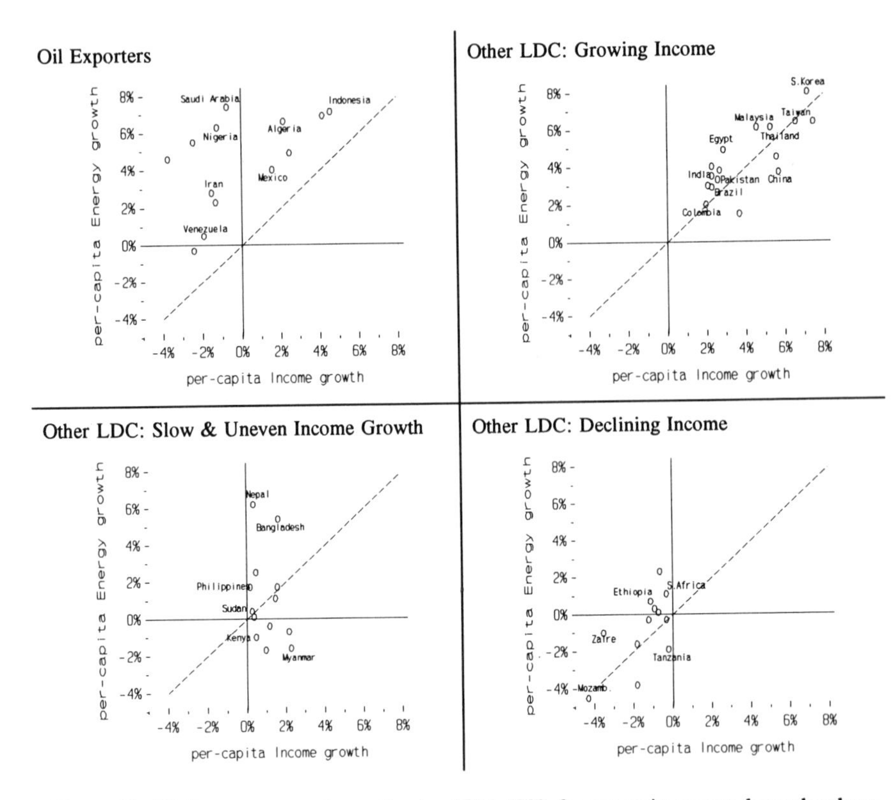

Figure 18 LDC average annual growth rates, 1971–1992, for per capita energy demand and per capita income.

We also analyze separately the Oil Exporters: Algeria, Bahrain, Ecuador, Gabon, Indonesia, Iran, Iraq, Kuwait, Libya, Mexico, Nigeria, Oman, Qatar, Saudi Arabia, United Arab Emirates, Venezuela.

Figures 17 and 18 present another contrast between the homogeneity of the OECD countries and the heterogeneity of the LDC countries in terms of energy income. These figures show, for the OECD and for the four LDC subgroups, respectively, the average annual rate of per capita energy demand growth vs per capita income growth over the period 1971–1992.

The OECD countries and country-groups[4] have exhibited relatively homo-

[4]This analysis is based on our analysis (2) of the OECD as 11 regions: the eight largest OECD countries—the US, Japan, Germany, France, Italy, the UK, Canada, and the Netherlands—and three aggregated regional groups—Europe North (Norway, Sweden, Finland, Denmark, Belgium, Luxembourg, Iceland, Ireland, Austria, Switzerland), Europe South (Spain, Portugal, Greece, Turkey), and Australia-New Zealand.

geneous performance not only in income growth but also in energy demand growth: from 0 to 2% (see Figure 17). In most of the OECD countries, demand grew more slowly than income, and in the US and UK it actually declined slightly. The most obvious exception is in Europe South—the group of countries with the lowest per capita incomes in the OECD—where energy demand increased more rapidly than income.

The four graphs in Figure 18 show the heterogeneity among the LDC, both in their per capita income growth and also in the relationship of per capita income growth to per capita demand growth. Although not shown, per capita oil demand is similarly heterogeneous. Only the most heavily populated countries are labeled in each graph.

Most of the Oil Exporters experienced negative growth in per capita income over this period. Their 1970s income growth was more than reversed by the oil price declines of the 1980s. Some of them, however, did experience growing per capita income; among these were Indonesia, Algeria, and Mexico. But regardless of whether their per capita income grew or declined, their growth in per capita energy (and oil) demand exceeded their income growth. In fact, most had positive growth in energy demand along with declining income.

In contrast, the Growing Income LDC experienced not only growth in per capita income but also growth in per capita energy (and oil) demand that was in rough proportion to income growth. Egypt, India, Malaysia, and South Korea experienced faster growth in energy demand than in income. But of the largest countries, only China (which is analyzed separately) had slower than equi-proportional growth.

The Declining Income countries include many that are less heavily populated. But several have populations in excess of 20 million: Zaire, Tanzania, Mozambique, Ethiopia, and South Africa. For these countries, the relationship between per capita energy demand growth and per capita income growth was not very close. South Africa and Ethiopia experienced growing per capita energy demand. In Zaire, per capita energy demand declined more slowly than per capita income, whereas in Tanzania, per capita energy demand declined more rapidly than per capita income.

The third subgroup, Slow and Uneven Income Growth, had a similarly low correlation between per capita growth in energy demand and income. Energy demand grew much more rapidly than income in Nepal, Bangladesh, and the Philippines. Other countries (Kenya, Myanmar) experienced modest declines in energy demand with slow growth in income.

Figure 19, virtually identical in structure to Figure 9, shows, for each of the three subgroups of the Other LDC, their 1970–1991 time-paths of demand vs income for energy, oil, transportation oil, and nontransportation oil, both as totals (left column) and per capita (right column).

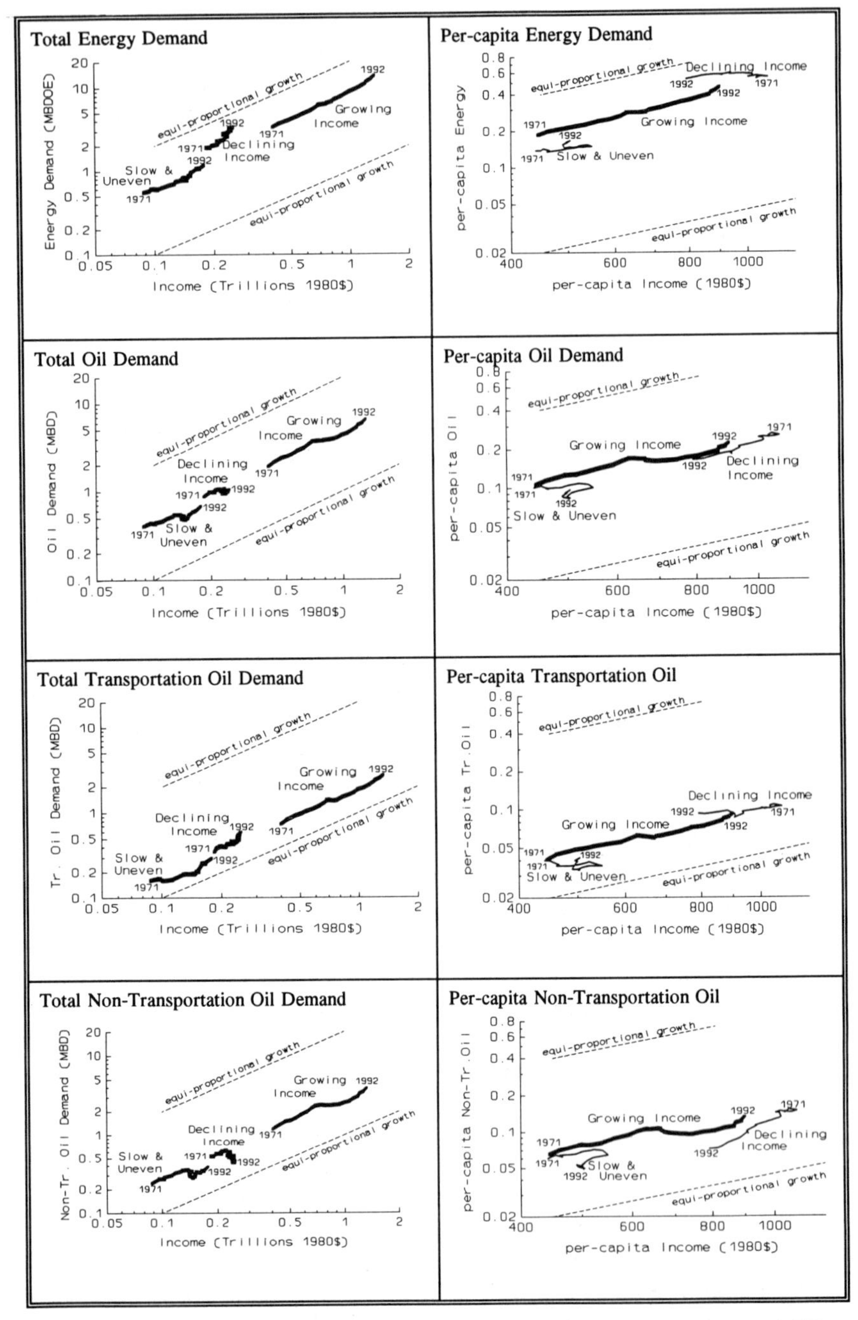

Figure 19 Other LDC, three sub-groups: energy and oil demand (MBD) vs real income (trillions 1980 dollars): per capita demand (tons) vs per capita income, 1971–1992; logarithmic scales (sources: IEA, UN).

The name of the Declining Income subgroup refers to its declining per capita income; its total income is rising over time, but less rapidly than population growth. The 1970 level of per capita income (above $1000) for this subgroup is more than double that of the other two subgroups. This difference can be attributed to the presence of two moderate-income countries (Argentina and South Africa), without whom the subgroup's per capita income in 1970 would be comparable to that of the other two subgroups.

The income-growth of the Growing Income subgroup is impressive. Total income tripled from about 0.4 trillion in 1970 to above 1.3 trillion in 1991. Per capita income doubled from approximately $450 in 1970 to approximately $900 in 1991. Energy and oil demand growth is most easily understood in this subgroup. We describe it in per capita terms. Energy demand grew slightly faster than income, and transportation oil demand grew about as fast as income. Nontransportation oil demand, which grew about as fast as income until the 1979–1980 price increases, declined until 1985, but has increased since the 1986 oil price collapse even faster than income growth. Compared with the OECD countries, the Growing Income subgroup experienced faster economic growth, and faster growth in energy demand and transportation oil demand. This group responded less to the oil price increases than did the OECD, and there was little evidence of imperfect price-reversibility: The price collapse of the 1980s stimulated demand growth, especially for nontransportation oil.

The Declining (per capita) Income countries were in some ways the opposite of the Growing Income subgroup. Their per capita income declined substantially, from above $1000 in 1970 to below $800 in 1991. Their total income grew slowly (less rapidly than population), from about $0.18 trillion in 1970 to $0.25 trillion in 1991. In per capita terms (moving right to left in the per capita graphs), their transportation oil demand remained relatively unchanged, but nontransportation oil fell more than twice as rapidly as income.

The third subgroup, Slow and Uneven Income Growth, experienced both increases and decreases in per capita income: It increased until 1981 but declined and stagnated thereafter. This subgroup's per capita demand experience was similar in some ways to each of the other two subgroups. When per capita income was increasing, energy and oil demand increased, although much more slowly than in the Growing Income countries. When per capita income declined, this subgroup cut back more on nontransportation oil than on transportation oil, as did the Declining Income countries.

A comparison of the Other LDC subgroup behavior in Figure 19 with the aggregated Other LDC behavior in Figure 9 is useful. For per capita nontransportation demand, the reversal of the 1980–1985 demand reduction in the Growing Income countries after the 1986 price collapse is masked when the Other LDC is aggregated, as in Figure 9. There, Other LDC per capita nontransportation oil demand appears to be imperfectly price-reversible after the

1986 price collapse. This imperfect price-reversibility results from the decline in the 1980s per capita income in not only the Declining Income countries but also Slow and Uneven Growth countries. These two subgroups responded by reducing per capita demand for nontransportation oil even faster than their per capita income declined. Thus, after the 1986 oil price collapse, the two subgroups with declining per capita incomes reduced nontransportation demand faster than their income declined, while the Growing Income countries were increasing their demand faster than their income growth. The aggregation of the three subgroups masks what happened to them separately. In particular, it masks the Growing Income countries' partial reversal of the nontransportation oil demand reductions of the early 1980s.

The effect of aggregation is different for transportation oil than for nontransportation oil, because the per capita demand behavior is different. For per capita transportation oil, declining-income countries reduced their demand less than proportionately to declines in per capita income, while the Growing Income countries increased demand about as rapidly as income. Combining countries with declining per capita incomes with the Growing Income countries will reduce the average per capita income growth but will not reduce per capita transportation oil demand growth proportionately. Hence the ratio of transportation oil growth to income growth will be higher for the aggregate than for the Growing Income countries. Thus, for transportation oil, aggregating declining per capita–income countries with growing per capita–income countries will cause an overestimate of the true effects of income growth.

We can also examine the LDC data econometrically, using the two demand specifications: the perfectly price-reversible (Equation 2a) and the imperfectly price-reversible (Equation 2b). Although we tried both specifications, we present only the results for the former, because there was little evidence of imperfect price-reversibility for the LDC. The econometric results appear in Table 3 for energy, transportation oil, and nontransportation oil. The table shows separate estimates for each of the three Other LDC subgroups, the total of all three Other LDC subgroups, the Oil Exporters, and the total of all LDC groups including the Oil Exporters, and China.

In all the LDC demand equations, we used the real price of crude oil, because product price data for most countries are inadequate. Using crude oil price is not satisfactory, because product prices need not move in the same fashion as crude oil prices, particularly when taxation and cross-subsidization play a major role in actual product price development. In oil-exporting countries, oil products are often sold domestically below export prices. Even many oil-importing countries show a pattern of taxing gasoline while cross-subsidizing other oil products such as kerosene and naphtha (see 16, p. 285). The effect of this specification error on the estimated elasticities is difficult to determine. If taxes or subsidies are constant over the estimation period, the use of crude

Table 3 Long-run Elasticities of Per-Capita Demand, for different groups of LDCs[a]

Groups of LDCs	Energy per capita		Transport oil per capita		Non-transport oil per capita	
	Price elasticity	Income elasticity	Price elasticity	Income elasticity	Price elasticity	Income elasticity
Other LDC: Growing Income	−0.12	1.20	−0.18	0.95	−0.09	1.04
Other LDC: Declining Income	0.05[b]	1.50[c]	0.07	0.73	−0.07[c]	3.60
Other LDC: Slow, Uneven Income Growth	−0.32	2.90	−0.29	2.00	NR[d]	NR[d]
All Other LDC excluding Oil Exporters	−0.08	1.40	−0.19	1.00	NR[d]	NR[d]
Oil Exporters	NR[d]	NR[d]	0.23[c]	3.50	0.10[b]	2.50
All LDC including Oil Exporters	−0.14	3.00	−0.18	2.50	−0.03[c]	2.00
China	0.03[b]	0.50[c]	0.18[c]	0.70[c]		

[a] Perfectly Price-Reversible: log Demand/Pop$_{.t}$ = $\alpha + \gamma$ log Income/Pop$_{.t}$ + β log Price$_t$ + ϕ log Demand/Pop$_{.t\text{-}1}$
[b] Coefficient had the wrong sign but was not statistically significant.
[c] Coefficient had the correct sign but was not statistically significant.
[d] The standard specification failed to yield meaningful results.

oil price might not affect the estimated price elasticities. However, taxes or subsidies are generally not constant over the estimation period, so estimated price elasticities may not be particularly reliable.

The results are best for the Growing Income countries, which is not surprising, given the relatively close correlation between income and demand growth. For both energy- and transportation-oil demand, price and income have the expected sign and are statistically significant. Unlike for the OECD, tests (not shown) gave little or no evidence of imperfect price-reversibility. The income elasticity was 1.20 for energy, 0.95 for transportation oil, and 1.04 for non-transportation oil.

For the other groups—Declining Income, Slow and Uneven Growth, and the Oil Exporters—the results were not nearly as good as for the Growing Income countries. In particular, the income elasticities were implausibly high, often indicating that per capita demand rises twice as rapidly as per capita income. This relationship suggests that their response to income growth is to increase demand perhaps twice as rapidly as in the Growing Income countries.

The results were least satisfactory for the Declining Income countries, which is perhaps not surprising. In the energy demand equation, price had the wrong sign and income was statistically insignificant. Results for transportation and nontransportation oil were similarly unsatisfactory. For the Slow and Uneven Growth countries, the results were also unsatisfactory. The income elasticity was surprisingly and misleadingly high.

If all three of these Other LDC subgroups were aggregated, the results would be largely determined by the experience of the Growing Income countries; they constitute a disproportionately large share of the total, two-thirds of population. However, such aggregated results would be contaminated by the irregular behavior of the other two subgroups; their declining income was not matched by proportional declines in transportation oil demand, and nontransportation oil declined more than proportionately.

Also unsatisfactory are the econometric results for the Oil Exporters, or for an LDC aggregate that includes both the Oil Exporters and the Other LDC. For China, the econometric results were reasonable, but not completely satisfactory. The low income-elasticity is certainly consistent with Figure 11 above.

In general, we see that the income elasticities are higher than those estimated for the OECD (compare with Table 2). This finding is consistent with the argument made earlier (regarding Figures 9 and 10) concerning the effect of economic development on the relationship between energy and income growth: The income elasticity declines as income reaches higher levels.[5]

We also see that the price elasticities are rather low in all instances and that demand is much less price-elastic than it was for the OECD. Part of the difference in estimated price elasticities may be a result of our use of the price of crude oil in the LDC demand equations, rather than the end-use price of oil products. But these relatively low estimates for the price elasticity are consistent with those described in the survey by Dahl (14) for total oil demand in the LDC: "The bulk of [long-run] price elasticities lies between –0.13 and –0.26....Income elasticities fall between 0.79 and 1.40" (14, p. 401). However, these price-elasticity estimates are lower than price-elasticity estimates for LDC oil-product demand in surveys by Dahl (15) and Bates (16). The survey by Dahl (15) is more complete, covering 35 different studies, with 232 estimates for gasoline alone. Summarizing such a wide range of estimates is difficult. In general, however, the transportation oil products (gasoline and diesel) have long-run elasticities of about –0.7 for price and 1.2 for income, whereas residual fuel oil was more price elastic and less income elastic. In Bates' survey (16, p. 409), the price elasticities of oil product demand in developing countries are also generally higher than those in Table 3, although they vary by country and by oil product. In other words, they are relatively high for gasoline but relatively low for LPG and kerosene (16, p. 490).

CONCLUSIONS

The responsiveness of energy demand to income growth is higher in the LDC than in the OECD: Energy demand grows faster than income in the LDC, but

[5]Dargay (18) estimated a nonconstant elasticity model for total energy demand for the UK. The results included both a declining income elasticity and imperfect price-reversibility.

slower than income in the OECD. Transportation oil demand has grown consistently, in all regions of the world, about as rapidly as income growth, and is less price-responsive than nontransportation oil demand.

The effect of the oil price increases of the 1970s on LDC energy and oil demand was not as great as its effect on OECD energy and oil demand. In the past two decades, OECD energy demand has grown much more slowly than income, as has OECD oil demand. OECD nontransportation-oil demand fell sharply after the 1979–1980 oil price increase, but that decline has not been reversed after the reversal of the oil price increases in the 1980s.

The Other LDC show much less evidence of imperfect price-reversibility than the OECD. A substantial part of the decline in Other LDC nontransportation demand in the early 1980s—for the Growing Income subgroup especially—was reversed by the 1986 oil price collapse. Since then, nontransportation oil demand has risen faster than income, and oil's share of energy has partially recovered.

Finally, the aggregation of the three subgroups of the Other LDC will distort the effects of true income growth on energy and transportation-oil demand and the effects of the oil price reductions of the 1980s on nontransportation oil demand. For transportation oil and energy, aggregating countries that have declining per capita incomes with countries that have growing per capita incomes causes an overestimate of the effects of true income growth. For nontransportation oil, the aggregation masks the Growing Income countries' post-1986 reversal of the demand reductions of the early 1980s.

Literature Cited

1. Hogan WW. 1993. OECD oil demand dynamics: trends and asymmetries. *Energy J.* 14(1):125–57
2. Dargay J, Gately D. 1995. The imperfect price-reversibility of non-transportation oil demand in the OECD. *Energy Econ.* 17(1):59–71
3. Griffin JM. 1979. *Energy Consumption in the OECD: 1980–2000.* Cambridge, MA: Ballinger
4. Pindyck RS. 1979. *The Structure of World Energy Demand.* Cambridge, MA: MIT Press
5. Hogan WW. 1986. Patterns of energy use. In *Energy Conservation,* ed. JC Sawhill, R Cotton, pp. 19–53. Washington, DC: Brookings Inst.
6. Dargay J, Gately D. 1994. Oil demand in the industrialized countries. *Energy J.* 15:39–67 (Special issue)

6a. OECD. 1993. *Energy Prices and Taxes, First Quarter 1993.* Paris: International Energy Agency

7. Traill B, Colman D, Young T. 1978. Estimating irreversible supply functions. *Am. J. Agric. Econ.* 60(3):528–31
8. Wolffram R. 1971. Positivistic measures of aggregate supply elasticities: Some new approaches—some critical notes. *Am. J. Agric. Econ.* 53:356–59

9. Gately D. 1993. The imperfect price-reversibility of world oil demand. *Energy J.* 14(4):163–82
10. Gately D. 1993. Oil demand in the US and Japan: Why the demand reductions caused by the price increases of the 1970s won't be reversed by the price declines of the 1980s. *Jpn. World Econ.* 5:295–320
11. Gately D. 1992. Imperfect price-reversibility of oil demand: asymmetric responses of US gasoline consumption to price increases and declines. *Energy J.* 13(4):179–207
12. Dargay J. 1992. The irreversible effects of high oil prices: empirical evidence for the demand for motor fuels in France, Germany, and the UK. In *Energy Demand: Evidence and Expectations,* ed. D Hawdon, pp. 165–82. London: Academic
13. Mory JF. 1993. Oil prices and economic activity: Is the relationship symmetric? *Energy J.* 14(4):151–61
14. Dahl C. 1993. A survey of oil demand elasticities for developing countries. *OPEC Rev.* XVII(4):399–419
15. Dahl C. 1994. A survey of oil product demand elasticities for developing countries. *OPEC Rev.* XVIII(1):47–86
16. Bates RW. 1993. The impact of economic policy on energy and the environment in developing countries. *Annu. Rev. Energy Environ.* 18:479–506
17. Chenery H, Robinson S, Syrquin M. 1986. *Industrialization and Growth: A Comparative Study.* Oxford: Oxford Univ. Press
18. Dargay J. 1992. Are Price and Income Elasticities of Demand Constant: The UK Experience. *Rep. EE16.* Oxford, UK: Oxford Inst. Energy Stud.

Annu. Rev. Energy Environ. 1995. 20:179–212

INTERNATIONALIZING NUCLEAR SAFETY: The Pursuit of Collective Responsibility

Jack N. Barkenbus
University of Tennessee, Knoxville, Tennessee 37996

Charles Forsberg
Oak Ridge National Laboratory, Oak Ridge, Tennessee 37831

KEY WORDS: energy, safety convention, nuclear power, international, IAEA

CONTENTS

ABSTRACT

The future of nuclear energy could depend upon the international infrastructure established to ensure the creation of a strong and uniform safety culture. Deliberations during the 1990s, leading to the recently promulgated International Nuclear Safety Convention, held out the prospect of both bolstering nuclear safety and gaining public recognition of the need to address transboundary safety concerns head-on. Unfortunately, the Convention that emerged from the deliberations constitutes little more than another form of technical assistance. The basis for an alternative, and more substantial, Convention is presented—one that would be based on the establishment and evaluation of

1056-3466/95/1022-0179$05.00

performance standards, the creation of a series of political "firebreaks," and the encouragement of nuclear power plant designs that minimize the catastrophic offsite consequences of accidents.

Introduction

After nearly half a century of development and utilization, nuclear power is now a major global energy source. Born in the shadow of the US military program, civilian nuclear power has emerged as a far-flung, multinational industry. This industry now has an identity, in nearly all countries that possess nuclear reactors, quite separate from any military applications. In 1993, nuclear power produced approximately 17% of all the electricity generated worldwide—a contribution similar to that of hydroelectric power. Overall, the world had seen approximately 7000 reactor years of nuclear plant operation at the end of 1994.

Despite the impressive maturation of nuclear power, the degree to which it will contribute to meeting the world's growing energy needs is very much in question. For numerous familiar and fundamental reasons, nuclear power may not be an energy source of choice—despite projections of enormous global electricity requirements and despite the fact that several other energy options are deficient in critical respects.

A prerequisite for a vigorous nuclear energy future is the creation of an extraordinarily dedicated and uniformly vigilant, safety culture throughout the nuclear industry itself—a culture so effective that Chernobyl-type accidents, or even less serious events, simply will not occur. Moreover, in order to change public perceptions about nuclear power, the nuclear industry must not only create a vigilant safety culture, it must also be seen as creating the institutional framework to support it over time. This review focuses on the nuclear power community's attempt to build an institutional pillar to support the necessary safety culture through the creation of the Convention on Nuclear Safety. It will be seen that the Convention that resulted from three years of international discussion and negotiation is a woefully inadequate tool for altering perceptions about nuclear power and for establishing the technical and institutional infrastructure necessary to manage this energy technology. A preferable framework, based on a far more substantial and ambitious Convention, is presented.

Nuclear Power Globally

While nuclear power is a global energy source, not all regions of the globe make equal use of it. North America is responsible for one third of the world's 337,000 megawatts of operating nuclear power capacity. Western Europe contributes another 35%, while states of the former Soviet Union and other Central and Eastern European states make up 14%. The Far East constitutes another 14% (with Japan making up nearly three quarters of this total). These figures, derived from the totals found in Table 1, illustrate that nuclear power

Table 1 Nuclear power status around the world, December 31, 1993[a]

	In operation		Under construction	
	Number of units	Total net MWe	Number of units	Total net MWe
Argentina	2	935	1	692
Belgium	7	5,527		
Brazil	1	626	1	1,245
Bulgaria	6	3,538		
Canada	22	15,755		
China	2	1,194	1	906
Cuba			2	816
Czech Republic	4	1,648	2	1,824
Finland	4	2,310		
France	57	59,003	4	5,815
Germany	21	22,559		
Hungary	4	1,729		
India	9	1,593	5	1,010
Iran			2	2,392
Japan	48	38,029	6[b]	5,645
Kazakhstan	1	70		
Korea, Republic of	9	7,220	7[b]	5,770
Lithuania	2	2,370		
Mexico	1	654	1	654
Netherlands	2	504		
Pakistan	1	125	1	300
Romania			5	3,155
Russian Federation	29	19,843	4	3,375
South Africa	2	1,842		
Slovak Republic	4	1,632	4	1,552
Slovenia	1	632		
Spain	9	7,101		
Sweden	12	10,002		
Switzerland	5	2,985		
Taiwan	6	4,890		
United Kingdom	35	11,909	1	1,188
Ukraine	15	12,679	6	5,700
United States	109	98,784	2	2,330
World total	430	337,718	55	44,369

[a] Source; International Atomic Energy Agency

[b] Japan and South Korea have the largest nuclear power plant construction programs. The scale of their programs is larger than suggested by this table because both countries have short power plant construction times, hence few power plants under construction at any one time but many reactors completed per decade.

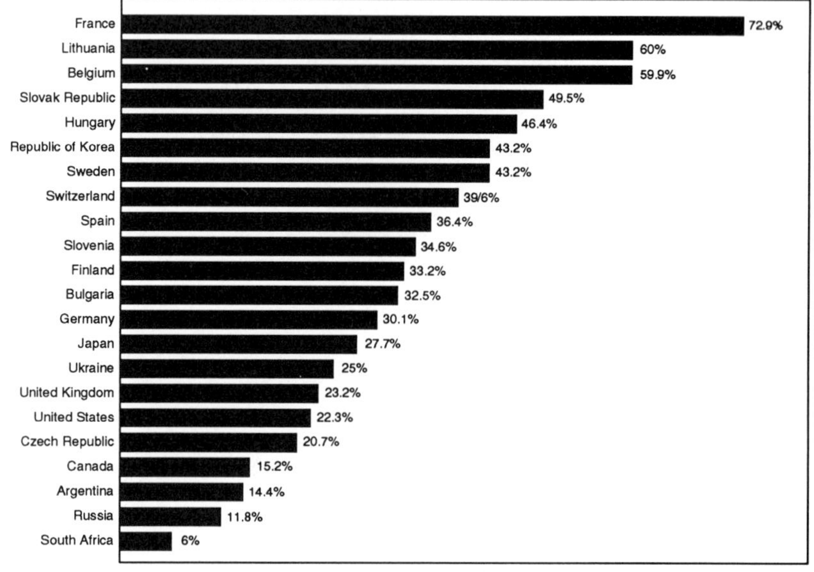

Figure 1 Nuclear share of electricity generation in selected countries

operates today within a small subset (30 countries) of the international community, and within the most affluent nations. Despite the impressive half century of growth, therefore, nuclear power's reach has not taken it across the globe. And if nuclear power is to play a significant role in meeting the enormous future energy requirements of developing countries, it needs to be introduced and accepted in countries that have little familiarity with it today.

A nation's on-line nuclear power capacity does not, of course, indicate that nation's dependence upon this source of power. Even though the United States is by far the largest generator of nuclear power in the world, with its 100-plus operating reactors, its electrical power system is so large that nuclear power constitutes just 22% of total electrical output. In contrast, Hungary has only four reactors, but the output from these sources constitutes almost 50% of that country's generated electricity. Nuclear power's contributions to national electricity supplies can be seen in Figure 1.

In the countries and regions most accustomed to nuclear power, further growth has reached an impasse. Twenty-one of the 30 countries operating nuclear power plants have no plans to build and operate additional plants. Fifty-five nuclear power plants are still officially in the construction "pipeline," but how many of these will actually be built remains to be seen. During 1993,

6 reactors moved from construction to operation, yet the official number of reactors designated as "under construction" fell from 72 to 55. We can expect further attrition. Notwithstanding excellent prospects for growth in a few countries today (primarily those in the Far East), the nuclear power industry is currently suffering stagnation (or is in, as one analyst has charitably put it, a period of "introspection") (1). Much has to transpire, therefore, if nuclear power is to fulfill the future role that its boosters envision.

An Optimistic Scenario for the Future

A strong case can be made for a revival of the nuclear power industry. The demand for the product nuclear energy currently produces—electricity—is likely to remain strong in virtually all countries for the foreseeable future. Because electricity is versatile and its end use is clean, its utilization is projected to expand faster than that of other energy sources (2). High growth rates are expected particularly in developing countries. Even in the industrial nations where one might expect saturation, and where aggressive conservation programs are being pursued, electricity is making inroads at the expense of other fuels (particularly oil) (3).

This growth in demand is taking place at a time when fully acceptable means of supplying electricity appear to be dwindling. Hydroelectric power, once seen as the "clean" energy option with enormous potential, is now being bashed consistently by nongovernmental organizations for its large-scale environmental impacts (4). Fossil fuels carry their own unfortunate baggage. Oil markets have proven extremely volatile and unpredictable. Coal is much more abundant and stable but unfortunately raises both near-term and long-term environmental concerns. Policy measures adopted to combat climate change fly in the face of an increasing role for coal. Natural gas is the most acceptable fossil fuel, but it is generally viewed solely as a transition fuel. Solar power options, either for direct or indirect (e.g. biomass) electricity generation, may play a growing role in the future, but because the technology is still in an early stage of development its role is likely to be small for some time. In short, nuclear power's competition demonstrates serious weaknesses in a time of robust electricity demand.

Evolving trends in nuclear power technology can also contribute to an optimistic scenario (5). The much-discussed movement to smaller, even modular, nuclear reactors provides the opportunity for a better match with small electric grids. The rule-of-thumb claim that a single power station should not produce more than 15% of a nation's electricity effectively ruled nuclear power out in many smaller countries when only 1000 MWe (and larger) reactors were on the market. Accompanying this movement to smaller reactors is a greater reliance on simplicity and passive safety features. Yesterday's reactor safety practices evolved piece-by-piece, fostered by intense regulatory oversight and incorporated incrementally by power plant designers. This approach has resulted in considerable complexity, and it relies for the maintenance of safety upon human and mechanical infallibility. New reactor designs are being created that are more forgiving of human mistakes. The added passive safety features will not obviate dedicated and diligent safety practices, but they will provide a higher level of comfort than we now enjoy about the consequences of human error. By so doing, they will permit the introduction of the technology into regions not now making use of it (6).

Under the right conditions, nuclear power can be economical. The standardized nuclear power plant designs of France have shown major economic advantages over the custom designs of American power plants. In Japan, shipyard construction of large modules that are later assembled into power plants has produced additional economic gains. Both approaches, however, require large construction programs and a willingness to restructure industrial organizations for the efficient production of nuclear power plants. Research on advanced reactors such as the gas-turbine Modular High Temperature Gas-Cooled Reactor (MHTGR) and the supercritical Light Water Reactor (LWR) indicate the potential for further major reductions in the cost of electricity generated from nuclear power (7).

A Pessimistic Scenario for the Future

While the scenario sketched above may convince some that a bright future awaits nuclear power, it overlooks the considerable liabilities that continue to plague this form of electricity production and that make it, for many, the energy option of last resort. Several factors have created this period of industry stagnation. The economic performance of nuclear reactors has been extremely variable, with some nuclear units providing the cheapest form of electricity and others the most expensive (8, 9). Some of the technological features that might make the nuclear option more acceptable (e.g. modularization) might also make it more expensive. Most states have decided to defer high-level waste disposal decisions in favor of monitored disposal and further research on geological disposal (10). Though this may be the prudent path today, it leaves a cloud of doubt and indecision over our ability to manage safely all phases of the nuclear power option. Furthermore, concerns over nuclear power's association with weapons proliferation, currently undiminished, would certainly intensify if nuclear energy technologies were to spread to nations that prefer to leave future military options open (11).

The public must be afforded a greater level of comfort about all of the issues noted above before industry stagnation can be replaced by growth. Many observers feel that the industry is incapable of overcoming these obstacles. And beyond this dim prospect looms far-reaching doubt about reactor safety. Prior to Chernobyl, this concern was hypothetical. The unprecedented accident in Ukraine, however, demonstrated that the threat was real. Claims that this accident resulted from an anomalous combination of egregious human error and unstable reactor technology can and have been made (12). But all efforts to return reactor accidents to the realm of the hypothetical must fail. Chernobyl, and its worldwide impact, cannot be erased from our collective consciousness.

It follows, therefore, that the foremost threat to nuclear power's future is yet another consequential reactor accident. As a truism oft-repeated within the nuclear power industry has it, "An accident anywhere in the world is an accident everywhere in the world." Regardless what idiosyncratic factors might cause another accident, such an event would tarnish the entire nuclear power industry. The political fallout from Chernobyl did indeed reach far beyond the borders of Ukraine, and of Central, Western, and Eastern Europe; it has spread over North America and the Far East as well (13). Reactor safety is a collective responsibility. Safety can no longer rest solely with national authorities, because the impact of an accident, both physically, psychologically, and politically, transcends political boundaries. What follows is a description of how the nuclear community has struggled over the past few years to create a regime based upon collective responsibility. It is not an exaggeration to say that the success of this effort could determine which scenario—optimistic or pessimistic—will prevail in the future.

Transboundary Issues

A host of challenges beset the state-based nature of our international system, and technologies such as nuclear power constitute one such challenge (14, 15). In fact, nuclear power may present the best known and most obvious technological challenge to state-based dominance. The international community has already responded boldly to collective responsibility covering one aspect of nuclear power: namely, implementing intrusive international oversight (safeguarding) over potential diversion of reactor fuel to weapons production. International inspection to deter and detect diversion of weapons-usable material from civilian to military applications has accompanied the spread of nuclear energy technology from its origins. The subsequent ratification of the Treaty on the Non-Proliferation of Nuclear Weapons (the NPT) further strengthened and legitimized international oversight. Recent events in Iraq and North Korea will likely intensify and bolster safeguarding practices (16).

The international approach to collective responsibility for nuclear safety, on the other hand, has been decidedly less aggressive. The same international organization that is responsible for safeguards—the International Atomic Energy Agency (IAEA)—has always been vested with the responsibility for promoting nuclear power globally. It has never been given the authority, however, to regulate or conduct mandatory inspections of nuclear power facilities to ensure safety. One may claim that the IAEA should not attempt to carry out promotional and regulatory responsibilities simultaneously. The fact is, however, that the international community has never granted authority to any international body for the purpose of regulatory oversight. States have doggedly maintained exclusive control and authority over nuclear power operations, despite other states' clear and legitimate transboundary concerns.

Public discomfort with state-based nuclear regulation escalated with the 1986 Chernobyl accident. The international community's "hands-off" approach had failed to prevent the accident at Chernobyl, and there were no guarantees that this accident would be the last (17). Ironically, it was the end of the Cold War and the introduction of information transparency between East and West that provided the next jolt to international confidence in nuclear power. Nuclear power within the Eastern Bloc had developed in relative isolation from the West, and as long as the East-West conflict dominated the international agenda there was little impetus for integration or cooperation. Chernobyl afforded the West a window through which to evaluate nuclear performance in the former Soviet Union, and the view was not reassuring. When Communism fell, the barriers to information transparency were gradually removed, revealing a nuclear power enterprise fraught with technical, institutional, and cultural deficiencies (18).

Figures in Table 1 reveal a heavy dependence upon nuclear power in the many states of the former Soviet Union and in Central and Eastern Europe. As of the end of 1993, 65 reactors operated in this region. The magnitude of the task involved in bringing these reactors up to international standards has subsequently overwhelmed the host states and the donor community. After the expenditure of tens of millions of dollars in multinational funding, the Office of Technology Assessment estimates that 40% of the operating reactors still present "serious safety concerns" (19). The General Accounting Office claims that all Soviet-designed reactors "pose significant safety risks" (20).

With few exceptions, the shutdown of deficient reactors is beyond what host governments feel is possible. Four units have been shut down in the former East Germany—but only because the new, unified Germany has excess electric generating capacity and the financial impacts were small. Two units have also been shut down in Armenia because they might be incapable of withstanding seismic shock; plans are under way, however, to bring one of the units back on-line. Teams of international experts have recommended additional shutdowns, the first back in 1991 when four nuclear units at Kozlodui, Bulgaria were found to have extremely serious structural and operational deficiencies. More recently, in 1994, international experts recommended that two of the still-operating units at Chernobyl, Ukraine be closed down. In both the Bulgarian and the Ukrainian cases, state officials declined to close the reactors, citing their citizens' dependency on the electricity generated at the sites, the high cost of using alternative fuels, and the financial burden accompanying the importation of electricity (21).

The problems associated with the fleet of reactors in this region are multifaceted. Designed and constructed in isolation from Western safety standards, many of the reactors lack even the most fundamental safety features found in the West. The earliest reactors (the so-called RBMK design) lack a robust containment shell and adequate fire protection systems, among other deficiencies. Under certain operating conditions these reactors are themselves unstable—witness the runaway core melt and explosion at Chernobyl. Fifteen of the 65 operating reactors in the region are of the RBMK design and are found in Russia, Ukraine, and Lithuania. The next generation of reactor design, the so-called VVER 440/230, suffered many of the same deficiencies, including poor instrumentation, limited capacity emergency core cooling systems, and inadequate seismic protection. Ten of these reactors are still operating in Bulgaria, Russia, and Slovakia. The 1970s generation of reactors, the VVER 440/213, constituted an improvement over the previous generations but still lacked a full, robust containment and adequate instrumentation. Fifteen of these reactors are still operating in the Czech Republic, Hungary, Russia, Slovakia, and Ukraine. Finally, the most recent generation of reactor, the VVER-1000, functions much as a Western pressurized-water reactor does and approximates the safety standards found in the West (22). Twenty-five of these reactors are operating across the region.

It would be misleading to imply that all, or even most, of the problems associated with nuclear power operations reside in the technical deficiencies of the reactors themselves. Unfortunately, the problem is much larger and results from the absence of what is commonly referred to as a necessary "safety culture" (23)—an operating ethic whereby individuals and organizations responsible for nuclear power operations demonstrate the fullest commitment to safety. Deficiencies in safety culture can result from any number of circumstances, such as inadequate financial resources, competing national objectives (e.g. emphasis on energy production), flagging technical capabilities, and human and organizational complacency. The absence of a safety culture is manifest as slipshod construction and maintenance practices, lackadaisical operating procedures, the absence of a quality control program, low worker morale, poor working conditions, delays in implementing corrective actions, and the absence of training opportunities. Additionally, a weak safety culture is characterized by a feeble regulatory body unable to impose its requirements upon operating entities under its jurisdiction (24).

Western efforts to deal with both the technical and institutional problems of the region formerly behind the "Iron Curtain" have been considerable. Bilateral initiatives date back to 1988 when the United States and the Soviet Union formed the Joint Co-ordinating Committee on Civilian Nuclear Reactor

Safety. This organization was superseded by the Group of 7 (Canada, France, Germany, Italy, Japan, the United Kingdom, and the United States) announcement in 1992 of a multimillion-dollar nuclear safety campaign in the region. As of mid-1994, $785 million had been pledged but only close to $60 million allocated. Western European funding for safety improvements in the region has been substantial, much of it going to improvement of conditions and operations in Bulgaria and Lithuania. US assistance has been largely directed toward Russia and Ukraine. Despite the influx of Western dollars and technical assistance, many believe that fundamental changes will require the influx of billions, not millions, of dollars. Frustration is rising in the region because although review committees from the West have proposed numerous changes, little follow-up financing has been forthcoming (25).

The need to create a uniformly excellent safety culture in the region surfaced well before the Chernobyl accident and before widespread recognition of problems in states under the control of the former Soviet Union. In fact, much of the discussion revolved around developing nations and how to insure their adoption of the norms and practices established in states more familiar with the use of nuclear power (26). Slower-than-anticipated diffusion of the technology to developing countries has somewhat muted the issue to date. Still, considering that future electricity use rates are expected to grow most robustly in the developing nations, the issue is likely to return as these countries once again evaluate nuclear power as an energy option.

Technical Assistance

Not all of the international community's efforts in collective responsibility for safety have been channeled through projects specifically directed toward nuclear power operations in the former Soviet Union or Central and Eastern Europe. It is worth reviewing other, more broad-based efforts as well.

The IAEA, while given no regulatory authority over nuclear power operations, has a growing technical assistance program located primarily within its Division of Nuclear Safety. This division receives a relatively small portion of the overall IAEA budget—approximately $17 million in 1992, or just a little over 7%—but its budget and responsibilities have grown significantly during the 1980s and 1990s. As Thompson has stated, "Safety and environmental protection were not high priorities in the Agency's early years, but, especially since the 1986 Chernobyl accident, have come to play a larger role in its activities" (27).

In the 1960s and 1970s emphasis within the Agency's nuclear power program centered on the production of consensus standards and guidance documents. The centerpiece of this approach was the Nuclear Safety Standard Program (NUSS), which involved the production of Agency publications covering codes of practices in all aspects of nuclear power operations. These publications were intended to serve as a frame of reference for safety—particularly for new states just entering the nuclear field. Over 60 reports have been produced under NUSS, which are continually updated as knowledge in the field increases. Other early safety activities included the production of specific-issue conventions covering transboundary safety issues—e.g. the Vienna Convention on Civil Liability for Nuclear Damage, adopted in 1963.

IAEA nuclear safety activities evolved during the 1980s owing in part to the repercussions from the Three Mile Island (TMI) accident in 1979 and, in greater degree, to the Chernobyl accident. These experiences made it clear that the Agency needed to go beyond the simple production of guidance documents and their distribution to national authorities. During the 1980s, therefore, the Agency created programs with a new and common thrust: the conduct of peer technical assistance reviews, carried out by multinational teams, intended to transfer experience and wisdom about successful industry practices. An emphasis on distribution of written documents changed to one on conducting site visits aimed at transfering knowledge and evaluating whether best industry practices were being carried out (28). It is important to point out, as the IAEA regularly does, that these programs have not been intended either to establish a new layer of regulation or to supplant national regulatory authority. The Agency possesses no authority with which to enforce compliance.

The major IAEA technical assistance programs, described briefly in Table 2, have been gathering momentum over the past decade and are now generally recognized as a legitimate and valuable contribution to international nuclear safety.

Technical Assistance—Nongovernmental Approaches

The nuclear power industry and the electric power industry have felt it imperative to launch their own technical assistance efforts, not relying upon the IAEA efforts for improved performance. While national authorities, the day-to-day IAEA contacts, may be held responsible for safety performance, the actual burden of achieving this safety falls upon nuclear power plant operators. It is to this significant segment of the industry that most nongovernmental technical assistance efforts are directed.

The foremost player in this respect is the World Association of Nuclear Operators (WANO), launched in 1987. WANO believes that the sharing of worldwide experience will enable operators to ensure safe operations at their individual facilities. The founding of WANO parallels the approach taken by the US domestic nuclear power utilities in 1983 in the wake of TMI, when the Institute for Nuclear Operations (INPO) was founded.

WANO is run primarily through four regional centers—in Atlanta, Moscow, Paris, and Tokyo—that serve primarily as communication hubs. A WANO Governing Board meets biannually, and a small coordinating center is located in London. WANO's goal is to strengthen utility capabilities through "com-

Table 2 IAEA technical assistance programs

Program	Description	States
Optional Safety Review (OSART)	Experts from numerous nations are pulled together in a team to spend 3 weeks on site evaluating specific nuclear power plant operations.	From 1983 to 1993 there had been 68 site missions conducted in 25 countries.
Assessment of Safety Significant Events Team (ASSET)	Multinational teams provide in-depth investigation of unusual incidents or events with safety significance.	There have been over 50 ASSET missions sent to 21 countries.
Engineering Safety Reviews (ESRs)	Multinational teams investigate plant design, plant aging, fire protection, and often engineering related issues.	At the beginning of 1993, 27 team missions had been conducted.
International Regulation Review Team (AERATE)	Multinational teams assemble to assess a nation's regulatory framework and practices.	At the beginning of 1993, only two nations—Brazil and Romania—had availed themselves of this service.
International Peer Review Service (IPERs)	Nations contribute information on nuclear events taking place within their borders.	26 nations now contribute to the database that holds reports of over 1000 events.
Incident Reporting System (IRS)	Multinational teams review the probabilistic safety assessments (PSAs) increasingly being used to determine where threats to operational safety reside.	Few have been carried out to date through this program established in 1989.

parison, communication, and emulation." It does this primarily through four programs (29):

1. the operating-experience information-exchange program—notifications of events taking place at individual reactor sites are compiled and distributed. An analysis of the events is undertaken.
2. the operator-to-operator program—exchanges of operating personnel take place through site visits, workshops and seminars.
3. the plant performance indicator program—operators are asked to evaluate their annual performance on the basis of 10 quantitative plant-performance indicators. These performance reviews are then sent to Atlanta for compilation and provide the basis for comparison.
4. the good practice program—documents setting forth good practices at individual plants are distributed in order to provide examples of exemplary performance.

WANO has had some success, primarily because it is a creation of the utilities and is a utility-to-utility support system. There is a utility culture worldwide with common interests. Advice from a manager or a technical expert based at a large Western power station is intrinsically credible to an Eastern European or Russian utility manager or technical expert. Operators in both world regions are responsible for producing electricity, assuring safety, and meeting conflicting demands. These common experiences assure a mutual credibility that is sometimes lacking in other information sources. The effectiveness of advice often depends as much upon its source as on its technical content.

Despite the millions of dollars being funneled to the former Eastern Bloc, and the increasing governmental and nongovernmental support for more broad-ranging technical assistance, public concern over reactor safety persists. If nuclear power is to have a signficant future, safety must be assured not only for the current fleet of 430 operating reactors but also for the reactors of new national entrants into the nuclear community. Such residual safety concerns, largely felt by several influential Western European states, provided impetus to the next step in collective responsibility: namely, the creation of an international nuclear safety convention.

Purposes of an International Nuclear Safety Convention

The disturbing revelations of safety deficiencies in Eastern Europe were at center stage when the IAEA hosted the International Conference on the Safety of Nuclear Power in September 1991, and when the IAEA General Conference met in the same month. With a strong push from the European community, a unanimously adopted resolution resulted from the General Conference on September 21, calling for the IAEA Director General to convene a group of experts to draft the structure and content of a possible nuclear safety convention. After the Director General presented the findings of the expert group to the IAEA's Board of Governors in February of 1992, he was given the go-ahead to convene subsequent working groups of experts who would provide the substance of a formal convention (30).

Rhetorical commitment to the principle of such a convention was nearly universal. Its appeal, however, rested on expectations and goals not equally shared by all states. At the very least, the convention was seen as providing the nuclear community with a window on what is transpiring within the existing nuclear power programs of all states.

The convention would open nuclear power programs to international scrutiny. All states hoped, more ambitiously, that the convention could provide the leverage by which the international nuclear community could assert its influence in creating a uniformly high safety culture. While this goal was shared by all, preferences for the means of accomplishing the goal were not shared.

Perceptions of the proper level and form of international intrusiveness differed. Finally, some states saw the convention as a significant confidence-building initiative that would create greater public trust in nuclear power and thereby facilitate its growth. IAEA Director General Hans Blix stated, "The future of nuclear power ... depends essentially on two factors: how well and safely it actually performs and how well and how safely it is perceived to perform" (31). Many believed the convention would enhance both of these objectives.

Once working-group deliberations got under way in 1992 and the drafting of the convention commenced, it became apparent that not all expectations and visions could be reconciled. At issue, among other things, were the scope of the convention, the level and specificity of the obligations incurred by signatories, the degree of independent or international oversight of state adherence to these obligations, and the openness or public scrutiny of future review sessions. During the period May 1992 through January 1994, no fewer than six Group of Experts meetings convened to hammer out a consensus. One participant in those meetings characterized them as "complex and difficult," and at times it seemed possible that the effort might be abandoned (32). A consensus was finally reached in 1994, and the Convention was adopted on June 17, 1994 by a Diplomatic Conference convened by the IAEA. Forty-nine countries have become signatories to the Convention, and it will go into effect 90 days after the twenty-second ratification is deposited with the IAEA. The Convention in its entirety can be found as Appendix A to this review (33).

The Convention

The Convention obligates states to maintain and demonstrate high safety standards in the operation of their nuclear power plants. Demonstration involves the presentation before regularly scheduled review conferences of reports illustrating how the incurred obligations are being fulfilled. Such international scrutiny at review conferences is referred to as a "peer review" process, implying the congregation of "equals" to review the reports and presentations. The Convention contains 35 articles subsumed under a Preamble and four chapters or sections: Chapter 1: Objectives, Definitions and Scope (Articles 1–3); Chapter 2: Obligations (Articles 4–19); Chapter 3: Meetings of the Contracting Parties (Articles 20–28); Chapter 4: Final Clauses and Other Provisions (Articles 29–35).

The Preamble highlights the importance of nuclear safety and reaffirms that responsibility for safety rests with the state where a nuclear power plant is located. While Chapter 1 clearly defines the scope of the Convention to be "nuclear installations" (i.e. civil nuclear power plants), the Preamble notes that future conventions could cover other parts of the nuclear fuel cycle, and even affirms the need for a waste management convention after international agreement has been reached on waste management safety fundamentals. Chapter 1

(Articles 1–3) also states that the key objective of its signatories is "to achieve and maintain a high level of nuclear safety worldwide through national measures and international cooperation."

Chapter 2 (Articles 4–19) speaks to the obligations or commitments parties to the Convention accept with their signature. These obligations are set forth as fundamental safety principles rather than detailed standards. The commitments can be summarized as follows: to ensure an adequate financial base of support; to create and maintain (through education and continuous training) an excellent personnel pool; to establish a firm legislative framework and an independent regulatory body capable of ensuring excellence in plant operation; to execute safety assurance and safety assessment programs—both prior to operation and on a continuing basis thereafter; to establish viable emergency evacuation plans; and to design reactors that will provide several levels of public protection, thereby implementing the "defense-in-depth" principle.

States also incur the obligation to produce a report detailing how the above requirements are being met, and to submit this report, for review, at regularly scheduled review conferences.

Chapter 3 (Articles 20–28) treats the review conference. The first review conference is to be held no later than 30 months after the Convention goes into force, and is to be preceded by a preparatory meeting that establishes the dates and procedures for subsequent review sessions. The intervals between review sessions are not to exceed three years. Signatory states are granted confidentiality with respect to their reports, and the content of the debates at the review sessions are to remain confidential. Signatories are to make public a summary report addressing issues discussed and conclusions reached at the review conference.

The IAEA has been designated the convenor or Secretariat for the review sessions. The IAEA can serve other functions or services but only as requested and funded by the signatories.

Chapter 4 (Articles 29–35) provides the terms under which the Convention can be amended.

A Peer-Review Convention

In its final manifestation, the Convention is only a slight departure from the nuclear community's norm of technical assistance. The Convention is simply another mechanism through which technical assistance can be rendered. Assistance was, in effect, the only element around which consensus could form. This outcome will not surprise scholars of international negotiations, who have noted that efforts to create international agreements in other arenas have produced lowest-common-denominator results—evidence of the operation of what one observer has called "the law of the least ambitious" program (34). As soon as the principles of "peer review" and "incentive basing" were accepted as the conceptual underpinnings of the deliberations, the eventual outcome (unless the Convention were to be abandoned entirely) was settled. As indicated by the term "peer review," the Convention's Review Conference is seen not as a forum in which to pass critical judgment, threaten sanctions, or take enforcement action, but rather as another opportunity to suggest ways that low achievers can improve upon their safety performance. Involvement of as many states as possible (the incentive basis) required a convention with few obligations and a minimum of evaluative and regulatory intrusion. Something more ambitious is implied in Article 6 of the Convention, which calls for the closing of nuclear installations that cannot be upgraded to meet safety standards and obligations. But even Article 6 asserts only that such closings should take place "as soon as practically possible." It continues, "The timing of the shutdown may take into account the whole energy context and possible alternatives as well as the social, environmental and economic impact."

The peer-review Convention constitutes a political victory for such states as the United States, France, Japan, and numerous other states fearful of prescriptive obligations, supranational regulation, and costly incumbrance. These states have expressed concern that a more prescriptive or regulatory convention would lead to key defections, thereby reducing the integration of the entire system. Moreover, it is sometimes claimed that the creation of more explicit and/or prescriptive standards would only establish an acceptable minimum level of performance and, in effect, discourage the achievement of excellence (35).

France, Japan, and the United States also argued for keeping the review process closed to public scrutiny. The argument for closed doors is that this environment encourages frank and open discussion among peers. When sessions are exposed to scrutiny from outside, participants will either play to the potential headlines or, fearing public retribution, conceal what they know.

The peer-review Convention's primary benefit resides in the information transparency produced by requiring states to report on their nuclear power activities. (Such reports are intended, of course, for the industry itself and not the general public). This information exchange helps in the assessment of levels of industry performance. In addition, as Convention advocates point out, peer pressure may in due time evolve into more prescriptive oversight.

The Rejected Prescriptive Approach

Irrespective of the eventual utility of the Convention, it is clear that something more ambitious was envisioned by many convention proponents when deliberations began. While there was virtually no support for the establishment of supranational command-and-control regulation, with enforcement provisions, numerous states argued for the establishment of measurable, technical standards by which state adherence to general principles could be judged. Most international agreements for the protection of the environment contain such measurable performance standards by which to assess whether parties to the agreements are in compliance. The Austrian representative to the Conference argued, for example, that the standards ultimately to be inscribed in the Convention should be based upon the international standards (codes and guides) the IAEA had produced in its NUSS program (36). Other states argued that not only should signatories pledge to create quality assurance programs, but they should also commit to specific and detailed quality assurance standards and practices. The fear expressed earlier that these standards would justify minimalist efforts was discounted by these states. Nothing in the Convention would prevent states from exceeding the standards in the pursuit of excellence.

In fact, in early working-group discussions the issue was not whether technical standards would be part of the effort but rather in which broad type of international convention they would be embedded. Two familiar models prevail in international deliberations. The first involves the creation of a "framework" convention in which fundamental principles are enumerated. States sign on to the framework convention with the expectation that it will be followed by specific protocols that provide detailed technical standards or goals. Each state then has the opportunity to sign and ratify these protocols as a subsequent and separate act. Perhaps the best-known example of this approach is the 1985 Vienna Convention and its subsequent 1987 Montreal Protocol, which set forth the terms governing the phase-out of substances harmful to the stratospheric ozone layer. The second model is the "comprehensive" convention in which general obligations and specific annexes are simultaneously negotiated and set forth in a single treaty. The Law of the Sea Convention is the most noteworthy example of this form of convention (37).

While much discussion early in the International Nuclear Safety Convention (INSC) negotiations centered on the relative merits of the two convention variants described above, neither was eventually chosen. Instead, an agreement on only the broadest of safety principles evolved, without commitment to the development of subsequent annexes or protocols—a manifestation of the lowest-common-denominator approach that often arises in contentious international negotiations (38).

Regardless of the level of detail to which states might commit themselves, a Review Conference would be quite a different undertaking were a more prescriptive approach eventually to be used. In the first place, state reports submitted to the conference would be structured for comparative purposes. During the negotiations toward the present Convention some states favored requiring state reports to address a uniform list of issues using quantitative performance indicators. Germany advocated that the evaluations, to be carried out by independent third parties, treat individual reactors (39). Using such an approach, missions would be dispatched to validate the safety claims reported, and states not living up to their obligations could be required either to upgrade their operations or shut the facility(ies) down. Advocates of the prescriptive approach also stress openness, feeling that both the reports themselves and the proceedings should be open to public scrutiny. Otherwise, they claim, the conference would be perceived as another old-boy club, where shortcomings are overlooked for the sake of harmony.

The IAEA's role under the current Convention is solely organizational: It serves as Secretariat of the Convention and its Depositary. During the deliberations some states envisioned a far more active agency, one that would establish the criteria by which safety is assessed, carry out third-party verification tasks (through mandatory OSART inspections), and provide an independent evaluation of state reports. Ultimately, these roles were rejected because states were reluctant to elevate international bodies in the nuclear power enterprise.

An Alternative Convention

The international nuclear power community has done little to focus worldwide attention on its new convention—and appropriately so. While proceeding in an incremental fashion is not only the norm for international environmental agreements but also the wise course, the Convention's lack of several key features will likely make its success problematic. First, the Convention lacks measurable performance or design standards; compliance with its spirit and letter cannot be assessed. Second, it lacks the information transparency between signatories necessary to enable full and fair compliance evaluations. Most of all, it lacks a 30–50-year vision of what it will take to maintain a greatly expanded nuclear power system free of major accidents. In place of such a vision the international nuclear power community has claimed that technical assistance (bolstered through a peer review process and supported through multinational funding) can deal with existing safety concerns and allow nuclear power technology to be used with confidence. By inference from the choices made in Convention negotiations, the community is betting that sustained peer pressure will cause signatories gradually to upgrade their safety performance; the community evidently believes that the Chernobyl experience will soon recede in public memory or be seen as an aberrant case.

However, the effectiveness of technical assistance has inherent limitations. Whether nations and power plant operators avail themselves of the assistance is optional: Poor performers are under no obligation to seek or accept assistance, and any lessons learned through such assistance may lapse in the absence of continued outside scrutiny and pressure. In fact, it may be the average and better safety performers, rather than poor performers, that benefit most from the assistance. Moreover, and more fundamentally, the Convention's reliance on technical assistance is based on the premise that because ignorance of good safety practices accounts for poor performance, the transfer of relevant knowledge must produce better performance. While ignorance is a contributing factor to poor performance, it is not obvious that its reduction will concomitantly increase performance excellence. A host of sociocultural, economic, and institutional factors bear on power plant performance, and these factors are not altered through technical assistance (40).

In short, the Convention is a very limited step toward acceptance of collective responsibility. Even if it were to function optimally, it is extremely unlikely to renew enthusiasm for nuclear power based on confidence that the world's reactors can be managed safely. As noted above, the nuclear power industry bears the burden of demonstrating to a skeptical public that this power-generation option poses no undue safety concerns. The Convention's agreement to deal with industry "laggards" behind the closed doors of a Review Conference is wholly inadequate to this task. Unless the public can come to know, through the reports of reputable media or nongovernmental organizations, that the nuclear power industry is indeed putting its house in order, a renewed wave of public support for the technology is unlikely.

The West's initial focus upon the near-term issue of unsafe nuclear power operations in Eastern and Central Europe had both positive and negative effects. On the positive side, it catalyzed international discussions on the transnational nature of the technology. It brought national delegates to the table to discuss an issue that had been of considerable concern since the Chernobyl accident. Having come to the table, however, these delegates failed to treat a more important long-term issue: the building of international confidence in the safe management of nuclear power technology, without which industry growth is neither desirable nor feasible. Extraordinary technologies require extraordinary management systems shaped by a common perception of standards, obligations, and responsibilities. By being unwilling to begin building the bedrock of such management systems the international nuclear community lost a most critical opportunity.

Supporters of the Convention cast it as roughly comparable to the international agreements that cover safety in air or maritime transportation (41); but in fact, the differences between the conventions in the two domains are more striking than the similarities. The degree of international institutionalization and prescriptiveness in the air and maritime safety conventions far exceeds that found in the International Nuclear Safety Convention. The International Maritime Organization (IMO) can fight ship-based pollution through the formulation of maritime design and operation standards, which cover navigational equipment, tank cleaning standards, and construction techniques (42). In addition, the IMO has provided the auspices for the drawing up of the Convention for the Prevention of Pollution from Ships (MARPOL) and the London Convention on the Prevention of Marine Pollution by Dumping of Wastes and

Other Matter (London Dumping Convention). Both of these conventions specify general state obligations and append a series of prescriptive and technical annexes (43).

Not only does the International Nuclear Safety Convention break no new paths in international cooperation, but it also fails even to match current levels of internationalization found in other realms where technology and the environment intersect. No one is suggesting the establishment of an intrusive international nuclear power regulatory regime, but the unwillingness of key national governments to create a genuine international infrastructure that fully supports global nuclear safety is all too evident.

The opportunity to reformulate the international nuclear safety convention may arise a year from now, or it may be a decade away. Should the political will to deal with the issue return, however, it would be important to have ready the framework of a more suitable convention. On the assumption that developing countries with enormous electricity requirements may seek to utilize nuclear power technologies, the time horizon for this alternative framework should be 30–50 years. The convention should require such states (even economically poor states) to recognize that the selection of nuclear power encumbers them with international obligations for safe operations.

This alternative convention would have three purposes. First, it would seek proactively to raise global operating performance levels. Second, it would establish a series of political "firebreaks," such that an accident in one part of the world would not tarnish the reputation and viability of the entire nuclear power enterprise. Third, the convention would explicitly encourage nuclear power plant designs that minimize catastrophic offsite consequences.

The first purpose of the alternative framework resembles that of the INSC but would elaborate upon measurable performance standards and the open monitoring of safety performance. It might envision, for example, the creation of a nongovernmental certification organization that would conduct performance reviews at all operating reactors (44). Such an organization could rank the operating safety of reactors, highlighting both high and low performance. Such rankings would carry no legal or regulatory status but would be open to public scrutiny.

The second purpose of an alternative framework, creation of political "firebreaks," is equally important. The negotiators of the current Convention attempted (with apparent success) to make it acceptable to as many states as possible, leaving no states operating nuclear power plants outside the bounds of the Convention. The price of success, however, was exceedingly high: an eviscerated convention with only amorphous and ill-defined obligations. A strong case can be made for quite a different approach. In order genuinely to strengthen the global nuclear power enterprise, the community should create a vigorous and demanding safety regime, even if such a convention is not signed by "outlier" states. Justification for this approach is twofold: First, as noted above, only a relatively small subset of the international community now operates nuclear power plants. Future demand for electricity may lead other states to contemplate the nuclear option. These states must recognize that by choosing the nuclear option they commit themselves to rigorous and extraordinary vigilance. By making entry costs higher than they are today, the international community can avoid future problems with states too far into reactor construction or operation to make necessary and expensive safety changes.

Second, the rejection of such a convention by a given state would not necessarily be bad, revealing that state as unwilling to obligate itself to the creation of the necessary safety culture. In the short term, such unwillingness might result in widely differing global performance standards; but nonsignatories would clearly be labeled as pariah states, and accidents in these states would not tarnish nuclear power operations among signatory states. The oft-repeated statement, "a nuclear accident anywhere is a nuclear power accident everywhere," would be put to rest (45).

The establishment of two blocs of states with differing performance standards is not desirable in the long term. Such a situation existed unacknowledged during the Cold War. Under the alternative convention, pressure could be brought to bear upon pariah states to alter their opposition to international obligations. Access to credit and to necessary technology could be made contingent upon signatory status. The message to all leaders and publics would be clear: Acceptance of nuclear power requires a commitment to operating excellence.

This message may have to be reinforced even among signatory states. The envisioned nongovernmental certification agency would no doubt uncover substandard operations. Lagging performers would be identified and their operations made visible. A second political firebreak would thus be created between those signatory states that live up to their obligations and those that do not.

The third purpose of such an alternative convention would be the creation of long-term incentives to develop nuclear power technologies that prevent nuclear accidents with catastrophic off-site consequences. Like those of chemical plant accidents, the consequences of nuclear plant accidents should be made a local instead of a national or international issue.

In the last decade, two technical approaches to the elimination of catastrophic nuclear power accidents have been identified: supercontainments and PRIME reactors (46). The first option is to build a box around the reactor capable of preventing catastrophic release of radioactivity. Groups in Russia, Germany, and Japan are currently examining underground siting of supercontained plants, and Japan may actually build such a project on a large scale.

Japan has a second incentive for siting reactors underground: These facilities are less vulnerable than surface-sited reactors to seismic disturbance.

The second technical approach to the avoidance of globally catastrophic nuclear power accidents is to design reactors incapable of such accidents—so-called PRIME (Passive, Resilient, Inherent, Malevolent, Extended Safety) reactors (47). Several PRIME concepts have been integrated in, for example, the Process Inherent Ultimate Safety (PIUS) reactor built by ABB-Atom of Sweden and the Modular High Temperature Gas-Cooled Reactor (MHTGR) invented in Germany now being tested in Russia, Germany, Japan, and the United States. The largest such program is currently in Japan. Finally, a few research reactors have been designed that meet the PRIME criteria.

The development of such technologies requires will and appropriate incentives. A rating of nuclear power plant designs in terms of safety could provide such an incentive. Investors and politicians alike have a strong preference for lower risk futures. For most nonnuclear technologies there now exist major-accident-safety rating systems, and these have played a significant role in the development of safety technologies (48, 49). The unique political history of nuclear power has prevented development of similar rating systems for nuclear power plants.

Conclusion

The international nuclear power community has taken a go-slow approach to the pursuit of collective responsibility for nuclear safety. The recently negotiated International Convention on Nuclear Safety lacks elements that allow for measured evolution as the nuclear community expands. The result will be 1. a continuing distrust of the technology and its institutions in those countries already familiar with nuclear power, leading to a growing marginalization of nuclear power; and/or 2. a failure to transfer critical international safety norms to developing countries seeking to adopt nuclear power. The current hiatus in the large-scale ordering of nuclear power plants has provided industry and government leaders an opportunity to reflect upon the technology's past and its long-term future. Thus far it has been an opportunity foregone.

APPENDIX A

CONVENTION ON NUCLEAR SAFETY

PREAMBLE

THE CONTRACTING PARTIES

(i) Aware of the importance to the international community of ensuring that the use of nuclear energy is safe, well regulated and environmentally sound;

(ii) Reaffirming the necessity of continuing to promote a high level of nuclear safety worldwide;

(iii) Reaffirming that responsibility for nuclear safety rests with the State where a nuclear installation is located;
(iv) Desiring to promote an effective nuclear safety culture;
(v) Aware that accidents at nuclear installations have the potential for transboundary impacts;
(vi) Keeping in mind the Convention on the Physical Protection of Nuclear Material (1979), the Convention on Early Notification of a Nuclear Accident (1986), and the Convention on Assistance in the Case of a Nuclear Accident or Radiological Emergency (1986);
(vii) Affirming the importance of international cooperation for the enhancement of nuclear safety by the use of existing bilateral and multilateral mechanisms and the establishment of this incentive Convention;
(viii) Recognizing that this Convention entails a commitment to the application of fundamental safety principles rather than detailed safety standards and that there are internationally formulated safety guidelines which are updated from time to time and so can provide guidance on contemporary means of achieving a high level of safety;
(ix) Affirming the need to begin promptly the development of an international convention on the safety of radioactive waste management as soon as the ongoing process to develop waste management safety fundamentals has resulted in broad international agreement;
(x) Recognizing the usefulness of further technical work in connection with the safety of other parts of the nuclear fuel cycle, and that this work may, in time, facilitate the development of current or future international instruments;

CHAPTER 1. OBJECTIVES, DEFINITIONS AND SCOPE

ARTICLE 1. OBJECTIVES

The objectives of this Convention are:
(i) to achieve and maintain a high level of nuclear safety worldwide through the enhancement of national measures and international co-operation including, where appropriate, safety-related technical co-operation;
(ii) to establish and maintain effective defenses in nuclear installations against potential radiological hazards in order to protect individuals, society and the environment from harmful effects of ionizing radiation from such installations;
(iii) to prevent accidents with radiological consequences and to mitigate such consequences should they occur.

ARTICLE 2. DEFINITIONS

For the purpose of this Convention:
(i) "nuclear installation" means for each Contracting Party any land based civil nuclear power plant under its jurisdiction including such storage, handling and treatment facilities for radioactive materials as are on the same site and are directly related to the operation of the nuclear power plant. Such a plant ceases to be a nuclear installation when all nuclear fuel elements have been

removed permanently from the reactor core and have been stored safely in accordance with approved procedures, and a decommissioning programme has been agreed to by the regulatory body.

(ii) "regulatory body" means for each Contracting Party any body or bodies given the legal authority by that Contracting Party to grant licenses and to regulate the siting, design, construction, commissioning, operation or decommissioning of nuclear installations.

(iii) "license" means any authorization granted by the regulatory body to the applicant to have the overall responsibility for the siting, design, construction, commissioning or operation of a nuclear installation.

ARTICLE 3. SCOPE OF APPLICATION

This Convention shall apply to the safety of nuclear installations.

CHAPTER 2. OBLIGATIONS

(A) General Provisions

ARTICLE 4. IMPLEMENTING MEASURES

Each Contracting Party shall take, within the framework of its national law, the legislative, regulatory and administrative measures and other steps necessary to implement its obligations under this Convention.

ARTICLE 5. REPORTING

Each Contracting Party shall submit for review, prior to each meeting referred to in Article 20, a report on the measures it has taken to implement each of the obligations of this Convention.

ARTICLE 6. EXISTING NUCLEAR INSTALLATIONS

Each Contracting Party shall take the appropriate steps to ensure that the safety of nuclear installations existing at the time the Convention enters into force for that Contracting Party is reviewed as soon as possible. When necessary in the context of this Convention, the Contracting Party shall ensure that all reasonably practicable improvements are made as a matter of urgency to upgrade the safety of the nuclear installation. If such upgrading cannot be achieved, plans should be implemented to shut down the nuclear installation as soon as practically possible. The timing of the shut-down may take into account the whole energy context and possible alternatives as well as the social, environmental and economic impact.

(B) Legislation and Regulation

ARTICLE 7. LEGISLATIVE AND REGULATORY FRAMEWORK

1. Each Contracting Party shall establish and maintain a legislative and regulatory framework to govern the safety of nuclear installations.

2. The legislative and regulatory framework shall provide for:
   (i) the establishment of applicable national safety requirements and regulations;
   (ii) a system of licensing with regard to nuclear installations and the prohibition of the operation of a nuclear installation without a license;
   (iii) a system of regulatory inspection and assessment of nuclear installations to ascertain compliance with applicable regulations and the terms of any license;
   (iv) enforcement of applicable regulations and of the terms of any licenses, including suspension, modification or revocation.

ARTICLE 8. REGULATORY BODY

1. Each Contracting Party shall establish or designate a regulatory body entrusted with the implementation of the legislative and regulatory framework referred to in Article 7, and provided with adequate authority, competence and financial and human resources to fulfill its assigned responsibilities.
2. Each Contracting Party shall take the appropriate steps to ensure an effective separation between the functions of the regulatory body and those of any other body or organization concerned with the promotion or utilization of nuclear energy.

ARTICLE 9. RESPONSIBILITY OF THE LICENSE HOLDER

Each Contracting Party shall ensure that prime responsibility for the safety of a nuclear installation rests with the holder of the relevant license and shall take the appropriate steps to ensure that each such license holder meets its responsibility.

(c) General Safety Considerations

ARTICLE 10. PRIORITY TO SAFETY

Each Contracting Party shall take the appropriate steps to ensure that all organizations engaged in activities directly related to nuclear installations shall establish policies that give due priority to nuclear safety.

ARTICLE 11. FINANCIAL AND HUMAN RESOURCES

1. Each Contracting Party shall take the appropriate steps to ensure that adequate financial resources are available to support the safety of each nuclear installation throughout its life.
2. Each contracting Party shall take the appropriate steps to ensure that for all safety related activities in or for each nuclear installation throughout its life sufficient numbers of qualified staff with appropriate education, training and retraining are available.

ARTICLE 12. HUMAN FACTORS

Each Contracting Party shall take the appropriate steps to ensure that capabilities and limitations of human performance are taken into account throughout the life of a nuclear installation.

ARTICLE 13. QUALITY ASSURANCE

Each Contracting Party shall take the appropriate steps to ensure that quality assurance programmes are established and implemented with a view to providing confidence that specified requirements for all activities important to nuclear safety are satisfied throughout the life of a nuclear installation.

ARTICLE 14. ASSESSMENT AND VERIFICATION OF SAFETY

Each contracting Party shall take the appropriate steps to ensure that:

(i) comprehensive and systematic safety assessments are carried out before the construction and commissioning of a nuclear installation and throughout its life. Such assessments shall be well documented, subsequently updated in the light of operating experience and significant new safety information, and reviewed under the authority of the regulatory body;

(ii) verification by analysis, surveillance, testing and inspection is carried out to ensure that the physical state of a nuclear installation and the operation of the installation continue to be in accordance with its design, applicable national safety requirements and with operational limits and conditions.

ARTICLE 15. RADIATION PROTECTION

Each contracting Party shall take the appropriate steps to ensure that in all operational states the radiation exposure to the workers and the public caused by a nuclear installation shall be kept as low as reasonably achievable and that no individual shall be exposed to radiation doses which exceed prescribed national dose limits.

ARTICLE 16. EMERGENCY PREPAREDNESS

1. Each contracting Party shall take the appropriate steps to ensure that there are on-site and off-site emergency plans that are routinely tested for nuclear installations and cover the activities to be carried out in the event of an emergency.
   For any new nuclear installation, such plans shall be prepared and tested before it commences operation above a very low power level.
2. Each contracting Party shall take the appropriate steps to ensure that, insofar as they are likely to be affected by a radiological emergency, its own population and the competent authorities of the States in the vicinity of the nuclear installation are provided with appropriate information for emergency planning and response.
3. Contracting Parties which do not have a nuclear installation on their territory, insofar as they are likely to be affected in the event of a radiological emergency at a nuclear installation in the vicinity, shall take the appropriate steps for the preparation and testing of emergency plans for their territory that cover the activities to be carried out in the event of such an emergency.

(d) Safety of Installations

ARTICLE 17. SITING

Each contracting Party shall take the appropriate steps to ensure that appropriate procedures are established and implemented:

(i) for evaluating all relevant site-related factors which are likely to affect the safety of a nuclear installation for its projected lifetime;

(ii) for evaluating the likely safety impact of a proposed nuclear installation on individuals, society and the environment;

(iii) for re-evaluating as necessary all relevant factors referred to under sub-paragraphs (i) and (ii) so as to ensure the continued safety acceptability of the nuclear installation;

(iv) for consulting Contracting Parties in the vicinity of a proposed nuclear installation, insofar as they are likely to be affected by that installation and upon request providing the necessary information to such Contracting Parties, in order to enable them to evaluate and make their own assessment of the likely safety impact on their own territory of the nuclear installation.

ARTICLE 18. DESIGN AND CONSTRUCTION

Each contracting Party shall take the appropriate steps to ensure that:

(i) the design and construction of a nuclear installation provides for several reliable levels and methods of protection (defense in depth) against the release of radioactive materials, with a view to preventing the occurrence of accidents and to mitigating their radiological consequences should they occur;

(ii) the technologies incorporated in the design and construction of a nuclear installation are proven by experience or qualified by testing or analysis;

(iii) the design of a nuclear reactor allows for reliable, stable and easily manageable operation, with specific consideration of human factors and the man-machine interface.

ARTICLE 19. OPERATION

Each contracting Party shall take the appropriate steps to ensure that:

(i) the initial authorization to operate a nuclear installation is based upon an appropriate safety analysis and a commissioning programme demonstrating that the installation, as constructed, is consistent with design and safety requirements;

(ii) operational limits and conditions derived from the safety analysis, tests and operational experience are defined and revised as necessary to identify safe boundaries for operation;

(iii) operation, maintenance, inspection and testing of a nuclear installation are conducted in accordance with approved procedures;

(iv) procedures are established for responding to anticipated operational occurrences and to accidents;

(v) necessary engineering and technical support in all safety related fields is available throughout the lifetime of a nuclear installation;

(vi) incidents significant to safety are reported in a timely manner by the holder of the relevant license to the regulatory body;
(vii) programmes to collect and analyze operating experience are established, the results obtained and the conclusions drawn are acted upon and that existing mechanisms are used to share important experience with international bodies and with other operating organizations and regulatory bodies;
(viii) the generation of radioactive waste resulting from the operation of a nuclear installation is kept to the minimum practicable for the process concerned, both in activity and volume, and any necessary treatment and storage of spent fuel and waste directly related to the operation and on the same site as that of a nuclear installation take into consideration conditioning and disposal.

CHAPTER 3. MEETINGS OF THE CONTRACTING PARTIES

ARTICLE 20. REVIEW MEETING

1. The Contracting Parties shall hold meetings (hereinafter referred to as "review meetings") for the purpose of reviewing the reports submitted pursuant to Article 5 in accordance with the procedures adopted under Article 22.
2. Subject to the provisions of Article 24 sub-groups comprised of representatives of Contracting Parties may be established and may function during the review meetings as deemed necessary for the purpose of reviewing specific subjects contained in the reports.
3. Each Contracting Party shall have a reasonable opportunity to discuss the reports submitted by other Contracting Parties and to seek clarification of such reports.

ARTICLE 21. TIMETABLE

1. A preparatory meeting of the Contracting Parties shall be held not later than six months after the date of entry into force of this Convention.
2. At this preparatory meeting the Contracting Parties shall determine the date for the first review meeting. This review meeting shall be held as soon as possible but not later than thirty months after the date of entry into force of this Convention.
3. At each review meeting the Contracting Parties shall determine the date for the next such meeting. The interval between review meetings shall not exceed three years.

ARTICLE 22. PROCEDURAL ARRANGEMENTS

1. At the preparatory meeting held pursuant to Article 21 the Contracting Parties shall prepare and adopt by consensus Rules of Procedure and Financial Rules. The Contracting Parties shall establish in particular and in accordance with the Rules of Procedure:
   (i) guidelines regarding the form and structure of the reports to be submitted pursuant to Article 5;

(ii) a date for submission of such reports;
(iii) the process for reviewing such reports;

2. At review meetings the Contracting Parties may, if necessary, review the arrangements established pursuant to subparagraphs (i)–(iii) above, and adopt revisions by consensus unless otherwise provided for in the Rules of Procedure. They may also amend the Rules of Procedure and the Financial Rules, by consensus.

ARTICLE 23. EXTRAORDINARY MEETINGS

An extraordinary meeting of the Contracting Parties shall be held:

(i) if so agreed by a majority of the Contracting Parties present and voting at a meeting, abstentions being considered as voting; or
(ii) at the written request of a Contracting Party, within six months of this request having been communicated to the Contracting Parties and notification having been received by the secretariat referred to in Article 28, that the request has been supported by a majority of the Contracting Parties.

ARTICLE 24. ATTENDANCE

1. Each Contracting Party shall attend meetings of the Contracting Parties and be represented at such meetings by one delegate, and by such alternates, experts and advisers as it deems necessary.
2. The Contracting Parties may invite, by consensus, any intergovernmental organization which is competent in respect of matters governed by this Convention to attend, as an observer, any meeting, or specific sessions thereof. Observers shall be required to accept in writing, and in advance, the provisions of Article 27.

ARTICLE 25. SUMMARY REPORTS

The Contracting Parties shall adopt, by consensus, and make available to the public a document addressing issues discussed and conclusions reached during a meeting.

ARTICLE 26. LANGUAGES

1. The languages of meetings of the Contracting Parties shall be Arabic, Chinese, English, French, Russian and Spanish unless otherwise provided in the Rules of Procedure.
2. Reports submitted pursuant to Article 5 shall be prepared in the national language of the submitting Contracting Party or in a single designated language to be agreed in the Rules of Procedure. Should the report be submitted in a national language other than the designated language, a translation of the report into the designated language shall be provided by the Contracting Party.

ARTICLE 27. CONFIDENTIALITY

1. The provisions of this Convention shall not affect the rights and obligations of the Contracting Parties under their law to protect information from disclosure. For the purposes of this Article, "information" includes, inter alia, (i) personal data; (ii) information protected by intellectual

property rights or by industrial or commercial confidentiality; and (iii) information relating to national security, or to the physical protection of nuclear materials or nuclear installations.

2. When, in the context of this Convention, a contracting Party provides information identified by it as protected as described in paragraph 1, such information shall be used only for the purposes for which it has been provided and its confidentiality shall be respected.
3. The content of the debates during the reviewing of the reports by the Contracting Parties at each meeting shall be confidential.

ARTICLE 28. SECRETARIAT

1. The International Atomic Energy Agency, (hereinafter referred to as the "Agency") shall provide the secretariat for the meetings of the Contracting Parties.
2. The secretariat shall:
   (i) convene, prepare and service the meetings of the contracting Parties;
   (ii) transmit to the Contracting Parties information received or prepared in accordance with the provisions of this Convention.

   The costs incurred by the Agency in carrying out the functions referred to in sub-paragraphs (i) and (ii) above shall be borne by the Agency as part of its regular budget.
3. The Contracting Parties may, by consensus, request the Agency to provide other services in support of meetings of the Contracting Parties. The Agency may provide such services if they can be undertaken within its programme and regular budget. Should this not be possible, the Agency may provide such services if voluntary funding is provided from another source.

CHAPTER 4. FINAL CLAUSES AND OTHER PROVISIONS

ARTICLE 29. RESOLUTION OF DISAGREEMENTS

In the event of a disagreement between two or more Contracting Parties concerning the interpretation or application of this Convention, the Contracting Parties shall consult within the framework of a meeting of the Contracting Parties with a view to resolving the disagreement.

ARTICLE 30. SIGNATURE, RATIFICATION, ACCEPTANCE, APPROVAL, ACCESSION

1. This Convention shall be open for signature by all States at the Headquarters of the Agency in Vienna from 20 September 1994 until its entry into force.
2. This Convention is subject to ratification, acceptance or approval by the signatory States.
3. After its entry into force, this Convention will be open for accession by all States.

4.
   (i) This convention shall be open for signature or accession by regional organizations of an integration or other nature, provided that any such organization is constituted by sovereign States and has competence in respect of the negotiation, conclusion and application of international agreements in matters covered by this Convention.
   (ii) In matters within their competence, such organizations shall, on their own behalf, exercise the rights and fulfill the responsibilities which this Convention attributes to States Parties.
   (iii) When becoming party to this Convention such an organization shall communicate to the Depositary referred to in Article 34, a declaration indicating which States are members thereof and which articles of this Convention apply to it, and the extent of its competence in the field covered by those articles.
   (iv) Such an organization shall not hold any vote additional to those of its Member States.
5. Instruments of ratification, acceptance, approval or accession shall be deposited with the Depositary.

ARTICLE 31. ENTRY INTO FORCE

1. This Convention shall enter into force on the ninetieth day after the date of deposit with the Depositary of the twenty-second instrument of ratification, acceptance or approval, including the instruments of seventeen states, each having at least one nuclear installation which has achieved criticality in a reactor core.
2. For each State or regional organization of an integration or other nature which ratifies, accepts, approves or accedes to this Convention after the date of deposit of the last instrument required to satisfy the conditions in paragraph 1, this Convention shall enter into force on the ninetieth day after deposit with the Depositary of the appropriate instrument by such a State or organization.

ARTICLE 32. AMENDMENTS TO THE CONVENTION

1. Any Contracting Party may propose an amendment to this Convention. Proposed amendments shall be considered at a review meeting or extraordinary meeting.
2. The text of any proposed amendment and the reasons for it shall be provided to the Depositary who shall communicate the proposal to the Contracting Parties promptly and at least ninety days before the meeting for which it is submitted for consideration. Any comments received on such a proposal shall be circulated by the Depositary to the Contracting Parties.
3. The Contracting Parties shall decide after consideration of the proposed amendment whether to adopt it by consensus, or in the absence of such

consensus, to submit it to a Diplomatic Conference. A decision to submit a proposed amendment to a Diplomatic Conference shall require a two-thirds majority vote of the Parties present and voting at the meeting, provided that at least one half of the Contracting Parties are present at the time of voting. Abstentions shall be considered as voting.

4. The Diplomatic Conference to consider and adopt amendments to this Convention shall be convened by the Depositary and held no later than one year after the appropriate decision taken in accordance with paragraph 3 of this Article. The Diplomatic Conference shall make every effort to ensure amendments are adopted by consensus. Should this not be possible, amendments shall be adopted with a two-thirds majority of all Contracting Parties.
5. Amendments to this Convention adopted pursuant to paragraphs 3 and 4 above shall be subject to ratification, acceptance, approval, or confirmation by the Contracting Parties and shall enter into force for those Contracting Parties having ratified, accepted, or approved or confirmed them on the ninetieth day after the receipt by the Depositary of the relevant instruments by at least three fourths of the Contracting Parties. For a Contracting Party which subsequently ratifies, accepts, approves or confirms the said amendments, the amendments will enter into force on the ninetieth day after that Contracting Party has deposited its relevant instrument.

ARTICLE 33. DENUNCIATION

1. Any Contracting Party may denounce this Convention by written notification to the Depositary.
2. Denunciation shall take effect one year following the date of the receipt of the notification by the Depositary, or on such later date as may be specified in the notification.

ARTICLE 34. DEPOSITARY

1. The Director General of the Agency shall be the Depositary of this Convention.
2. The Depositary shall inform the Contracting Parties of:
   (i) the signature of this Convention and of the deposit of instruments of ratification, acceptance, approval or accession, in accordance with Article 30;
   (ii) the date on which the Convention enters into force, in accordance with Article 31;
   (iii) the notifications of denunciation of the Convention and the date thereof, made in accordance with Article 33;
   (iv) the proposed amendments to this Convention submitted by the Contracting parties, the amendments adopted by the relevant Diplomatic Conference or by the meeting of the Contracting Parties, and the date of entry into force of the said amendments, in accordance with Article 32.

ARTICLE 35. AUTHENTIC TEXTS

The original of this Convention of which the Arabic, Chinese, English, French, Russian and Spanish texts are equally authentic, shall be deposited with the Depositary who shall send certified copies thereof to the Contracting parties.

Literature Cited

1. Hohenemser C, Goble RL, Slovic P. 1992. Nuclear power. In *The Energy-Environment Connection*, ed. JM Hollander, p. 135. Washington DC/Covelo, CA: Island. 414 pp.
2. World Energy Council. 1993. *Energy for Tomorrow's World*, pp. 50–57. New York: St. Martin's. 320 pp.
3. Schurr S. 1990. *Electricity in the American Economy: Agent of Technological Progess*. New York: Greenwood. 443 pp.
4. Mikesell RF, William L. 1992. *International Banks and the Environment*. San Francisco: Sierra Club Books. 302 pp.
5. Forsberg CW, Weinberg A. 1990. Advanced reactors, passive safety, and acceptance of nuclear energy. *Annu. Rev. Energy* 15:205–29
6. Douglas J. 1994. Reopening the nuclear option. *EPRI J.* 19(6):6–17
7. Oka Y, Koshizuka S, Jevremovic T, Okano Y, et al. 1995. System design of a Direct Cycle Supercritical-Water-Cooled Fast Reactor. *Nucl. Tech. 109(1):1–10*
8. Hinman GW, Lowinger TC. 1987. A comparative study of Japan and United States nuclear enterprise: industry structure and construction experience. *Energy Syst. Policy* 11:205–29
9. Lester RK, McCabe MJ. 1993. The effect of industrial structure on learning by doing in nuclear power plant operation. *Rand J. Econ.* 24(3):418–38
10. US General Accounting Office. 1994. *Nuclear Waste: Foreign Countries' Approaches to High-Level Waste Storage and Disposal.* GAO/RCED-94-172. Washington DC: USGPO. 59 pp.
11. Fischer D. 1994. *Towards 1995: The Prospects for Ending the Proliferation of Nuclear Weapons.* Aldershot, UK: Dartmouth. 257 pp.
12. Taylor, JJ. 1987. Chernobyl and its legacy. *EPRI J.* 12(4):5–21
13. Frankena F. 1992. *Implications of the Chernobyl Nuclear Accident for Policy and Planning*. Chicago: Council of Planning Libraries. 26 pp.
14. Walker RBJ, Mendlovitz SH, eds. 1990. *Contending Sovereignties: Redefining Political Community*. Boulder/London: Lynne Rienner. 189 pp.
15. Elkins D. 1995. *Beyond Sovereignty: Territory and Political Economy in the Twenty-First Century*. Toronto: Univ. Toronto Press. 308 pp.
16. International Atomic Energy Agency. 1994. *Against the Spread of Nuclear Weapons: IAEA Safeguards in the 1990s*. Vienna: Int. Atomic Energy Agency. 65 pp.
17. Barkenbus J. 1987. Nuclear power safety and the role of international organization. *Int. Org.* 41(3):475–90
18. Rosen M. 1993. Energy for development. *IAEA Bull.* 35(4):34–44
19. US Office of Technology Assessment. 1994. *Fueling Reform: Energy Technologies in the Former East Bloc*. OTA-ETI-599, p. 32. Washington DC: USGPO. 205 pp.
20. US General Accounting Office. 1994. *Nuclear Safety: International Assistance Efforts to Make Soviet-Designed Reactors Safe*. GAO/RCED-94-234, p. 2. Washington DC: USGPO. 43 pp.
21. International Atomic Energy Agency. 1993. *International Assistance to Upgrade the Safety of Soviet-Designed Nuclear Power Plants*. Vienna: Int. Atomic Energy Agency. 71 pp.
22. US Department of Energy. 1988. *Department of Energy's Team Analysis of Soviet Designed VVERs*. DOE/NE-0086. Washington DC: USGPO. 55 pp.
23. International Nuclear Safety Advisory Group. 1991. Safety culture. In *Safety*

Series No. 75-INSAG-4. Vienna: Int. Atomic Energy Agency

24. Ostrom L, Wilhelmsensen C, Kaplan B. 1993. Assessing safety culture. *Nucl. Safety* 34(2):163–72
25. US Office of Technology Assessment. 1994. See Ref. 19
26. Rosen M. 1977. The critical issue of nuclear power plant safety in developing countries. *IAEA Bull.* 19:12–16
27. Thompson G. 1992. *Strengthening the International Atomic Energy Agency.* Work. Pap. 6. Cambridge, MA: Inst. Resour. Secur. Stud.
28. Yaremy E, Hide K. 1992. A more vigorous approach to IAEA safety services. *IAEA Bull.* 34(2):15–23
29. Anderson, SJ. 1991. International commercial nuclear reactor safety. Hearing before the Senate Subcommittee on Nuclear Regulation (Senate Hearing 102-276), July, 25:85–91
30. deLa Fayette L. 1993. International environmental law and the problem of nuclear safety. *J. Envir. Law 5(1):31–69*
31. deLa Fayette L. 1993. See Ref. 30
32. US General Accounting Office. 1993. *Nuclear Safety: Progress Toward International Agreement to Improve Reactor Safety.* GAO/RCED-93-153. Washington DC: USGPO. 11 pp.
33. International Atomic Energy Agency. 1994. Convention on nuclear safety. *Inf. Circ. (INFCIRC/449)* July 5. 13 pp.
34. Sand PH. 1991. International cooperation: the environmental experience. In *Preserving the Global Environment,* ed. J. Mathews, pp. 236–78. New York: W.W. Norton
35. Group of Experts on a Nuclear Safety Convention. 1993. US Governmental Comments on an International Nuclear Safety Convention. Submitted to the Int. Atomic Energy Agency. April 5. 9 pp.
36. Group of Experts on a Nuclear Safety Convention. Submitted to the Int. Atomic Energy Agency, April 5. 77 pp.
37. Sebenius J. 1994. Towards a winning climate coalition. *Negotiating Climate Change,* ed. I Mintzer, JA Leonard, pp. 277–320. Cambridge: Cambridge Univ. Press
38. Susskind L. 1994. *Environmental Diplomacy: Negotiating More Effective Global Agreements.* Oxford: Oxford Univ. Press. 180 pp.
39. Group of Experts on a Nuclear Safety Convention. 1993. Draft Elements for a Nuclear Safety Convention. Submitted to the Int. Atomic Energy Agency, February 23. 35 pp.
40. Moray N, Huey B. 1988. *Human Factors Research and Nuclear Safety.* Washington DC: Natl. Res. Counc., Natl. Acad. Sci.
41. Jankowitsch O, Flakus FN. 1994. International Convention on Nuclear Safety: a legal milestone. *IAEA Bull.* 36(3):36–40
42. Carroll, J, ed. 1988. *International Environmental Diplomacy: The Management and Resolution of Transfrontier Environmental Problems.* Cambridge: Cambridge Univ. Press. 274 pp.
43. deLa Fayette, L. 1993. See Ref. 30, p. 44
44. Hinchberger B. 1993. Non-governmental organization: the third force in the Third World. In *Green Globe Yearbook 1993,* ed. Fridtjof Nansen Inst., Norway, pp. 45–54. Oxford: Oxford Univ. Press
45. Rees J. 1994. *Hostages of Each Other: The Transformation of Safety Since Three Mile Island.* Chicago: Chicago Univ. Press. 238 pp.
46. Hafele W. 1990. Energy from nuclear power. *Sci. Am.* 263(3):136–44
47. Forsberg CW, 1991. *Worldwide Advanced Nuclear Power Reactors with Passive and Inherent Safety.* ORNL/TM-11907. Oak Ridge, TN: Oak Ridge Natl. Lab.
48. Mitchell R. 1993. International oil pollution of the oceans. In *Institutions for the Earth: Sources of Effective International Environmental Protection,* ed. PM Hass, RO Keohane, MA Levy, 5: 183–248. Cambridge, MA: The MIT Press. 448 pp.
49. Chuse R, Eber SM. 1984. *Pressure Vessels: The ASME Code Simplified.* NY: McGraw-Hill. 6th ed.

Annu. Rev. Energy Environ. 1995. 20:213–32

ASIA-WIDE EMISSIONS OF GREENHOUSE GASES

Toufiq A. Siddiqi
Program on Environment, East-West Center, Honolulu, Hawaii 96848

KEY WORDS: carbon dioxide, methane, energy use, biomass, rice fields

---

CONTENTS

Abstract

Emissions of principal greenhouse gases (GHGs) from Asia are increasing faster than those from any other continent. This is a result of rapid economic growth, as well as the fact that almost half of the world's population lives in Asian countries. In this paper, we provide estimates of emissions of the two principal greenhouse gases, carbon dioxide (CO_2) and methane (CH_4), from individual countries and areas. Recent literature has been reviewed for emission estimates for individual sources, such as carbon dioxide from cement manufacture, and methane from rice fields. There are very large uncertainties in many of these estimates, so several estimates are provided, where available.

The largest anthropogenic source of CO_2 emissions is the use of fossil fuels. Energy consumption data from 1992 have been used to calculate estimated

1056-3466/95/1022-0213$05.00

emissions of CO_2 from this source. In view of the ongoing negotiations to limit future greenhouse gas emissions, estimates of projected CO_2 emissions from the developing countries of Asia are also provided. These are likely to be 3 times their 1986 levels by 2010, under "business as usual" scenarios. Even with the implementation of energy efficiency measures and fuel switching where feasible, the emissions of CO_2 are likely to double within the same time period.

INTRODUCTION

The rapid rate of industrialization and population growth in much of Asia has been accompanied by large increases in emissions of greenhouse gases from the region. During the past few decades, Asia's share of global emissions has increased rapidly. Carbon dioxide emissions caused by the use of fossil fuels, for example, now exceed those from North America and from the European Union, as shown in Figure 1. (Asia, as defined in this paper, includes the central Asian republics of the former Soviet Union, the Middle East, and Turkey). On a per capita basis, emissions from the developing countries of Asia are still only a fraction of those from the more industrialized countries (Figure 2). Therefore, increasing emissions from the Asian countries are to be expected for several decades.

The Framework Convention on Climate Change (FCCC) has been signed by more than 160 countries, and took effect in early 1994. The signatories are required to inventory greenhouse gases (GHGs) and identify policy options for stabilizing or reducing future emissions. The signatories to the FCCC met

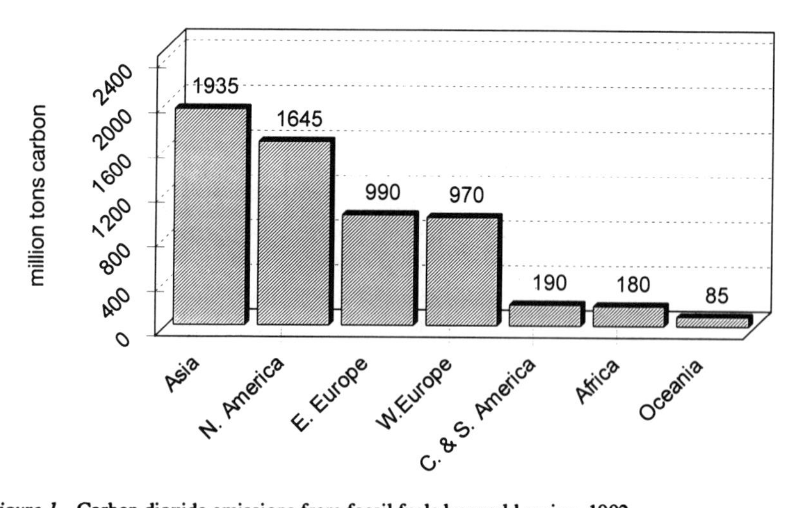

Figure 1 Carbon dioxide emissions from fossil fuels by world region, 1992.

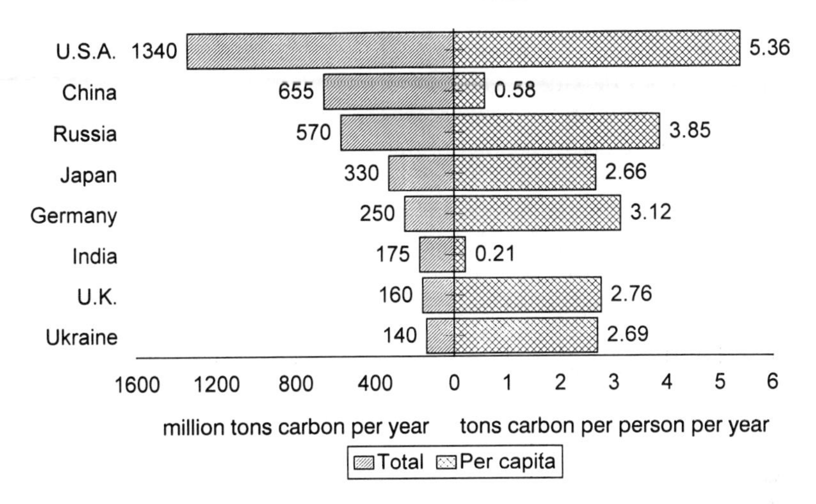

Figure 2 **Total and per capita emissions of carbon dioxide from fossil fuel use for countries with the largest total emissions, 1992.**

in Berlin in March and April, 1995, but were unable to agree on emission limits or a timetable to achieve them. This situation is different from the one that resulted in the Montreal Protocol for reducing emissions of substances that reduce the ozone layer, where it was possible to agree on a complete phase-out of such substances by the industrialized countries by 1997, and by the developing countries a decade later.

A number of factors make the setting of emission limits for GHGs difficult. These factors include:

1. The current recessions in many industrialized countries and the concern that limits on greenhouse gas emissions would exacerbate economic problems;
2. The concern by developing countries that emission limits would lower their economic growth rates;
3. The disparity in per capita emissions between industrialized and developing countries, and the fact that most anthropogenic greenhouse gas emissions result from activities undertaken by industrialized countries;
4. The view in some industrialized countries that most emissions increases will come from developing countries and that stabilizing or reducing their own emissions is pointless as long as developing countries give no indication of doing so.

The economic and political aspects of limiting greenhouse gas emissions have been discussed in a number of excellent reviews (1–5), and we shall not pursue that subject further here.

This paper presents current estimates of the emissions of greenhouse gases, principally carbon dioxide (CO_2) and methane (CH_4), from the countries of Asia, the continent with the most rapid increase in such emissions. For many of the countries, detailed estimates are under way, and the results will be available within a year or two. For other countries, particularly Japan, China, India, and the Republic of Korea, several estimates have already been made. These are summarized in the relevant sections below.

Each of the greenhouse gases has many anthropogenic sources of emissions (6–8), and the uncertainty in emissions estimates varies greatly with each source. Whereas emissions of CO_2 from the combustion of fossil fuels can usually be accurately estimated to within 5%, uncertainties exceeding 50% are not uncommon for other sources. Uncertainties in estimates of methane emissions from rice fields or coal mines, for example, often exceed 50%. A great deal of additional research is needed to reduce these uncertainties. Some research has begun, but a major effort needs to be made to reduce the uncertainties at a country-specific level. This would assist the countries greatly in assessing the potential for reducing emissions from different sectors, and designing reduction strategies that are cost-effective.

CARBON DIOXIDE EMISSIONS

Carbon dioxide is the largest contributor to greenhouse warming, and the combustion of fossil fuels is the largest anthropogenic source of CO_2 (9). The Carbon Dioxide Information Analysis Center (CDIAC) publishes annual data indicating trends in atmospheric concentrations, as well as emissions from industrial use from various regions of the world, as well as from the larger emitting countries (10). We begin our discussion with CO_2 emissions from fossil fuel use in Asia.

Fossil Fuel Combustion

Annual estimates of CO_2 emissions from fossil fuel combustion and cement manufacturing, since 1950, for several Asian subregions, as well as for many individual countries, are provided in CDIAC reports (11) and are now available on computer disk. Estimates of emissions in more recent years are provided in the biennial reports published by the World Resources Institute (12). More detailed information on several countries is provided in a number of conference proceedings (13–15).

A study commissioned by the Economic and Social Commission for Asia and the Pacific (ESCAP, 16, 17) was the first to provide not only estimated emissions for a base year (1986), but also projected future emissions under alternate scenarios for each of the developing countries in Asia and the Pacific (excluding the Middle East and the central Asian republics that were a part of

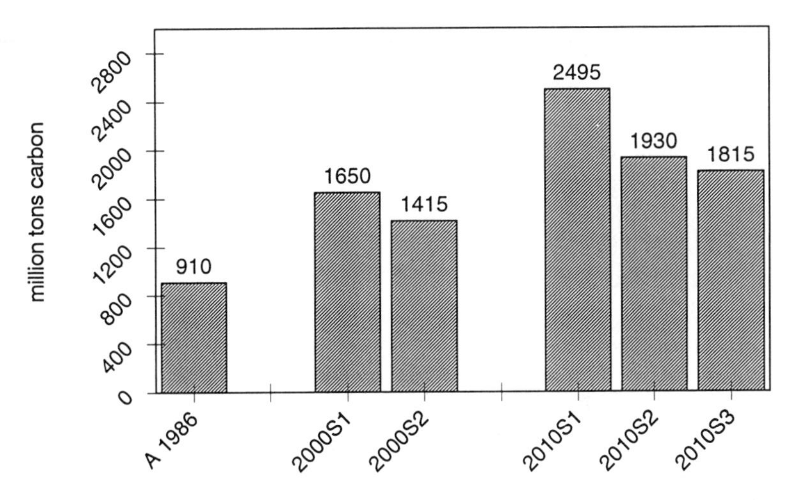

Figure 3 **Recent and projected emissions of carbon dioxide from Asian developing countries.**

alternate scenarios for each of the developing countries in Asia and the Pacific (excluding the Middle East and the central Asian republics that were a part of the USSR at that time). The study used the following scenarios to estimate future emissions:

1. Business As Usual, meaning no special efforts to reduce future emissions;
2. Conservation, meaning efforts to reduce emissions by improving the efficiency of energy use and thereby reduce the actual amount of energy used;
3. Conservation + Fuel Switching, meaning that, in addition to the measures taken in scenario 2, the use of coal and oil would be reduced wherever feasible and replaced by natural gas, hydro- or nuclear power, and renewable energy sources.

The estimated CO_2 emissions from the use of fossil fuels for 1986 for developing Asian countries that were members of ESCAP are shown in Figure 3. The projections for 2000 and 2010 are also shown in Figure 3 for the different scenarios. Because of the long lead time required for fuel switching, scenario 3 was not included for the year 2000. The totals in Figure 3 include emissions from the Pacific islands, but not from Japan, the Middle East, and the central Asian republics. The contribution of the Pacific islands to global emissions of CO_2 is small, amounting to 2.1 million metric tons carbon (MtC) in 1986, 3.0–3.5 MtC in 2000, and 3.4–4.9 MtC in 2010.

The ESCAP study showed that CO_2 emissions from the developing countries of Asia could almost double between 1986 and 2000 and reach three times the

1986 emissions by 2010 under the Business As Usual scenario. Even with the implementation of measures to improve the efficiency of energy use and fuel switching, the 2010 emissions would be almost double those of 1986. Most of the absolute increase would result from anticipated continued growth in coal use in China and India, which derive most of their energy from this source, and where alternate sources of energy could not replace the large amounts required in the 20-year time horizon.

The present author has also estimated the emissions of CO_2 from fossil fuel use in each of the Asian countries during 1992 (Table 1). The energy consumption data are from the United Nations (18), except for Taiwan (19). The data contain some ambiguities regarding the extent to which the coal, oil, and natural gas are used for purposes other than combustion, such as the production of lubricants or tar. Therefore, the estimated emissions shown in Table 1 may be somewhat high, but the overestimation is not expected to exceed 5% in most cases.

The factors for converting energy supplied by coal, oil, and natural gas into carbon emitted into the atmosphere vary somewhat from country to country (11, 20). Since separate emission factors for each of the Asian countries are not available, the emission factors for China (21), the country with the highest emissions, have been used for all the countries in Asia, except for Japan, the second largest emitter, for which a somewhat larger emission factor for coal (26.2 gC Gj^{-1}) has been used, based on the OECD Pacific estimate by the International Energy Agency (IEA) (20).

For all countries except India, 98% of the fuel's carbon content is assumed to be emitted into the atmosphere as carbon dioxide. The Central Fuel Research Institute of India has estimated (22) that an average of 90% of carbon in the coals is converted into CO_2 [this low value may be a result of the very large ash content (30–40%) in Indian coals], and this estimate was used to calculate India's CO_2 emissions from coal in Table 1.

Some of the other estimates (23–29) for CO_2 emissions for countries with substantial emissions are shown in Table 2. A number of additional estimates are expected to become available during the next year.

Cement

In Asia as a whole, cement manufacturing contributes about 4% of the total CO_2 emissions from industrial processes (12), the remainder coming from the use of fossil fuels and gas flaring. The total CO_2 emissions due to cement manufacturing from the Asian countries, excluding the central Asian republics of the former USSR, amounted to 77 million tons (Mt) in 1991 out of a world total of 162 Mt. China is by far the largest emitter, with Japan, India, and the Republic of Korea also contributing substantially. The emissions of CO_2 from cement manufacturing in the major emitting countries in Asia are given in Table 3.

Biomass Combustion

The uncertainties in estimating emissions of CO_2 from biomass combustion are much larger than those for fossil fuels or cement manufacturing. Uncertainties about the amount of biomass burned, as well as the conditions under which it is burned, are relatively large. Biomass usually includes fuelwood, agricultural wastes, and animal wastes. The actual amounts of these items that undergo combustion in each country are not known accurately. The amount burned in stoves, the efficiency of combustion of the stoves, and the amounts burned in the open are also subject to large uncertainties.

The United Nations has compiled estimated amounts of fuelwood and bagasse (from sugarcane) used for energy purposes in the Asian countries (18). These estimates have been converted into energy supplied, in terms of million metric tons of coal equivalent (Mtce), using the conversion factors given in Reference 18. The energy supplied by these sources is given in Table 4 for the subregions and for the countries where the total amount of energy supplied by these biomass sources exceeds 100 thousand metric tons (100 ktce or 0.1 Mtce). Figures given separately for charcoal are included in the fuelwood column, with a conversion value of 0.986 tce/ton of charcoal.

The total amount of biomass in all forms used for energy, as well as that used as fuelwood, has been estimated for most countries in a report published by the Food and Agriculture Organization of the United Nations (FAO) (28). These estimates are based on FAO's own database, as well as data from the UN Statistical Office and detailed studies by the Biomass Users Network. For developing countries with no published data, FAO assumed that the annual per capita consumption of biomass is 1 ton of wood equivalent (twe) for rural inhabitants and 0.5 twe for urban inhabitants.

Estimates of the amount of energy derived from all forms of biomass in several developing countries during 1987 have been compiled by Hall (29). These estimates were converted into tons of carbon emitted as CO_2 using the conversion factor of 0.636 metric tons of carbon per ton of coal equivalent (20), with the usual provisos regarding large uncertainties in the estimates. The results for the Asian countries are given in Table 5.

Estimates of biofuels consumption in several Asian countries have also been provided by Myers and Leach (30). Recent studies on China (21) and India (22) estimate the total amount of energy supplied by biomass at 266 and 176 million tons coal equivalent respectively during 1990.

Many countries have not made estimates of the emissions of CO_2 from biomass combustion because they maintain that these figures are about equal

Table 1 Carbon dioxide emissions from fossil fuel consumption in Asia (1992)[a]

		Fossil fuel consumption (Mtoe)			Carbon dioxide emissions from fossil fuels[b] (million tons carbon)			
Subregion	Country/area	Coal and lignite	Oil	Nat. gas and LNG	Coal and lignite	Oil	Nat. Gas and LNG	Total
East Asia	China	544.5	109.8	14.7	594.0	96.2	9.0	699.2
	Hong Kong	5.5	4.2		6.0	3.7	0.0	9.7
	Japan	81.4	209.9	52.5	94.4	183.9	32.1	310.4
	Korea, DPR	60.7	4.6		66.2	4.0	0.0	70.3
	Korea, Rep. of	25.3	78.5	5.7	27.5	68.8	3.5	99.8
	Macao		0.4		0.0	0.3	0.0	0.3
	Mongolia	1.9	0.6		2.1	0.5	0.0	2.6
	Taiwan[c]	16.5	28.6	2.0	18.0	25.1	1.2	44.3
					Total for Subregion			1236.6
Southeast Asia	Brunei		0.8	2.4	0.0	0.7	1.5	2.2
	Cambodia	0.0	0.2		0.0	0.1	0.0	0.1
	Indonesia	3.9	27.2	19.2	4.2	23.8	11.8	39.8
	Lao PDR	0.0	0.1		0.0	0.1	0.0	0.1
	Malaysia	1.6	14.6	7.0	1.8	12.8	4.3	18.9
	Myanmar	0.0	0.6	0.9	0.0	0.5	0.6	1.1
Philippines		1.3	11.8		1.5	10.4	0.0	11.8
	Singapore	0.0	16.5		0.0	14.4	0.0	14.4
	Thailand	4.1	23.2	7.2	4.5	20.3	4.4	29.2
	Vietnam	2.4	2.9	0.0	2.6	2.6	0.0	5.2
					Total for Subregion			122.9
South Asia	Afghanistan	0.0	0.3	0.2	0.0	0.3	0.1	0.4
	Bangladesh	0.2	2.1	4.7	0.2	1.8	2.9	4.9
	Bhutan	0.0	0.0		0.0	0.0	0.0	0.0
	India	143.9	52.8	10.9	144.2	46.2	6.7	197.1
	Iran, Isl. Rep.	1.4	46.3	23.3	1.6	40.6	14.3	56.4
	Maldives		0.0		0.0	0.0	0.0	0.0
	Nepal	0.1	0.3		0.1	0.3	0.0	0.3

	Sri Lanka	0.0	1.6		0.0	1.4	0.0	1.4
					Total for Subregion			279.2
Central Asia	Armenia	0.1		1.7	0.1	0.0	1.1	1.2
	Azerbaijan	0.0	0.4	10.8	0.0	0.4	6.6	7.0
	Kazakhstan	52.6	6.0	17.2	57.4	5.2	10.5	73.1
	Kyrgyzstan	1.7	1.5	1.8	1.9	1.3	1.1	4.3
	Tajikistan	0.1	0.0	1.7	0.1	0.0	1.0	1.1
	Turkmenistan	0.2	0.5	10.4	0.2	0.5	6.4	7.1
	Uzbekistan	4.8	5.1	38.5	5.3	4.5	23.5	33.3
					Total for Subregion			127.0
Middle East and West Asia	Bahrein	0.0	0.7	4.8	0.0	0.6	3.0	3.6
	Cyprus	0.0	1.4		0.0	1.3	0.0	1.3
	Iraq	0.0	14.0	2.8	0.0	12.2	1.7	14.0
	Israel	3.5	8.3	0.0	3.8	7.3	0.0	11.0
	Jordan	0.0	3.4		0.0	3.0	0.0	3.0
	Kuwait[d]	0.0	3.1	2.4	0.0	2.7	1.5	4.2
	Lebanon	0.0	3.5	0.0	0.0	3.1	0.0	3.1
	Oman	0.0	1.6	1.7	0.0	1.4	1.4	2.4
	Qatar	0.0	1.8	10.7	0.0	1.6	6.6	8.2
	Saudi Arabia[d]	0.0	35.9	32.1	0.0	31.4	19.6	51.1
	Syrian Arab R.	0.0	9.6	1.8	0.0	8.4	1.1	9.5
	Turkey	15.4	20.7	4.2	16.8	18.1	2.6	37.5
	United Arab E.	0.0	6.7	23.8	0.0	5.9	14.5	20.4
	Yemen	0.0	2.9	0.0	0.0	2.5	0.0	2.5
					Total for Subregion			171.7
					Total for Asia			1937.4

[a] Sources: The energy data are for 1992, and taken from the *United Nations' Energy Statistics 1992* (18) unless indicated otherwise.
Mtoe = million metric tons oil equivalent.

[b] The carbon dioxide emissions were calculated by the present author using the following conversion factors: Coal: g-C Gj − 1, 24.8; tons-C toe − 1, 1.091. Petroleum products: kg-C Gj − 1, 19.9; tons-C toe − 1, 0.876. Natural Gas: kg-C Gj − 1, 13.9; tons-C toe − 1, 0.612. Please see the text for a discussion of these.

[c] From Reference 19.

[d] Includes part of Neutral Zone.

Table 2 Carbon dioxide emissions estimates from fossil fuel consumption in selected Asian countries

Country	Emissions (Mt-C)	Estimate for Year	Reference
China	660.5	1991	23
China	614.5	1990	21
India	182.2	1991	23
India	153.0	1989–1990	22
India	167.5	1989–1990	24
Indonesia	32.9	1990	22
Iran, Isl. Rep. of	52.9	1991	23
Japan	285.7	1991	23
Japan	322.7	1992	25
Korea, DPR	64.2	1991	23
Korea, Rep. of	67.7	1991	23
Korea, Rep. of	63.0	1990	26
Pakistan	17.6	1990–1991	27
Philippines	10.1	1990	22
Saudi Arabia	53.0	1991	23
Thailand	23.2	1991	27
Vietnam	5.3	1990	22

Table 3 Carbon dioxide emissions estimates from cement production in selected Asian countries

Country	Emissions (Mt-C)	Estimate for Year	Reference
China	33.7	1991	23
China	25.5	1990	21
India	6.8	1991	23
India	6.3	1989–1990	22
Indonesia	1.7	1990	22
Iran, Isl. Rep. of	2.0	1991	23
Iraq	0.7	1991	12
Israel	0.4	1991	12
Japan	12.1	1991	23
Korea, DPR	2.2	1991	23
Korea, Rep. of	4.6	1991	23
Malaysia	1.0	1991	12
Pakistan	1.0	1991	12
Philippines	0.6	1991	12
Saudi Arabia	1.6	1991	23
Thailand	2.5	1991	12
Turkey	3.5	1991	12
Vietnam	0.9	1990	22

Table 4 Carbon dioxide emissions from selected[a] biomass consumption in some Asian countries, 1992

		Fuelwood		Bagasse		Subtotal	CO_2
Subregion	Country/Area	million m3	million tce[b]	million tons	million tce[b]	biomass (Mtce)	emissions[c] (Mt-C)
East Asia	China	203.77	67.24	23.11	6.10	73.35	46.65
	Japan	0.17	0.15	0.44	0.12	0.27	0.17
	Korea, Dem. P.R.	4.18	1.38			1.38	0.88
	Korea, Rep. of	2.40	0.79			0.79	0.50
	Mongolia	1.35	0.45			0.45	0.28
Southeast Asia	Cambodia	5.71	1.88			1.88	1.20
	Indonesia	146.28	48.27	7.66	2.02	50.29	31.99
	Lao Peop. Dem. Rep.	4.13	1.36			1.36	0.87
	Malaysia	9.16	3.02	0.34	0.09	3.11	1.98
	Myanmar	18.63	6.15	0.36	0.10	6.24	3.97
	Philippines	35.04	11.56	6.03	1.59	13.16	8.37
	Thailand	60.84	20.08	10.43	2.75	22.83	14.52
	Vietnam	25.16	8.30	1.28	0.34	8.64	5.50
South Asia	Afghanistan	5.67	1.87			1.87	1.19
	Bangladesh	5.90	1.95	3.35	0.88	2.83	1.80
	India	257.79	85.07	45.22	11.94	97.01	61.70
	Iran, Islamic Rep. of	2.51	0.83	0.72	0.19	1.02	0.65
	Nepal	17.11	5.65	0.07	0.02	5.66	3.60
	Pakistan	24.38	8.05	8.15	2.15	10.20	6.49
	Sri Lanka	9.10	3.00	0.22	0.06	3.06	1.95
Central Asia	Kazakhstan	0.55	0.18			0.18	0.12
Middle East	Lebanon	0.48	0.16			0.16	0.10
	Turkey	9.75	3.22			3.22	2.05
	Yemen	0.32	0.11			0.11	0.07
				Total for Asia		309.26	169.69

[a] Data on agricultural and animal wastes are not included here.
[b] Conversion factors (18): Fuel wood: 0.33 tce/m3; Bagasse: 0.264 tce/t; tce = tons of coal equivalent.
Sources: The biomass energy data are from the United Nations (18).
[c] The carbon dioxide emissions have been calculated by the present author using the conversion factor: 0.636 t-C/tce (20).

Table 5 Energy derived from all biomass in selected Asian countries, 1987

Country	Biomass energy[a] (Mtoe)	CO_2 emissions[b] (Mt-C)
Bangladesh	36.3	33.9
China	221.1	206.2
India	203.4	189.7
Indonesia	63.2	59.0
Malaysia	15.8	14.7
Nepal	10.4	9.7
Pakistan	29.7	27.7
Philippines	20.4	19.0
Sri Lanka	4.3	4.0
Thailand	4.9	4.6

[a] The energy data are from Hall (29)
[b] The carbon dioxide emissions have been calculated by the present author using the conversion factor: 0.636 t-C/tce (20).

balance for each country is to be calculated. India is one of the few countries where detailed estimates have been made (24), and such emissions amounted to about 148 MtC during 1989–90.

Land Use Changes

Changes in land use such as the switch from forests to agriculture have been a major source of carbon dioxide emissions for centuries (31). The pace of such change has accelerated in Asia during the past several decades, owing to population increases, as well as demand for fuelwood, timber, and residential land. The rate at which land conversions have been taking place in individual countries and regions is the source of much controversy. Thus the available data should be considered very tentative. Table 6 shows estimated CO_2 emissions (12) caused by land use changes in 1991 in several Asian countries. Estimates of such emissions in a few Asian countries are also provided in Reference 22.

METHANE EMISSIONS

Global anthropogenic methane emissions rank second only to carbon dioxide emissions in their contribution to greenhouse warming. The Asian countries (excluding the central Asian republics of the former USSR) are estimated to have emitted about 120 million tons (Mt) of methane from human activities, out of a global total of about 250 Mt emitted in 1991 (12). More than half of

Table 6 Estimated emissions of carbon dioxide due to land use changes in selected Asian countries, 1991

Country	Emissions (Mt-C)
Bangladesh	1.9
Bhutan	1.1
Cambodia	9.3
India	5.7
Indonesia	89.9
Lao PDR	9.8
Malaysia	30.0
Myanmar	32.7
Nepal	2.1
Pakistan	2.6
Philippines	30.0
Sri Lanka	1.0
Thailand	24.8
Vietnam	9.0

Source: World Resources Institute (12)

this figure is attributed to rice production, with other major contributions coming from livestock, coal mining, solid wastes, and oil and gas production. The estimates compiled by the World Resources Institute (WRI) (12) for the Asian countries are shown in Table 7. As mentioned earlier, estimates of methane emissions are subject to very large uncertainties (6, 32), particularly for emissions from rice production, coal mines, and natural gas distribution. These are discussed in the relevant sections that follow.

Rice Fields

The flux of methane (defined as milligrams of methane per square meter per hour, mg/m^{-2} h^{-1}) varies enormously, depending on several factors such as agricultural practices, soil conditions, time of day and year, and rice varieties. Thus the estimates of methane emissions from entire countries, and from the world as a whole, are subject to very large uncertainties. The Intergovernmental Panel on Climate Change (IPCC) (6) estimates global methane emissions from rice paddies range from 20–150 Mt (recently revised to a range of 20–100 Mt). Their report also gives revised estimates of methane fluxes in several Asian countries, which range (in units of mg/m^{-2} h^{-1}) from 0.1–27.5 for India, 7.8–60.0 for China, 0.4–16.2 for Japan, and 3.7–19.6 for Thailand.

An earlier review by Khalil and Rasmussen (32) of 11 global methane budgets found a range of 18–280 teragrams (Tg). Estimates by Bachelet &

Table 7 Estimated emissions of methane from anthropogenic sources in selected Asian countries, 1991

Country	Methane emissions					
	Rice production	Livestock	Solid waste	Coal mining	Oil & Gas production	Total
	(teragrams or million metric tons methane per year)					
Afghanistan	0.1	0.2	0.0	0.0	0.0	0.3
Bangladesh	5.0	1.0	0.2		0.1	6.2
China	19.0	5.4	0.9	14.0	0.3	39.5
India	19.0	12.0	2.3	1.8	0.2	35.3
Indonesia	5.1	0.7	0.5	0.1	0.7	7.0
Iran, Islamic Rep.	0.3	0.6	0.3	0.0	0.4	1.6
Iraq	0.1	0.1	0.3		0.0	0.5
Japan	1.4	0.3	1.9	0.1	0.0	3.7
Korea, DPR	0.3	0.1	0.1	0.8		1.3
Korea, Rep. of	0.6	0.1	0.3	0.3		1.2
Lao PDR	0.2	0.1	0.0			0.3
Malaysia	0.3	0.0	0.1		0.3	0.6
Mongolia		0.3	0.0	0.0		0.3
Myanmar	3.0	0.4	0.1	0.0	0.0	3.5
Nepal	0.6	0.4	0.0		0.2	3.2
Pakistan	1.1	1.9	0.0	0.0	0.2	3.2
Philippines	1.5	0.2	0.2			1.9
Saudi Arabia		0.1	0.2		0.6	0.9
Sri Lanka	0.5	0.1	0.0			0.6
Syrian Arab Rep.		0.1	0.1			0.2
Thailand	5.3	0.4	0.1	0.0	0.1	5.9
Turkey	0.0	0.7	0.3	0.0	0.0	1.1
United Arab Emr.		0.0	0.1		0.4	0.5
Vietnam	3.4	0.3		0.1	0.0	3.7

Source: World Resources Institute (12).

[a] Many of these estimates are subjected to very large uncertainties. For example, for the two largest emitting countries, China and India, other estimates (21, 22) give total emissions of 28 and 16 Tg per year respectively.

Neue (33) of methane emissions from rice fields in Asia using several techniques suggest total emissions of about 60 Tg.

Perhaps the best known example of the wide differences in estimates in the Asian context is for emissions from rice fields in India. The US Environmental Protection Agency (USEPA) (34) estimates that India's emissions of methane amounted to 16 Tg (=Mt) during 1990, whereas the Indian Agricultural Research Institute has estimated (22, 24) these emissions at 4.0 Tg and Sinha (35) at 0.83–1.22 Tg. In the case of China, scientists at the Chinese Academy of Agricultural Sciences have estimated (36) total CH_4 emissions from rice fields in China at 11.3 Tg, whereas the USEPA estimate is 21 Tg. All of these studies suggest that extrapolations of fluxes from Europe and North America to specific Asian countries may not be useful, and that detailed measurements in different parts of larger countries would be required to improve the accuracy of the estimates.

Enteric Fermentation and Animal Wastes

Livestock are a major source of methane, owing directly to enteric fermentation, as well as manure emissions. Global emissions from the former source are estimated (34) to be in the range of 65–100 Tg, and 10–18 Tg from the latter. According to the WRI estimate (12), about one third of all methane emissions from livestock is from Asian countries.

The USEPA (34) has estimated China's emissions of CH_4 from enteric fermentation at 6 Tg, and from livestock manure at 2 Tg. The former number is fairly close to the 5.5 Tg estimated in the study for the Asian Development Bank and the State Science and Technology Commission of China (ADB/SSTC) (21), which includes estimates of animal populations in individual provinces, as well as emission factors for the different animals.

For India, the Council for Scientific and Industrial Research (CSIR) estimate (24) for CH_4 emissions from enteric fermentation from all animals is 6.9 Tg, about 30% less than the 10 Tg estimated by USEPA (34). The differences in the estimates for China and India are considerably smaller than for emissions from rice fields.

Coal Mines

Coal mines are a substantial source of methane emissions in the major coal-producing countries of the world. China and India are the largest coal producers in Asia, the former producing about 1,100 Mt per year. Estimates of CH_4 emissions from China's coal mines vary by a factor of about 3. The review by IEA (37) discusses the assumptions underlying estimates of 5.3, 7.6, 15.3, and 16.1 Tg per year made during the 1990–1992 period. The ADB/SSTC study estimated (21) the emissions at 5.3 Tg, based on the methane content of seams in the major coal fields. This is lower than the range of 9.5–16.6 Tg estimated by USEPA (34). Their estimate of India's emissions of 0.4 Tg agrees with the ADB study (22).

Oil and Gas

The natural–gas and oil production and distribution system is believed to be the largest source of anthropogenic methane in Russia (34), but its contribution to emissions in the Asian countries is estimated at only about 3.2 Tg (12). Estimates of contributions from the major emitting countries are provided in Table 7. The largest source of uncertainty in these estimates, owing to lack of

Table 8 Estimated emissions of methane from anthropogenic biomass combustion in selected Asian countries[a]

Country/Region	Emissions (Tg/year)
Bangladesh	0.8
China	7.3
India	2.6
Indonesia	2.9
Mayanmar	1.0
Pakistan	0.7
Thailand	0.9
Vietnam	0.8
Other Asia	2.9

[a] Source: USEPA (34)

data, is leakage from natural–gas pipeline systems, which can reach 5% of the throughput in some of the older systems.

Biomass Burning

Worldwide anthropogenic emissions of methane from biomass combustion are estimated in the range of 20–80 Tg per year. Estimates for many of the countries in Asia are not available, but those provided by USEPA (34) for several countries for 1990 are summarized in Table 8.

Solid Wastes

Urban wastes are the largest source of methane emissions in many industrialized countries, including the US (34) and Japan (38). They are of growing importance in the larger and more developed countries of Asia, including China and India. Emissions in China are estimated (21) in the range of 0.6–2.0 Tg per year, in India at 2.4 Tg per year (22), and in Thailand at 0.3 Tg per year (34). Japan's emissions are estimated in the range of 0.4–0.8 Tg per year (38).

In many Asian countries, agricultural wastes are likely to contribute more to methane emissions than urban wastes. In China, for example, emissions from agricultural wastes are estimated at 2–4 Tg per year (21), about 2–3 times the emissions from urban wastes.

OTHER GREENHOUSE GASES

The other major (6) long-lived greenhouse gases are nitrous oxide (N_2O), chlorofluorocarbons (CFCs), and carbon tetrachloride (CCl_4). Anthropogenic sources are estimated to contribute about 1–6 Tg of nitrogen per year in the form of N_2O. Combustion of fossil fuels and biomass is the largest source of N_2O. Estimated emissions from individual countries are not yet available in most cases, but initial estimates have been made for a few, including China, India and Japan (38). The estimated emissions for China (21) lie in the range of 0.2–0.6 Tg- Nitrogen (N) per year. Approximately one third to one half this amount is estimated to be from fossil fuel combustion. The total for India is estimated (22) at approximately 0.04 Tg - N per year. Most of these emissions are from fossil fuel combustion, with fertilizer use also contributing approximately 20% of the total. Again, these estimates are subject to very large uncertainties.

Almost all the Asian countries have signed the Montreal Protocol for the elimination of ozone-depleting substances, and are committed to the phaseout of CFCs during the next 12 years. In the interim period, emissions from Asia have become a significant part of the global total because the industrialized countries are phasing out the use of these substances by 1997 and their current emissions are declining rapidly.

While the total amounts of the various CFCs used in each country are fairly well known, the amount that actually escapes into the atmosphere in any country is less well determined. CFCs are used for various purposes, principally refrigeration, foams, and aerosols. For some uses, such as aerosols, the release into the atmosphere is instantaneous. For other uses, the release may be delayed several years. Taking these variations into account, emissions of CFC-11, CFC-12, and HCFC-22 from India are estimated at 1308, 8, and 1020 metric tons respectively during 1989–1990 (24).

Estimates of CFC emissions by countries that have not reported them have been calculated (39) using a linear relationship between Gross Domestic Product and total consumption of CFCs. The uncertainties based on this method are only one half as large as those using the relationship between CFC consumption and electric power use. Estimates of CFC use for all Asian countries are provided for the year 1986.

CONCLUSION

Emissions of greenhouse gases from the Asian countries are already a large fraction of anthropogenic global emissions, and their share is expected to increase in the years ahead. Thus any international agreement to slow down the rate of increase, eventually stabilize, and even reduce greenhouse gas emissions would require the commitment and cooperation of the developing countries of Asia, as well as the industrialized nations on all continents, if it is to be effective.

For the formulation of longer-term strategies for addressing global climate change concerns, needed inputs include estimates of current emissions of greenhouse gases in each of the major emitting countries, technological and policy options for reducing future emissions, and the economic, social, and environmental costs of implementing such options. Several regional and national studies to undertake these estimates have been initiated during the past few years by international and regional organizations such as the Global Environment Facility (40), the United Nations Development Programme (41), the World Bank (42), and the Asian Development Bank (21, 22). Some of the work is also supported by individual countries such as the US (43), various European countries, Australia, Canada, and Japan.

The results from some of these studies became available during 1994 (21, 22), and others will be available soon. Many of the Asian countries have also initiated their own research and analysis programs to address global climate change issues. All these efforts will assist in defining the full scope of the problem and the options available for addressing it. The negotiations that preceded and followed the United Nations Conference on Environment and Development (UNCED) conference (44) in Rio de Janeiro and the First Conference of the Parties to FCCC made clear that financial resources will be a major constraint in the implementation of strategies for reducing emissions of greenhouse gases. An equitable arrangement between the industrialized and developing countries of the world is essential for mobilizing resources.

In some countries, infrastructure development, including the establishment of organizations with the needed scientific and technical capabilities, is also required for the successful implementation of policies to reduce future emissions of greenhouse gases. The needed institutions and the pool of technically qualified persons are growing at a fast rate in most Asian countries. Consequently, even though these factors are important at present, they are not expected to be long-term constraints on the Asian countries' ability to cooperate actively with the rest of the international community in addressing the challenges and opportunities presented by global climate change.

ACKNOWLEDGMENTS

The author would like to thank Amrita Achanta, Lin Erda, Dina Kruger, Najib Murtaza, RK Pachauri, Kirk Smith, Mingxing Wang, and Zongxin Wu for making available some of the data cited in the text. The views and interpretations, of course, are solely those of the present author.

Literature Cited

1. Org. Econ. Coop. Dev. 1992. *Convention on Climate Change: Economic Aspects of Negotiations.* Paris: OECD. 97 pp.
2. Intergov. Panel Climate Change. 1994. *Climate Change: Policy Instruments and Their Implications. Proc. Tsukuba Workshop IPCC Work. Group. III.* Tsukuba, Jpn: Cent. Glob. Environ. Res., Natl. Inst. Environ. Stud., Environ. Agency Jpn. 407 pp.
3. Achanta AN, ed. 1993. *The Climate Change Agenda: An Indian Perspective.* New Delhi: Tata Energy Res. Inst. 305 pp.
4. Hayes P, Smith KR, eds. 1993. *The Global Greenhouse Regime: Who Pays?* London: Earthscan/Tokyo: UNU. 382 pp.
5. Nakicenovic N, Nordhaus WD, Richels R, Toth FL, eds. 1994. *Integrative Assessment of Mitigation, Impacts, and Adaptation to Climate Change. Proc. Workshop held at IIASA.* Laxenburg: Int. Inst. Appl. Syst. Anal. 669 pp.
6. Houghton JT, Callander BA, Varney SK, eds. 1992. Climate Change 1992. *Suppl. Rep. IPCC Sci. Assess.,* Cambridge UK: Cambridge Univ. Press, 200 pp.
7. Lashof DA, Tirpak DA, eds. 1990. Policy Options for Stabilizing Global Climate. *Rep. Congr. Off. Policy, Plan. Eval.,* US Environ. Prot. Agency, Washington, DC
8. German Bundestag. 1991. *Protecting the Earth: A Status Report with Recommendations for a New Energy Policy,* Vols. 1 , 2, Bonn: German Bundestag. 672 pp. 1008 pp.
9. Intergov. Panel Climate Change. 1991. *Climate Change: The IPCC Response Strategies.* Washington, DC: Island. 272 pp.
10. Boden TA, Kaiser DP, Sepanski RJ, Stoss FW, eds. 1994. *Trends '93: A Compendium of Data on Global Change.* Oak Ridge, TN: Carbon Dioxide Info. Anal. Cent., Oak Ridge Natl. Lab.
11. Marland G, Boden TA, Griffin RC, Huang SF, Kanciruk P, Nelson TR. 1989. *Estimates of CO_2 Emissions from Fossil Fuel Burning and Cement Manufacturing, Based on the United Nations Energy Statistics and the US Bureau of Mines Cement Manufacturing Data.* Oak Ridge, TN: Carbon Dioxide Info. Anal. Cent., Environ. Sci. Div., Oak Ridge Natl. Lab. 712 pp.
12. World Resour. Inst. 1994. *World Resources 1994–95: A Guide to the Global Environment.* New York: Oxford Univ. Press. 400 pp.
13. Streets DG, Siddiqi TA, eds. 1990. *Responding to the Threat of Global Warming. Proc. Conf. sponsored by Argonne Natl. Lab. and Environ. Policy Inst., East-West Cent., Honolulu.* Argonne, IL: Environ. Assess. Info. Sci. Div., Argonne Natl. Lab.
14. Environ. Agency Jpn. 1991. *The Asian-Pacific Seminar on Climate Change. Proc. Conf., Nagoya, Jan.* Tokyo: Environ. Agency
15. Jones BP, Wheeler EF, eds. 1992. *Greenhouse Research Initiatives in the ESCAP Region: Energy., Proc. Conf., Bangkok, Aug. 1991.* Canberra: Austr. Bur. Agric. Res. Econ. 325 pp.
16. Econ. Soc. Comm. Asia Pacific. 1991. *Energy Policy Implications of the Climatic Effects of Fossil Fuel Use in the Asia Pacific Region.* Bangkok: ESCAP, UN
17. Siddiqi TA, Foell WK, Hills P, Keesman AT, Nagao T, Torok SJ. 1991. Climate change and energy scenarios in Asia-Pacific developing countries. *Energy* 16:1467–88
18. UN. 1994. *1992 Energy Statistics Yearbook.* New York: UN. 492 pp.
19. Br. Petroleum. 1994. *BP Statistical Review of World Energy.* London: Br. Petroleum. 36 pp.
20. Int. Energy Agency. 1991. *Greenhouse Gas Emissions: The Energy Dimension.* Paris: Org. Econ. Coop. Dev. 199 pp.
21. Asian Dev. Bank. 1994. *National Response Strategy for Global Climate Change: People's Republic of China,* ed. TA Siddiqi, DG Streets, Wu Zongxin, He Jiankun. Manila: Off. Environ., Asian Dev. Bank. 390 pp.
22. Asian Dev. Bank. 1994. *Climate Change in Asia, Executive Summary and Volume on India.* Manila: Off. Environ., Asian Dev. Bank.
23. Marland G, Andres RJ, Boden TA. 1994. Global, regional, and national CO_2 emissions. In *Trends '93: A Compendium of Data on Global Change,* ed. TA Boden, DP Kaiser, RJ Sepanski, FW Stoss, pp. 505–84. Oak Ridge, TN: Carbon Dioxide Info. Anal. Cent., Oak Ridge Natl. Lab.
24. Mehra M, Damodaran M. 1993. Anthropogenic emissions of greenhouse gases in India. In *Climate Change Agenda: An Indian Perspective,* ed. A Achanta,

pp. 10–32. New Delhi: Tata Energy Res. Inst.
25. Kudo H. 1994. Status quo and subjects of commitments to global warming. *Energy Jpn.* 130:46–58
26. Lee H. 1991. Energy outlook and environmental implications for Korea. *Energy* 16:1489–93
27. Asian Dev. Bank. 1993. *Climate Change in Asia: Vol. 1. South Asia,* ed. A Qureshi, D Hobbie. Manila: Off. Environ., Asian Development Bank
28. Woods J, Hall DO. 1994. *Bioenergy for Development.* Rome: Food Agric. Org. UN
29. Hall DO. 1992. *Biomass.* Background Pap. World Dev. Rep. 1992. Off. Vice President, Dev. Econ., Washington, DC: World Bank
30. Meyers S, Leach G. 1989. *Biomass Fuels in the Developing Countries: An Overview.* Berkeley, CA: Appl. Sci. Div., Lawrence Berkeley Lab., Univ. Calif.
31. Kammen DM, Smith KR, Rambo AT, Khalil MAK, eds. 1994. Preindustrial human environmental impacts: Are there lessons for global change science and policy? *Chemosphere* 29: No. 5 (Special issue)
32. Khalil MAK, Rasmussen RA. 1990. Constraints on the global sources of methane and an analysis of recent budgets. *Tellus* 42:229–36
33. Bachelet D, Neue HU. 1993. Methane emissions from wetland rice areas of Asia. *Chemosphere* 26:219–37
34. Adler MJ, ed. 1994. International Anthropogenic Methane Emissions: Estimates for 1990. *Rep. Congr. Off. Policy, Plan. Eval.,* US Environ. Prot. Agency, Washington, DC
35. Sinha SK. 1995. Global methane emissions from rice paddies: Excellent methodology but poor extrapolation. *Curr. Sci.* In press
36. Lin E, Dong H, Li Y. 1994. Methane emissions of China: Agricultural sources and mitigation options. In *Non-CO_2 Greenhouse Gases,* ed. J van Ham et al. pp. 405–410
37. Smith IM, Sloss LL. 1992. *Methane Emissions from Coal.* London: IEA Coal Res. 28 pp.
38. Environ. Agency Jpn. 1994. *Global Environmental Research of Japan in 1993.* Tokyo: Res. Info. Off., Glob. Environ. Dep., Environ. Agency
39. McCulloch A, Midgley PM, Fisher DA. 1994. Distribution of emissions of chlorofluorocarbons (CFCs) 11, 12, 113, 114 and 115 among reporting and non-reporting countries in 1986. *Atmos. Environ.* 28:2567–82
40. Glob. Environ. Facil. 1994. *Q. Oper. Rep.,* GEF, Washington, DC
41. UN Dev. Prog. and Asian Dev. Bank. 1994. *Asian Least-Cost Greenhouse Gas Emission Reduction Plans.* Reg. Bur. Asia Pac., UNDP, New York/Off. Environ., ADB, Manila
42. World Bank. 1994. *Monthly Report on World Bank-Implemented Global Environment Facility Operations.* Glob. Environ. Coord. Div., Environ. Dep., World Bank, Washington, DC
43. US Country Stud. Prog. 1993. *Support for Climate Change Studies.* US Country. Stud. Prog., Washington, DC
44. UN. 1992. *Adoption of Agreements on Environment and Development. UN Conf. Environ. Dev., Rio de Janeiro.* New York: UN

Annu. Rev. Energy Environ. 1995. 20:233–64

INTERNATIONAL ENVIRONMENTAL LABELING

Albert Mödl
AT&T Global Information Solutions Deutschland GmbH, Ulmer Straße 160,
D-86135 Augsburg, Federal Republic of Germany

Ferdinand Hermann
AT&T Corporate Environmental, Health and Safety - Europe, Ulmer Straße 160,
D-86135, Augsburg, Federal Republic of Germany

KEY WORDS: ecolabel, environmental label, labeling initiatives, national environmental-labeling programs, international environmental-labeling programs

CONTENTS

ABSTRACT

Environmentally acceptable products are continuing to gain importance on the market. This review covers programs that are targeted to mark products or services that are less environmentally harmful than others that serve the same purpose. Several environmental-labeling programs were created in the late 1980s and early 1990s. To illustrate the principles of such labeling systems, various programs are presented in detail. The programs were chosen to represent a wide spectrum of characteristics: stringent and less demanding, well-established and recently created, national and multinational, and from various

1056-3466/95/1022-0233$05.00

parts of the world. Similarities and differences, as well as advantages and disadvantages, are discussed.

INTRODUCTION

For many decades industrial goods have been produced to satisfy consumer demands irrespective of the products' real or potential impact on the environment. At the end of a product's lifetime, the "throw-away" society simply replaces the product with a new one. With unlimited resources and ever-increasing demand, this practice would ultimately lead to unlimited waste. The so-called oil crisis in the early 1970s increased awareness of the limits on energy resources and of the need to reduce pollution and protect the environment. Since that time, saving energy and resources has become both an industrial and a private concern. Saving energy both during manufacture, which reduces costs, and during product use is a marketing advantage. Therefore, industry became more interested in these environmental issues.

Environmental consciousness has increased in many areas of our society. The need for recycling initiatives and waste minimization is strong, as shown by the following examples: (*a*) Within the European Union about 50 million tons of used packaging material are produced each year (1); (*b*) In 1992, electronic equipment waste in Germany (excluding the federal states of the former Democratic Republic of Germany) reached 1.2 million tons (2), an average of approximately 200 kg per inhabitant per year.

The need to treat people, the environment, industrial processes and the interactions among them as a whole is becoming more and more evident. Future products, services, and processes must be designed to be sustainable and to protect the environment, both as they are made and throughout their lifetime. The complex of interactions among industry, the environment, and people, based on the principle that an industrial system cannot be viewed in isolation from its surrounding systems, but rather in accordance with them, is termed industrial ecology (3, 4).

One concept of industrial ecology is called design for the environment (DFE), introduced in 1990 (5). The idea of DFE is to take environmental issues into consideration in the concept and design stages of a product, long before the product impacts the environment (6–8). Everybody involved in product development, manufacture, and marketing should aim to reduce environmental impacts throughout a product's lifetime (9). In the electronics industry, the DFE methodology is currently being implemented (10, 11).

In order to enable DFE to function in practice, a tool is needed to determine, control and reduce the impact of processes and products on the environment. An analytical tool used for this purpose is called life-cycle assessment (LCA) (12–14). LCA implies that the environmental friendliness of a product should be judged not only by the impacts of its use, but also by the impacts of its raw materials, production, manufacturing, distribution, recycling, and final disposal (13). Thus, LCA assesses environmental impact from "cradle to grave."

The first step in LCA is the inventory stage in which the different life phases of a product are determined and the environment-related inputs and outputs are identified. The second step is life-cycle environmental impact analysis in which a product's impact on the environment is investigated and assessed. Assessing environmental impact is difficult because it requires ranking product characteristics. Life-cycle improvement analysis is the last step in LCA.

Since most companies are not involved in all phases of the life of a product or its constituents, LCA is complicated, lengthy, and requires collaboration. Product or process complexity often makes a complete LCA impossible. However, even an incomplete LCA can yield considerable improvements in a product's environmental soundness while simultaneously improving cost and per- formance issues (15–17).

Often, consumers do not have the information they need to determine whether a product is environmentally friendly. Increasingly environmentally-conscious consumers now want clear criteria for making product purchase decisions, and they should be able to select a product on the basis of its environmental impact. This is the reason green indicators, or environmental labels, were introduced. Environmental labels enable consumers to make informed product comparisons and choices by marking products or services that are less harmful to the environment than similar products or services without labels (14).

Environmental labels should be based on a serious and comprehensive LCA and should be bestowed by an independent organization. Many countries have created or are developing environmental labels based on life-cycle assessment or on selected life-cycle concerns. In order to keep the labeling process practicable and manageable, the various environmental-labeling schemes have set different priorities. This report reviews national, multinational, and supranational environmental-labeling initiatives and programs. As of this writing (late 1994), most countries have presented their programs in brochures. Comprehensive papers about environmental labeling have been published by the United States Environmental Protection Agency (18, 14) and others (19, 20).

COMMON FEATURES OF ENVIRONMENTAL LABELS

Definition

In principle, any labels describing or identifying environment-related characteristics of products or services can be considered environmental labels. "Environmental label" is a generic term encompassing expressions such as "green label" or "ecolabel." Environmental labels can be divided into two categories:

1. Seals of approval, which are awarded by a neutral and independent organization after certain critical requirements have been met, identify products or services as less harmful to the environment than similar ones without the seal (see Introduction);
2. Information labels, which are provided either directly by the manufacturer or by a commissioned party, list individual characteristics without giving an overall approval of the product as environmentally benign or preferable. These labels include information disclosure labels, such as report cards and hazard/warning labels, which are required by law.

A more subtle classification is given by the US Environmental Protection Agency (EPA) (18). In the following, the term "environmental label" shall designate only seals of approval as defined above, based on specific award criteria. Therefore, neither environmental claims certification programs, such as the one created in 1990 by Scientific Certification Systems (SCS) USA (21), or the Environmental Choice Australia program (effective 1991– May 1994), nor programs targeted exclusively at one environmental aspect [such as EPA's Energy Star Computer program (22)] are included in this review.

Goal of an Environmental Label

The primary goal of environmental labeling is to mark products or services as environmentally preferable to similar products or services without labels so consumers can make decisions based on the product's or service's environmental impact. In addition, an environmental label should assure the consumer that the product bearing it has met strict and extensive environmental requirements set by an independent organization. Because consumers are now environmentally conscious, products without an environmental label will have a disadvantage on the market. Thus, the market advantage created by environmental labeling provides a clear incentive for developing environmentally sound products or processes. Consequently, environmental labeling implicitly aims at stimulating market forces and improving the environmental performance of products or services.

Ecological Aspects and Criteria

The distinction between ecological aspects and ecological criteria is important. Ecological aspects (or environmental fields) referring to a product are the areas in which a product influences the environment. The ecological criteria are the requirements a product must fulfill and the means to judge the ecological impact of the product within the corresponding environmental areas.

Certain ecological criteria must be met before a product is eligible for an environmental label. In many cases, not all aspects apply to a given product. Therefore, products are assigned to certain product groups which have a fixed set of environmental aspects. In general, the ecological aspects of environmental labels can be categorized as follows: consumption of resources; use of hazardous substances; emissions into or pollution of air, water, and soil; energy efficiency; generation of noise; waste relevance; economic efficiency.

Environmental friendliness is assessed by investigating the fulfillment of the criteria within the environmental fields in all stages of the product life cycle. The comprehensive assessment system can in principle be represented as a matrix listing the aspects as a function of the stages of the life cycle (see also Figure 3).

Assigning weighting factors to the various ecological aspects is a convenient way to compare the environmental friendliness of individual products (23). This rating provides a quantitative scoring system and an overall green index or environmental impact assessment index. An environmental label should not be awarded on the basis of this index, however, because this mean value may be misleading. Single characteristics that are particularly harmful to the environment could be masked by several other features that are environmentally benign. Therefore, each individual requirement must be met separately before an environmental label is awarded.

THE ENVIRONMENTAL-LABELING PROGRAMS

An overview of the various environmental-labeling programs is given in Table 1 in chronological order, along with the number of the figure in which the labels are depicted and the corresponding references.

The first environmental-labeling program was established by Germany in 1977. Other countries did not begin similar programs until the late 1980s. In the following section, only several selected programs from the complete list in Table 1 will be described individually. In the Discussion section, similarities and differences among the programs will be evaluated and discussed.

Supranational and Multinational Initiatives and Programs

ISO/TC 207 ENVIRONMENTAL MANAGEMENT The International Organization for Standardization (ISO), a Geneva-based worldwide federation of national standards bodies is developing environmental-management and product-design standards in a supranational initiative. In 1991, ISO established the Strategic Advisory Group on Environment (SAGE) which was superseded in 1993 by a technical committee (TC) for environmental management called ISO/TC 207. The goal of this committee is to standardize environmental-management tools and systems (50). ISO/TC 207 (environmental management) comprises

Table 1 Environmental labelling programs (in chronological order)

Country/countries	Program name	Date of creation	Number of figure showing label	References
Germany	Environmental Label ("Blue Angel")	1977	1a	24–27
Canada	Environmental Choice	1988	1b	28, 29
Nordic countries (Sweden, Norway, Finland, Iceland)	Nordic Environmental Label ("Swan")	1989	1c	30
Japan	Eco Mark	1989	1d	31, 32
Sweden	Good Environmental Choice	1990	1e	33
USA	Green Seal	1990	1f	34, 35
USA	Scientific Certification Systems[a]	1990	—	21
New Zealand	Environmental Choice	1990	1g	36
France	NF Environnement	1991	1h	37
Austria	Environmental label	1991	1i	38
India	Ecomark	1991	2a	39
Australia	Environmental Choice Australia[a,b]	1991	—	—
European Union	EU eco-label	1992	2b	40–44
The Netherlands	Ecolabel ("Stichting Milieukeur")	1992	2c	45
Singapore	Green Labelling Scheme	1992	2d	46
Korea	Eco-Mark	1992	2e	47
USA	EPA Energy Star Computers[a]	1992	—	22
Spain	Aenor Medioambiental	1993	2f	48
The Czech Republic	Eco-Label	1994	2g	49

[a] not an environmental labelling program in the strict sense
[b] discontinued as of May 1st, 1994

six subcommittees and one working group. The various subcommittees are divided into as many as five working groups.

The environmental-labeling subcommittee (SC3) is charged with standardizing first-party labeling practices (self-declaration) and guiding principles for third-party certification programs. The three working groups within SC3 are developing guiding principles for environmental-labeling practitioner programs, environmental-labeling standards for manufacturer self-declaration, and general principles for all environmental labeling, respectively. The international standards (a draft version) adopted by the technical committees of ISO are to be approved by the member bodies before their acceptance as international standards by the ISO Council.

EUROPEAN UNION ECO-LABEL The European Union eco-labeling program (40) is the multinational environmental-labeling initiative of the European Union (EU), which entered into force on March 23, 1992. The EU was then called the European Community. Currently (1994), the EU consists of twelve coun-

NORDIC ENVIRONMENTAL LABEL

a. b. c.

Figure 1 (*a*) The German Environmental Label (unofficially called Blue Angel); (*b*) The Eco-Logo[M], the environmental label of Canada's Environmental Choice[M] Program; (*c*) The label of the Nordic environmental-labeling system; (*d*) The Eco Mark symbol of Japan's environmental-labeling program; (*e*) The label of the Swedish Good Environmental Choice program; (*f*) The Certification Mark of the environmental-labeling organization Green Seal, USA (*g*) The certification mark of Environmental Choice New Zealand; (*h*) The emblem of France's environmental-labeling program NF-Environnement; (*i*) The Austrian environmental label.

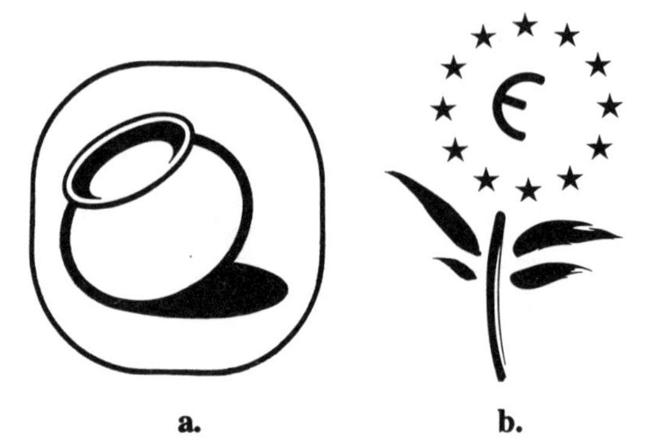

a. b. c.

d. e. f.

g.

Figure 2 (*a*) India's EcoMark; (*b*) Eco-label of the European Union; (*c*) The Dutch environmental label, owned by Stichting Milieukeur; (*d*) The environmental label of The Singapore Green Labelling Scheme; (*e*) The label of Korea's Eco Mark System; (*f*) The Spanish environmental label AENOR -MEDIOAMBIENTAL; (*g*) The emblem of the National Eco-Labeling Program of the Czech Republic.

Environmental fields	Product life-cycle				
	Pre-production	Production	Distribution (including packaging)	Utilization	Disposal
Waste relevance					
Soil pollution and degradation					
Water contamination					
Air contamination					
Noise					
Consumption of energy					
Consumption of natural resources					
Effects on eco-systems					

Figure 3 Indicative assessment matrix for the EU eco-labeling program.

tries: Belgium, Denmark, France, Germany, Greece, Ireland, Italy, Luxemburg, The Netherlands, Portugal, Spain, and the United Kingdom. In 1995, Austria, Finland, and Sweden will join the EU, making it a league of 15. In a referendum held in late 1994, Norway voted against joining the Union. Switzerland, although not an EU member, is working on participating in the EU eco-labeling program.

The Commission, EU's executive branch, has the obligation of ensuring that the member states implement EU laws. As a Council Regulation, the EU eco-labeling program must be implemented in each member state and is currently being introduced gradually (41); the use of the label, however, is not compulsory. National programs can continue to exist alongside the EU scheme. The EU eco-label, a stylized flower, is reproduced in Figure 2b. The program (40) is targeted at consumer goods, not at the overall environmental performance of manufacturers or companies, and it explicitly excludes food, drink, and pharmaceuticals. The eco-label cannot be awarded to dangerous substances or preparations.

The conditions for awarding the label are defined within product groups. The specific ecological aspects and criteria for each product group are established by using a cradle-to-grave approach. The life-cycle assessment for the product group is based on the indicative assessment matrix shown in Figure 3.

Several EU member states are taking the lead in developing proposals for the ecological criteria for several product groups (41, 42). Twenty-six individual product groups are now being assessed (see Table 2).

In each member state, a competent body (i.e. an appropriate independent and neutral authority) is designated to run the EU eco-labeling program. The

Table 2 Member states and product groups for which they are developing ecological criteria for the EU eco-labelling system

Member state	Product groups
Denmark	copying paper; writing paper; toilet paper[a]; kitchen rolls[a]; materials for building insulation; textiles
France	paints and varnishes; batteries and accumulators; shampoos
Germany	detergents; dish washing agents; household cleaning agents
Italy	packaging; refrigerators and freezers; ceramic tiles
The Netherlands	shoes; cat litter
United Kingdom	dishwashers[a]; washing machines[a]; hairsprays[a]; hairstyling agents; anti-aspirants; deodorants; light bulbs[a]; soil improvers[a]; growing media

[a] work for developing proposals for the eco-labelling criteria completed

application process for an eco-label is as follows (see Figure 4). The manufacturer or importer must apply to the competent body of the member state in which the product is manufactured or first marketed or into which the product is imported from a third country. The application must include all required certification and documents. The competent body then assesses the product's environmental performance using the relevant ecological criteria and decides whether to award the eco-label. If the body decides to award the label, it must notify the EU Commission which in turn informs the competent bodies of all other EU member states. If no reasoned objections are raised within 30 days following the dispatch of the notification to the Commission, the competent body may award the EU eco-label. If such objections are raised and cannot be resolved informally within 45 days, the Commission submits the proposed award to the EU Regulatory Committee.

When an eco-label is awarded, the competent body and the applicant sign a contract governing its use. Applicants are required to cover the application-processing costs and to pay a fee for use of the label. The length of time the label may be used depends on the validity of the environmental criteria for the relevant product group. Generally, this period is three years from the date of adoption of the criteria. The expiration date of any individual contract is the end of the validity period, so the later contracts are concluded during the period of validity, the shorter they will be (43).

The first EU eco-labels were awarded in November 1993 for washing machines. Currently, only seven products (all washing machines) bear the label. So far, the Commission has adopted ecological criteria for only two product groups: washing machines and dishwashers. The member states have approved the criteria for kitchen rolls, toilet paper, and soil improvers, and a final adoption by the Commission is expected soon. The member states are expected to vote on the ecological criteria for detergents, paints, and varnishes in the near future.

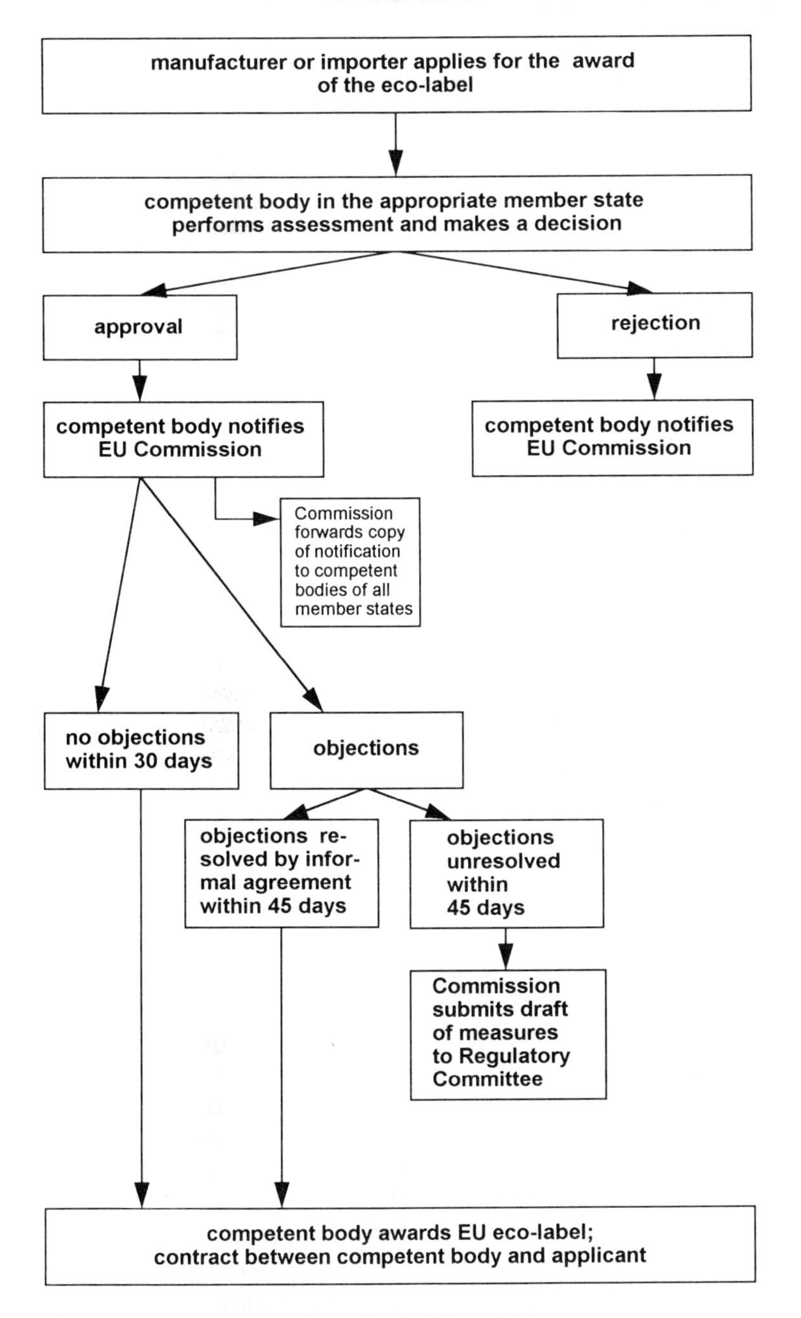

Figure 4 The steps of the process of awarding the EU eco-label.

The EU environmental-labeling program is not without critics. The application and standardization of a cradle-to-grave assessment has been a continuing source of conflict. The lack of consistent guidelines for the LCA has been criticized as well. Concern that the practice of awarding the EU eco-label could lead to political negotiation is also an issue. In December 1993, the European Commission established the Group des Sages to advise on the role of LCA in the EU eco-labeling program. This group issued an LCA guidance document (44) to facilitate the determination of ecological criteria. In spite of the criticism it has received, the European eco-labeling program continues to make progress, albeit at a moderate pace.

NORDIC ENVIRONMENTAL-LABELING SYSTEM The Nordic Council of Ministers decided in November 1989 to introduce a common, voluntary, Nordic environmental-labeling system (30). Finland, Sweden, Norway, and Iceland participate in this program (Denmark does not take part but participates in the EU's labeling program). The multinational program is administered in these countries by national secretariats (The Finnish Standards Institution (SFS), the Swedish Standards Institution (SIS), the Norwegian Foundation for Environmental Labeling and the Ministry of Environment, Iceland) working under the Nordic Coordinating Body, which is responsible for the common rules for Nordic environmental labeling. The Nordic system has been a forerunner in many ways, particularly for the EU's eco-labeling initiative. In fact, the organization called Swedish Standards Institution (SIS) Environmental Labelling has cooperated with the EU and participated in expert groups for the preparation of criteria for the EU eco-labeling program. The similarities between the Nordic environmental-labeling program and the EU eco-labeling program are pronounced.

The label of the Nordic environmental-labeling program, which represents a swan, is depicted in Figure 1c. The words "Nordic environmental label" appear in the relevant language above the label. A brief description of how the product is less burdensome for the environment is below the label.

The labeling system is based on the following steps: defining product groups, identifying the most important environmental aspects of a product group, and establishing environmental criteria. As of August 1994, the ecological criteria for 24 product groups had been approved, the criteria for 8 more were under review, and the criteria for 10 more were under development (see Table 3).

In the determination of the ecological criteria for a product group, the entire life cycle of the product is assessed. Notably, the criteria also include rules for inspection and testing and for the submission of documentation. The coordinating body determines the product groups to be considered. One country takes the lead in developing the criteria, which are circulated for comments within

Table 3 Status of the criteria for the product groups of the Nordic environmental labelling system[a]

Product groups with approved criteria	
building materials: chipboard, fibre board and gypsum board	household and toilet paper
button cell batteries	lawnmower
chain lubricants	light sources
cleaning agents for cars	marine engines
composters	machine dishwashing agents
copying machines	newsprint paper
dishwashing machines	oilburners and oilburner/boiler combinations
envelopes	processed fine paper products
fine paper for copying and printing	rechargeable batteries
furniture and kitchen units	textile detergents
glues	toner cartridges for printing and copying
guidelines for packaging	washing machines
Criteria sent out for review	
all purpose cleaners (N)	paints and lacquers (N)
diapers (F)	refrigerators, freezers (S)
floor covering (N)	toilet bowl cleaners (F)
magazine paper (F)	textiles (S)
Criteria under development	
coffee filters/vacuum cleaner bags (F)	hand dishwashing agents (S)
computers and printers (F)	hand towels (F)
floor cleaning agents (N)	insulation material (S)
graphic products (F)	packaging paper (S)
hair care products (S)	soap (N)

[a] The letter in parentheses marks the lead country: N: Norway, F: Finland, S: Sweden.

the other Nordic countries before being finalized by the Nordic coordination body. The criteria are valid for two to three years.

Manufacturers or importers in Sweden, Norway, Finland or Iceland who apply for the Nordic environmental label submit their application, along with detailed documentation and proof of tests by independent laboratories, to the national environmental-labeling organization in their own country, just as in the EU eco-labeling program. Applicants from other countries apply in the country that has drawn up the relevent criteria document. The individual countries manage the application and approval process. The competent labeling organization has the right to carry out periodic inspection after the label has been awarded. The fee for the label license consists of a fixed application fee and an annual fee which is related to product turnover. This regulation is similar to that in the EU eco-labeling program. An environmental-labeling license awarded in a Nordic country also entitles the licensee to use the environmental label in the other Nordic countries.

Product life cycle: / Environmental aspects:		Production (1)	Use	Disposal
Hazardous substances		O	●	● (4)
Emissions of pollutants into	air	O	● (2)	●
	water	O	-/-	-/-
	soil	O	-/-	-/-
Noise emissions		O	●	-/-
Waste	avoidance/ reduction/ recycling	O	● (3)	● (5)
Sparing use of resources (energy, water, raw material)		-/-	●	-/-
Fitness for use			●	
Safety			●	

KEY:
● requirements considered
O requirements not included
-/- not relevant

EXPLANATION:
(1) import only
(2) ozone, dust and hydrocarbons
(3) recommendation to use recycled paper
(4) separate recycling of selenium
(5) non-reusable toner cartridges

Figure 5 Environmental-label test chart for environmentally-sound photocopying equipment from the German environmental-labeling program (see Reference 25).

Whereas the EU's eco-labeling program is still in its infancy, the Nordic environmental-labeling system is more advanced and has become very successful. Currently, more than 650 products bearing the Nordic environmental label are on the market. Finland and Sweden joined the EU in 1995 (when this article went to press), so coordination of the EU eco-labeling program with the Nordic system will be needed.

Examples of National Initiatives and Programs

GERMANY: ENVIRONMENTAL LABEL ("BLUE ANGEL") In 1977, the Minister of the Interior of the Federal Republic of Germany and the Ministers for Environmental Protection of the Federal States of Germany decided to introduce an environmental label (24, 25). By the end of 1978 the first six sets of criteria for the label had been established. Since 1986, the Minister for the Environment, Nature Conservation, and Nuclear Safety, rather than the Minister for the Interior, has been responsible for environmental issues.

The German environmental label is shown in Figure 1a. The label is officially named "Umweltzeichen" (Environmental Label) and consists of the Environmental Emblem of the United Nations, a blue ring with a laurel wreath and a blue figure with arms spread wide in the center. The label has been dubbed "Blue Angel" because of this figure. At the bottom is a brief description of the reason the label was awarded (e.g. "because made of 100% scrap paper" or "because of low noise").

The assessment of environmental friendliness is based on a wholistic view which takes into account all aspects of environmental protection including resource conservation (24). The labeling system is voluntary, positive, and relative. Products marked with the environmental label are not allowed to show safety or usability restrictions. The environmental label is a "soft instrument" (25) of German environmental policy. The label fosters environmentally-benign product characteristics.

The ecological criteria are not determined for individual products but for product groups, each of which has its own LCA matrix. Safety and usability are also taken into account. Figure 5 shows the environmental aspects relevant to assessing the environmental friendliness of copiers (25).

During the development of environmental aspects and criteria, the product-specific environmental characteristics are of primary interest. For paper products, for example, the most important factor is the percentage of recycled paper contained in the product; for lawn mowers it is noise emission. Thus, the German environmental label for a product in a certain product group is awarded on the basis of only one or two main characteristics that decisively lessen its impact on the environment. The environmental label for the product group "copiers" (see Figure 5), for example, bears the caption: "because: low emission and waste reducing." The labeling program's emphasis on only a few aspects has been repeatedly criticized. However, the criteria that refer to these aspects are often very detailed and demanding compared to the average characteristics of similar products or to minimum legal requirements (26). Criteria for one product group are valid for three years. Shorter periods can be set if the state-of-the-art in a certain product group is expected to advance very quickly.

Germany's environmental-labeling program presently comprises 78 product groups. More than 3900 separate products bear the "Umweltzeichen." Approximately 15% of these products come from foreign manufacturers (24). The product groups include lacquers, batteries, products made from recycled plastics, toner cartridges, and drums for laser printers. Only recently have the criteria for environmentally acceptable workstation computers been established (27). The other product groups are mainly related to heating and cooling systems, building material, construction equipment, automotive products, packaging, and paper products.

Three institutions are involved in the environmental-label award procedure:

1. The Jury Umweltzeichen (Environmental Label Jury). This is an independent group of representatives from industry, business, and trade unions,

as well as consumer, environmental-protection and nature-conservation organizations. The Jury selects the product groups that should have environmental labels and decides on both the environmental criteria and how compliance must be demonstrated.

2. The RAL Deutsches Institut für Gütesicherung und Kennzeichnung e.V. (German Institute for Quality Assurance and Labeling). This group is a nonprofit organization heavily involved in both developing ecological criteria for product groups and awarding the label. Applications for an environmental label are submitted to this institute, which also manages the awards. RAL checks whether the requirements are met, consults the necessary organizations and finally awards the label and signs a contract with the manufacturer regarding label use.
3. The Umweltbundesamt (Federal Environmental Agency). This agency reviews and examines new proposals (see below) and submits them, together with its opinion, to the Environmental Label Jury. It is in charge of drafting conditions for awarding the environmental label and participates in developing the final criteria. This agency also informs RAL of its opinions regarding individual environmental label applications.

The German environmental-labeling procedure encompasses two stages: (*a*) the development of ecological criteria, and (*b*) the processing of individual applications. Proposals to award an environmental label to a product that does not fit into an existing product group can be submitted by the general public to the Federal Environmental Agency. This "new proposal" must explain how the product is distinctly different from other products made for the same purpose in terms of environmental friendliness. The further processing of this new proposal is shown in Figure 6. The ecological criteria are established in a thorough process which may take up to two years to complete. If a product group and award criteria already exist for a particular product, the manufacturer can apply directly to RAL for an environmental label. The flow chart in Figure 7 shows the application process.

Proposals for new product groups are processed free of charge by the Federal Environmental Agency. RAL charges an environmental label–application fee for a product within an existing product group. Following completion of a contract between the manufacturer and RAL, a graduated annual fee is paid to RAL (25), the amount based on the estimated annual turnover of the product. The signed contracts are valid at most until the end of the validity period for the award criteria. The environmental label is also available to foreign manufacturers, subject to the same conditions as German companies.

CANADA'S ENVIRONMENTAL CHOICE[M] PROGRAM Canada's Environmental Choice Program, created in 1988, runs under the auspices of the Canadian

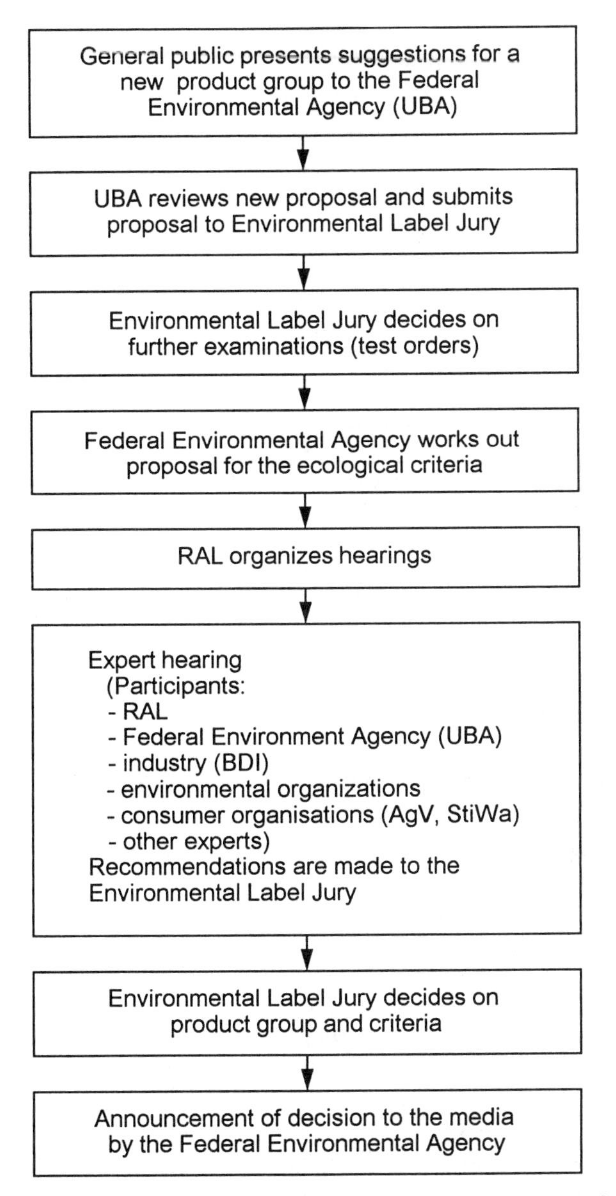

Figure 6 Stage 1 of the German environmental-labeling program: development of criteria for a new product group (see Reference 25). Abbreviations: UBA: Umweltbundesamt (Federal Environment Agency); RAL: RAL Deutsches Institut für Gütesicherung und Kennzeichnung e. V. (German Institute for Quality Assurance and Labeling); BDI: Bundesverband der Deutschen Industrie e. V. (Federation of German Industries); AgV: Arbeitsgemeinschaft der Verbraucherverbände e. V. (Consumers' Association); StiWa: Stiftung Warentest (Stiftung Warentest Foundation).

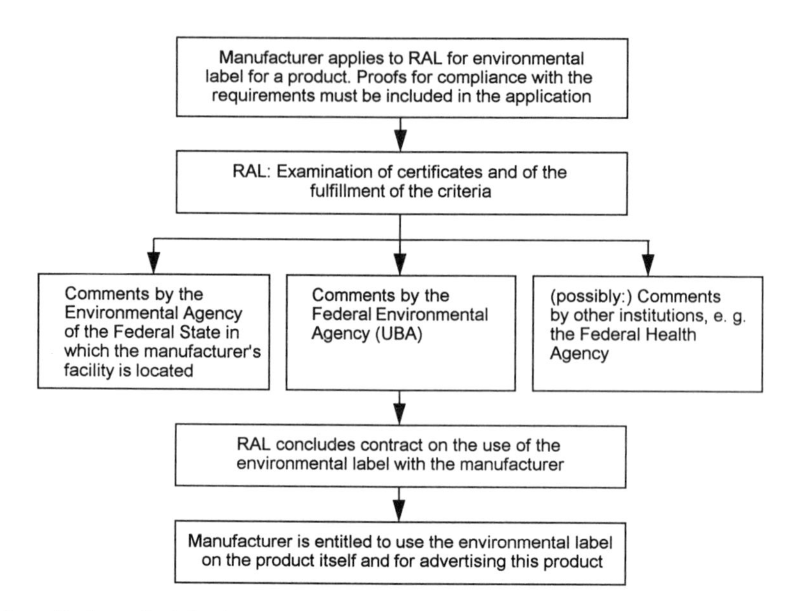

Figure 7 Stage 2 of the German environmental-labeling program: processing of an individual application for an environmental label for a product for which a product group and award criteria have been adopted (see Reference 25; for abbreviations, see Figure 6).

government. The program is voluntary, and manufacturers and purveyors are encouraged to improve their products and services so that they qualify. A product or service may be certified if, for example, it is made or offered in a way that improves energy efficiency, reduces hazardous by-products or uses recycled materials (28). Certified products or services are labeled with the EcoLogo^{M} (see Figure 1b). The logo consists of a maple leaf symbol composed of three stylized doves and is accompanied by a brief statement of the reason for qualification. The words "environmentally friendly" are intentionally avoided on the label so consumers will not mistakenly consider products with the EcoLogo completely harmless to the environment.

Canada's environmental-labeling program is operated by the Environmental Choice Program office within Environmental Canada, Canada's federal environmental agency. An independent Technical Agency is under contract to assist in verifying and testing products and services. The development of ecological criteria is overseen by the Environmental Choice Program Board. This 16-member board of independent volunteers is appointed by the Minister of the Environment who is responsible for the operation and activities of the Environmental Choice Program.

The development of ecological criteria for product groups is based on LCA.

The procedure is as follows (28). First, the Environmental Choice Program Board identifies potential product groups. Then the Environmental Choice Program staff, assisted by several experts, drafts criteria for consideration by the Board. Representatives from industry, environmental groups, government, and universities, as well as independent scientific and technical advisors participate. (The criteria include the requirement that the product must continue to meet all applicable safety and performance standards.) The public is allowed a 60-day review period, after which the Board may revise the criteria. The Board submits the final recommended criteria to the Minister for the Environment for approval.

When the criteria for a product group are promulgated, purveyors, manufacturers and importers of relevant products or services can apply for the environmental label (29) through the Environmental Choice Program office, which instructs the Technical Agency to start the verification procedures. These include product testing, an inspection of the location where the product or service is manufactured or processed, including the quality control system, and a review of product compliance with the appropriate criteria. After completing the verification process, the Technical Agency makes a recommendation to the Environmental Choice Program.

If all requirements are met, the Environmental Choice Program and the applicant sign a license agreement on use of the EcoLogo. While the contract is valid, the Technical Agency annually verifies guideline compliance. The licensee pays the cost of the verification procedures and an annual license fee determined by the gross annual sales of the certified product or service.

As of August 1994, criteria were established for 29 different product groups. Six product categories are under discussion, and guidelines for 11 product groups are under development. The product groups include automotive fluids, such as engine oil, fuels, and coolants; paints; adhesives; detergents; paper goods; batteries; toner cartridges; and building materials. Currently, 1467 products have been awarded the environmental label by the Canadian Environmental Choice Program. The certified products, however, do not yet span the spectrum of available product groups. Rather, the certified products are concentrated in a few groups: water- and solvent-based paints (~88%) and fine paper from recycled paper (7%).

JAPAN: ECO MARK PROGRAM Japan's Eco Mark Program was founded in 1989 and is administered by the Japan Environment Association (JEA) under the authority of the Environment Agency. The Eco Mark Secretariat within the Japan Environment Association handles general affairs. The JEA comprises two advisory bodies: the Eco Mark Promotion Committee and the Eco Mark Expert Committee. The Eco Mark Promotion Committee determines product categories and establishes ecological criteria. This group consists of specialists from administrative agencies and consumer groups, as well as environmental conservation specialists. The Eco Mark Expert Committee is responsible for judging products and for approving the use of the Eco Mark. This committee is comprised of experts in environmental impact assessment and in reducing damage to the environment.

The general idea behind Japan's environmental-labeling initiative is somewhat different from that of other labeling programs. The overall objective is (31) to preserve valuable natural resources, pursue sustainable development, create an ecological lifestyle and, ultimately, an environmentally-sound society.

The label of the Eco Mark Program, the Eco Mark symbol (see Figure 1d), represents the desire to protect the earth with one's own hands. The phrase "Friendly to the Earth" appears at the top of the symbol, and at the bottom the specific environmentally-friendly characteristics of the product are listed. Only products sold in Japan, irrespective of their country of manufacture, are allowed to bear the Eco Mark.

In addition to the basic requirement of being less detrimental to the environment, the following requirements must be met (31) by the products in all categories: (*a*) proper antipollution measures must be applied in the manufacturing process; (*b*) products must effectively conserve energy and resources during use; (*c*) product disposal must not cause additional problems; (*d*) product quality and safety must comply with the related laws, standards, and criteria; and (*e*) the product must be priced competitively.

The ecological criteria for a particular product group are set by the Eco Mark Promotion Committee in consultation with the Expert Committee and revisions are carried out regularly. Often the requirements for a product group are not very stringent and are based on one property rather than a complete LCA. In such cases the products fulfill the general requirement that they reduce the burden on the environment in other ways. Typical products in this context are (31) fine mesh strainers for kitchen sinks [which can be awarded the Eco Mark "for cleaner water" (32)], and magazines and books on environmental problems [which have the Eco Mark "Green publication" (32)].

The creation of a new product group can be proposed to the Eco Mark Secretariat. The procedure for establishing a new product group is depicted schematically in Figure 8 (see Reference 31). The application process for the environmental label for a product in an existing product group of the Eco Mark Program is shown in the diagram in Figure 9.

The Japanese Eco Mark Program does not charge an application fee, but the applicant has to pay for necessary third-party testing. In the final contract, the annual licensing fee for the Eco Mark is based on the product's retail price, not product turnover, which makes fee-handling easier. As of August 1994, Japan's Eco Mark program had 60 different product groups and 2418 indi-

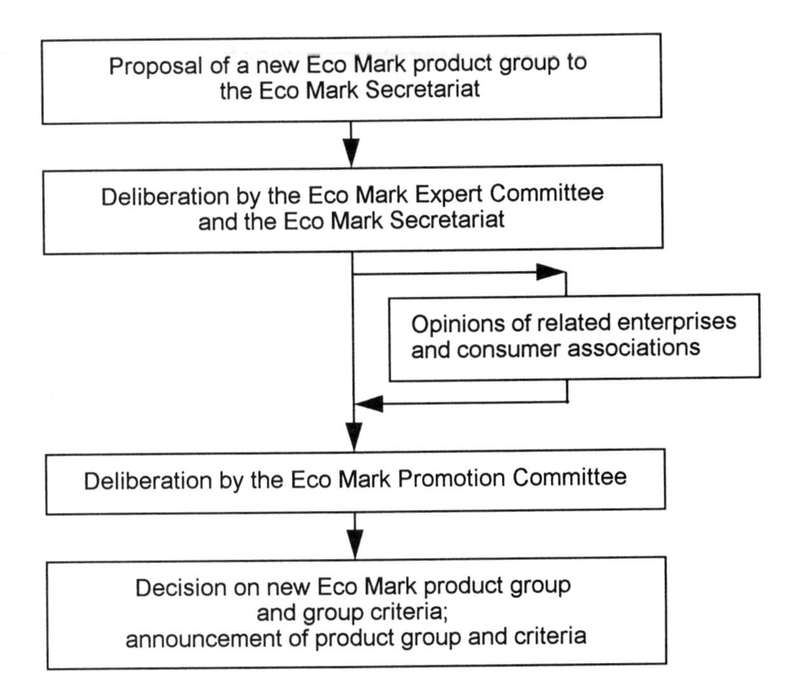

Figure 8 The process of establishing a new product group in Japan's Eco Mark program.

vidually certified products. Most of these products either keep the water and air clean, conserve water and energy, preserve greenery, are reyclable or reusable, or are made from recycled or natural materials. Many product groups relate to household items. Recently, product categories for the electronics industry were established, such as low-waste printers for business machines, replacable ink cartridges, and ribbon cassettes.

A product's impact on the environment is judged during the life-cycle stages, but not necessarily during every stage for each product group. Some products impose less of a load on the environment during specific life-cycle stages (31) such as manufacturing (e.g. unbleached coffee filters), use (e.g. water heating systems that use solar energy), and disposal (e.g. biodegradable oils). For this reason, the Japanese labeling program is not based on a complete LCA.

Products can also merit the environmental label because they foster environmentally-friendly behavior, even if they have no other environmentally-friendly properties. For instance, composting containers can be awarded the Eco Mark only because they serve the purpose of composting garbage. (In most other environmental-labeling programs, these containers would not be

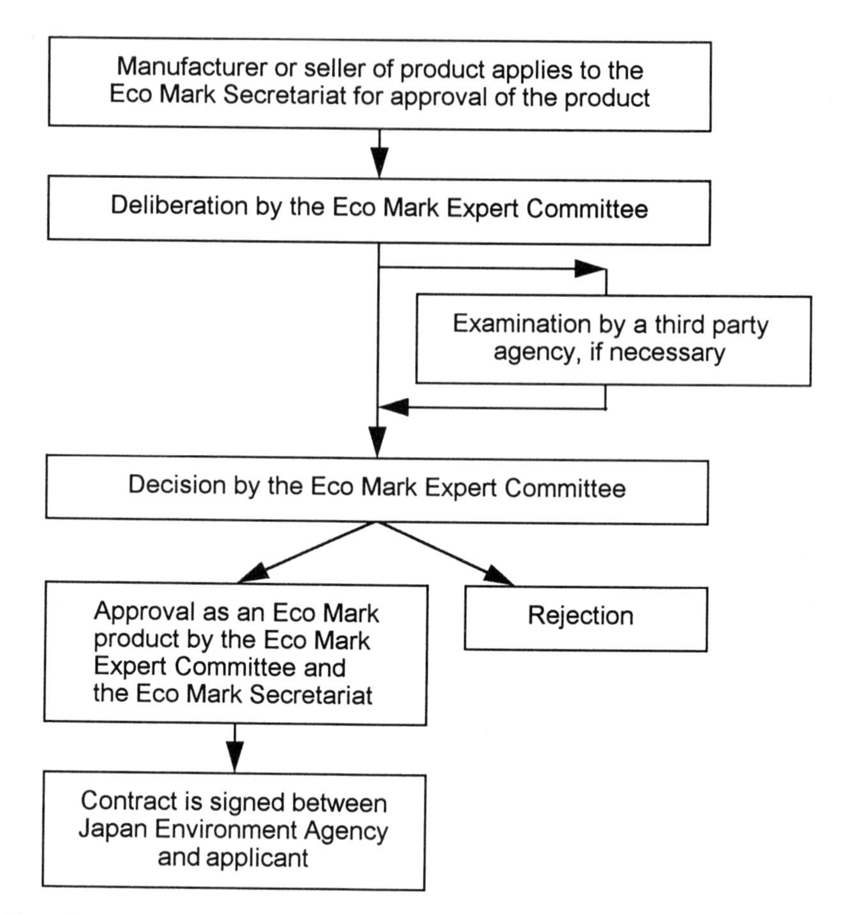

Figure 9 Flow chart of product-approval process for Japan's Eco Mark.

eligible for an environmental label because the characteristic of "representing a composting container" is common to all other products serving the same purpose.) Some criteria of the Japanese Eco Mark Program are obviously not very demanding. This approach, however, reflects the goal of making consumers more environmentally conscious and encouraging them to help preserve the environment.

SWEDEN'S GOOD ENVIRONMENTAL CHOICE Besides the Nordic environmental-labeling system, Sweden has a second eco-labeling program called Good Environmental Choice (Bra Miljöval) (33), a nongovernmental initiative. The Swedish Society for Nature Conservation (Naturskyddsföreningen), in conjunction with the three major retailers in Sweden (which control about 75% of Sweden's grocery business), founded the Good Environmental Choice program in 1990. Since 1992, Good Environmental Choice has been managed by a committee equally represented by the Swedish Society for Nature Conservation and the retail trade.

The strategy of the labeling program is to focus on daily consumption, that is, goods that are bought frequently, and to create dynamic criteria that require only small improvements at a time. Market forces can react very quickly as a result of this policy and the environmental criteria can be broadened or made more stringent quickly. Requirements are not set so high that most products would have no chance of meeting them and getting a label because then consumers would continue to buy products that are less environmentally friendly.

Unlike most other environmental-labeling systems, Sweden's Good Environmental Choice program does not label the products themselves, but rather the appropriate supermarket shelves, because of the tight collaboration with retailers. Since 1992, however, labeling the product itself has also been possible. Sweden's Good Environmental Choice label is yellow and features a peregrine falcon in the center (see Figure 1e).

In developing environmental criteria, a process jointly financed by the Society for Nature Conservation and the retail trade, the Good Environmental Choice committee decides which product groups to take into account (33). These groups are then studied to determine which ecological aspects should be considered. A proposal for the criteria is drawn up by a consultant, based on scientific findings and on the input of manufacturers and retailers. The criteria are reviewed by the Society's panel of experts, and manufacturers are given the chance to express their opinions. The finalized criteria are ratified by the Society's head of conservation and the environment, and the committee decides on the date the criteria take effect (usually six months later).

In the interim, the Society investigates the products (assisted by the information from the manufacturers) and determines which products comply with the criteria. The Society spot-tests the contents of the approved products for adherence to the manufacturer's claims. The list of approved products is forwarded to the retailers who label the shelves with the Falcon mark when the criteria come into effect.

The Committee that awards the Nordic environmental label does not allow one product to carry both the Nordic's swan label and Good Environmental Choice's falcon label. However, the Nordic environmental label may be on the product while the Good Environmental Choice label is on the shelf. Shelf-labeling is free for the manufacturers. If they want to put the environment label directly on their products, they pay a one-time fee, which is substantially lower for additional products labeled after the first. Environmental criteria have been established within the Good Environmental Choice program for the following product groups: batteries, shampoos, toilet and sanitary cleaners, diapers, washing powders and liquids, dish detergents, cleaning detergents, stain removers and bleaches, and paper products.

USA: GREEN SEAL Green Seal is an independent (nongovernmental) nonprofit environmental-labeling organization in the US. Its environmental-labeling program (created in 1990) awards a seal of approval called Green Seal Certification Mark (see Figure 1f).

The Green Seal organization is led by a Board of Directors composed of leaders from business, government, national environmental and consumer organizations, and other public interest groups (34). Under contract with Green Seal, Underwriters Laboratories, Inc., a product testing and factory inspection organization, tests the products and monitors compliance with requirements.

Green Seal selects product groups for which to evaluate certification criteria. Proposals for product groups are accepted from the public or from manufacturers. The ecological aspects taken into account include energy efficiency, recycled content, significant reduction or elimination of an environmental risk, and reduction of sources of potential pollution. The environmental criteria are determined by Green Seal with input from manufacturers, trade associations, federal, state, and regional regulatory agencies, and environmental and other public interest groups. In place of a complete quantitative life-cycle assessment, the environmental impacts are evaluated and the criteria are then focused on the most important of those (35). A product cannot be certified, for example, if it is toxic or contains carcinogens. After the initial criteria for a certain product group are prepared, they are publicly circulated for comment before the final criteria are established. The criteria are reviewed every three years.

To obtain the Green Seal Certification Mark, the manufacturer of a product applies to Green Seal, which evaluates the suitability of the product for certification. Eligible products are tested, and the main manufacturing facility is inspected for compliance with the environmental criteria. Most of this work is carried out by Underwriters Laboratories, Inc. (If a product group and the corresponding criteria have not yet been defined for the product in question, Green Seal will develop them.) If the product meets all requirements, the manufacturer is authorized to use the label on the product and in product advertising. Green Seal conducts annual monitoring, including product tests and factory inspections, to ensure that the product continues to meet the environmental criteria.

Applicants to the program must pay an application fee and cover the costs of evaluating the product's adherence to the environmental criteria and of the annual monitoring procedures. The costs associated with developing the environmental criteria of a new product group must also be paid by the applicant.

Such costs are partially credited to the applicant to cover the annual monitoring fees or are partially reimbursed when additional manufacturers apply for an environmental label in this new product group.

To date, the environmental criteria for the following product groups have been established: tissue paper; re-refined engine oil; compact fluorescent lamps; water-efficient fixtures; printing and writing paper; household cleaners; paper towels and napkins; coated printing paper; paints; windows; window films; newsprint; reusable utility bags; refrigerator-freezers, freezers, ovens and ranges; and washing machines, dryers, and dishwashers. Environmental criteria for the following product groups have been proposed but not yet adopted: sealants and caulking compounds; adhesives; recycled toner cartridges; water heaters; and room air conditioners.

INDIA'S ECOMARK The Indian ecolabeling program (39), called EcoMark, was created in 1991 by the Indian government. The emblem, an earthen pot, is shown in Figure 2a.

A technical panel (composed of representatives from the government, the scientific community, consumer and environmental groups and industry), set up by the Ministry of Environment and Forests, is in charge of identifying product groups and developing the environmental criteria. The ecological criteria include energy conservation, wise use of natural and nonrenewable resources, reduction and utilization of waste, biodegradability and others. Only the few most important criteria are taken into account for the EcoMark.

The product groups selected and the environmental criteria defined by the technical panel are reviewed by the Ministry of Environment and Forests, which also takes into consideration the comments from various interest groups. The Ministry then makes the criteria public. Applications for the EcoMark are submitted to The Bureau of Indian Standards (BIS), the certifying agency for the program. BIS tests, assesses, and certifies the product and awards the EcoMark. Compliance with the requirements is monitored thereafter. The applicant bears the cost of the assessment.

As of June 1993, 19 product groups had been identified. The criteria for four groups (soaps, detergents, papers, and paints) have been made public. Initial criteria have been drafted for 13 more product groups, and criteria for two more groups (pesticides and drugs) are currently being developed.

THE SINGAPORE GREEN LABELLING SCHEME The Singapore Green Labeling Scheme (46) was launched in May 1992 by Singapore's Ministry of the Environment. The labeling program is administered by the Secretariat of The Singapore Green Labelling Scheme, which is provided by the Waste Minimization Department of the Ministry for the Environment. Both local and foreign companies can participate in this voluntary program. The program does not apply to foods, beverages, or pharmaceutical services and processes. The environmental label for this program is called Green Label; its logo consists of a green leaf, surrounded by the words "Green Label SINGAPORE" (see Figure 2d). The Green Label is accompanied by a brief description of the reason for which the label was awarded (e.g. Recycled Paper).

The product groups are chosen and defined by the Secretariat of the program, which takes suggestions from industry and the public into account. The Secretariat sets up a technical working group to draft ecological criteria for the product group. The final criteria are developed by an advisory committee with members from industry, retailing, academia, statutory organizations, and environmental groups. The ecological criteria are based on the few product-group characteristics most relevant to environmental issues, rather than an LCA. For the office-automation paper-product group, for example (46), the only requirement is that the paper contain more than 50% recycled paper. Other groups, such as modular compact fluorescent lamps, must meet only three requirements.

In the criteria development process, the Secretariat relays feedback from industry and the public to the advisory committee which reviews the input and finalizes the ecological criteria for the product group. Once the criteria are approved by the Approving Board, which is headed by the Permanent Secretary (Environment), manufacturers and distributors may apply to the Secretariat for the Green Label. The applicant may be required to submit product samples for testing to laboratories accredited by the Singapore Laboratory Accreditation Scheme (SINGLAS). (The applicant must pay for this testing.) The applicant then submits the laboratory results to the Secretariat which conducts its own appraisal. If the product is approved, the applicant signs a three-year agreement with the Secretariat regarding use of the Green Label.

As incentive for companies to quickly put on the market products that meet recently stipulated requirements, cost advantages are granted (46). For example, annual fees for the first five years are waived if the application for a product is made within one year of the release of the final qualifying criteria. The Ministry of Environment currently bears the cost of administering the Singapore Green Labelling Scheme, but self-financing is a goal.

As of October 1994, the Singapore Green Labelling Scheme comprised 14 product groups including recycled paper (e.g. computer and photocopier paper), batteries (e.g. carbon-zinc and alkaline), compact fluorescent lamps, and detergents and washing machines. The product categories currently under consideration include correction fluid/tape, personal care products, paints, personal computers, lithium/rechargeable batteries, and solar cell products.

KOREA'S ECO MARK Korea's Eco Mark System, established in June 1992, is administered by the Korean Environmental Preservation Association under the guidance of the Ministry of the Environment. The main purpose of the program (47) is to promote products that are made from recycled materials or can easily be recycled; that is, to encourage consumers to buy environmentally friendly products. The Eco Mark emblem bears the slogan "cleaner and greener" and the words "Eco Mark Committee" (see Figure 2e).

Suggestions for product groups can be submitted by the general public to the Eco Mark Committee (consisting of representatives from environmental and consumer organizations, industry, government, and technical and legal experts). This committee selects product groups and sets criteria in collaboration with the Examination Committee, and the Ministry of the Environment announces the final product groups and requirements. The ecological criteria are not very comprehensive; each product group has only one criterion. For example, clutch facings must be free of asbestos and recycled paper goods must contain more than 50% recycled paper.

Applications for the Eco Mark are submitted to the Eco Mark Committee. The Examination Committee evaluates the product qualities and criteria, and the Eco Mark Committee uses the results to decide whether or not to grant the label. The royalty fee for using the Eco Mark is linked to the product's price and total sales volume.

As of December 1993, 141 products in 21 product groups had been granted permission to use the Eco Mark. The product categories include primarily products containing large amounts of recycled material, such as paper, plastic and packing products, and construction materials. Other product groups include sprays that do not contain chlorofluorocarbons, unbleached towels, energy-saving fluorescent lamps, and biodegradable lubricants and oils.

THE CZECH REPUBLIC The National Eco-Labeling Program of the Czech Republic was initiated in 1993 by the Czech Republic and entered into force in April 1994. The eco-label (49) is granted by the Minister of the Environment who is advised by the Board of the Czech National Program. The board is composed of experts from the government and from environmental and consumer organizations. The program label includes the text "Ecology preserving product" (see Figure 2g). Both domestic and imported products are granted access to the program, and participation is voluntary. The program does not include foods, beverages, or pharmaceuticals.

Product group proposals may be submitted to the Board by anyone. The ecological criteria are prepared by an ad hoc group of experts, considered by the Board, and recommended to the Minister for the Environment for approval. The program stipulates that a cradle-to-grave assessment of environmental impact (i.e. including production, utilization, and disposal) should be used as much as possible in ecological criteria development (49). Criteria for each product group are valid for two years.

An agency established by the Minister of the Environment to administer producers' applications examines together with test laboratories and the Czech Environmental Inspection the product's and production technology's compliance with the criteria. Applicants pay an application fee and, after the eco-label has been awarded, a certain percentage of the annual product turnover. Contracts for using the labels are valid for two years.

Currently, the Czech environmental-labeling program encompasses six product groups: thermal insulation made from recycled paper, water-based paints, phosphate-free detergents, low-emission gas heaters for homes, and biodegradable lubricating oil for chain saws. As of October 1994, two products from the thermal-insulation-made-from-recycled-paper group and the water-based-paints group bore the Czech eco-label, with four more products soon to follow. Product groups in preparation are printing paper and toilet paper made from recycled paper, liquid household detergents, and water-based glues.

DISCUSSION

Although existing environmental-labeling programs differ considerably in detail, they share many features. A common goal is to label products that are environmentally preferable to others of the same kind. Participation in all programs is voluntary, and almost all are created by governments and administered by governmental organizations (typically the Department of Environment) or other independent federal or state institutions. (The only non-governmental labeling organization that awards seals of approval that comply with the definition given at the beginning of this article is Green Seal in the United States.) All program labels are emblems that are easily discernible and identifiable. The Austrian label was even created by the well-known artist Friedensreich Hundertwasser. Generally, products themselves are labeled; only the Swedish program also labels the corresponding store shelves.

The structure of most labeling systems can be divided into two parts: defining product groups and the corresponding ecological criteria; and the application, approval and awarding process. The more thorough the process of establishing the ecological criteria, the stronger the environmental-labeling program (see below). The criteria are usually reviewed every three to five years.

Applications for an environmental label frequently must include proof of compliance with the requirements, and labels are generally awarded for a limited time (typically two to three years). In most programs, the annual license fees for using the label are approximately 0.15–0.50% of annual product turnover, not to exceed a maximum amount. In the Japanese program, however, the annual fee is based on the product's retail price, which eliminates the bureaucracy from fee determination. In the EU, the annual fee is fixed to 0.15%

of the corresponding annual turnover, without an upper limit. This arrangement may actually prevent manufacturers that expect a large annual turnover from applying for the label, because the relative license costs do not decrease with increasing volume. One way of encouraging manufacturers to apply for the environmental label (as demonstrated by Singapore) is to waive the license fees if the application is made within a short time after the promulgation of the ecological criteria.

Many product groups are common to the various labeling programs: paints and varnishes; paper products; batteries; detergents and household cleaners; electrical appliances; and office equipment. Therefore, cooperation between the responsible labeling organizations could reduce the effort required to establish and update ecological criteria for these product groups.

Although they are similar in many ways, the environmental-labeling programs are also notably different. For example, Japan's Eco Mark program is the only environmental-labeling program that explicitly aims at creating an ecological life-style and an environmentally-sound society. Emphasis on public awareness of environmental preservation is naturally more important in this program than pushing the limits of certain ecological criteria.

The strategy of Sweden's Good Environmental Choice program is also notable. By concentrating on goods that are consumed quickly and in large quantities, Sweden has laid a strong foundation that should enable it to make a positive impact relatively quickly. By dynamically strengthening the criteria, Sweden has also lowered entry hurdles, and manufacturers are anxious to maintain the label even when the requirements become more stringent.

Some European countries (e.g. France and Spain) skillfully combine their national environmental-labeling program with that of the EU. This cooperation is achieved by having the same members on the corresponding national and European labeling committees and by developing ecological criteria for product groups that can be used by both programs.

The most prominent differences among the environmental-labeling programs are in the ecological criteria for the product groups, which make some programs far more demanding than others. The most demanding labeling programs are based on a complete LCA, especially those that take the ecological aspects of the preproduction phase into account. The EU's eco-label program is a typical example. Thorough development of ecological criteria takes much time and effort for the organization in charge. Industry, on the other hand, has the difficult task of creating products that meet the requirements. Although the EU's labeling program was created in 1992, only seven products (all in one product category) have been awarded the label to date. The stringent requirements of these demanding eco-labeling programs may in principle be desirable, but they also prevent many products from acquiring labels.

Other programs (e.g. India's and Korea's) are based on one or a few requirements which can be fulfilled comparatively easily. In Korea's Eco Mark System, the requirement for recycled paper goods, for instance, is that they contain more than 50% recycled paper. In these programs only the criteria most relevant to environmental impact are taken into consideration.

The most successful labeling systems are currently those with criteria between the two extremes mentioned above. This category includes the programs of Austria, Canada, Germany and New Zealand. These programs do not assess the complete life cycle (often preproduction is not considered) but use a reasonable and effective set of ecological criteria. For these reasons the programs are well accepted by both industry and consumers and contribute considerably to environmental protection.

Environmental-labeling programs must be established in such a manner that they are accepted and supported by industry as well as consumers. Australia's Environmental Choice, an environmental claims certification program, for example, was not supported by industry and was discontinued in 1994 after only three years.

Because national and multinational programs sometimes exist side by side, one might be inclined to question the need for national programs. For multinational industries delivering to the world market, multinational environmental-labeling programs would be desirable. However, multinational programs are difficult to create and to manage. Decision-making is slow because inputs are required from the various member states. Furthermore, a national label for the country in which the product is primarily sold will produce the greatest economic advantage. In addition, requirements often have higher priority in one country or region than another, depending on, for example, geographic location, degree of industrialization, standard of living and consumer behavior. For some countries thermal insulation for houses and energy efficient heating systems are very important whereas other countries may need to concentrate on reducing water pollution. Even small countries, such as the Czech Republic, have created their own environmental-labeling programs because of both their concern for the environment and their desire for economic gain.

Environmental-labeling programs need to balance economic and ecological concerns and must accommodate the dynamics of development and consumption cycles in our society. During the past five years, public awareness of environmental concerns and environmental labels has proliferated and will continue to motivate industry to develop and offer products, processes, and services that are more environmentally benign.

Acknowledgement

The authors would like to extend their gratitude to Mr. Thomas Graedel for his review of the manuscript and to the many colleagues from AT&T organizations in various countries on several continents for providing the authors with information and with contacts at national environmental agencies.

Literature Cited

1. Van Goethem A. 1993. Packaging Waste-The Regulatory Framework in the Twelve EU Member States. *Features from Europe Environment,* p. 1. Brussels: Eur. Info. Serv.
2. *Elektronikschrott-Entsorgung/Recycling.* 1993. Heft 1. Impuls-Stiftung für den Maschinenbau, den Anlagenbau und die Informationstechnik, Frankfurt a. Main
3. Graedel TE, Allenby BR, Linhart PB. 1993. Implementing industrial ecology. *IEEE Tech. Soc. Mag.* Spring:18–26
4. Graedel TE, Allenby BR. 1994. *Industrial Ecology.* Englewood Cliffs, NJ: Prentice-Hall
5. Frosch RA, Gallopoulos NE. 1989. Strategies for manufacturing. *Sci. Am.* 261(3):144–52
6. Sekutowski J. 1991. Design for environment. *MRS Bull.* June:3
7. Allenby BR. 1991. Design for environment: a tool whose time has come. *SSA J.* Sept:5–9
8. Glantschnig WJ. 1992. Design for environment (DFE): a systematic approach to green design in a concurrent engineering environment. *Proc. 1st Int. Congr. Environ. Conscious Design Manuf.* Boston, MA
9. US Congr., Off. Technol. Assess. 1992. *Green Products by Design: Choices for a Cleaner Environment, OTA-E-541.* Washington, DC: US Gov. Print. Off.
10. Billatos SB, Nevrekar VV. 1994. Challenges and practical solutions to designing for the environment. *Proc. Design Manufacturability Conf. Natl. Design Eng. Conf., Chicago,* DE-Vol. 67:49–64. New York: Am. Soc. Mech. Eng.
11. Berko-Boateng V, Azar J, de Jong E, Yander GA. 1993. Asset recycle management-a total approach to product design for the environment. *Proc. IEEE Int. Symp. Electron. Environ.,* pp. 19–31, *Arlington, VA*
12. Fava J, Denison R, Jones B, Curran M, Vigon B, et al, eds. 1991. A technical framework for life-cycle assessment. *Proc. wrkshp. Smuggler's Notch, VT, Aug. 1990.* Washington, DC: Soc. Environ. Toxicol. Chem. (SETAC)
13. Consoli F, Allen D, Boustead I, Fava J, Franklin W, et al, eds. 1993. Guidelines for life-cycle assessment: a "code of practice." Proc. wrkshp. Sesimbra, Portugal, Mar. 1993. Pensacola, FL: Soc. Environ. Toxicol. Chem. (SETAC)
14. US EPA, Off. Pollut. Prev. Toxics. 1993. *The use of life cycle assessment in environmental labeling, EPA/742-R-93-003.* Washington, DC: EPA
15. Fiksel J, Wapman K. 1994. How to design for environment and minimize life cycle cost. *Proc. IEEE Int. Symp. Electron. Environ., San Francisco,* pp. 75–80. Piscataway, NJ: Inst. Electr. Electron. Eng.
16. Keoleian GA, Glantschnig WJ, McCann W. 1994. Life cycle design: AT&T demonstration project. *Proc. IEEE Int. Symp. Electron. Environ., San Francisco,* pp. 135A–135H. Piscataway, NJ: Inst. Electr. Electron. Eng.
17. Environmental consciousness: a strategic competitiveness issue for the electronics and computer industry. 1993. *Comprehensive Rep.: Anal. Synth., Task Force Rep./Append.* US Dep. Energy/Microelectron. Comput. Technol. Corp. (MCC), Austin, TX
18. US EPA, Office of Pollution and Prevention of Toxics. 1993. *Status report on the use of environmental labels worldwide, EPA 742-R-9-93-001,* Washington, DC: EPA
19. Haynes D. 1990–1991. Harnessing market forces to protect the Earth. *Issue Sci. Technol.* Winter:46–51
20. Smith TT. 1993. Understanding European environmental regulation. *Rep. 1026,* pp. 15–20. Conf. Board, New York
21. Scientific Certification Systems Information Package. 1994. Oakland, CA: Sci. Certification Syst.
22. US EPA, Air and Radiation. 1994. *Purchasing an Energy Star^{SM} computer, EPA 430-K-94-006.* Washington, DC: EPA

23. Graedel TE, Allenby BR, Comrie PR. 1995. Matrix approaches to abridged life cycle assessment. *Environ. Sci. Tech.* 29(3):134A–39A
24. Fed. Environ. Agency. Information sheet on the Environmental Label. 1994. Berlin: Umweltbundesamt
25. *The Environmental Label Introduces Itself.* 1992. Bonn: RAL Dtsch. Inst. Gütesicher. Kennzeichn. e. V. [German Inst. Quality Assurance Labeling]/Berlin: Umweltbundesamt [Fed. Environ. Agency]
26. *Umweltzeichen. Produktanforderungen, Zeichenanwender und Produkte (Environmental label. Product requirements, label users and products).* 1993. Sankt Augustin, Germany: RAL Dtsch. Inst. Gütesicher. Kennzeichn. e. V. [German Inst. Quality Assurance Labeling]
27. *Basic Criteria for the Award of the Environmental Label. Environmentally acceptable Workstation Computers RAL-UZ 78.* 1994. Sankt Augustin, Germany: RAL Dtsch Inst Gütesicher. Kennzeichn. e. V. (German Inst. Quality Assurance Labeling)
28. *Environmental Choice.* n.d. Information sheet by Environmental Choice, Environ. Canada: Ottawa, Canada
29. Application Information. 1992. Environmental Choice, Environ. Canada: Ottawa, Canada
30. *Official environmental labeling in the Nordic countries.* Stockholm: SIS Environmental Labeling-Swedish Stand. Inst.
31. *The Eco Mark Program.* 1989. Tokyo: Eco Mark Secretariat, Japan Environ. Assoc.
32. *Certification Criteria for Eco Mark Products.* 1994. Tokyo: Eco Mark Secretariat, Japan Environ. Assoc.
33. *Re-mark-able ecolabeling and The Swedish Society for Nature Conservation.* 1993. Göteborg, Sweden: Swedish Soc. Nature Conserv.
34. Green Seal information brochure. 1994. Washington, DC: Green Seal
35. US EPA, Off. Pollut. Prev. Toxics. 1993. *Status report on the use of environmental labels worldwide, EPA 742-R-9-93-001,* pp. 71–75. Washington, DC: EPA
36. *Environmental Choice New Zealand, Some Questions Answered.* 1992. Telarc New Zealand, *Doc. EC 003,* Issue 4, Auckland, NZ
37. *Information File on: -The NF-Environnement Mark and the European Ecolabel in France, -The NF-Environnement Mark, -The European Ecolabel.* 1993. Assoc. Française de Normalisation (AFNOR), Paris
38. Sedlar C, Raneburger J, Plankensteiner B, Kohlmann M. 1993. *Das österreichische Umweltzeichen (The Austrian Environmental Label).* Wien, Austria: Umweltbundesamt (Fed. Environ. Agency)
39. *EcoMark.* 1993. New Delhi, India: Ministry Environ. Forests
40. Council Regulation (EEC) 1992. 880/92 of 23 March 1992 on a Community eco-label award scheme. *Off. J. Eur. Communities* L99:1–7, 11.4
41. *Commission information on eco-labelling.* 1993. Iss 1. Brussels: Directorate-General XI, Comm. Eur. Communities.
42. *Commission information on eco-labeling.* 1993. Iss. 2. Brussels: Directorate-General XI, Comm. Eur. Communities.
43. *European Union Eco-Label Award Scheme. Information for Applicants.* 1994. Brussels: Directorate-General XI, Eur. Comm.
44. Groupe des Sages: Udo de Haes HA, Bensahel JF, Clift R, Fussler CR, Griesshammer R, Jensen AA. 1994. *Guidelines for the application of life-cycle assessment in the EU eco-labelling programme.* Leiden
45. *The Dutch Ecolabel, Added Value for Your Product and the Environment.* 1994. The Hague, The Netherlands: Stichting Milieukeur
46. *Make Green Label Your Choice.* 1994. Singapore: Ministry Environ.
47. *Environmental labeling in Korea: The Eco Mark.* 1993. Republic of Korea: Ministry Environ.
48. *Marca AENOR-MEDIO AMBIENTE.* 1994. Madrid: AENOR Associación Espanola de Normalización y Certificación
49. *Environmental labeling in the Czech Republic.* 1994. Prague: Ministry Environ. Czech Republic
50. *ISO MEMENTO.* 1994. pp. 32–33. Geneva: Int. Org. Standardization

Annu. Rev. Energy Environ. 1995. 20:265–300

ATMOSPHERIC EMISSIONS INVENTORIES: Status and Prospects

J.M. Pacyna
Norwegian Institute for Air Research, Kjeller, Norway

T.E. Graedel
AT&T Bell Laboratories, Murray Hill, New Jersey 07974

KEY WORDS: acid rain, global warming, ozone depletion, photochemical smog, sulfur oxides, nitrogen oxides, ammonia, methane, carbon monoxide, carbon dioxide, chlorofluorocarbons, trace metals, greenhouse gases

---

CONTENTS

1056-3466/95/1022-0265$05.00

ABSTRACT

Emissions inventories are the basis for many studies of the relationships between human activities and the environment, especially computer-model studies of the present and future, and are used by the scientific community to link theory to field observations. Within the policy community, they provide some of the most important information needed to formulate international environmental agreements. In North America, Europe, parts of Asia, and a few smaller regions, detailed emissions inventories are available for the principal greenhouse gases, acid-related species, and selected toxics and natural emittants. Internationally-generated, global-scale, gridded emissions inventories have only recently been released for general use. International agreements on emissions reductions are constrained by philosophical, technological, and financial differences among nations, but have nonetheless been shown to be of substantial value in decreasing the emissions of chemical species that clearly cause environmental degradation.

INTRODUCTION

Historically, terrestrial and marine biological processes, volcanic eruptions, and other actions of nature have generated emissions of numerous trace constituents. These emissions, in combination with atmospheric chemical and photochemical processes, have changed the atmospheric chemical composition over time. The magnitude and temporal frequency of these perturbations has been significantly increased by the releases of pollutants as a result of human activities such as energy generation, production of industrial goods, vehicular traffic, and waste disposal. These emissions have affected and will continue to affect human health and the global, regional, and local environment.

Emissions inventories are fundamental to scientific investigations of atmospheric pollution and to the design and implementation of policy response options. Various models have been developed to simulate the physical, chemical, and hydrometeorological processes that result in environmental changes. Information on emissions is a key parameter in these models. The model results are important for making policy decisions on the reduction of pollutants that enter the environment from various anthropogenic sources. In this way, emissions numbers are often linked to economic models developed at a country or regional level. Therefore, two major groups use emissions data: modelers and

Table 1 Typical characteristics of emissions inventories developed for different constituencies

	Inventory purpose	
Characteristic	Reporting and compliance	Scientific research
Spatial resolution	Appropriate legal entities (county, province, etc)	Geographical grid (150 km by 150 km, 1° by 1°, etc)
Spatial coverage	Appropriate legal entities	Geographical coverage as needed
Temporal resolution	Annual averages	Time-dependent values (seasonal, hourly, etc)
Chemical detail	Little	Modest to substantial
Included processes	Anthropogenic	Anthropogenic and natural
Perspective	Measurement-based	Estimation-based
Developer	Governments	Scientists
Data input	Self-reporting	Evaluation by external research teams
Reporting period	Annual	As needed

policymakers. Other emissions-data users include regulators, technologists, policy analysts, and advocacy groups, as well as business and the public.

The emissions inventory needs of these different constituencies vary. Those inventories developed for scientific research tend to have detailed spatial and temporal resolution and significant chemical detail (Table 1). They are developed by scientists and include those data needed as input for computer models of the atmosphere. For example, they include oceans and polar regions and natural as well as anthropogenic emissions. In contrast, emissions inventories generated because of the legal obligations of reporting and compliance generally relate to political feasibility and cost of compliance. They are characterized by average emissions fluxes rather than time-resolved fluxes, self-reporting (perhaps with guidance), little chemical detail (important to models but expensive to acquire), and a focus on a specific political entity or entities.

Ultimately, of course, the scientific and policy communities both aim to determine accurate rates and magnitudes of specific chemical compound emissions into the atmosphere. Hence, their approaches should produce identical results once factors such as spatial resolution and approaches to time-averaging are taken into account. However, the fact that all source emissions cannot be quantified, as well as the continual change in the number and types of sources, means that emissions inventories, no matter how well done, are "never right and never finished," but rather are continually aiming at a moving target. In addition, scientists and policymakers approach the inventory task with different tools, resources, and options. Because inventories for policy purposes generally represent conceptually simplified versions of scientific inventories (although completing them may be much more detailed and labor-intensive), we base our inventory development presentation in this paper largely on the scientific approach.

BUILDING EMISSIONS INVENTORY DATA SETS

Definition of Emissions

Atmospheric emissions arise from both physico-chemical and biological processes. If such processes involve sources related to socioeconomic activities (e.g. combustion of fuels to produce energy) the emissions are defined as anthropogenic, in contrast to natural processes occurring without significant human interference.

An atmospheric emissions inventory is a database of information on: (*a*) emissions measurements, which determine (*b*) emissions factors (i.e. how much of emittant A is produced from a source of type B per unit time), (*c*) number of individual emissions sources, (*d*) activity statistics (i.e. number of hours per month of smelter operation, number of miles driven per year), and (*e*) emissions estimates derived from the above data (1).

In addition, a satisfactory emissions database includes information on the accuracy of emissions estimates and/or measurements, reporting procedures, and verification procedures. A good emissions inventory is transparent; that is, it not only includes the final emissions estimates but also the data from all stages of analysis, so that anyone can recompute the emissions if activity levels or the numbers or types of sources change.

For consideration of global and regional issues, inventories and emissions estimates are structured hierarchically. For example, regions are aggregated into countries and countries into groups of countries in international inventories. Individual emissions sources are aggregated into larger groupings, such as activities, subsectors (or subcategories), and sectors (or categories).

In the majority of cases, various emissions measurements form the basis of emissions estimation procedures. One exception is a calculation procedure involving chemical analyses of input and output materials. Measured data can be used for emissions estimates directly or indirectly. Data are used directly from stack tests performed by means of continuous or spot monitoring at individual sites, for example. Data are used indirectly in emissions measurements whenever this information is transformed into an emissions factor or included in a special calculation procedure, such as a chemical mass balance. The emissions factor is often defined as the amount of a given material (gases, particles, vapor, chemical compounds, etc) generated during the consumption of a unit of raw materials (fossil fuels, ores, etc) or the production of a unit of industrial goods (electricity, nonferrous and ferrous metals, etc). For emissions from transportation, emissions factors are defined as the amount of a given material generated per distance driven.

Major Steps of Emissions Determination

The initial steps of emissions determination include the selection of (*a*) substances to be inventoried, (*b*) source categories, (*c*) determination procedures, (*d*) source resolution, and (*e*) source activity data. Emissions inventories are typically grouped into two source categories: engineering and economic. In the former, the source category split is based on emissions generation processes (e.g. the burning of coal), while in the latter the split relates to major socio-economic activities (e.g. the production of motorcycles). An example of the engineering approach used jointly in the UN Economic Commission for Europe (ECE) European Monitoring and Evaluation Programme (EMEP) and the Commission of the European Communities (CEC) CORINAIR Programme is presented in Table 2.

Depending on the circumstances, sources can be treated in emissions inventories individually or collectively. The source-by-source approach is used for point sources such as power plants, refineries, waste incinerators, and airports for which site-specific activity and emissions data are available. The trend is for more sources to be treated in this way as legislative requirements extend to more source types and pollutants. The collective approach is predominantly used for source types comprising many small emitters for which the emissions conditions are similar. These source types are often called area sources. The collective approach is also used to estimate emissions from line sources, such as vehicle emissions from road transport, railways, and inland navigation.

Table 2 CORINAIR/EMEP common source sector split[a]

Source sector	Cd	Hg	As	Cr	Cu	Ni	Pb	Zn	Se
Public power, cogeneration and district heating	X	XX	XX	X	XX	XX	X	X	XX
Commercial, institutional and residential combustion	X	XX	X	X	X	X	X	X	X
Industrial combustion	X	XX	X	X	X	X	X	X	X
Production processes	XX	XX	XX	XX	XX	X	XX	XX	X
Extraction and distribution of fossil fuels					X	X	X	X	
Solvent use									
Road transport	X				X	X	XX	X	
Other mobile sources and machinery	X				X	X	XX	X	
Waste treatment and disposal	XX	XX	X	X	X	X	X	XX	X
Agriculture			X					X	
Nature		X	X		X		X	X	X

[a]XX = Major source category.

Source activity data, which is the frequency with which sources generate emissions, are crucially important in establishing an accurate emissions inventory. In general, activity data should be linked to the emissions generation process as closely as possible. For example, emissions characteristics for combustion processes depend on the type of fuel. Thus, activity data must be reported separately for different fuels, not as total energy consumption or production. Special measurement campaigns might be needed whenever relevant activity data are not available or not readily accessible.

In the final step, emissions factors, source counts, and source activity data are used to calculate emissions. All emissions inventories ultimately seek to compute emission rates from a formula of the type

$$E_{\mathrm{i},\Delta t, A_{x,y}} = \sum_{t} \sum_{A_{x,y}} \sum_{\mathrm{s}} N_{\mathrm{s},A_{x,y}} \mathrm{U}_{\mathrm{s},A_{x,y},t} \Phi_{\mathrm{s,i},t}$$

where $E_{\mathrm{i},\Delta t,A_{x,y}}$ is the emission rate of species i over time interval Δt over an area specified by the coordinates x and y, $N_{s,A_{x,y}}$ is the number of sources of type s in area $A_{x,y}$; $\mathrm{U}_{s,A_{x,y},t}$ is the time-dependent usage given to source s in area $A_{x,y}$; $\Phi_{\mathrm{s,i},t}$ is the flux of species i from source s at time t. Flux terms are sometimes called emissions factors.

Verification of Emissions Data

Programs and approaches to verify or validate emissions data must respond to the rigorous demands being placed on emissions inventories. Concepts for emissions inventory verification are being reviewed and modified, particularly within the UN ECE Task Force on Emission Inventories (2). Since emissions data are inherently estimates, deriving statistically meaningful quantitative error bounds for inventory data is generally impossible. Frequently, however, ranges that bound the likely minimum and maximum for an emissions estimate can be provided or a quantitative data quality parameter can be developed to assess the relative confidence that can be associated with various estimates.

Efforts to improve emissions estimation reliability are focused on three major topics: the accuracy of activity data, the level of detail of procedures, and the quality of emissions factors.

The accuracy of the activity data relates to the availability of satisfactory statistics for all sources within socioeconomic activities considered in a given source-category grouping. Data related to natural processes may also be required. And, crucially, the factors used to represent emissions generation processes from these various activities and processes must be well determined. Various plausibility checks can be performed, such as comparing emissions factors for similar sources in different methodologies and inventories.

More advanced concepts of verification of emissions estimates include a thorough examination of the following activities:

1. documentation of data quality (quality assurance activities);
2. application of the completed inventories in: the assessments of the specific air-quality problems in an area, the air-quality modeling activities, and the regulatory activities including the design, implementation, and tracking of the effects of air-quality control strategies;
3. comparison of alternative estimates;
4. uncertainty estimates; and
5. techniques that make direct comparisons between emissions estimates and some other known quantity that is related either directly to the emissions source or indirectly to the underlying process that results in emissions (2).

Spatial Resolution of Emissions Data

Spatial resolution of emissions data is needed when the data are to be used as input to air-quality models. Many potentially useful allocation data are available from public statistics, but these always relate to administrative regions. However, modelers need allocation into rectangular grid systems. The statistical data are thus transformed into the grid cells using information on the geographical location of point sources, as well as allocation parameters that assume that the emissions determined collectively (emissions from area and line sources) can be regarded as homogeneously distributed within the individual governmental units. Examples of allocation parameters include population density, number of industrial employees, industrial energy use, length of traffic lanes, and ratio of arable land to total land area. The decision on the appropriate allocation or transformation method depends largely on the share of emissions already determined and allocated on a point-source basis.

Species Resolution of Emissions Data

The amount of chemical detail needed in an inventory depends on inventory use. Scientists generally require a significant amount. For example, photochemical oxidant models require the specification of many individual nonmethane volatile organic compounds (NMVOCs), because of differences in the reactivity and chemical degradation of these species. However, other photochemical models require various other NMVOC species, depending on the complexity of their chemical schemes.

Similarly, toxic compound models may require the specification of individual chemical forms of mercury and other heavy metals or the specification of various persistent organic compound species because of the species' differences in bioaccumulation, biomagnification, and toxicity. An example of the species resolution of mercury emissions in Europe at the end of the 1980s is presented in Figure 1 (3) as it might be provided to a scientific model. Such detail is merely an aggravating and expensive complexity for policymakers,

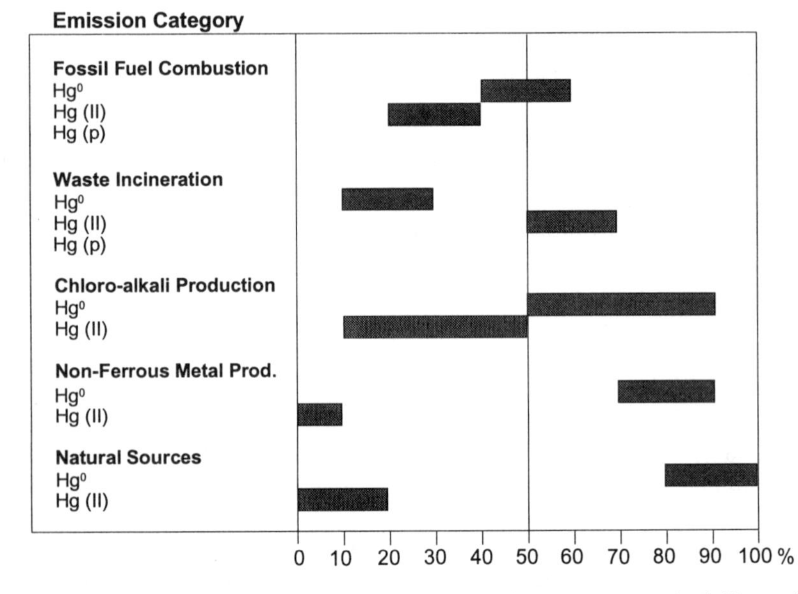

Figure 1 Species resolution of mercury emissions from various source categories in Europe (JM Pacyna, unpublished).

because legal constraints on emissions are generally crafted on a grouped compound basis.

Examples of Emissions Estimation Methodologies

The precise approach to inventory generation, the methodology, can often be clearly specified and reference data can be provided when one is dealing with a single governmental unit. For example, various guidebooks, emissions factor handbooks, and other documents describe methodologies for estimating emissions at regional and national levels. One of the first compilations of emissions factors was made at the US Environmental Protection Agency (EPA). The document, often called AP-42 (4), has been widely used in North America and Europe. An electronic version, the ClearingHouse for Inventories and Emission Factors (the CHIEF) recently became available (5).

Governments with common interests can often establish standard methodologies on a regional basis. A good example of such an approach is the CEC CORINAIR program, established to assist the CEC member countries in preparation of their emissions inventories (6). The CORINAIR85 inventory addressed national emissions as well as spatial distribution to each of 470 statistical regions of the Community. The project generated a source-sector nomenclature for socioeconomic activity as well as emissions source sectors, subsectors, and activities; a handbook of default emissions factors; and a computer software package for data input and the calculation of regional, sectoral, and national emissions estimates. After CORINAIR85 was completed, the CEC decided to support further development of the CORINAIR methodology and computer system so an emissions inventory for 1990, CORINAIR90, could be produced. An emissions factor manual has been developed to help estimate emissions of trace elements and persistent organic pollutants which are then deposited into the North Sea (7).

The UN ECE Task Force on Emission Inventories, in cooperation with the CEC, is preparing the Emission Inventory Guidebook, designed to provide an up-to-date comprehensive summary of emissions inventory methodology for each pollutant and source (8). Each chapter presents information on emissions of various pollutants from a given source category. The chapters have a common format, a crucial feature of the guidebook designed to ensure that users can readily locate and understand the essential aspects of the technical information. Supplementary documents, including various computer programs that estimate emissions from mobile sources and sources with temperature-driven emissions, such as field-applied pesticides, are also provided. The computer-based guidebook makes use of appropriate software, screen interfaces, and bulk storage media.

When methodologies are needed for the construction of global inventories, one must devise common practices for governmental units with widely differing financial resources, technical and administrative staffs, and levels of commitment to inventory assessment. Simply reporting and compiling numbers furnished by the participating governmental units invariably produces inventories containing data generated under differing assumptions and with varying degrees of precision. The policy challenge is to devise ways for such a group of governmental participants to generate reliable inventories with a mutually agreed-upon framework.

An example of such a global effort is the preparation of a compendium of methods to calculate emissions of greenhouse gases. This activity is being carried out under the auspices of the Organization for Economic Cooperation and Development (OECD), in cooperation with the International Energy Agency (IEA) and other international organizations, at the request of the Intergovernmental Panel on Climate Change (IPCC) (9). The resulting IPCC Guidelines for National Greenhouse Gas Inventories consist of three volumes: the Greenhouse Gas Inventory Reporting Instructions, the Greenhouse Gas Inventory Workbook, and the Greenhouse Gas Inventory Reference Manual.

The guidelines provide default methods and assumptions for characterizing the major sources and sinks of greenhouse gases. Users can choose methods most suited to their needs and capabilities. The guidelines also discuss weaknesses in existing methodologies for estimating greenhouse gas emissions and identify technical areas in which additional work is needed to develop better methods. In general, the IPCC guidelines present minimum requirements for estimating national emissions, but allow users to apply more detailed methods when possible.

The IPCC effort on greenhouse gas inventories is also an interesting illustration of the manner in which inventory methodologies are developed and implemented for international accords. Developing countries found the Guidelines too cumbersome and complex for their needs, although several European countries also missed the deadline for filing their reports. Another difficulty with the Guidelines is that they require countries to provide data on sinks (i.e. rates of surface loss) of atmospheric species and sources, yet we do not have the science to reliably generate these data.

These examples demonstrate the ways in which cooperative inventories have been accomplished; they also illustrate the problems inherent in achieving consistent and accurate reporting of emissions on a regional or global basis from diverse governmental units. The inevitable time delays and incomplete reporting, combined with the somewhat different needs of the scientific community (see Table 1), have led many researchers to bypass governmental units and perform their own emissions estimates. These inventories are often either the only ones available or the most reliable, particularly for global estimates. The most comprehensive such effort is the Global Emissions Inventory Activity (GEIA), a component of the International Global Atmospheric Chemistry Program (9a). GEIA is a volunteer association of some 60 scientists from more than 15 countries, and its teams are developing and updating global gridded inventories for virtually all the principal atmospheric emittants. Although the accuracy of such efforts cannot in principle approach that of coordinated and well-funded government efforts, they are often the best that can be hoped for over short to moderate time scales, especially from a research scientist's perspective.

TYPES OF EMISSION INVENTORIES

Inventories may, of course, be prepared for any gaseous or particulate species emitted into the air, but the principal foci of most inventory efforts are those substances that are the cause of many current and potential environmental problems. These problems include: (*a*) acidification (sulfur and nitrogen compounds), (*b*) global warming/climate change (greenhouse gases, particles), (*c*) air-quality degradation (sulfur and nitrogen compounds, photooxidants, heavy metals, persistent organic compounds, radionuclides), (*d*) damage and soiling of buildings and other structures (sulfur and nitrogen compounds, photooxidants), and (*e*) stratospheric ozone depletion (CFCs).

Table 3 Examples of types of inventories[a]

Designator	Geog Area	Spatial Resolution	Temporal Resolution
Supporting Inventories			
Vegetation	Global	1° × 1°	NA
Land cover	Global	1° × 1°	NA
Effects Inventories			
Acidification (SO_2)	Global	1° × 2°	Hourly
Air toxics (metals)	UK	None	Annual
Atmos. visibility (NO_x)	Global	80 km × 80 km	Hourly
Climate (CO_2)	Global	5° × 5°	Annual
Event Inventories			
Volcanic emissions	Global	None	NA
War-related emissions	Kuwait	Country	NA
Time Scale Inventories			
Historical	USA	States	Annual
Future projections	Global	Countries	NA

[a] Adapted from (50), where citations of the specific inventories are given.

Atmospheric emissions inventories may be of many types, depending on the uses for which they are intended. Some typical inventories are listed in Table 3, and the characteristics of the inventory types are discussed below.

Supporting Inventories

The first group of inventories are so-called supporting inventories: those that are needed for the construction of various emissions inventories although they are not in themselves atmospheric in nature. One set of these inventories, sometimes termed primary supporting databases, consists of information needed to estimate emissions from natural processes, e.g. nitric oxide from soils or volatile hydrocarbons from vegetation. Examples of this subset of supporting inventories include databases on topography, vegetation, and soil types.

Secondary supporting databases are used for anthropogenic rather than natural emissions estimates and provide geographic information on specific human-related activities such as the number of sources in an area and their time-dependent emissions. The most crucial data sets are perhaps political unit boundary locations (because many data are known by country, region, or area), land use, and human population. Other inventories of interest are those for various agricultural and industrial activities.

Keeping primary and secondary databases current is a prerequisite to developing and maintaining accurate emissions inventories. The United Nations and many individual government units provide annual updates on production and consumption statistics for various agricultural and energy activities. Updating the gridded databases should be straightforward as long as political boundaries and activity locations within the countries remain the same, although in practice the resources necessary to perform periodic updating are often limited or nonexistent.

Effects Inventories

Effects inventories are directed toward species that contribute to a specific environmental impact such as ozone depletion or photochemical smog. The inventories may be global or they may refer to a specific continent, country, or region. The most common of all the atmospheric inventories—single species emissions inventories—are included in this category.

Event Inventories

This group of inventories encompasses emissions from specific events. They are produced as needed, rather than on a regular basis. Examples of such inventories include, but are not limited to, volcanic emissions and atmospheric emissions related to war. Event inventories are often less accurate than other types, but are important in studying the atmosphere's response to sudden perturbations. A great deal of data on volcanic emissions has been gathered during the last few decades by research groups around the world (10). Emissions rates of various pollutants for individual volcanoes can vary by more than two orders of magnitude. Among these pollutants are sulfur, mercury, and several trace metals (arsenic, cadmium, selenium, and zinc).

Examples of war-related emissions inventories are those prepared for black carbon and aerosols emitted from the Middle East region during the 1991 Iraqi conflict.

Historic and Prospective Inventories

This group of inventories is composed of those inventories related to a time other than the present. Such inventories may refer to specific periods in the past or they may be future emissions scenarios. The former are used for studies related to historical atmospheric chemistry, such as comparisons with sediment core deposition data. The latter are used in connection with predictive studies of atmospheric chemistry and of air-quality or emissions-reduction policies.

EMISSIONS OF GREENHOUSE GASES

In this section, we present a review of major sources and fluxes of carbon dioxide, methane, nitrous oxide, chlorofluorocarbons and other halocarbons, and carbon monoxide. The review is based on emissions inventory practices within international programs and organizations, particularly those of the OECD/IPCC and GEIA.

Emissions of Carbon Dioxide

The short-run carbon cycle in the environment consists of three main components: the atmosphere, the oceans, and land vegetation. The current net flows, as well as the stocks, sinks, and sources are shown in Figure 2 (11).

The carbon content in the atmosphere was quite constant during the last 10,000 years (12), but major changes have been observed since the industrial revolution. The total carbon burden in the atmosphere has increased from 600 to 760 billion tonnes in 1992. At present, as much as 7 billion tonnes of carbon enters the atmosphere annually. The net flux is estimated at about 3.2 billion tonnes. Combustion of fossil fuels is by far the main anthropogenic source of carbon dioxide. The total emissions of carbon dioxide from fossil fuel burning, cement manufacturing, and gas flaring was estimated at 6.2 billion tonnes carbon in 1992 (13). These emissions have been allocated by a GEIA project team to a global 1° by 1° grid (RJ Andres, G Marland, I Fung, & E Matthews, private communication, 1995).

Along with fossil fuel use, land use changes are also sources of carbon

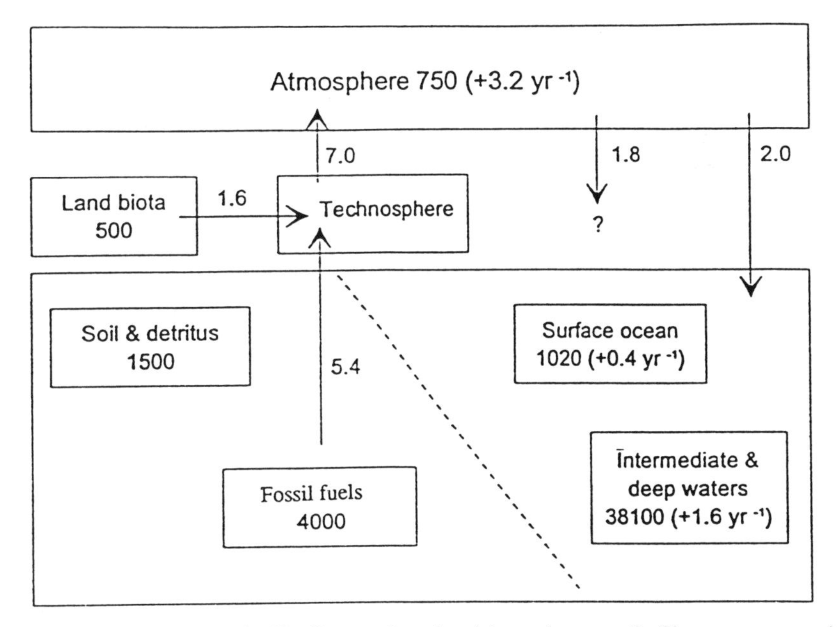

Figure 2 The carbon cycle: Net flows and stocks, sinks, and sources (in Gigatonnes per year). Boxes indicate stocks; arrows indicate flows (11).

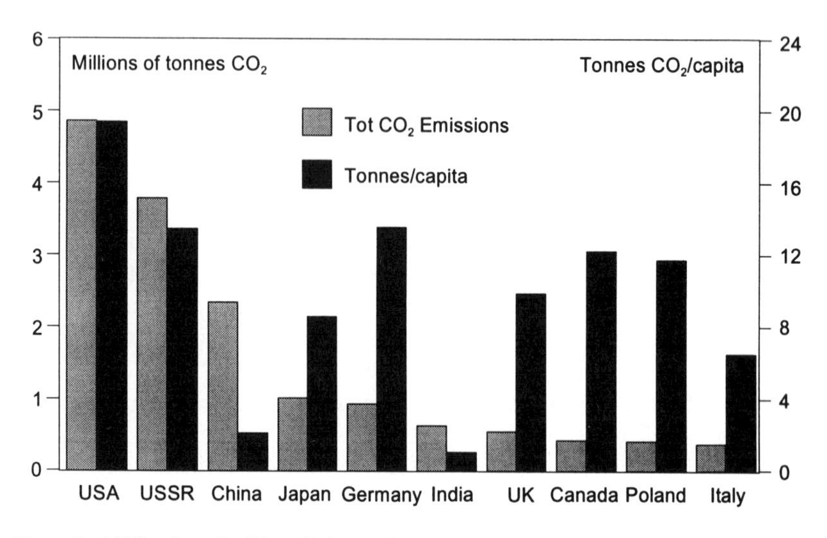

Figure 3 **1989 carbon dioxide emissions estimates for selected countries (13).**

dioxide emissions. In certain countries, these changes, especially forestation changes, are among the most important factors in carbon dioxide net emissions. In New Zealand, for example, reforestation has become the dominant source of reductions in carbon dioxide emissions.

Carbon dioxide emissions estimates for 1989 in several countries that burn fossil fuels are presented in Figure 3. The United States, the former Soviet Union, and China head the list of major emitters of carbon dioxide to the atmosphere, contributing about half of the world emissions from fossil-fuel burning.

The importance of coal combustion as a source of greenhouse gases will most likely grow with time, because the world's coal reserves are very large in comparison with those of petroleum and natural gas. In addition, coal is generally the least expensive reliable source of energy.

Emissions of Methane

Methane is an important greenhouse gas that accounts for about 15% of the current greenhouse forcing, based on model calculations (15). Methane sources are numerous, diverse, and geographically dispersed. Additionally, methane is generally emitted by complex and highly variable biological and industrial systems whose inputs and outputs cannot be translated easily into emissions. Therefore, precise measurement of emissions from individual locations is difficult. A preliminary estimate of methane emissions from various sources

Table 4 Global methane emission source fluxes (16) in million tonnes C year^{-1}

Source	Methane
Fossil fuel	
Coal mining (no post mining)	35-50
Natural gas	25-42
Transmission loss > vented/flares loss	
Biomass burning	45
(Burning of forest/grass, agricultural waste)	
Natural wetlands	100-115
High latitude (bogs, tundra)	30-35
Low latitude (swamps, alluvial)	70-80
Rice cultivation	40-70
Animals—mainly domestic ruminants	80
Termites	20
Oceans and coastal sediments	10
Landfills	20-25
Waste water treatment and animal waste	30
Fuel combustion	28
Other minor (industrial, residential waste burning, peat mining, geothermal)	
TOTAL	442-524

is shown in Table 4 (16). The total emissions estimate of 442–542 million tonnes carbon agrees well with the estimate presented by the IPCC [440–640 million tonnes carbon (15)], although there are some differences for individual source categories.

Natural wetlands generate the largest fraction of methane emissions, followed by animals, rice cultivation, and biomass burning. The area and type of wetlands, as well as general seasonal dynamics of water and temperature, are important when assessing this source. Methane production from animals is affected by quantity and quality of feed, body weight, age, and activity level, and therefore varies among animal species, as well as among individuals of the same species. Many factors affect the production, transport, and release of methane in flooded rice fields, among them temperature, water status, fertilizer application, soil properties, and plant phenology.

Emissions of Nitrous Oxide

Nitrous oxide is generated by a variety of biogenic sources such as natural soils and oceans (17). Many anthropogenic sources of the gas are also known, including fertilized fields, animal nitrogen excretion, postburn effects of land

Table 5 Global nitrous oxide source fluxes (51) in million tonnes N $year^{-1}$

Source	Emission flux
Natural soils	7.0
Oceans	3.6
Animal waste	1.0
Cultivated soils	0.4
Tropical forest conversion	0.4
Mobile sources	0.3
Adipic acid production	0.3
Nitric acid production	0.2
Fuelwood combustion	0.1
Biomass burning	0.1

use changes, fossil fuel combustion, trash incineration, traffic, and some industrial activities. A global inventory has also been prepared for N_2O emissions from biomass burning (18). All of this information was used by GEIA experts to develop a global emissions inventory for nitrous oxide, shown in Table 5 (18a). The total emissions range from 12.3–22.8 million tonnes of N_2O-N, with more than half from natural sources.

Emissions of Chlorofluorocarbons (CFCs) and Other Halocarbons

Unlike carbon dioxide and methane, CFCs and halons are all manufactured; they do not occur in nature. Since they were introduced in the 1930s, CFCs have become widely used for refrigeration, air-conditioning (AC), aerosol propellants, production of foam packing and insulation, and as solvents. Halons are used in fire extinguishers. At present, the total emissions of CFCs are about 0.8 million tonnes chlorine annually. The emissions trends for CFC-11 and CFC-12 during the period from 1931 to 1989 are shown in Figure 4. The production and release of CFCs has been declining in the past few years as a result of the Montreal Protocol agreements to limit and eventually stop production of these compounds. The resulting decreases in global releases have been reflected in time-series measurements at ground-based observing sites (20).

Emissions of Carbon Monoxide

The major sources of carbon monoxide emissions include incomplete combustion, particularly in engines and furnaces, where emissions are affected by type of fuel, technological parameters of combustion, and the type and efficiency of control equipment. Other sources of carbon monoxide emissions include:

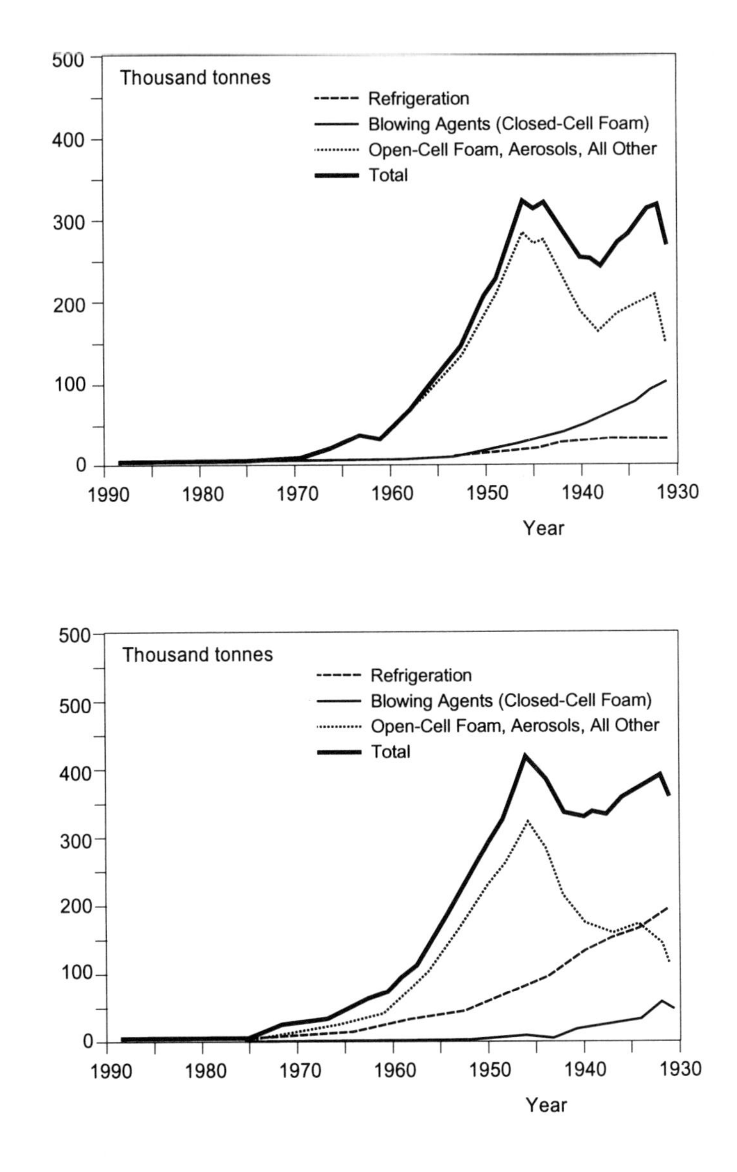

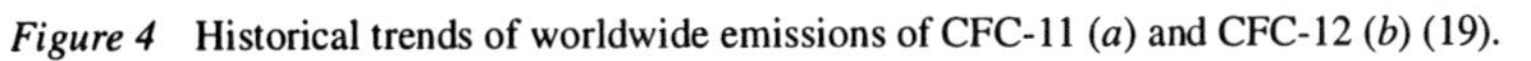

Figure 4 Historical trends of worldwide emissions of CFC-11 (*a*) and CFC-12 (*b*) (19).

Table 6 Global and tropical budget of atmospheric carbon monoxide (19) in million tonnes CO year^{-1}

Budget item	Global	Tropics (30°S-30°N)
Sources		
Technological sources	440 ± 150	—
Biomass burning	700 ± 200	600 ± 200
Vegetation	75 ± 25	60 ± 20
Ocean	50 ± 40	25 ± 20
CH_4 oxidation	600 ± 200	400 ± 150
NMHC oxidation	800 ± 400	500 ± 200
Total production	2700 ± 1000	1600 ± 600
Sinks		
Oxidation by OH	2000 ± 600	1200 ± 400
Uptake by soils	250 ± 100	70 ± 35
Flux into stratosphere	110 ± 30	80 ± 20
Total destruction	2400 ± 750	1400 ± 450

incineration and open burning of biomass and wastes, particularly agricultural wastes and grasslands; industrial production without combustion, such as steel manufacturing; and extraction and distribution of natural gas that contains carbon monoxide.

Carbon monoxide plays a minor role as a greenhouse gas: Its total emissions are about two orders of magnitude lower than those of carbon dioxide. Preliminary estimates of the global and tropical budget of atmospheric carbon monoxide are presented in Table 6 (19).

EMISSIONS OF ACIDIC COMPOUNDS AND PHOTOOXIDANTS

Acidification of precipitation and the formation of photooxidants in the atmosphere are recognized as regional environmental phenomena. Major contributors to the formation of acid rain and photooxidant buildup are emissions of sulfur and nitrogen species and of volatile organic compounds. These emissions are reviewed here.

Emissions of Sulfur and Nitrogen Species

The majority of the anthropogenic emissions of sulfur oxides into the atmosphere is in the form of sulfur dioxide. Other sulfur compounds, including sulfates, sulfuric acid, and nonoxygenated compounds are also emitted, but are of less importance.

Almost all solid and liquid fuels contain sulfur. The combustion of sulfur-

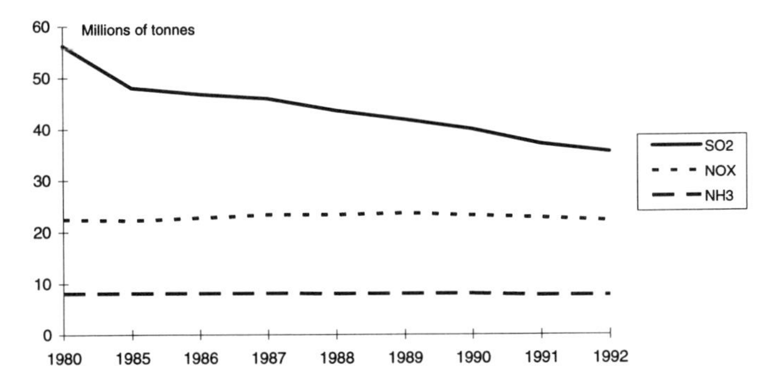

Figure 5 Historical trends of emissions of sulfur and nitrogen oxides and ammonia in Europe (21).

containing fuels and wastes is by far the main source of sulfur oxide emissions into the air. Other sources include processing of sulfur-containing materials (e.g. roasting or sintering of ores), production of sulfur compounds, (e.g. sulfuric-acid manufacturing), using sulfur compounds to produce other industrial goods (e.g. cellulose production with the use of the sulfide process), and processes of sulfur removal (e.g. from liquid and gaseous fuels). The sulfur content of fuels and sulfur retention in ashes, along with the type and efficiency of flue-gas desulfurization installations, are the major parameters affecting the amount of emissions from combustion processes.

Combustion of fuels in stationary and mobile sources is also a predominant source of nitrogen oxide emissions (excluding nitrous oxide). Three major mechanisms are responsible for the formation of nitrogen oxides during combustion: fixation of atmospheric nitrogen in the combustion air (thermal nitrogen oxides), conversion of chemically bound nitrogen in the fuel (fuel nitrogen oxides), and interaction of reactive radicals in the primary combustion zone (prompt nitrogen oxides). Similar processes result in the formation of nitrogen emissions during waste combustion. The amount of emissions is highly dependent on the parameters related to fuel or waste characteristics (e.g. nitrogen content), the apparatus (e.g. design), the operating conditions, and the type and efficiency of control equipment. Other sources of nitrogen oxide emissions include: production of nitrogen-containing chemicals (e.g. nitric acid), open thermal processes involving oxygen (e.g. production of steel in open-hearth and electric arc furnaces), and use of nitrogen compounds (e.g. for surface treatment).

In Europe, the most comprehensive inventories of sulfur and nitrogen emissions are those assembled by the Cooperative Programme for Monitoring and Evaluation of the Long Range Transmission of Air Pollutants in Europe

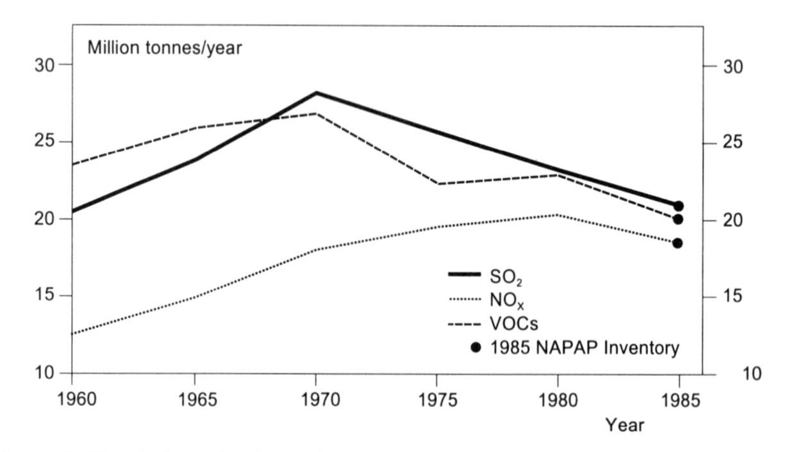

Figure 6 Historical trends of emissions of sulfur and nitrogen oxides and volatile organic compounds in the United States (22).

(EMEP) (21), operated under the UN Economic Commission for Europe (ECE) Convention on Long Range Transboundary Air Pollution. National total and disaggregated emissions data are collected from the Parties to the Convention. Based on these emissions data, two EMEP meteorological centers calculate budgets for airborne acidifying and photochemical compounds across Europe. The results of the calculations form the scientific basis for discussions of emissions reduction protocols within the Convention. The first EMEP emissions inventories included only sulfur dioxide, and the information provided by the countries was rather incomplete. More recently, nitrogen oxides, NMVOCs, and ammonia emissions rates have also been collated or estimated by the EMEP centers. The trends of total emissions of sulfur (decreasing) and nitrogen dioxide (approximately constant) are presented in Figure 5.

The most comprehensive emissions inventories for the United States and Canada have been compiled by the National Acid Precipitation Assessment Program (NAPAP) (22). Among the aims of this program, which was reauthorized under the 1990 Clean Air Act, was to determine what reductions in the deposition for various compounds were needed to prevent adverse ecological effects. The decreasing trends of total emissions of sulfur and nitrogen oxides in the United States are presented in Figure 6.

In Asia, SO_x and NO_x emissions inventories for 25 countries have been developed for several reference years in the 1970s and 1980s (23). The information was compiled on a provincial and regional basis for China and India and on a national basis for all other countries. Recently, these results were apportioned to a 1° by 1° grid (24).

The European, North American, and Asian emissions inventories were used by a GEIA project team, along with information on emissions in other regions of the world, to prepare global emissions inventories for SO_x and NO_x gridded to 1° by 1° (CM Benkovitz, MT Scholtz, J Pacyna, L Tarrason, J Dignon, et al, personal communication, 1995). The spatial distribution of sulfur oxides emissions from this work is shown in Figure 7. The emissions map demonstrates that sulfur emissions are highest throughout North America, Europe, India, and southeastern Asia, and they are virtually absent in much of northern South America, northern Africa, central Asia, and Australia.

Natural sources of emissions are often spatially quite distinct from anthropogenic sources. This trait is demonstrated by Figure 8, a gridded global emissions inventory for oxides of nitrogen emissions from soils (26). The highest emissions occur in northern South America, central Africa, and the southeastern United States, reflecting emissions owing to agriculture, grasslands, and tropical rain forests.

Emissions of Ammonia

Biological degradation processes within animal husbandry wastes, commercial fertilizers, sewage treatment, and landfills are the major anthropogenic sources of ammonia emissions to the atmosphere. Of particular importance is the decomposition of nitrogen compounds in domestic animal wastes during housing of the animals, application and storage of manure, and the period when animals are out to pasture. Other sources of ammonia include the manufacture of certain nitrogen compounds such as ammonia and artificial fertilizers, handling of ammonia in cooling installations, and combustion of solid and liquid fuels and wastes.

Information on ammonia emissions in the European countries, collected within EMEP, reveals a stable pattern with time, as presented in Figure 5. A preliminary global emissions inventory for ammonia is summarized in Table 7 (26a). The total estimate of 75 million tonnes of ammonia nitrogen agrees well with an earlier estimate of 62.5 million tonnes NH_3-N [including 52.7 million tonnes NH_3-N from anthropogenic sources (27)], and a more recent inventory on a 10° by 10° grid (28). An updated emissions inventory is now being prepared on a 1° by 1° grid by a GEIA project team.

Emissions of Nonmethane Volatile Organic Compounds

Nonmethane volatile organic compounds are often inventoried as a group, though scientific research teams generally require some degree of speciation breakdown. Emissions sources of NMVOCs are numerous. They include incomplete combustion (particularly in engines); incineration and open burning of wastes and biomass; industrial production processes, mostly in the chemical and petrochemical industries; end-use of products and solvents; processing of

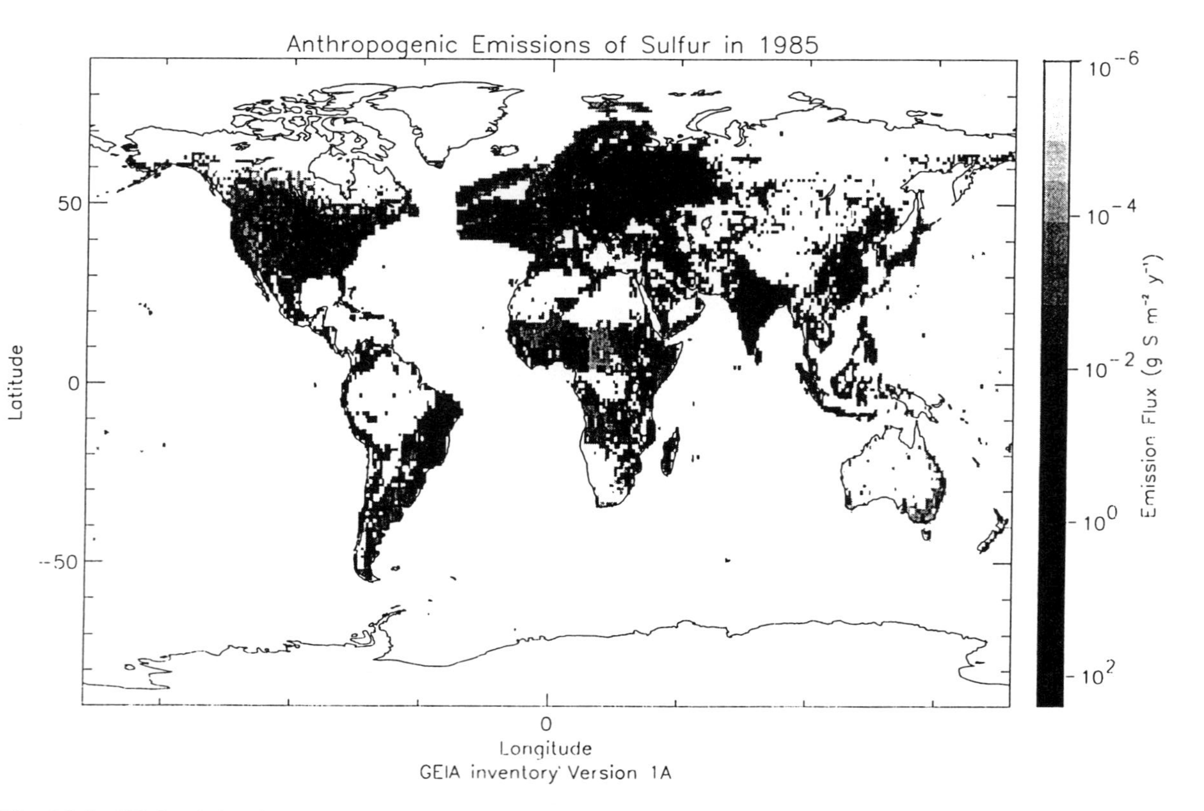

Figure 7 The global gridded emissions inventory map (1985 epoch) for anthropogenic oxides of sulfur (CM Benkovitz, MT Scholtz, J Pacyna, L Tarrason, J Dignon, et al, Personal communication, 1995).

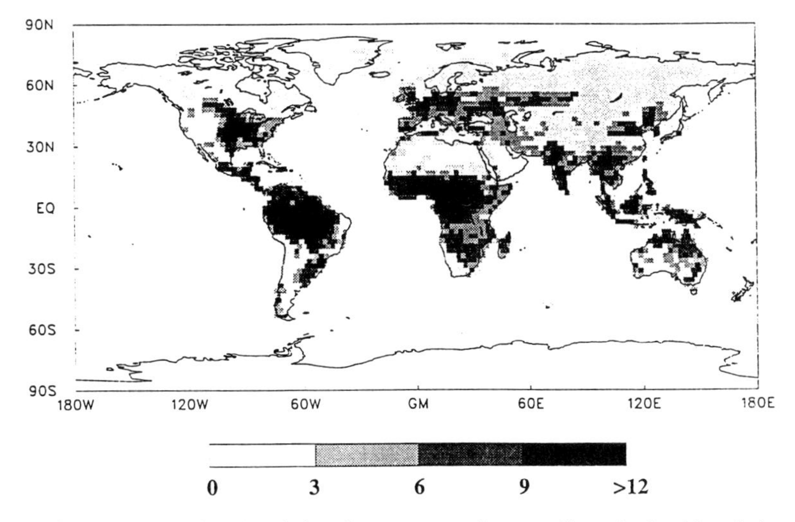

Figure 8 **The global gridded emissions inventory map for naturally-emitted oxides of nitrogen [mmole N (m^{2}year)$^{-1}$](26).**

Table 7 Global ammonia source fluxes (52) in million tonnes N year^{-1}

Source of ammonia	Emission
Natural soils	10
Wild animals	
Biomass burning	5
Coal combustion	2
Domestic animals	32
Fertilizer application	9
Fertilizer production	
Seas and oceans	13
Human excretion	4
Sewage sludge	
Automobiles	0.2
Agricultural crops	—
Total on global scale	75

Table 8 Anthropogenic NMVOC source fluxes (16)

	Developed countries			Developing countries			World		
Activity	Tg product	kg Mg^{-1}	Tg NMVOC	Tg product	kg Mg^{-1}	Tg NMVOC	Tg product	Tg NMVOC	%
Fuel production/ distribution									
Petroleum	1,300			1,800			3,100	8.0	6.0
Natural gas	1,700			450			2,150	2.0	1.5
Oil refining	2,050		3.2	900	2	1.8	2,950	5.0	3.5
Gasoline distribution	600	3	2.0	130	3	0.4	730	2.5	2.0
Fuel consumption									
Coal	150	6	0.9	450	6	2.7	600	3.5	2.5
Wood	250	16	4.0	1,300	16	20.8	1,550	25.0	18.0
Crop residues (incl. waste)	650	7	4.6	1,400	7	9.8	2,050	14.5	10.0
Charcoal				20	120	2.4	20	2.5	2.0
Dung cakes				400	7	2.8	400	3.0	2.0
Road transport								36.0	25.0
Chemical industry	300	5	1.5	30	10	0.3	330	2.0	1.5
Solvent use	15	1,000	15.0	5	1,000	5.0	20	20.0	14.0
Uncontrolled waste burning								8.0	5.0
Other								10.0	7.0
Total								142	100

various organic products using solvents; evaporation during storage, transfer, and handling operations of volatile organics (such as vehicle fuels); evaporation from cars; and fermentation in the production of food and alcoholic beverages. Obviously, many parameters, such as temperature, fuel composition, and combustor type, affect the amount of NMVOC emissions from these sources.

EMEP and NAPAP inventories contain information on NMVOC emission fluxes in Europe and North America, respectively. These data are presented in Table 8 along with estimates of NMVOC emissions from sources in other parts of the world (16). The downward emissions trend of NMVOCs in the United States is presented in Figure 6.

Species profiles for major NMVOC source categories are now being prepared, particularly within the UN ECE *Atmospheric Emission Inventory Guidebook* (8). The spatial resolution of global emissions of anthropogenic NMVOCs on a 1° by 1° grid is under development by a GEIA project team to complement an earlier GEIA inventory on emissions of natural NMVOCs (29).

EMISSIONS OF TOXIC COMPOUNDS

Emissions of two major groups of toxic compounds are discussed here: trace metals and persistent organic pollutants (POPs). Recent studies on the behavior of trace metals and POPs in the environment have shown that many of these compounds have the potential to create serious problems because of their toxicity and bioaccumulation in various environmental compartments.

Emissions of Trace Metals

High temperature processes such as coal and oil combustion in electric power stations and heating and industrial plants; gasoline combustion; roasting and smelting of ores in nonferrous metal smelters; melting operations in ferrous foundries; refuse incineration; and kiln operations in cement plants emit various trace metals into the atmosphere. The amounts of trace metal emissions depend on: the contamination of fossil fuels and other raw materials by these compounds, trace metal physico-chemical properties affecting the compounds' behavior during industrial processes, the technology of industrial processes, and the type and efficiency of control equipment.

The first attempt to estimate atmospheric emissions of trace metals from anthropogenic sources in Europe was completed at the beginning of the 1980s (30). This survey included information on emissions of 16 major trace metals. In recent years, the emissions estimates have been updated. The latest estimates of As, Cd, Cr, Cu, Hg, Ni, Pb, Se, and Zn emissions in Europe are presented in Table 9 (31). A spatial distribution of the As, Cd, Hg, Pb, and Zn emissions estimates in Europe is also available within the EMEP grid system of 150 km by 150 km (32).

Pursuant to the requirements of the 1990 Clean Air Act Amendment, an interim toxic emissions inventory has been developed for the continental United States. Preliminary results of this work include the geographical distribution and source type analysis (33) and are based on the 1985 NAPAP inventory (now being updated; see References 34–36). Metal emissions data for major sources in Canada are presented in References 36 and 37.

The first quantitative worldwide estimate of the annual industrial input of 16 trace metals into the air, soil, and water has been published by Nriagu and Pacyna (38). The study used information collected during the estimation of atmospheric emissions of trace metals in Europe and North America. Recently, the global emissions inventory of lead was updated for 1989. The estimates show that lead emissions are between 150,000 and 209,000 tonnes annually, with 62% from gasoline combustion and 26% from nonferrous metal produc-

Table 9 Heavy metal source fluxes in Europe in tonnes year^{-1}

Source category	As (1982)	Cd (1989)	Cr (1979)	Cu ((1983)	Hg (1987)	Ni (1979)	Pb (1989)	Se[b] (1979)	Zn (1982)
1. Stationary fuel combustion	710	330[a]	2,780	1,880	405	12,050	3,830	373	3,290
2. Non-ferrous metal industry	3,660	764	—	1,360	29	1,780	12,960	13	26,700
3. Iron and steel production	230	53	15,400	870	2	340	3,900	—	9,410
4. Gasoline combustion	—	—	—	—	—	1,330	46,720	—	—
5. Other sources	370	—	720	940	290	500	1,820	34	5,190
Total	4,970	1,147	18,900	5,050	726	16,000	69,230	420	44,590

[a] including "other sources"
[b] Se emissions on particles

tion (39). The upper range of these estimates is presented in Table 10. The emissions in 1989 were dominated by leaded gasoline use in Europe and Asia, with strong contributions from metal production activities in the same regions.

Emissions of Persistent Organic Pollutants (POPs)

Ambient air contains many volatile and semivolatile organic compounds (VOCs and SOCs). The distinction between the two groupings is that VOCs are present in air solely as vapors, whereas SOCs consist at least partly of airborne particles. Persistent organic pollutants are generally SOCs because of their high molecular weights. Among the various classes of POPs, the greatest emphasis has been placed on estimating emissions of polycylic aromatic hy-

Table 10 Global lead source fluxes, maximum scenario, tonnes year^{-1}

Continent	Gasoline combustion	Fossil fuel combustion	Non-ferrous metal production	Waste incineration	Cement production	Steel and iron	Total
Africa	12,316	628	4,421	—	85	86	17,536
Asia	44,006	4,209	21,323	207	867	3,713	74,325
Australia	2,000	411	2,775	62	12	118	5,378
Europe	47,579	3,477	13,041	537	641	4,278	69,553
North America	14,192	993	7,613	2,498	177	1,316	36,789
South America	8,796	195	5,422	—	98	555	15,066
Total	128,889	9,913	54,595	3,304	1,880	10,066	208,647

drocarbons (PAHs), polychlorinated biphenyls (PCBs), polychlorinated dibenzo-p-dioxins (PCDDs), polychlorinated dibenzofurans (PCDFs), and several pesticides.

The source categories that generate the largest amounts of PAHs, PCBs, PCDDs, and PCDFs include industrial processes, stationary fuel combustion, solid waste incineration, agricultural uses of POPs, transportation, and various open sources.

Emissions of some of the above-mentioned POPs have been inventoried in Europe (32) and North America (40). One conclusion from these inventories was that the uncertainties are substantial and difficult to quantify. This problem is largely a result of the poor quality of emissions data, which are often incomplete and highly variable among sources, even within the same class of POPs. Proprietary restrictions also make the generation of reliable pesticide use and production figures difficult. Nevertheless, the global usage and emissions of selected POPs have been preliminarily quantified (41). The cumulative global usages of toxaphene, DDT, hexachlorocyclohexane (HCH), and lindane were estimated at 450,000, 1,500,000, 550,000, and 720,000 tonnes, respectively. The overall usage of toxaphene and of DDT from 1950–1993 was estimated at 1,330,000 and 2,600,000 tonnes, respectively. The compounds are still legally used in several countries and the suggestion has been made that illegal use is occurring in other countries.

QUALITY ASSESSMENTS OF EMISSION INVENTORIES

Although assigning precise quality metrics to existing emissions inventories is difficult, expert analysts generally can provide qualitative merit ratings. Such an activity is particularly useful in prioritizing additional emissions inventory efforts. Merit-rating assessments were performed by project leaders in the Global Emissions Inventory Activity several years ago (42). We present in Table 11 an updated version of those ratings for global emissions inventories for several species or groups of species. In our opinion, the only species for which total global fluxes are reasonably well-determined are the CFCs, carbon dioxide, the sulfur oxides, and the nitrogen oxides. On a regional basis, the total fluxes of those species plus carbon monoxide but minus CFCs also appear safisfactory for the more developed countries.

More detailed spatial resolution of inventory totals is problematic. For most of North America, Western Europe, and Japan, the spatial resolution of the emissions information can be regarded as good for carbon dioxide, carbon monoxide, nitrogen oxides, and sulfur oxides. Otherwise, detailed regional inventories are available for very few species and very few regions, and the total fluxes and spatial resolutions of those that do exist require refinement. On a global basis, the spatial resolution of many inventories is poor or non-

Table 11 Status assessments of emissions inventories

		Spatial resolution		Temporal resolution	
Species	Global flux	Specific regions	Overall	Specific regions	Overall
CO_2	G[a]	G	P[b]	G	G
CO	F[c]	G	P	F	NI[d]
CH_4	F	F	P	P	P
VOC	P	F	P	F	P
PAH	P	P	NI	NI	NI
Chlorinated HC	F	F	P	NI	NI
NO_x	G	G	F	F	P
N_2O	P	F	P	F	F
NH_3	P	F	NI	F	NI
CFC	G	P	P	P	F
SO_2	G	G	F	F	P
Reduced S	P	F	P	F	P
HCl	P	F	NI	NI	NI
HF	NI	P	NI	NI	NI
Radon	F	P	P	NA[e]	NA
TPM	P	F	P	NI	NI
SO_4^{2-}	NI	P	NI	P	NI
Metals	F	F	F	NI	NI
Soot	P	P	P	NI	NI

[a] G: good; [b] P: poor; [c] F: fair; [d] NI: no inventory; [e] NA: not applicable.

existent. The temporal resolution is generally no better. All in all, present emissions inventories are unsatisfactory for most species surveyed. In nearly all cases, however, improvements are being made as appropriate effort is expended in data acquisition and assimilation.

The spatial scale and chemical nature of an inventoried compound has an important effect on the quality of emissions data. Contamination of the environment by selected trace metals such as thallium, tin, or chromium is generally limited to a certain region, and accurate emissions information in such cases is needed for that region alone. Other pollutants, such as mercury, are much more widely dispersed by air motions. For these pollutants, and for highly reactive species, accurate gridded emissions data on the global scale are needed. Some pollutants, such as CO_2, are long-lived in the atmosphere, but are only weakly reactive or unreactive; their longevity makes their emissions surveys more critical than surveys of short-lived pollutants, but minimizes the need for high spatial resolution. In many such cases, the quality of the resulting inventory can be improved by time or space averaging without loss of usefulness.

INTERNATIONAL AGREEMENTS ON EMISSION REDUCTIONS

About 170 treaties, most of them drafted in the past 20 years, are designed to safeguard the environment (43). Many of these treaties deal with emissions reductions. The 1987 Montreal Protocol is one of the best known agreements to reduce emissions on a global scale. This protocol was prepared to halt the destruction of the ozone layer by CFCs; its latest amendments ban CFC production in the industrial world after January 1, 1996. Members of the Montreal Protocol are forbidden to purchase CFCs or products containing them from nations that have not agreed to the treaty. These provisions were a significant factor in convincing more than 100 countries to sign the agreement.

More controversial is the agreement to ameliorate climate change, discussed by more than 150 governments at the UN Conference on Environment and Development (UNCED) in Rio de Janeiro in June 1992. The final version of this treaty urges countries to stabilize their carbon dioxide emissions, as well as emissions of other greenhouse gases not controlled by the Montreal Protocol, at 1990 levels by the year 2000. The version of the treaty that required—not urged—this stabilization was not acceptable to the United States or to over 100 developing countries. This example illustrates how difficult it is to obtain agreement on environmental issues when international decisions must be unanimous (43).

Various regional treaties are also designed to reduce emissions to the air. The UN ECE, encompassing all the countries of Europe and North America, as well as Israel and the Asian Republics of the former Soviet Union, has been the only permanent intergovernmental forum for economic cooperation. In the past three years, the membership of ECE has risen from 34 to 54 countries. In March 1983, the Convention on Long-range Transboundary Air Pollution entered into force, 90 days after it was ratified by 24 countries. By May 1, 1994, the convention had been ratified by 38 parties. The Convention constitutes a framework within which the member countries identify the problems posed by transboundary air pollution and accept their responsibility to undertake appropriate abatement action (44).

Although the ECE Convention covers all types of air pollutants, it gave priority in the first phase to the abatement of acid rain triggered primarily by emissions of sulfur compounds. In July, 1985, the Protocol on the Reduction of Sulphur Emissions or Their Transboundary Fluxes By at Least 30% was signed in Helsinki (the so-called Helsinki Protocol). This reduction from 1980 levels of emissions was expected to be achieved by 1993 at the latest. However, it is still too early to discuss the effectiveness of this agreement because not all countries have yet reported their 1993 and 1994 emissions data to the Convention. A new Protocol on Further Reduction of Sulphur Emissions After

1993 was adopted in June 1994. This new agreement was negotiated to obtain emissions reductions of the 1980 levels increasing from 30% to 87% by the year 2010. In addition, requirements have been set for emissions controls at certain stationary combustion sources and for the sulfur content of petroleum products.

To moderate acid rain and ozone formation, the ECE Convention also provided a platform for emissions reduction protocols on nitrogen oxides and volatile organic compounds (VOCs). In October 1988, the Protocol concerning the Control of Emissions of Nitrogen Oxides or Their Transboundary Fluxes was signed in Sofia, Bulgaria (the so-called Sofia Protocol). The countries agreed to reduce or control emissions "so that these, at the latest by 31 December 1994, do not exceed their national annual emissions of nitrogen oxides or transboundary fluxes of such emissions in the calendar year 1987." Alternatively, a country may choose any year prior to 1987 as a base year, provided that the average annual national level of emissions between 1987 and 1996 does not exceed the 1987 level (44).

In November 1991, in Geneva, the Protocol concerning the Control of Emissions of Volatile Organic Compounds or Their Transboundary Fluxes was adopted (the so-called Geneva Protocol). The emissions target set out in the basic obligations of the Protocol is to reduce national annual VOC emissions by at least 30% by 1999, using either 1988 levels as a basis or any other annual level between 1984 and 1990. However, alternatives to these obligations are provided for either specific regions or situations in which the national annual emissions of VOCs in 1988 were below 500,000 tonnes, 20 kg per inhabitant, and 5 tonnes per square kilometer (44).

Separate legal measures are discussed under the ECE Convention to reduce emissions of POPs and heavy metals, because these groups of compounds have been found to be subject to transboundary transport and to have detrimental effects on human health and the environment. The effectiveness of environmental agreements can be hindered by various factors, such as the requirement for unanimity mentioned previously. However, these agreements are the most viable instruments available to achieve emissions reductions. In the case of CFCs, the results of the international agreements are already obvious.

FUTURE EMISSIONS SCENARIOS

Various forecasting models are used by decision makers to predict future emissions of air pollutants. These models are based on mathematical equations and formulas that represent the relationship between the emissions and the prospective activities that may result in changes of emissions. Devising and developing satisfactory emissions projections is very difficult. The accuracy of forecasting models depends on the accuracy of the model inputs (which are themselves projections of the future, e.g. economic growth models) and the key model parameters, which summarize future relationships between inputs and model outputs. As a consequence, model forecasts can be highly uncertain.

Various factors need to be considered in order to construct emissions scenarios that are as reliable as possible. All major emitters, both existing and potential, must be identified and their future activity levels projected. The economic factors that form the basis for operating and investment decisions, such as fuel prices, interest rates, and cost of control equipment, need to be projected and considered. The same considerations apply to the rate of technological progress. Phasing in new emissions regulations should also be considered. The demographic composition of the population (e.g. wealthier people buying more goods and using more energy than poorer people) is also taken into account in some emissions forecasting models.

Approaches to forecasting emissions through the application of various models have been summarized and discussed by the US EPA in connection with the National Acid Precipitation Assessment Program (45). Several models that forecast emissions are available for the electric utility sector; they use historical experience and expected future trends to make predictions about future capacity utilization rates and emissions rates. Alternatively, the rates are forecast on the basis of simulations of electric utility behavior, such as plant dispatching, under various economic factors (46).

Much less effort has been made to forecast emissions from other energy-consuming sectors, including the industrial sector, the transportation sector, and residential and commercial boilers. Such models exist, but often employ less-sophisticated modeling techniques (45). Emissions scenarios for the transportation sector are generally based on models designed to forecast aggregate fuel consumption, using information on changes in vehicle fleet registration and per-vehicle utilization. Some of these models use time-trended vehicle registration or fuel sales data to derive information on travel mileage or fuel consumption for further use in emissions forecasting. Often, the forecast models for transportation emissions are developed on a local scale, e.g. urban or metropolitan, in contrast to utility-emissions models.

Emissions forecasting can be done on a local, regional, or global basis to suit various purposes. Thus, following the results of reviews on national strategies and environmental policies within the UN ECE region, the Convention on Long Range Transboundary Air Pollution has requested its members to report on projections for the years 2000, 2005, and 2010 for national emissions of sulfur, nitrogen oxides, ammonia, NMVOCs, and carbon dioxide. The Convention recommended that the projection be based on current national plans for emissions reduction. Some members will probably also make projections for methane, carbon monoxide, persistent organic pollutants, and heavy metals. The UN ECE Task Force on Emission Projections has been established to help prepare the emissions forecasts (47).

Guidelines on estimation and reporting of emissions projections within the ECE Convention have been prepared with a focus on two scenarios. The first scenario covers current reduction plans, reflecting the politically determined intention to reach specific targets. The second scenario is the baseline emissions projection, which reflects the state of legal/regulatory emissions constraints in place as of the end of the year prior to the reporting deadline.

For international purposes, common key factors, which are exogenous to models for emissions projections, should be made consistent. These factors include energy prices, international economic growth, international regulations, and technological evolution.

Various global emissions scenarios have also been prepared. The IPCC has used various approaches to forecast future emissions of greenhouse gases (48). In 1990, global models were used to develop four emissions scenarios which were then used to prepare the future warming scenarios. These original reference scenarios were: (*a*) Business as Usual, (*b*) Low Emissions, (*c*) Control Policies, and (*d*) Accelerated Policies. In 1992, the original reference scenarios were updated with new information, and six new reference scenarios based on varied rates and patterns of development were generated. These scenarios were designed for one purpose: to help evaluate the environmental/climatic consequences of a nonintervention scenario, e.g. no action to reduce greenhouse gas emissions. In 1994, the 1992 scenarios were assessed. A new approach consisting of a set of purposes for scenarios and criteria to satisfy those purposes was proposed. The following purposes were suggested (49):

1. to help evaluate the environmental/climatic consequences of nonintervention (already mentioned above);
2. to help evaluate the environmental/climatic consequences of intervention to reduce greenhouse gas emissions;
3. to help examine the feasibility and costs of mitigating greenhouse gases from different regions and economic sectors and over time (including the examination of driving forces of emissions and sinks to identify which of these forces can be influenced by policies); and
4. to help negotiate possible emissions reductions for different countries and geographic regions.

The mixed scientific/policy nature of the scenarios makes the separation of scientific and policy objectives difficult. The scenarios can be used by the scientific community to explore future states of global climate forcing, but they must first be gridded and interpreted if they are to serve as input to computer models. For those involved in policy-making, the scientific results based on the scenarios can be used to assess the consequences of various emissions levels on climate and on society (49). Obviously, policymakers need the scientific results, just as the scientific community must rely on the policymakers to develop reasonable scenarios.

FINAL REMARKS

Emissions inventories are never completely accurate, because unsurveyed or inadequately described sources are always present. Inventories are also never finished, because society moves ever on, building new sources of emissions, controlling the emissions of others, and ceasing the operation of still others. If emissions inventories are thus never correct and never finished, are they worth preparing? The obvious answer is yes, and the value of the inventories lies not in the inventories themselves, but in the uses to which they are put. These inventories serve as the basis for models of local, regional, and global environmental quality in the past, present, and future. They are the foundation for reporting and compliance activities. Without emissions inventories, model calculations that are crucial for both scientific and political activities could not be performed.

Rapid strides have been made in the past decade or so in the scope and accuracy of emissions inventories. Many current local and regional inventories are thought to be accurate to perhaps 10%, although quantifying accuracy remains a seemingly intractable problem. As 2D and 3D regional and global models have appeared on the scene, gridded inventories have been generated to support them. Such inventories are now available for many of the important atmospheric species.

An increasingly important development of the past decade has been the creation of organizations that act as "umbrella groups" under which teams of experts can develop comparable, comprehensive emissions inventories and scenarios. These teams are now working to improve the accuracy of emissions estimates from the less-developed world, particularly in much of Africa, Asia, and South America. Scientists from these regions are now active participants in many emissions inventory projects, and their involvement will greatly improve the reliability of information pertaining to those areas of the world.

All of science is exemplified by the triumvirate activities of estimation, modeling, and measurement. If everything in the world were known, a perfect emissions inventory would form the basis for computer model calculations that would predict certain levels of environmental impacts, and those impacts would be confirmed by measurement. Our knowledge is never complete, of course, and so inaccuracies in our understanding are revealed by discrepancies between emissions-driven model results and measurement results. Attempts to resolve these discrepancies eventually lead to improved understanding of

Earth's physical and chemical workings. In this scientific enterprise, emissions inventories have played, and will continue to play, a central role.

ACKNOWLEDGMENTS

This work is a contribution of the Global Emissions Inventories Activity (GEIA) project of the IGBP International Global Atmospheric Chemistry (IGAC) program. Financial support from the Norwegian Research Council (Norges Forskningsrad) and the Norwegian Institute for Air Research (NILU) is greatly appreciated. Some information presented in this chapter has been taken from JM Pacyna's contribution to *Topics in Atmospheric and Interstellar Physics and Chemistry,* CF Boutron, ed., Les Editions de Physique, Les Utis, France, with permission from the editor.

Literature Cited

1. McInnes G, Pacyna JM, Dovland H, eds. 1992. *Proc. 1st Meet. Task Force Emission Inventories, EMEP/CCC Rep. 4/92.* Lillestrom, Norway: Nor. Inst. Air Res.
2. Mobley JD, Saeger M. 1993. *Proc. 2nd Meet. Task Force Emission Inventories, EMEP/CCC Rep. 8/93,* pp. 137–53. Lillestrom, Norway: Nor. Inst. Air Res.
3. Pacyna JM, Munch J, Keeler GJ. 1993. *Proc. Emission Inventories Issues,* pp. 133–46. Pittsburgh, PA: Air & Waste Manage. Assoc., VIP 27
4. US Environ. Prot. Agency. 1973. Compilation of air pollutant emission factors. *US EPA AP-42.* EPA, Res. Triangle Park, NC
5. US Environ. Prot. Agency. 1992. The clearinghouse for inventories and emission factors. *EPA 454-N-92-020.* EPA, Res. Triangle Park, NC
6. Bouscaren R. 1990. *Proc. Workshop Int. Emission Inventories, EMEP/CCC Rep. 7/90,* pp. 93–105. Lillestrom, Norway: Nor. Inst. Air Res.
7. van der Most PFJ, Veldt C. 1992. Emission factors manual. *TNO Rep. 92-235.* TNO Environ. Energy Res., Apeldoorn, The Netherlands
8. McInnes G, ed. 1994. *The Atmospheric Emission Inventory Guidebook.* Brussels: Eur. Environ. Agency Task Force (Draft)
9. Org. Econ. Coop. Dev. 1994. *Greenhouse Gas Inventory Reference Manual. IPCC Draft Guidelines for National Greenhouse Gas Inventories.* Paris: OECD

9a. Graedel TE. 1992. The IGAC activity for the development of global emissions inventories. In *Emission Inventory Issues in the 1990s, Spec. Conf. Proc.,* pp. 140–47. Pittsburgh: Air & Waste Manage. Assoc.

10. Zoller WH. 1984. Anthropogenic perturbation of metal fluxes into the atmosphere. In *Changing Metal Cycles and Human Health,* ed. JO Nriagu, pp. 27–42. Berlin: Springer-Verlag
11. Siegenthaler U, Sarmiento JL. 1993. Atmospheric carbon dioxide and the ocean. *Nature* 365:119–25
12. Sundquist ET. 1993. The global carbon dioxide budget. *Science* 259:934–41
13. Marland G, Andres RJ, Boden TA. 1994. In *Trends '93: A Compendium of Data on Global Change, Rep. ORNL/CDIAC-65,* ed. TA Boden, DP Kaiser, RJ Sepanski, FW Stoss, pp. 505–84. Oak Ridge, TN: Oak Ridge Natl. Lab.
14. Deleted in proof
15. US Environ. Prot. Agency. 1990. Methane emissions and opportunities for control. Workshop results Intergovern. Panel Clim. Change. *EPA/400/9-90/007 Rep.,* EPA, Washington, DC
16. Bouwman AF, ed. 1993. *Rep. 3rd Workshop Glob. Emissions Inventory Activ.*

(GEIA), Jan. 31–Feb. 2, Amersfoort, The Netherlands

17. Bouwman AF, Fung I, Matthews E, John J. 1993. Global analysis of the potential for N_2O production in natural soils. *Glob. Biogeochem. Cycles* 7:557–97
18. Hao WM, Liu MH, Crutzen PJ. 1990. Estimates of annual and regional releases of CO_2 and other trace gases to the atmosphere from fires in the tropics, based on the FAO statistics for the period 1975-1980. In *Fire in the Tropical Biota. Ecological Studies 84,* ed. JG Goldhammer, pp. 440–62. Berlin: Springer-Verlag

18a. Bouwman AF, Van der Hoek KW, Olivier JGJ. 1995. Uncertainties in the global source distribution of nitrous oxide. *J. Geophys. Res.* 100:2785–800

19. Cicerone RJ. 1988. In *The Changing Atmosphere,* ed. FS Rowland, ISA Isaksen, p. 49. New York: Wiley
20. Cunnold DM, Fraser PJ, Weiss RF, Prinn RG, Simmonds PG, et al. 1994. Global trends and annual releases of CCl_3F and CCl_2F_2 estimated from ALE/GAGE and other measurements from July 1978 to June 1991. *J. Geophys. Res.* 99:1107–26
21. Tuovinen J-P, Barrett K, Styve H. 1994. Transboundary acidifying pollution in Europe: calculated fields and budgets 1985–93. *EMEP/MSC-W Rep. 1/94,* Meteorol. Synthesizing Centre-West, Nor. Meteorol. Inst., Oslo, Norway
22. National Acid Precipitation Assessment Program. 1990. Technologies and other measures for controlling emissions: performance, costs and applicability. *NAPAP Rep. 25,* NAPAP, Washington, DC
23. Kato N, Akimoto H. 1992. Anthropogenic emissions of SO_2 and NO_x in Asia: emissions inventories. *Atmos. Environ.* 26A:2997–3017
24. Akimoto H, Narita H. 1994. Distribution of SO_2, NO_x, and CO_2 emissions from fuel combustion and industrial activities in Asia with 1° × 1° resolution. *Atmos. Environ.* 28:213–25
25. Deleted in proof
26. Yienger JJ, Levy H II. 1995. Empirical model of global soil-biogenic NO_x emissions. *J. Geophys. Res.* 100:In press

26a. Schlesinger WH, Hartley AE. 1992. A global budget for ammonia. *Biogeochemistry* 15:191–211

27. Pacyna JM. 1989. Atmospheric emissions of nitrogen compounds. In *The Role of Nitrogen in the Acidification of Soils and Surface Waters,* ed. JL Malanchuk, J Nilsson. *Nordic Counc. Ministers, NORD 1898:92 Rep.,* Copenhagen, Denmark
28. Dentener F, Crutzen PJ. 1994. A three-dimensional model of the global ammonia cycle. *J. Atmos. Chem.* 19: 331–69
29. Guenther A, Hewitt CN, Erickson D, Fall R, Geron C, et al. 1995. A global model of natural volatile organic compound emissions. *J. Geophys. Res.* 100: 8873–92
30. Pacyna JM. 1984. Estimation of the atmospheric emissions of trace elements from anthropogenic sources in Europe. *Atmos. Environ. 18:41–50*
31. *Pacyna JM. 1994.* Emissions of heavy metals in Europe. In *Proc. EMEP Workshop Eur. Monit., Model., Assess. Heavy Metals Persistent Organ. Pollut., May 3–6.* Beekbergen, The Netherlands
32. Axenfeld F, Munch J, Pacyna JM, Duiser JA, Veldt C. 1992. Test-emissionsdatenbasis der spurenelemente As, Cd, Hg, Pb, Zn und der speziellen Organischen Verbindungen y-HCH (Lindan), HCB, PCB und PAK fur Modellrechnungen in Europa. *Umweltforschungsplan des Bundesministers fur Umwelt Naturschutz und Reaktorsicherheit, Luftreinhaltung, Forschungsbericht 104 02 588,* Dornier GmbH Rep. Friedrichshafen, Germany
33. Benjey WG, Coventry DH. 1992. *Geographical distribution and source type analysis of toxic metal emissions.* Presented at Int. Symp. Measurement Toxic Related Air Pollut., May 3–8, Durham, NC
34. Voldner E, Smith L. 1989. *Production, usage and atmospheric emissions of 14 priority toxic chemicals. Rep. Joint Water Qual. Board/ Sci. Advis. Board/ Int. Air Qual. Advis. Board Int. Joint Comm.* Presented at Workshop Great Lakes Atmos. Depos., Oct. 29–31, 1986, Ottawa
35. US Environ. Prot. Agency. 1993. Locating and estimating air emissions from sources of mercury and mercury compounds. *Rep. EPA-454/R-93-023,* EPA, Res. Triangle Park, NC
36. US Environ. Prot. Agency. 1993. Locating and estimating air emissions from sources of cadmium and cadmium compounds. *Rep. EPA-454/R-93-040,* EPA, Res. Triangle Park, NC
37. Jacques AP. 1987. Summary of emissions of antimony, arsenic, cadmium, copper, lead, manganese, mercury, and nickel in Canada. *Environ. Anal. Branch Rep.,* Conserv. Prot. Div., Environ. Can., Ottawa
38. Nriagu JO, Pacyna JM. 1988. Quanti-

tative assessment of worldwide contamination of air, water and soils with trace metals. *Nature* 333:134–39

39. Pacyna JM, Shin BD, Pacyna P. 1993. Global emissions of lead. *Rep. Atmos. Environ. Serv.*, Environ. Can., Toronto
40. Johnson ND, Scholtz MT, Cassaday V, Davidson K. 1992. MOE toxic chemical emission inventory for Ontario and eastern North America. *Final Rep. Air Res. Branch, Ontario Ministry Environ., ORTECH Int., Rep. P92-T61-5429/OG,* Mississauga, Canada
41. Voldner E. 1993. Global usage and emission of selected persistent organochlorines. In *Proc. 4th Int. Workshop Glob. Emissions Inventories,* Nov. 30–Dec. 2, Boulder, CO: GEIA Data Manage. Org.
42. Graedel TE, Bates TS, Bouwman AF, Cunnold D, Dignon J, et al. 1993. A compilation of inventories of emissions to the atmosphere. *Glob. Biogeochem. Cycles* 7:1–26
43. French HF. 1994. Making environmental treaties work. *Sci. Am.* 271:94–97
44. United Nations Economic Commission for Europe. 1994. *Convention on Long-range Transboundary Air Pollution.* ECE, Environ. Human Settlements Div., Geneva
45. National Acid Precipitation Assessment Program. 1990. Methods for modeling future emissions and control costs. *Acid Deposition: State Sci. Technol. Rep. 26,* NAPAP, Washington, DC
46. Placet M, Streets DG, Williams ER. 1986. Environmental trends associated with the fifth national energy policy plan. *ANL/EES-TM-323,* Argonne Natl. Lab., Argonne, IL
47. United Nations Economic Commission for Europe. 1993. Emission projections. *Doc. EB.AIR/WG.5/R.39,* UN ECE, Geneva
48. Intergovernmental Panel Climate Change. 1990. Global climate change. *Rep. UN Panel Climate, Intergovern. Panel Climate Chang., Ministry Environ.,* Oslo, Norway (In Norwegian)
49. IPCC 1994. Climate Change 1994. Radiative forcing of climate change and an evaluation of the IPCC IS92 emission scenarios, ed. JT Houghton, LG Meira Filho, J Bruce, Hoesung Lee, BA Callander, et al. *Rep. Working Groups I/III Intergovern. Panel Clim. Change, part of IPCC Spec. Rep. 1st sess. Conf. Parties UN Framework Conv. Clim. Change,* Cambridge, UK: Cambridge Univ. Press

Annu. Rev. Energy Environ. 1995. 20:301–24

THE ELIMINATION OF LEAD IN GASOLINE

V. M. Thomas
Center for Energy and Environmental Studies, Princeton University, Princeton, New Jersey 08544-5263

KEY WORDS: unleaded gasoline, exposure, valve-seat recession, tetra-ethyl lead, MTBE, refinery costs, catalytic converters, benzene

---

CONTENTS

Abstract

Due to the health consequences of lead exposure, as well as to the introduction of catalytic converters, many countries have reduced or eliminated use of lead additives in motor gasolines. But in many other countries, leaded gasoline remains the norm. In these countries there is often confusion about the health significance of gasoline lead, the ability of cars to use unleaded gasoline, and the costs of unleaded gasoline. This chapter shows that leaded gasoline is a major source of human lead exposure. All cars, with or without catalytic

1056-3466/95/1022-0301$05.00

converters, and with or without hardened exhaust valve seats, can use unleaded gasoline exclusively. Unleaded gasoline typically costs on the order of $0.01 more per liter than leaded gasoline to produce. Recent concerns about benzene exposure from unleaded gasoline have been addressed through choice of gasoline formulation and other measures.

INTRODUCTION

The health effects of lead and the dynamics of exposure to lead from gasoline are better understood than those of almost any other pollutant. A growing number of countries have seen a successful transition to unleaded gasoline. Yet in most countries, leaded gasoline is still used, and in many developing countries unleaded gasoline is unavailable. When governments consider reducing lead levels in gasoline, confusion and argument often surround questions about the health significance of lead in gasoline, the ability of the vehicle fleet to use unleaded gasoline, and the costs of alternatives to lead. This review examines and clarifies the points of confusion, focusing on exposure to lead from gasoline and the ability of cars to use unleaded gasoline.

STATUS OF THE USE OF UNLEADED GASOLINE

As of 1993, an estimated 70,000 tons of lead were added to gasoline worldwide (1). As shown in Tables 1a and 1b, about half of the worldwide total of lead used in gasoline is used in the former USSR, eastern Europe, and the Far East. The remainder is used, in approximately equal amounts, in western Europe, Africa, the Middle East, and the Americas. In many countries, unleaded gasoline is unavailable.

Octel, Ltd. (88% owned by Great Lakes Chemical and 12% owned by Chevron) is almost the sole producer of lead gasoline additives worldwide. Ethyl Corporation, until recently the other major producer, stopped production of these additives in 1994. However, Ethyl continues to sell lead additives, buying its supplies from Octel (2). The other producers are facilities in Germany (less than 4000 tons lead per year) (3) and in Dzerzhinsk, Russia (about 5000 tons lead per year) (V Prozerov, personal communication). Total annual sales of lead additives are on the order of $1 billion.

The antiknock properties of tetra-ethyl lead (TEL) were discovered in the United States by Thomas Midgley and colleagues at the General Motors Research Laboratory in 1921 (4). (Nine years later, in the same laboratory, Midgley also developed chlorofluorocarbons.) Commercially introduced in the United States in the 1920s, leaded gasoline soon became standard (5). Owing to the introduction of catalytic converters and a better understanding of the risks of lead exposure, leaded gasoline was phased out in the United States

Table 1a Estimated use of leaded motor gasoline in Europe and the Middle East

Country	Motor gasoline consumption[a] (10^9 liters per year)	Lead content of leaded gasoline (g liter^{-1})[b]	Total added lead (tons per year)	Leaded gasoline percent share (year)
Western Europe[c]				
France	27	0.15	2,400	59 (1993)
Italy	19	0.15	2,200	76 (1993)
United Kingdom	33	0.15	2,300	47 (1993)
Former Yugoslavia	3.6	0.5	1,800	98 (1991)
Spain	11	0.15	1,600	94 (1992)
Germany	42	0.15	690	11 (1993)
Turkey	4.4	0.15	650	90 (1994)[d]
Portugal	1.9	0.40[e]	600	79 (1993)
Greece	3.3	0.40[e]	380	77 (1993)
Switzerland	5.0	0.15	260	35 (1992)
Belgium	3.7	0.15	240	43 (1993)
Luxembourg	5.2	0.15	240	31 (1993)
Netherlands	4.7	0.15	180	25 (1993)
Norway	2.4	0.15	140	40 (1993)[f]
Finland	2.7	0.15	120	30 (1992)
Ireland	1.2	0.15	110	62 (1993)
Denmark	2.2	0.15	79	24 (1993)
Sweden	5.7	0.15	9	1 (1994)[g]
Austria	3.5	—	0	0 (1993)
Other	0.2	0.15	30	
Total	170		14,000	
Eastern Europe and Former USSR				
Former USSR	100	0.2[h]	10,000[i]	60 (1994)[j]
Romania	2.8	0.6	1,700	
Poland	3.7	0.15	490	88 (1993)[k]
Former Czechoslovakia	2.3	0.15	330	97 (1991)[l]
Hungary	2.1	0.15[k]	240	75 (1993)[m]
Total	120		13,000	
Middle East				
Saudi Arabia	9.3	0.4	3,700	100
Iraq	4.4	0.4	1,700	100
Iran	8.1	0.19	1,500	100
Kuwait[n]	1.2	0.53	620	100
UAR	1.3	0.4	530	100
Syria	1.5	0.24	360	100
Israel[o]	2.1	0.15	310	<100
Quatar	0.4	0.4	75	47 (1992)
Other	4.6	0.4	1,900	
Total	31		11,000	
World Total	950		70,000	

[a] Aviation gasoline, used by aircraft with piston engines, is not included. In the United States, the lead in aviation gasoline is typically 0.5–0.8 g liter^{-1} and about 10^9 liters per year of aviation gasoline is used, for total lead emission of about 840 tons per year. Data are taken from Reference 98. [b] Data from Reference 8, unless otherwise noted. [c] Unless otherwise noted, from Eurostat, Brussels (1994 data). [d] Reference 99. [e] Reference 100. [f] Reference 101. [g] Reference 12. [h] Reference 102. [i] V Prozerov, personal communication. [j] Y Minaev, personal communication. [k] Reference 11. [l] Reference 81. [m] B Levy, personal communication. [n] Reference 103. [o] Reference 104.

Table 1b Estimated use of leaded motor gasoline in Europe and the Middle East

Country	Motor gasoline consumption[a] (10^9 liters per year)	Lead content of leaded gasoline (g $liter^{-1}$)[b]	Total added lead (tons per year)	Leaded gasoline percent share (year)
North America				
Mexico[c]	26	0.07	1,300	70 (1993)
United States[d]	430	0.026	100	1 (1993)
Canada	32	—	0	0 (1993)
Total	467	—	1,400	
Central and South America and Caribbean				
Venezuela[e]	7.7	0.37	2,600	90 (1993)
Ecuador	1.5	0.84	1,200	>95 (1993)[f]
Argentina[g]	4.5	0.2	900	70 (1994)[h]
Peru	1.2	0.84	920	91 (1992)
Chile	1.6	0.42	660	99 (1992)
Puerto Rico	2.9	0.13	380	?
Virgin Islands	0.23	1.12	260	?
Trinidad and Tobago	0.36	0.4[i]	140	100
Colombia[c]	6.4	—	0	0
Brazil	17	—	0	0
Suriname	0.06	—	0	0
Antigua	0.02	—	0	0
Other	8	0.4	2,700	
Total	53		9,700	
Africa				
Nigeria	6.4	0.66	4,200	100
South Africa	6.4	0.4	2,600	100
Algeria	2.6	0.6	1,700	100
Libya	2.0	0.8	1,600	100
Egypt[j]	2.0	0.35	700	100 (1994)
Other	6.4	0.4	2,600	100
Total	26		13,000	
Far East and Oceania				
Australia	17	0.3	2,700	53 (1993)[k]
China[l]	30	0.03	900	100 (1994)
India[m]	4.8	0.15	700	100 (1994)
New Zealand	2.5	0.45	640	57 (1994)[n]
Thailand	3.7	0.15	440	79 (1993)[e]
Malaysia	3.8	0.15	300	50 (1994)[o]
Taiwan[p]	4.9	0.026	70	55 (1993)[q]
Sri Lanka[r]	0.23	0.2	50	100 (1993)
Singapore	0.58	0.15	35	40 (1993)[e]
Hong Kong	0.35	0.15	32	62 (1991)[s]
Japan	44	—	0	0
South Korea	3.8	—	0	0
Other	15	0.4	6,000	
Total	130		12,000	
World Total	950		70,000	

[a] See Footnote a, Table 1a. [b] Data from Reference 8, unless otherwise noted. [c] M Barbiux, personal communication. [d] J Caldwell, personal communication. [e] Reference 11. [f] Reference 105. [g] Reference 106. [h] Reference 107. [i] Ramlel, personal communication. [j] Reference 108. [k] Reference 67. [l] Reference 109. [m] Reference 110. [n] Reference 111. [o] Reference 112. [p] Reference 113. [q] T-N Wu, personal communication. [r] Reference 114. [s] Reference 81.

beginning in the 1970s and now accounts for less than 1% of gasoline sales. The use of leaded gasoline for highway vehicles will be banned in the United States as of January 1, 1996.

In Japan, reduction of lead in gasoline began in the 1970s, after reports of high blood lead concentrations in Tokyo (6, 7). More recently, leaded gasoline was eliminated in Canada, Brazil, Colombia, Austria, South Korea, and Sweden (8–12). In addition, Suriname and Antigua have reportedly eliminated leaded gasoline (8). In the European Union, the lead content of gasoline is limited to 0.15 g $liter^{-1}$, and all new cars are required to have catalytic converters, which require the use of unleaded gasoline (13).

The Soviet Union was the first country to restrict use of lead in gasoline: By 1967, its sale was banned in Moscow, Leningrad, Kiev, Baku, Odessa, and tourist areas in Caucasia and the Crimea (14). This action was apparently prompted by Soviet research on the effects of low-level lead exposure (15). Use of leaded gasoline continued elsewhere in the country, and the former USSR is now apparently the largest user of leaded gasoline additives.

The following countries—all developing nations—allow the highest concentrations of lead ($\geq$ 0.8g $liter^{-1}$): Aruba, Bahamas, Barbados, Belize, Benin, Burkina, Burundi, Cape Verde Islands, Central African Republic, Chad, Cuba, Curacao, Dominican Republic, Ecuador, Equatorial Guinea, Fiji, Guinea, Guinea-Bissau, Haiti, Honduras, Indonesia, Ivory Coast, Jamaica, Lebanon, Libya, Macao, Madagascar, Mali, Maritius, Marshall Islands, Myanmar (Burma), Nauru, New Caledonia, Norfolk Islands, Panama, Papua, Paraguay, Peru, Rwanda, Sahara West, Saint Martin, Seychelles, Sierra Leone, Solomon Islands, Somalia, Uganda, Virgin Islands, West Samoa, and Zimbabwe.

RELATIONSHIP OF LEAD USE IN GASOLINE TO LEAD IN BLOOD

The US Public Health Service has said that lead exposure is the greatest environmental health threat to children (16). Figure 1 shows the effects of lead exposure at various levels. Prenatal, postnatal, and perinatal blood lead concentrations of 5–15 μg dl^{-1} have been associated with detrimental effects on infant growth and development (17). Most studies report a 2- to 4-point average IQ deficit for each increase of 10–15 μg dl^{-1} in blood lead within the range of 5–35 μg dl^{-1}, and no threshold has become apparent for this effect.

One of the main points of controversy has been the relationship of lead in gasoline to blood lead levels. Gasoline lead, emitted to the air, falls back to the earth and contaminates soil, urban dusts, and crops. Thus, this lead is not only inhaled, but is also ingested as lead deposited on soil and dust, food crops and pasture land. Exposure to gasoline lead depends not only on the amount of lead used locally in gasoline, but also on traffic patterns, diets, food sources,

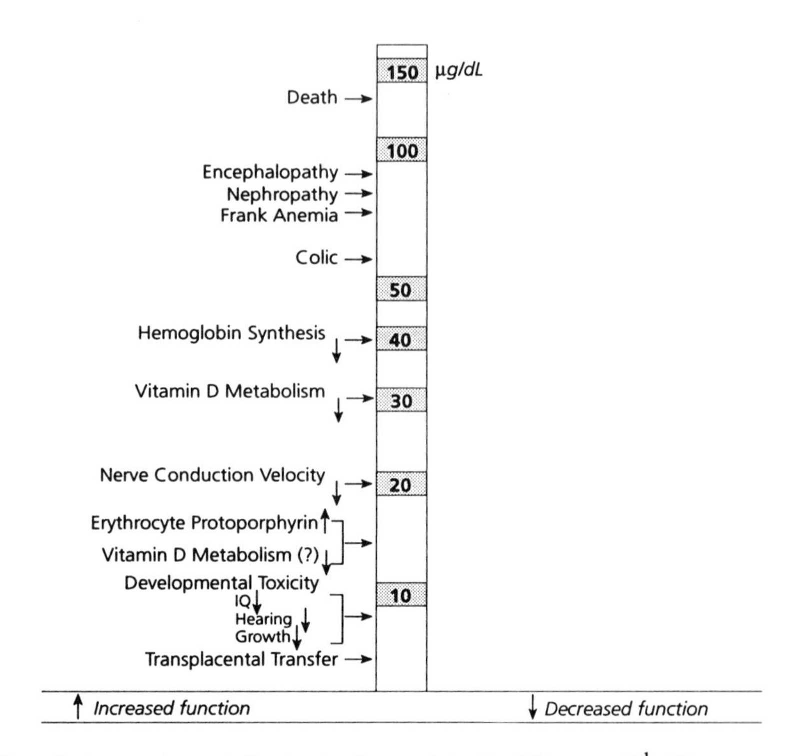

Figure 1 Lowest observed effect levels of inorganic lead in children (µg dl^{-1}) (18).

and personal habits. Exposure to lead found in paint, plumbing, cosmetics, eating or drinking utensils, soldered food cans, some industries, and drinking water complicates measurements of leaded gasoline's contribution to blood lead levels.

Overall, human exposure to lead is currently two to three orders of magnitude greater than it was in the preindustrial period (19). How much does leaded gasoline contribute to blood lead levels? Below I review data on changes in population blood lead levels as use of lead in gasoline has decreased, calculations of the expected contribution of lead in gasoline to lead in blood, isotopic studies of the contribution of gasoline lead to blood lead, and studies claiming that gasoline lead is only a minor contributor to blood lead levels.

NHANES

The most extensive studies of blood lead concentrations are NHANES (National Health and Nutrition Examination Study) II and NHANES III. NHANES II surveyed 27,801 people aged 6 months to 74 years living in 64 areas of the

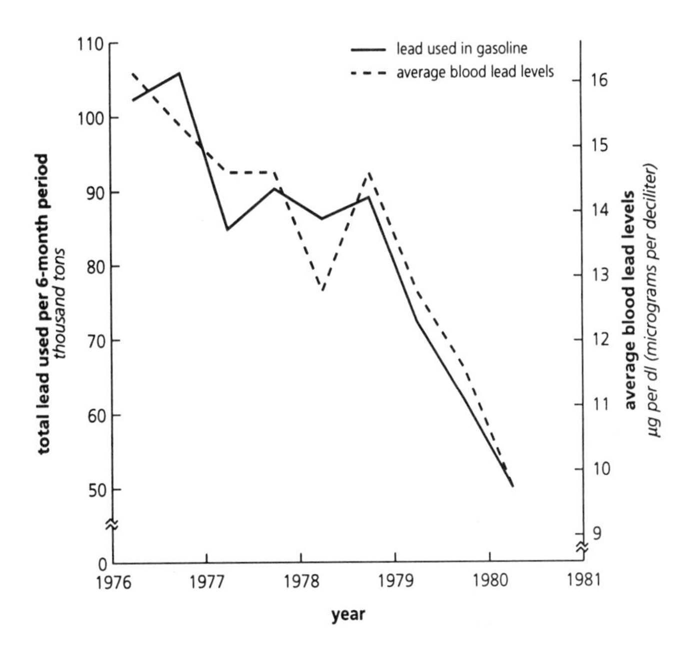

Figure 2 Decreases in blood lead values and amounts of lead used in gasoline during 1976–1980 (18).

United States. The results, in Figure 2, show US use of lead in gasoline and average blood lead levels in the United States between 1976 and 1980 (20). The figure shows a 50% drop in use of lead in gasoline, coincident with a 30% drop in blood lead levels (21). Lead from gasoline was the major source of air emissions of lead (22), and the reduction of lead in gasoline is generally accepted as the cause of most of the reductions in blood lead levels.

Correlation, of course, does not prove causation. Several factors could have caused the downward trend. Some authors have suggested that the data could be explained by either the reduced use of lead-soldered food cans or by errors in sampling and statistical analysis (23). A time-dependent sampling or statistical error does not plausibly explain the data; the NHANES II data were adjusted via regression for the effects of race, sex, region of the country, season, income, and degree of urbanization. Further studies, such as NHANES III, confirm that blood lead levels continued to decrease significantly. By 1988–1991, when the use of lead in US gasoline had been almost completely eliminated, the average blood lead levels of persons aged 1–74 years in the United States had dropped to 2.8 $\mu g\ dl^{-1}$ (24).

Reduced use of lead-soldered food cans might explain some of the decrease in blood lead levels. Although 90% of all US-manufactured food cans were still lead soldered as of 1979 (25), the canning industry had already started programs to reduce the lead content of canned foods, especially infant formula, evaporated milk, and baby foods (26, 27).

Exposure Calculation

A great deal of effort has gone into developing models for the prediction of lead exposure. Although a review of this body of work is beyond the scope of this chapter, the results are used here as a rough estimate of the magnitude of blood lead levels expected from use of lead in gasoline. In Table 2, the standard parameters for estimating lead transport and exposure are used to estimate typical adult exposure to gasoline lead through inhalation, dust and soil ingestion, and food ingestion (17, 28, 29). Total exposure for children would be of a similar magnitude, with a larger contribution from soil and dust ingestion and smaller contributions from inhalation and food.

In this calculation, exposure through inhalation can be estimated with the most confidence. Estimating the contribution from food is much more difficult, because this value depends on atmospheric transport and deposition, retention of lead in the food chain, and what people eat. The US EPA has estimated that in the United States, each μg m^{-3} in air (for 139 urban sites, based on the average of the maximum quarterly average air lead concentration) contributes about 30 μg per day of atmospheric lead to adult diets (30).

Estimating exposure to atmospheric lead via ingestion of soil and dust is also complex. The average amount of soil or dust ingested has been measured relatively well, particularly for children, through studies of fecal concentrations of various indigestible soil minerals. But the concentration of lead in soil or dust depends on atmospheric transport and deposition, the time scale of soil

Table 2 Estimated typical adult lead exposure per microgram per meter cubed of lead in air[a]

Medium	Gross intake	Lead in medium per μg m^{-3} in air[b]	Percent absorbed	Lead Absorbed (μg per day per μg m^{-3} in air)
Soil and Dust	0.05 g per day[c]	>100 μg g^{-1}	10–15%[d]	>0.5–0.75
Food		~30 μg per day	10–15%	3–4.5
Inhalation	20 m^3 per day[e]	1	30–50%[d]	6–10
Total				>9.5–15

[a] Air lead in terms of maximum quarterly average concentrations.
[b] Data from References 29, 30.
[c] Data from Reference 29.
[d] Data from Reference 17.
[e] Data from Reference 115.

mixing and erosion, dust retention on surfaces, and the personal habits of the population. Based on studies of air, dust, and soil lead levels in several locations (28), the US EPA has estimated that each $\mu g\ m^{-3}$ of lead in air contributes roughly 100 $\mu g\ g^{-1}$ or more to lead found in dust or soil (29).

The sum of these contributions is an absorption of roughly 10–15 μg per day for each microgram per meter cubed of lead in air. At this level of exposure, each microgram per day corresponds to about 0.5 $\mu g\ dl^{-1}$ in blood (28), and thus each microgram per meter cubed of lead in air corresponds to roughly 5–8 $\mu g\ dl^{-1}$ in adult blood lead. The exposure values for individuals and population subgroups can vary significantly from this estimated typical exposure.

This estimated exposure can be compared with the measured reductions in blood lead levels as the use of lead in gasoline declined. National data on urban air lead concentrations are only available since 1979. Maximum quarterly average urban air lead concentrations fell from 0.8 $\mu g\ m^{-3}$ in 1979 to about 0.6 $\mu g\ m^{-3}$ in 1980 (31). Because gasoline lead is the major source of air lead emissions, most of the air lead can be attributed to lead from gasoline. According to the model above, gasoline would have contributed 4–6 $\mu g\ dl^{-1}$ of blood lead in 1979, and 3–5 $\mu g\ dl^{-1}$ in 1980. In fact, Figure 2 shows that average blood lead levels were about 12 $\mu g\ dl^{-1}$ in 1979 and 10 $\mu g\ dl^{-1}$ in 1980. Because of other significant sources of lead exposure, such as lead-based paint and lead-soldered food cans, it is not surprising that the estimated contribution of gasoline lead is less than the total lead exposure.

The exposure estimate in Table 2 also predicts a fall in blood lead levels of 1–1.6 $\mu g\ dl^{-1}$ between 1979 and 1980. In fact, they fell by approximately 2 $\mu g\ dl^{-1}$ over this time period. Given the rough nature of the exposure estimate, this is fairly good agreement. Thus, exposure analysis indicates that reductions in gasoline lead account for at least a large fraction of the drop in blood lead levels shown in Figure 2.

Isotopic Analysis

The contribution of lead in gasoline to lead in blood can be directly measured with isotopic techniques. The stable isotopes of lead, especially 204, 206, 207, and 208, are present in different ratios in different lead ores. If the lead used in gasoline has sufficiently different isotopic ratios than the other sources of lead exposure, then by measuring the isotopic ratios of lead in the blood, gasoline, and other sources, we can determine the fraction resulting from lead in gasoline. In Turin, Italy, the lead used in gasoline was switched to a lead with isotopic ratios significantly different from the other local sources. Before, during, and after the switch, lead isotope ratios were measured in blood and air. The results showed that blood lead attributable to gasoline lead was approximately 5 $\mu g\ dl^{-1}$ for persons living in Turin and about 3 $\mu g\ dl^{-1}$ for persons living in the surrounding area. Inhaled lead accounted for approximately half of the exposure to lead from gasoline for residents of the city; outside of Turin, the inhaled contribution fell to less than 10% of the total exposure to gasoline lead (32, 33).

Claims That Blood Lead Levels Are Uncorrelated With Leaded Gasoline Use

Despite these studies, some have claimed that the declines in blood lead levels are "for the most part, unrelated in a significant way to lead in gasoline usage" {34, p. 19; see also (23)}. Authors have cited instances of falling blood lead levels while gasoline lead remained constant as evidence that leaded gasoline does not contribute significantly to blood lead levels. For example, in Christchurch, New Zealand, when the manufacture of many types of lead-soldered cans (including beer cans), and the sale of lead-based paints, stopped, average blood lead levels decreased significantly (36). However, this finding does not show that the contribution of leaded gasoline is insignificant, but rather that other sources of lead exposure are also significant.

Octel, Ltd., the lead additive manufacturer, claims that "recent UK Government studies demonstrated conclusively that the reduction of lead levels in gasoline...had no direct relationship to the decline in population blood lead levels" (37, see Executive Summary). In fact, in the UK a 50% drop in use of lead in gasoline (38) resulted in a 20% drop in blood lead levels (39).

A more egregious approach has been to use old data on blood lead levels to try to argue that blood lead levels in the United States were decreasing during the time that gasoline lead use was increasing. Figure 3 shows the published reports of blood lead levels in the United States since the 1930s (34, 40–46), plotted with data on the use of lead in gasoline (5). These data have been used to argue that blood lead levels were uncorrelated with use of lead in gasoline: "The rise in gasoline lead usage was particularly dramatic from about 1940 to 1970, as blood lead levels were decreasing" (34, p. 10). A version of this figure was later published with the statement that (39, p. 22):

> it is clear that blood lead levels have been declining steadily in the USA for the past 60 years, even during the period up to the mid-1970s when consumption of leaded gasoline underwent rapid increase. It is now generally accepted that: lead from gasoline is a very minor contributor to lead in the body....

This interpretation of Figure 3 is deeply misleading. In fact, many of the measurements of lead concentrations in biological and environmental samples were incorrect until the mid-1960s (47–50). Reported blood lead levels may have been off by an order of magnitude; reports of the lead concentration of seawater, now recognized to be in the parts per trillion range, were wrong by several orders of magnitude. According to a recent study by the US National Research Council on the measurement of lead exposure (17, p. 191):

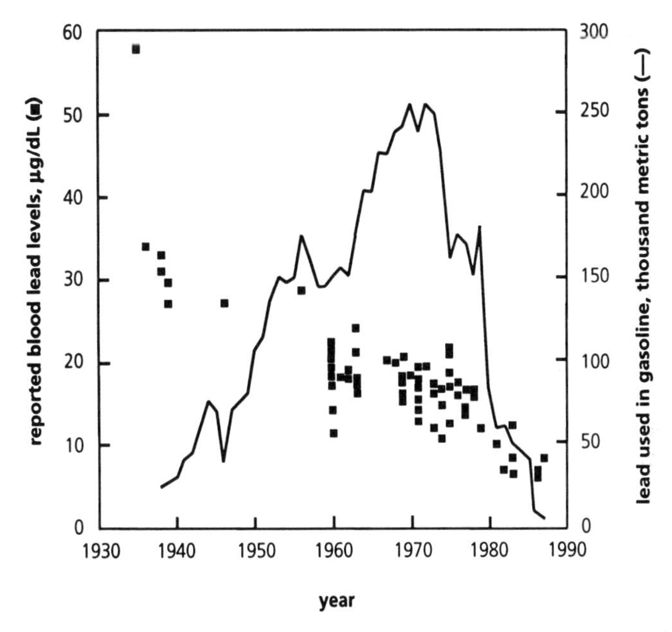

Figure 3 Amount of lead used in gasoline (*solid line*) and published reports of blood levels for the United States (34).

> Ultraclean techniques have repeatedly shown that previously reported concentrations of lead can be erroneously high by a factor of several hundred (Patterson and Settle, 1976 [51]). The flawed nature of some reported data was initially reported in oceanographic research: reported concentrations of lead in seawater have decreased by a factor of 1000 because of improvements in the reduction and control of lead contamination during sampling, storage, and analysis (Bruland, 1983) [51a].... Similar decreases in concentrations of lead in biologic materials have been reported by laboratories that have adopted trace metal clean techniques....One study revealed that lead concentrations in some canned tuna were 10,000 times those in fresh fish, and that the difference had been overlooked for decades because all previous analyses of lead concentrations in fish were erroneously high (Settle and Patterson, 1980 [52]). Another study demonstrated that lead concentrations in human blood plasma were much lower than reported (Everson and Patterson, 1980 [53])....Previous blood lead measurements cannot be corroborated now, because no aliquots of samples have been properly archived.

Patterson & Settle, leaders in the development of ultraclean trace analysis of lead, reported in 1974 that (51):

> Most present analytical practices for lead cannot reliably determine lead concentrations in...blood. The unreliability of lead analyses is caused by a universal lack of familiarity with the extent, sources, and control of industrial lead contamination during sample collection, handling, and analysis. As a consequence, the great

> mass of published lead data...is associated with gross positive errors and the error noise in lead concentration data below a few ppm obscures the meaning of most work dealing with lead at these concentration levels.

By the mid-1970s, procedures for contamination avoidance and quality assurance had been widely adopted, and the trend in the more recent data in Figure 3 can be interpreted as a true reduction in blood lead levels in the United States. However, the trend in the data up to the 1960s reflects improved measurements and cannot be interpreted as a reflection of the true values.

DO SOME CARS NEED LEADED GASOLINE?

The main purpose of lead additives is to increase the octane level of the gasoline. A secondary role of lead additives is lubrication of the exhaust valves. Cars said to require leaded fuel have exhaust valve seats made of cast iron or soft steel, and without leaded gasoline, the valve seats can become worn with heavy and long use. In principle, if wear is severe enough, engine failure could result (54–57). Concern about valve-seat recession has posed a significant barrier to the elimination of leaded gasoline in Europe, Australia, and elsewhere.

Cars with catalytic converters cannot use leaded gasoline because lead poisons the catalyst. Consequently, all new cars sold in countries requiring catalytic converters have hardened valve seats made of hardened steel or another hard material. Finding information on which companies are currently manufacturing cars without hardened valve seats and where these cars are sold is difficult. However, as of 1989, the following companies were manufacturing automobiles without hardened valve seats for sale in Europe: Austin Rover, Citroen, Dacia, Fiat, Ford, FSO, Honda, Mercedes Benz, Reliant, Renault, Saab, and Skoda (58).

Laboratory studies confirm that under prolonged, severe driving conditions, valve-seat recession will occur if unleaded gasoline is used in engines without hardened valve seats. However, such tests also show that only very low concentrations of lead are needed to prevent valve wear. Even under severe engine conditions, all studies agree that 0.05 g liter^{-1} is sufficient to prevent valve-seat recession (55, 59, 60). In contrast, current concentrations of lead in gasoline in Europe are 0.15 g liter^{-1}, three times the amount needed to protect against valve-seat recession, even under severe conditions.

But more importantly, all studies have shown that, under normal driving conditions (real cars, real roads, and nonprofessional drivers), valve-seat recession is not a problem in practice. These results, discussed in more detail below, can be explained by two factors. First, lead additives have harmful as well as beneficial effects on the automobile. Lead additives include halogenated compounds added as lead scavengers to prevent excess build-up of lead compounds within the engine. These halogens also form corrosive compounds, which degrade exhaust valves, spark plugs, mufflers, and exhaust pipes (61). Thus, the use of leaded gasoline, while decreasing the risk of valve-seat recession, increases the risk of other maintenance problems.

Second, valve wear depends on engine operating conditions and increases when engines run for extended periods at high speeds (55). In laboratory tests of valve wear, valves are often subjected to many hours of constant operation at moderate or high speeds; consequently, valve recession in laboratory studies or specialized road tests is more pronounced than would be expected under normal driving conditions.

The most extensive study of the car maintenance consequences of using unleaded gasoline in cars designed for leaded gasoline was a 5-year study of 64 matched pairs of cars (62). One vehicle in each pair used leaded gasoline; the other used unleaded exclusively. The mileage on the cars averaged more than 24,000 km per year. Maintenance costs for the leaded-gasoline users were greater in terms of spark-plug replacement and exhaust-system repairs (muffler, exhaust pipe, etc), whereas those for the unleaded-gasoline users were greater in terms of engine repairs (valve reconditioning, cylinder-head replacements, etc.) Valve reconditioning—the grinding of exhaust valve seats and replacement of valves—was performed more often on cars using leaded gasoline, whereas cylinder heads were replaced more often on the cars using unleaded gasoline, because of excessive valve-seat wear. Although cylinder-head replacement is a more expensive operation than valve reconditioning, the total fuel-related maintenance costs (including spark plugs, exhaust-system repairs, and valve repairs) were greater for the leaded-gasoline users ($0.0012 km^{-1}) than for cars using unleaded gasoline ($0.0008 km^{-1}), even though all of these cars were designed for leaded gasoline. Total maintenance costs, including tires, lubrication, oil filters, brakes, etc, were an order of magnitude greater than the fuel-related maintenance costs alone, and averaged about $0.008 km^{-1} for the cars using unleaded gasoline, and about $0.009 km^{-1} for the cars using leaded gasoline. The authors concluded that the maintenance costs of using leaded vs unleaded gasoline in cars designed to use leaded gasoline differ little, and the use of unleaded gasoline may even be slightly beneficial. Valve-seat recession in the cars using unleaded gasoline was counterbalanced by other maintenance problems in cars using leaded gasoline, and in any case, the cost of nonfuel-related maintenance costs far outweighed the fuel-related costs.

The US Army ran a major three-year study in which unleaded fuel was used in light-duty cars and trucks; heavy-duty trucks, tractors, jeeps, and tactical and combat vehicles; and some motorized heavy equipment. No untoward maintenance problems could be attributed to the use of unleaded gasoline (63).

The US Postal service carried out a similar experiment. In 42 months operating 1562 Ford heavy-duty trucks with unleaded gasoline, 1.2% of them experienced valve-seat problems, less than the failure rate (engine failure from all causes, including valve problems, using leaded gasoline) from the truck manufacturer's warranty data (61). Three other studies, conducted about the same time, gave similar results (64–66).

Despite all this evidence, car manufacturers have nevertheless recommended that car drivers in Europe and Australia use leaded gasoline every third to sixth fill for cars without hardened valve seats (58, 67). These recommendations are increasingly ignored in western Europe (G Muller, personal communication).

Chevron has reported that "valve seat recession has not been seen as a problem by Chevron in passenger car or commercial operations" (68). Moreover, in an advertising campaign, Caltex, an overseas subsidiary of Standard Oil and Texaco, guarantees that no car using their unleaded gasoline will experience problems such as valve-seat recession; this gasoline contains no special additives to prevent valve-seat recession (69; L Burke, personal communication).

Lead-additive substitutes designed to protect engines from valve-seat recession (61) are used in Europe and South America. The additional cost of one such compound (which is based on phosphorus and cannot be used with catalytic converters) is approximately \$0.001 liter^{-1} of gasoline (E Squire, personal communication).

COST OF ELIMINATING LEAD FROM GASOLINE

Another factor in the decision to eliminate lead from gasoline is cost, which is often cited as a reason against switching to unleaded gasoline. Octel claims that "it is possible, but extremely expensive to remove lead from gasoline" (70, p. 4). To replace the octane provided by lead, either the gasoline production must be changed, through various refinery options, or other additives must be used, such as the oxygenate methyl tertiary butyl ether (MTBE). The cost difference between leaded and unleaded gasoline depends on the details of the refinery, the gasoline feedstock, as well as many choices concerning refinery reconfiguration, gasoline properties including octane number, and the use of additives.

Cost of Lead Additive

The octane added by a given amount of lead additive (TEL) varies somewhat with the composition of the gasoline (71). Figure 4, a typical octane response curve, shows that the rate of increase of octane value decreases as the amount of lead additive increases. (All octane numbers presented here are the average of the research and motor methods for determining octane.)

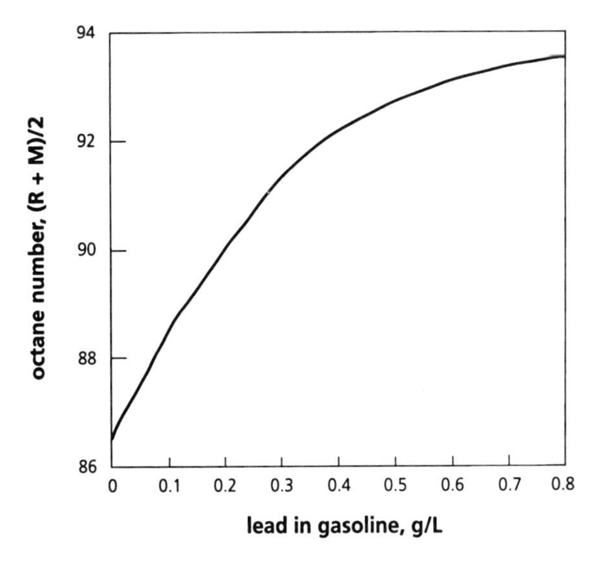

Figure 4 Octane response curve (71).

Currently, TEL sells for about \$7 kg^{-1} (2). With approximately 0.3 g lead per gram TEL, the cost of lead additive is \$0.02 g^{-1} lead. As shown in Figure 4, the first 0.1 g $liter^{-1}$ lead added to a gasoline with 87 octane provides about two octane numbers; this addition would cost about \$0.002 $liter^{-1}$ of gasoline.

Cost of MTBE

Octane can also be increased by adding organic compounds such as methanol, ethanol, MTBE, and ethyl tertiary butyl ether (ETBE). These additives have octane values ranging from 110 to 120. The wholesale price of MTBE is about \$0.16 $liter^{-1}$, and it has an octane value of 110. MTBE displaces gasoline, so the net cost of MTBE is the cost of MTBE minus the cost of gasoline. For a wholesale cost of gasoline of about \$0.11 $liter^{-1}$, the net cost of MTBE is \$0.16–0.11 = \$0.05 $liter^{-1}$ of MTBE. The MTBE equivalent of the 0.1 g $liter^{-1}$ leaded gasoline, discussed above, would provide a two–octane number increase. To increase the octane of an 87 octane gasoline by two octane numbers requires about 0.1 liter MTBE per liter gasoline. The net cost of the 0.1 liter MTBE would be \$0.005 $liter^{-1}$ of gasoline. This is 2.5 times the cost of the equivalent amount of lead additive, estimated above. The cost difference between the gasoline with 0.1 g $liter^{-1}$ lead and the gasoline with the equivalent amount of MTBE is \$0.003 $liter^{-1}$ gasoline (2).

Costs of Refinery Options

The basic refinery options include catalytic reforming, which produces higher octane aromatics (such as benzene, xylene, and toluene) and isoparaffins; isomerization, which transforms normal paraffins into higher-octane isoparaffins; and alkylation, by which lower-molecular-weight olefins react with isoparaffins to form higher-molecular-weight isoparaffins.

A transition from 0.66 g liter^{-1} leaded gasoline, which is typically found in many developing countries, to unleaded gasoline has been estimated to cost approximately \$0.005–0.01 liter^{-1} (72). Similarly, Octel estimates the cost of one octane number from refining to be about \$0.002 liter^{-1} (39).

For example, Thailand is making the transition from leaded gasoline with 0.45 g liter^{-1} lead to unleaded gasoline using both refinery improvements and MTBE. With investment costs amortized over 12 years, discounted at 12% per year, the capital cost is \$0.010 liter^{-1}, the operating cost \$0.0025 liter^{-1}, and the cost of MTBE (11%) \$0.0030 liter^{-1} (73). In Malaysia, the production cost for unleaded gasoline is 4% more than it is for 0.15 g liter^{-1} leaded gasoline (74). In the United States, the cost per octane number was about \$0.0025 liter^{-1} in the 1980s (61). In Germany, the cost of producing unleaded gasoline, vs that of 0.15 g liter^{-1} leaded gasoline, is \$0.01 liter^{-1} (3).

In summary, refinery improvements and MTBE are roughly three times as expensive as TEL. The additional cost of unleaded gasoline is about \$0.01 liter^{-1}. However, benefits, as well as costs, must be considered. The US EPA estimated that the car-maintenance benefits of using unleaded gasoline (leaded gasoline decreases the lifetime of spark plugs and exhaust systems and requires more frequent oil changes) alone are greater than the cost of eliminating lead in gasoline. When the health benefits of unleaded gasoline are included, the benefits are estimated to exceed the costs by a factor of three (61).

OTHER HEALTH AND ENVIRONMENTAL ASPECTS OF GASOLINE FORMULATIONS

Health and environmental concerns associated with the use of gasoline, include emissions of carbon monoxide and ozone-forming chemicals, the emission of toxic substances such as benzene, the potential health effects of the oxygenate MTBE (76), and issues of energy consumption and global warming. Any reformulation of gasoline, including the decision to eliminate lead in gasoline, can affect all of these factors. Here, I discuss two of these issues: catalytic converters and benzene.

Catalytic Converters

Catalytic converters reduce emissions of hydrocarbons, carbon monoxide, and nitrogen oxides. These devices require unleaded gasoline because lead poisons the catalyst. As catalytic converters become increasingly widespread, the use of leaded gasoline will of necessity be phased out.

However, the belief that unleaded gasoline can only be used in cars with catalytic converters is not uncommon, even among vehicle salesmen and government administrators (77, 78). In India, for example, the Environment Ministry said that "the introduction of un-leaded petrol would have to be synchronized with the development of vehicles with new technology and new engine designs which could entail colossal expenditure. The engines would also have to be fitted with catalytic converters" (79, p. 9). This misperception is not surprising, given Octel's claim that "unleaded gasoline should only be used in cars with catalytic converters" (70, p. 8). In fact, unleaded gasoline can be used in any car, regardless of whether or not the vehicle has a catalytic converter. As detailed above, the only potential risk is valve-seat recession in cars with unhardened exhaust valve seats, and this problem has not been reported in practice.

Benzene

As mentioned above, increasing the concentrations of benzene and other aromatic hydrocarbons in gasoline is one way to increase the gasoline's octane rating. Benzene is a carcinogen (80), and the content of benzene and other aromatics must be considered when choosing a manufacturing process for unleaded gasoline. In the United States, the benzene content of gasoline is limited to 1% volume; in the European Community and Australia, it is limited to 5% (81). Especially when used in a car lacking a catalytic converter, gasoline with higher aromatics will produce higher emissions of aromatics such as benzene.

Octel, the manufacturer of TEL, has made the benzene emissions of unleaded gasoline the focus of its international effort to market leaded gasoline. According to John Little, Managing Director of Octel, "We have launched a worldwide effort to promote the dangers of using unleaded gasoline, particularly in cars that are not equipped with catalytic converters...For the past five years we have been visiting developing countries to promote the use of leaded gasoline and to warn against the use of unleaded in non-catalyzed cars" (82). This corporation has actively campaigned in the United Kingdom, Israel, Egypt, Hong Kong, Italy, the Philippines, New Zealand, Australia, and elsewhere (70, 83–89). Octel claims that use of unleaded gasoline in cars without catalytic converters may increase exposure to benzene. Based on this claim, Octel argues that only cars with catalytic converters should use unleaded gasoline.

But human exposure to benzene has little to do with whether or not the cars have catalytic converters. Typical exposure to benzene results primarily from evaporation from the gasoline tank and not from tailpipe exhaust (90, 91).

Exposures can be especially high when the car is parked in a garage attached to the house, in which case benzene fumes seep into the house (92). Exposures in parking garages or at gasoline stations can also be high. Moreover, the most significant source of benzene exposure is cigarette smoke. So although the aromatic content of gasoline is an important consideration, the resulting exposure to benzene bears little relation to the presence or absence of catalytic converters.

Moreover, the benzene and aromatic content of unleaded gasoline is not necessarily higher than the benzene content of leaded gasoline. To counter the confusion caused by Octel's advertising, Shell Australia has issued press releases to the effect that the benzene and aromatic contents of its gasolines have not increased with the reduced use of TEL (93).

Energy

The energy implications of the transition to unleaded gasoline are determined by several factors, including the extent to which new components are added to the gasoline to furnish the octane formerly provided by lead additives; the energy and extra petroleum that may be required to make these new components; the potentially higher energy density of unleaded gasoline; and whether, if octane levels drop below the octane requirement of a car, the ignition timing must be adjusted to reduce the car's octane requirement.

First consider the case in which not all of the octane provided by lead additives is replaced in the unleaded gasoline, so that the resulting fuel has a lower octane level. Because the lead additives have essentially no fuel value, the reduction in octane number has no direct energy implications. However, because reduced octane is widely believed to result in reduced fuel efficiency (39), a more detailed discussion follows.

The main purpose of adding lead to gasoline is to prevent engine knock. Although the octane required to prevent knock depends on several factors, the basic parameter is compression ratio: Each unit increase in compression ratio requires approximately three octane numbers (94); and each unit of increase raises energy efficiency by approximately 3% for practical fuels and engines (95). The net result is that each unit of octane number provides the potential for a 1% increase in fuel efficiency. However, reducing the octane number of the gasoline does not reduce the fuel efficiency of the car. The compression ratio is a property of the car's engine. For a given car, increasing the octane will not in itself change energy efficiency (unless, as mentioned below, the octane is provided by components that increase mileage).

When the gasoline octane is reduced below the octane requirement of the car, the car's octane requirement can be decreased somewhat by retarding the ignition timing. Spark retard of approximately 10°, a typical recommendation to compensate for an octane requirement reduced by three octane numbers (58), can result in a decreased fuel efficiency of about 1% (96).

A different set of considerations pertains to the energy required for increased octane in unleaded gasoline. Refinery production of unleaded gasoline requires more crude oil, and more energy to run the refinery components, than would be needed for leaded gasoline of the same octane rating. The petroleum industry has estimated that a refinery-produced, seven–octane number increase (equivalent to about 0.3 g $liter^{-1}$ lead additive) requires an approximately 1.0–1.5% increase in crude oil (97). However, the resulting gasoline may have a higher energy density or, in the case of oxygenates, may provide more efficient combustion, resulting in increased mileage (97). Exxon estimated that in the United States, unleaded gasoline would have 1.0–1.5% greater energy efficiency than the octane-equivalent gasoline with 2.5 g $liter^{-1}$ lead (61). These estimates indicate that the 1.0–1.5% energy penalty resulting from increased octane in unleaded gasoline may be counterbalanced by a similar increase in mileage.

In summary, if the octane of leaded gasoline is replaced with high-octane gasoline components or additives, ~1% more energy may be required to produce the gasoline, but the resulting fuel may provide a roughly similar amount of additional energy, for a net change in energy use of considerably less than 1%. The lead additive itself has no energy value, so the removal of lead, with a resulting drop in octane, results in no change in energy use or energy efficiency. If, however, the octane is reduced below the octane requirement of a given car, and if the car's ignition timing is adjusted to reduce its octane requirement, that car's energy efficiency may decrease slightly.

CONCLUSIONS

The use of leaded gasoline is gradually being phased out worldwide, because of increasing recognition of the health risks of lead exposure and the introduction of catalytic converters, which require unleaded gasoline. However, as each country considers the issues of leaded gasoline, the same questions are revisited—those concerning health risks, technical feasibility, and costs.

As to health effects, many independent studies have produced a scientific consensus that blood lead levels of 10–15 $\mu g\ dl^{-1}$ are associated with reduced intelligence and effects on growth and development of children. Because of the multiple pathways of exposure to lead from gasoline, a precise quantification of the average contribution of gasoline lead to blood lead is not available. But between 1976 and 1990, as use of lead in gasoline in the United States fell from near its peak to almost zero, average blood lead levels in the United States decreased from 16 $\mu g\ dl^{-1}$ to 3 $\mu g\ dl^{-1}$. Calculations of expected exposure indicate that much of this decrease can be attributed to the reduced use of leaded gasoline.

The problem of valve-seat recession has often been cited as a barrier to the use of unleaded gasoline in cars without hardened valve seats. Although valve-seat recession can be induced in laboratory experiments and specially designed road tests, this problem does not arise in practice. Normal driving does not provide the extreme conditions most conducive to valve-seat recession, and any effects seem to be small and outweighed by corrosion caused by the halogenated additives used as scavengers in most leaded gasolines.

In general, the transition to unleaded gasoline will require refinery improvements and perhaps use of other additives such as MTBE. The components of unleaded gasoline (and all other gasolines) should be carefully considered in light of other potential health and environmental risks. The use of catalytic converters, however, is an entirely separate issue: If a gasoline does have a high benzene (or other aromatic) content, evaporation—from the tank, not through the tailpipe—will be the typical primary source of exposure. Hence, catalytic converters are unlikely to make much difference in the prevention of benzene exposure.

The cost of the transition to unleaded gasoline must be calculated on a case-by-case basis, but experience in the developing and developed countries indicates that the net cost increase is about \$0.01 liter^{-1}.

ACKNOWLEDGMENTS

The author thanks Sharmon Anciola for invaluable assistance with research and an anonymous reviewer for detailed and thoughtful comments.

Literature Cited

1. Organization for Economic Cooperation and Development. 1994. *Lead based gasoline additives*. Presented at OECD Workshop Lead Products, Toronto
2. Ottenstein R, Tsuei A. 1994. *Ethyl Corporation*. New York: CS First Boston
3. Muller G. 1994. *Use of lead antiknock additives in Germany*. Presented at OECD Workshop Lead Products, Toronto
4. Nickerson SP. 1954. Tetraethyl lead: a product of American research. *J. Chem. Educ.* 31:560–71
5. Nriagu JO. 1990. The rise and fall of leaded gasoline. *Sci. Total Environ.* 921: 13–28
6. Japan Environmental Agency. 1988. *Motor Vehicle Pollution in Japan*. Tokyo: Automotive Pollution Control Division, Air Quality Bureau. 3rd ed.
7. Kimura M. 1974. Changing to lead-free gasoline. *J. Jpn. Petrol. Inst.* 18(3):166–74 (In Japanese)
8. Octel, Ltd. 1992. *Worldwide Survey of Motor Gasoline Quality 1991*. London: Octel
9. International Environment Reporter. 1991. Colombia: production, sale of leaded gasoline eliminated, state-owned

oil company says. Washington, DC: Bur. Natl. Aff.
10. International Environment Reporter. 1993. Austria: gradual abolition of leaded fuel likely to result in lower emissions, higher costs. Washington, DC: Bur. Natl. Aff.
11. Natural Resources Defense Council and Cape 2000. 1994. *Four in '94: Assessing National Actions to Implement Agenda 21: A Country-by-Country Progress Report.* Washington, DC: Nat. Res. Def. Counc. Cape 2000
12. Swedish Environmental Protection Agency. 1994. EU drivers have taken up lead-free. *Current Environ.* 5:9 (In Swedish)
13. International Environment Reporter. 1994. European Union: unleaded gasoline takes a larger share of EU market, statistics agency reports. Washington, DC: Bur. Natl. Aff.
14. Danielson L. 1970. Gasoline containing lead. *Ecol. Res. Comm. Bull. No. 6,* 45 pp.
15. Gusev MI. 1960. *Limits of Allowable Lead Concentrations in the Air of Inhabited Localities,* Book 4, ed. VA Ryazanov. Moscow: Comm. Determination of Allowable Atmospheric Concentrations of Atmospheric Pollutants Allied with the Chief State Sanitary Inspectorate of the USSR; Transl. distributed by the US Dep. Commerce, Off. Tech. Serv.
16. US Department of Health and Human Services. 1991. *Preventing Lead Poisoning in Young Children.* Washington, DC: US Dep. Health Hum. Serv.
17. National Research Council. Committee on Measuring Lead in Critical Populations. 1993. *Measuring Lead Exposure in Infants, Children, and Other Sensitive Populations.* Washington, DC: Natl. Acad. Press
18. Agency for Toxic Substances Disease Registry. 1988. *The Nature and Extent of Lead Poisoning in Children in the United States: A Report to Congress.* Atlanta: ATSDR
19. Patterson CC, Ericson J, Manea-Krichten M, Shirahata H. 1991. Natural levels of lead in *Homo sapiens sapiens* uncontaminated by technological lead. *Sci. Total Environ.* 107:205–36
20. Annest JL, Pirkle JL, Makuc D, Neese JW, Bayse DD, Kovar MG. 1983. Chronological trend in blood lead levels between 1976 and 1980. *New Engl. J. Med.* 308(23):1373–77
21. US Environmental Protection Agency. 1983. *Air Quality Criteria for Lead,* Vol. 2. *EPA-600/8-83-028A.* Washington, DC: Environ. Prot. Agency
22. US Environmental Protection Agency. 1993. *National Air Pollutant Emission Trends 1900–1992. EPA-454/R-93–032.* Washington, DC: Environ. Prot. Agency
23. Elwood PC. 1986. The sources of lead in blood: a critical review. *Sci. Total Environ.* 52:1–23
24. Pirkle JL, Brody DJ, Gunter EW, Kramer RA, Paschal DC, et al. 1994. The decline in blood lead levels in the United States: the National Health and Nutrition Examination Surveys (NHANES). *J. Am. Med. Assoc.* 272(4):284–91
25. Capar SG. 1990. Survey of lead and cadmium in adult canned foods eaten by young children. *J. Assoc. Off. Anal. Chem.* 73(3):357–64
26. Jelinek CF. 1982. Levels of lead in the United States food supply. *J. Assoc. Off. Anal. Chem.* 65(4):942–46
27. Schaffner RM. 1981. Lead in canned foods. *Food Technol.* Dec:60–64
28. US Environmental Protection Agency. 1989. *Review of the National Ambient Air Quality Standards for Lead: Exposure Analysis Methodology and Validation. EPA-450/2-89-011.* Washington, DC: Environ. Prot. Agency
29. US Environmental Protection Agency. 1994. *Guidance Manual for the Integrated Exposure Uptake Biokinetic Model for Lead in Children. EPA-540/R-93/081.* Washington, DC: Environ. Prot. Agency
30. Flegal AR, Smith DR, Elias RW. 1990. Lead contamination in food. In *Advances in Environmental Science and Technology,* Vol. 23, *Food Contamination from Environmental Sources,* ed. JO Nriagu, MS Simmons, pp. 85–120. New York: Wiley
31. US Environmental Protection Agency. 1990. *National Air Quality and Emissions Trends Report, 1988. EPA 450/4-90-002.* Washington, DC: Environ. Prot. Agency
32. US Environmental Protection Agency. 1983. *Air Quality Criteria for Lead,* Vol. 3. *EPA-600/8-83-028A.* Washington, DC: Environ. Prot. Agency
33. Facchetti S. 1984. Isotopic Lead Experiment. European Community. *EUR. 8352 EN. MF. 84456EN. CD No. 84 005 EN A.*
34. International Lead and Zinc Research Organization. 1991. *Lead in Gasoline: Environmental Issues.* Research Triangle Park, NC: ILZRO
35. Deleted in proof
36. Hinton D, Coope PA, Malpress WA, James ED. 1986. Trends in blood lead

levels in Christchurch (NZ) and environs. *J. Epidemiol. Commun. Health* 40:244–48
37. Octel, Ltd. ca. 1993. Position paper on leaded gasoline. London: Octel
38. Jones KC, Symon C, Taylor PJL, Walsh J, Johnston AE. 1991. Evidence for a decline in rural herbage lead levels in the U.K. *Atmos. Environ. A* 25:361–69
39. Larbey RJ. 1994. Issues surrounding the use of lead in gasoline—energy, economic and environmental. *Sci. Total Environ.* 146/147:19–26
40. McMillen JH, Scott GH. 1936. Spectrographic studies of lead in human blood. *Proc. Soc. Exp. Biol. Med.* 35(3): 364–65
41. Smith FL, Rathmell TK, Marcil GE. 1938. The early diagnosis of acute and latent plumbism. *Am. J. Clin. Pathol.* 8:471–508
42. Kaplan E, McDonald JM. 1942. Blood lead determinations as a health department laboratory service. *Am. J. Public Health* 32:481–86
43. Kehoe RA, Cholak J, Story RV. 1970. A spectrochemical study of the normal ranges of concentrations of certain trace metals in biological materials. *J. Nutr.* 19:579–92
44. Letonoff TV, Reinhold JG. 1940. Colorimetric determination of lead chromate by diphenylcarbazide: application of a new method to analysis of lead in blood, tissues and excreta. *Ind. Eng. Chem. Anal. Ed.* 12(5):280–84
45. Kehoe RA. 1947. Exposure to lead. Conference on lead poisoning. *Occup. Med.* 3(2):156–71
46. The Working Group on Lead Contamination. 1965. *Survey of Lead in the Atmosphere of Three Urban Communities.* Washington, DC: US Dep. Health, Educ. and Welfare, Public Health Serv.
47. Patterson CC. 1965. Contaminated and natural lead environments of man. *Arch. Environ. Health* 11:344–60
48. Keppler JF, Maxfield ME, Moss WD, Tietjen G, Linch AL. 1970. Interlaboratory evaluation of the reliability of blood lead analysis. *Am. Ind. Hyg. Assoc. J.* 31:412–29
49. Pierce JO, Koirtyohann SR, Clevenger TE, Lichte FE. 1976. *The Determination of Lead in Blood: A Review and Critique of the State of the Art, 1975.* New York: Int. Lead Zinc Res. Org.
50. Lob M, Desbaumes P. 1976. Lead and Criminality. *Br. J. Ind. Med.* 33:125–27
51. Patterson CC, Seattle DM. 1974. *The reduction of orders of magnitude errors in lead analyses of biological materials and natural waters by evaluating and controlling the extent and sources of industrial lead contamination introduced during sample collecting and analysis.* Presented at Mater. Res. Symp., 7th, Natl. Bur. Stand., Gaithersburg, MD
51a. Bruland KW. 1983. Trace elements in sea-water. In *Chemical Oceanography,* Vol. 8, ed. JP Riley, R Chester, pp. 157–220. London: Academic
52. Settle DM, Patterson CC. 1980. Lead in Albacore: Guide to Lead Pollution in Americans. *Science* 207:1167–76
53. Everson J, Patterson CC. 1980. "Ultraclean" isotope dilution/mass spectrometric analyses for lead in human blood plasma indicate that most reported values are artificially high. *Clin. Chem.* 26(11):1603–7
54. Williams CG. 1937. Factors influencing the wear of valve seats in internal combustion engines. *Engineering* 143 (3715):357–58, 143(3719)475–76
55. Schoonveld GA, Riley RK, Thomas SP, Schiff S. 1986. Exhaust valve recession with low-lead gasolines. *Soc. Automot. Eng.* No. 861550
56. Godfrey D, Courtney RL. 1971. Investigation of the mechanism of exhaust valve seat wear in engines run on unleaded gasoline. *Soc. Automot. Eng.* No. 710365
57. Giles W. 1971. Induction hardening makes exhaust-valve seats wear less with non-leaded fuel. *Automob. Eng.* 79(6):33–37
58. Autodata, Ltd. 1989. *Unleaded Petrol Information Manual.* Maidenhead, UK: Autodata. 3rd ed.
59. McArragher S, Clarke L, Paesler H. 1994. *Protecting engines with unleaded fuels. Shell selected paper.* London: Shell Int. Petrol.
60. Garbak JA, Grinnell GE. 1987. Effects of using unleaded and low-lead gasoline and gasoline containing non-lead additives on agricultural engines designed for leaded gasoline. *Soc. Automot. Eng.* No. 871622
61. US Environmental Protection Agency. 1985. Costs and Benefits of Reducing Lead in Gasoline. Final Regulatory Impact Analysis. EPA-230–05–85–006. Washington, DC: Environ. Prot. Agency
62. Wintringham JS, Felt AE, Brown WJ, Adams WE. 1972. Car maintenance expense in owner service with leaded and nonleaded gasolines. *Soc. Automot. Eng.* No. 720499
63. Tosh JD. 1976. Performance of army engines with unleaded gasoline. *AFLRL-82, Southwest Res. Inst., San Antonio, Tex., NTIS #AD-A032075*

64. Gray DS, Azhari AG. 1972. Saving maintenance dollars with lead-free gasoline. *Soc. Automot. Eng.* No. 720084
65. Schwochert HW. 1969. Performance of a catalytic converter that operates with nonleaded fuel. *Soc. Automot. Eng.* No. 690503
66. Crouse WW, Johnson RH, Reiland WH. 1971. Effect of unleaded fuel on lubricant performance. *Soc. Automot. Eng.* No. 710584
67. Commonwealth Environmental Protection Agency. 1994. *Lead Alert: A Guide for the Motor Trades.* Canberra: Commonw. Environ. Prot. Agency
68. Chevron. 1985. Motor gasolines. *Chevron Res. Bull. Richmond, CA*
69. Tiam HL. 1992. An unleaded guarantee for motorists. *Bus. Times* (Singapore), Oct. 29
70. Gidlow DA. 1994. *Lead in gasoline.* Presented at OECD Workshop Lead Products, Toronto
71. Owen K, Coley R. 1990. *Automotive Fuels Handbook.* Warrendale, PA: Soc. Automot. Eng.
72. Michalski GW, Cunningham RE, Stevens CA, Lyons JM. 1994. *Lead in Gasoline: A Refinery Perspective.* Presented at OECD Workshop Lead Products, Toronto
73. World Bank. 1994. Thailand: mitigating pollution and congestion impacts in a high growth economy. *Country Econ. Rep.*
74. World Bank. 1993. Malaysia: managing costs of urban pollution. *Country Econ. Rep.*
75. Deleted in proof
76. US Environmental Protection Agency. 1993. *Assessment of Potential Health Risks of Gasoline Oxygenated with Methyl Tertiary Butyl Ether (MTBE). Off. Res. Dev. EPA/600/R-93/206.* Washington, DC: Environ. Prot. Agency
77. Lobban M, Valere E. 1993. Caribbean: environment chokes despite switch to unleaded gas. *Interpress Serv.* Aug. 22
78. Vehicle manufacturers want petrol plan delay. 1994. *Bangkok Post,* Jan. 20
79. Singh GK. 1994. Government finds unleaded gasoline use 'impractical.' *The Pioneer* (Delhi), Jun. 29, p. 9
80. Mehlman MA. 1992. Dangerous and cancer causing properties of products and chemicals in the oil refining and petrochemical industry. VIII. Health effects of motor fuels: carcinogenicity of gasoline—scientific update. *Environ. Res.* 59:238–49
81. The Oil Companies' European Organization for Environmental and Health Protection (CONCAWE). 1992. Motor vehicle emission regulations and fuel specifications—1992 update. *CONCAWE Rep. No. 2/92*
82. Little, JS. 1994. *Great Lakes Chemical Corporation: Octel overview.* Presented at Annu. CS First Boston Chem. Conf., 6th, New York
83. UK firm claims research strips away unleaded petrol's green status. 1993. *The Financial Times,* Nov. 12
84. Ben Shaul D. 1994. Compounding pollution. Jerusalem *Post,* Jan. 24
85. Why breathing is killing us. 1994. *Al Ahram Weekly,* Jan. 24
86. Carter J. 1993. Lead-free petrol linked to cancer-causing agent. South China *Morning Post,* Dec. 19
87. Call to test cancer risk of petrol lead. 1994. Sydney *Morning Herald,* Apr. 27
88. International Environment Reporter. 1994. Group asks commission to charge Italy with breaking EU lead-in-gas directive. Washington, DC: Bur. Natl. Aff.
89. Unleaded petrol and cancer: asbestos, cigarettes. Now benzene. (Advertisement). 1995. *Evening Post,* Wellington, NZ, Feb. 25
90. Wallace LA. 1989. Major sources of benzene exposure. *Environ. Health Perspect.* 82:165–69
91. Wallace L. 1993. A decade of studies of human exposure: what have we learned? *Risks Anal.* 13(2):135–39
92. Lansari A, Streicher JJ, Huber AH, Crescenti JH, Zweidinger RB, Duncan JW. 1993. Dispersion of uncombusted auto fuel vapor within residential and attached garage microenvironments. In *Measurement of Toxic and Related Air Pollutants,* pp. 52–57. Pittsburgh: Air Waste Manage. Assoc.
93. Shell Australia, Ltd. 1994. Media release. Nov. 22
94. Downstream Alternatives Inc. 1992. *Changes in Gasoline,* Vol. 2. *Technician's Manual.* Bremen, IN: Downstream Alternatives
95. Heywood JB. 1988. *Internal Combustion Engine Fundamentals,* p. 842. New York: McGraw-Hill
96. Obert EF. 1973. Fig. 14–44b. *Internal Combustion Engines and Air Pollution,* p. 552. New York: Harper & Row
97. Baker RE, Shelef M. 1985. Engine requirements for fuels and lubricants: a petroleum industry response. *ChemTech* 15(8):504–12
98. US Department of Energy. 1991. *International Energy Annual 1991. Energy Information Administration. DOE/EIA-012(91).* Washington: US Gov. Print. Off.
99. Somunkiranoglu F. 1995. Statement

presented at Int. Workshop Phasing Lead Out of Gasoline, Washington, DC
100. Organization for Economic Cooperation and Development. 1993. *Lead. Environmental Directorate. Risk Reduction Monogr.* No. 1. Paris: OECD
101. "Statoil's New Lead-Free Petrol." 1993. *Eur. Energy Rep.*, March 19
102. Russian Standard GOST 2084–77. 1987. "Motor Gasoline." Technical Conditions. Dec. 7
103. Janssen N. 1994. Cars must switch for unleaded Kuwait by '97. *Moneyclips,* March 10
104. Foner HA. 1993. Lead pollution in Israel. *Water Sci. Tech.* 27(7–8):253–62
105. "Ecuador: alarming levels of lead poisoning in Quito." 1993. *Interpress Serv.*, Sept. 27
106. Argentina: Petrol Plan—YPF. 1993. *Lloyd's List,* Aug. 21
107. Scott S. 1994. *Latin America: saying no to lead.* Decision Brief. Cambridge, MA: Cambridge Energy Res. Assoc.
108. Medhat Y. 1995. Statement presented at Int. Workshop on Phasing Lead Out of Gasoline, Washington, DC
109. Wang X-C. 1995. Statement presented at Int. Workshop on Phasing Lead Out of Gasoline, Washington, DC
110. Uniyal M. 1992. India: cities choking on vehicle pollution. *Interpress Serv.*, Dec. 5
111. NZ Government to Review Leaded Petrol Policy. 1994. *Xinhua News Agency,* Aug. 10
112. Fernandez J. 1994. Malaysia: Gov't Promotes Unleaded Gasoline; Sales Reach 50%. *Kuala Lampur Star*, Oct. 14
113. Taiwan: Govt to Advance Mandatory Unleaded Gas by Two Years. 1992. *China Economic News Serv.*, Dec. 15
114. Chandrasekera D. 1993. Using of non-leaded petrol. In *The Engineer* 13(5):3. Colombo: Soc. Sri Lanka Eng.
115. US Environmental Protection Agency. 1989. *Exposure Factors Handbook. EPA 600/8-89/043.* Washington, DC: US Environ. Prot. Agency

Annu. Rev. Energy Environ. 1995. 20:325–86

DETERMINANTS OF AUTOMOBILE USE AND ENERGY CONSUMPTION IN OECD COUNTRIES

Lee Schipper
Lawrence Berkeley Laboratory, Berkeley, Califiornia 94720

KEY WORDS: transport, travel, energy use, CO_2 emissions

CONTENTS

ABSTRACT

Energy use is associated with environmental problems and other externalities arising from personal transportation. In this article, we review trends in the use of the car and other modes of personal transportation in 10 OECD countries

1056-3466/95/1022-0325$05.00

from 1970 to 1992. We analyze changes in energy use for these activities and discuss underlying components of, and causes for, these changes. We compare differences between the United States and the European countries studied, concluding that most of the variations arise because of differences in total transport activity and modal choice, not because of the energy efficiency of each mode. We analyze more closely differences in activity, which is dominated by the automobile, relating automobile use to differences in fuel prices and car taxation, in patterns of mobility, in demographic patterns, and in geographical factors like land use or place of residence. We conclude with a focus on one externality of transportation, the carbon dioxide emissions from travel. We note that these emissions are rising in all the countries studied. We suggest that policies aimed at stemming this rise must be integrated with other policies related to other problems of transportation, many of which are perceived to be more important than that of CO_2 alone.

INTRODUCTION

This article reviews underlying trends that have boosted energy use for transportation and travel in OECD countries. We also show how these trends have affected emissions of CO_2. We focus on the automobile, but provide important information on factors affecting energy use of complementary modes of transportation. We discuss some of the policy instruments that have affected mobility and energy use. We speculate briefly on how changes in some policies might affect these variables in the future, concluding that only a broad framework that integrates concerns for CO_2 with strategies to solve other problems related to transportation can be successful.

This article summarizes many of the findings of the study "Sustainable Transportation: The Future of the Automobile in an Environmentally Constrained World," which aims to understand how travel and energy use for travel (and freight) are changing, how these changes may affect the environment, and how the environmental problems may in turn affect future travel and freight activity. During its first three years, the study focused on personal transportation in countries in the Organization for Economic Cooperation and Development (OECD), but we acknowledge that problems facing rapidly growing transportation demand are manifest in developing countries and in the transitional economies of Central and Eastern Europe.

The countries in the study include the United States, Japan, the former West Germany, France, Italy, Great Britain (here called the EU-4), and the four Nordic countries Sweden, Finland, Denmark, and Norway (the NO-4). Figure 1, using data both from this study and from the International Energy Agency (1), shows total energy use for transportation in all OECD countries and transportation's share in total primary energy use for all purposes in all OECD

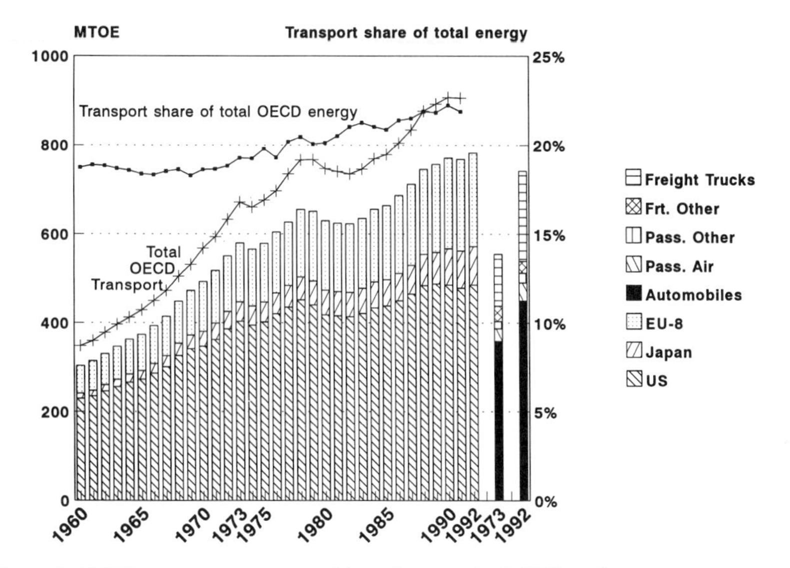

Figure 1 OECD transport energy use with use by countries in LBL study.

countries. We then show total transportation energy use (as reported to the IEA) for the ten countries included here, and for 1992, the total as we have analyzed it for those countries, splitting uses into travel and freight components as well. It can be seen that the countries under study consume more than 80% of all IEA energy use for transportation, and the travel-related sectors account for around two-thirds of that total. By comparing our 1992 figures for the ten countries included here and those provided by the IEA, it can be seen that we have omitted some energy uses (i.e. bunkers for international shipping and air travel, and some miscellaneous use of heavy land vehicles, including military vehicles). Note that the transportation share of total energy use has risen over time.

BACKGROUND: WHY WORRY ABOUT THE AUTOMOBILE?

A multitude of challenges associated with ever-increasing motorization, mobility, traffic, energy use, and pollution from road vehicles confront authorities and citizens of OECD countries and major developing countries as well (2–7). The litany of problems that must be addressed includes traffic safety, congestion, noise, and many kinds of pollution resulting from the use of cars—especially those leading to the formation of photochemical smog and greenhouse gas emissions from transportation fuels. The pollution generated by the manufacture of vehicles and the use, disposal, or recycling of transportation equipment cannot be ignored. The problems associated with obtaining a stable and secure supply of transportation fuels continue to be a concern. More broadly, transportation planners and policymakers have confronted a variety of problems associated with access to growing, sprawling cities, including segregation of land uses and competition for land, and access to vital services for those who cannot drive or cannot afford the costs of private or often even public transportation.

Although the automobile stands at the center of many of these problems, not all of them involve only the car or even road traffic. Some of the problems (air pollution, greenhouse gas emissions, oil importation) are related to the nature of fuel. Indeed, it appears that actions designed to remedy transportation-related problems could have as great an impact on fuel use as those aimed only at fuel, through provoking changes in travel itself. We emphasize, however, that it is neither travel nor the automobile per se that is a problem, but rather the externalities that arise in transportation itself. Understanding how to remedy these problems demands an analysis of their underlying determinants, particularly those forces that cause the overall volume of travel to grow. For that reason, we devote considerable space to discussing trends in overall travel and automobile use as well.

Deluchi (8) has begun to systematically evaluate many of the costs of using automobiles (and other vehicles) in the United States and has discovered wide variation in valuations of their "costs" per kilometer of automobile travel, or in the case of fuels, per liter of fuel consumed. He has tabulated a large number of costs that are paid by drivers (of trucks and cars), by society as a whole through funding of roads and other transportation-related services, and by specific groups in society through environmental costs or externalities.

Deluchi's calculations (currently being completed) show a range of social costs—monetary and non-monetary externalities, as well as direct, paid monetary costs—resulting in motor vehicle users paying 84% of the total monetary cost of transport, although not necessarily in proportion to the costs they actually incur. The share of total social costs (monetary costs plus non-monetary externalities) is larger, and users pay 68–80% of these costs. The externalities work out to approximately $0.06–$0.12 vehicle-kilometer, or veh-km, but these include both those caused by trucks and buses (most likely high per veh-km) and those caused by cars (much lower per veh-km). These results lie within the range of costs reviewed by Kaageson (9) and those estimated for the health costs of vehicle-source air pollution in the Los Angeles region (10). The total unpaid costs of transportation are large, albeit much less than the total private benefits of transportation and the privately (or publicly) paid user costs for roads, vehicles, fuel, and so on.

Some of these externalities, such as precursors to photochemical smog or noise, are being reduced by specific technologies (emission controls or walls, respectively); others, through certain regulations; and still others, through certain taxes in some countries. One of Deluchi's surprising findings was that road dust (literally particulates stirred up by moving vehicles as well as those breaking away from tires) may constitute a major health hazard. Breaking these costs into their marginal components and then estimating them at specific locations and times (important for pollution, congestion, and noise) are difficult. But the social cost of driving a noisy vehicle in a residential neighborhood on a congested and polluted day in Los Angeles is greater than driving the same vehicle in an industrial district at 2 A.M. before a rainstorm. The key challenge is to find ways of reducing the unpaid externalities from transportation, not to stop or limit transportation per se.

MIND THE GAP: QUANTIFYING ENERGY USE FOR TRANSPORTATION

Figure 2 shows per capita travel by mode in the United States, Japan, the EU-4 and the NO-4; Figure 3 shows the smooth time trend for the total from 1970 to 1991. Figure 4 shows the energy use associated with activities in each of these modes. The main sources of data from each country are listed in an appendix.

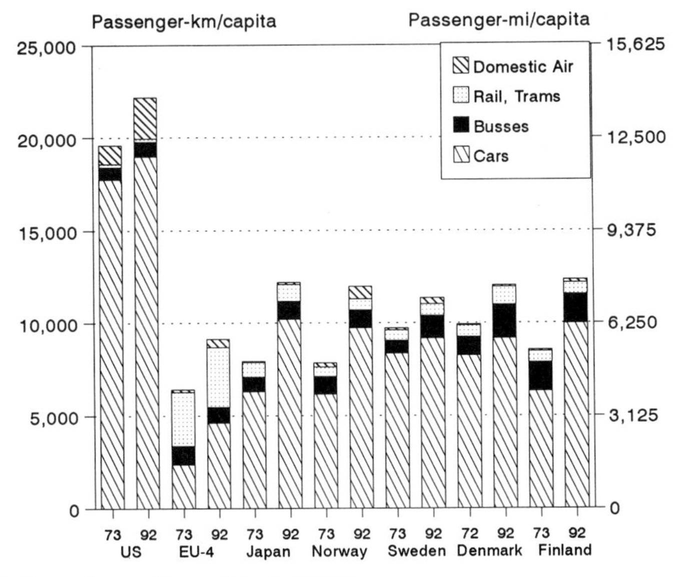

Figure 2 Per capita travel by mode in the OECD.

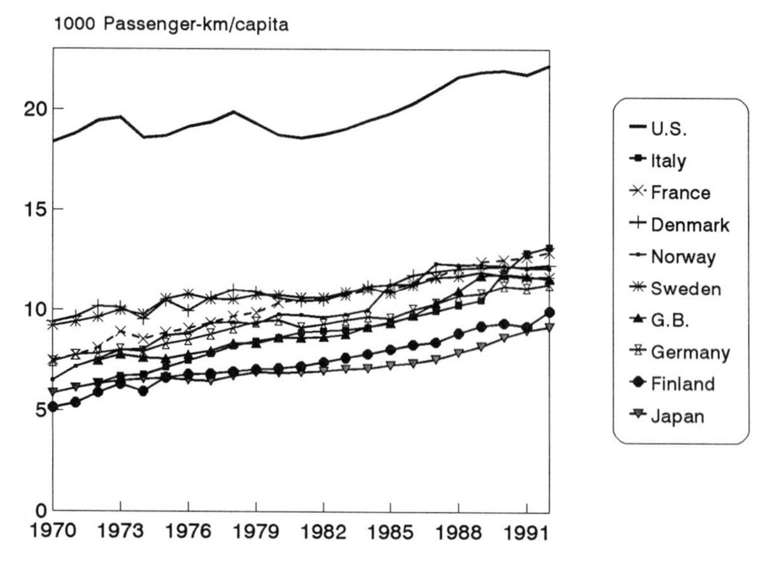

Figure 3 OECD per capita travel, all modes, 1970–1992.

The calculations behind Figures 2 through 4 are the first published in an international context and narrow many key uncertainties in the gap in how we understand the link between fuel use and transportation activity. Relying on national data sources provided by a large network of government, academic, and private experts, we collected and analyzed data that describe the structure of travel (and freight) activity, fleets of vehicles, infrastructure, and energy use in nearly a dozen OECD countries from 1970 to 1992. The references for national data on vehicles, transportation activity, and energy use are listed in an appendix. In addition, we have undertaken our own studies of Denmark (11a), Sweden (11b), and more recently Japan (12).

Vehicles, Activity, and Energy Use for Travel: Focus on the Car

At the center of examining the problems of transportation and its resulting air pollution is an analysis of how energy is used by vehicles. In the initial phase of this study, we confronted the vicious circles of interdependent data and truly circular calculations lying behind "data" on travel and freight activity, fuel use, and fuel economy of each transportation mode. Behind these figures lie careful tabulations of gasoline and diesel fuel (also LPG and natural gas) for each mode of road traffic (including motorcycles, waterborne passenger traffic, and other minor modes not shown in Figures 2 or 4), a split of energy use for

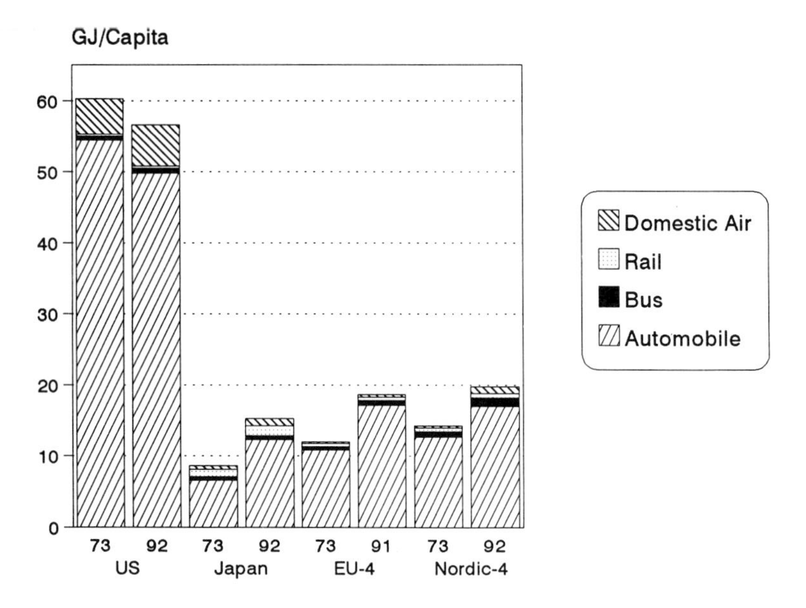

Figure 4 Per capita energy use for passenger travel. EU-4 includes Great Britain, France, Italy, and West Germany. Nordic-4 includes Norway, Denmark, Sweden, and Finland.

domestic rail and water traffic into passenger and freight shares, and a determination of the domestic share of fuel used for air travel.

Car Ownership and Characteristics

Figure 5 shows per capita car ownership from 1970 to 1992. Schipper et al (13) noted that "car" has no unique definition for two important reasons. First, a significant number of light trucks (vans, etc.) are used as cars in the United States and Denmark, and have made up 10–25% and 5–7%, respectively, of the light-duty passenger vehicle populations in those countries. These are included in our tabulations. For the United States these figures are derived from three important surveys taken at various intervals since the 1960s. For Denmark and Britain a small number of vans are counted in traffic and travel statistics as personal transportation vehicles. These cannot be separated from cars and are included as well. For Sweden and Norway, these vehicles are registered as trucks (for tax purposes) and are not included herein, but make up less than 2% of the passenger vehicle stock. For France, Japan, and Germany, we believe the numbers of light trucks used as passenger vehicles are even smaller. For Japan, by contrast, we do count the considerable number of mini-cars.

Second, the number of cars registered in a country fluctuates seasonally (15,

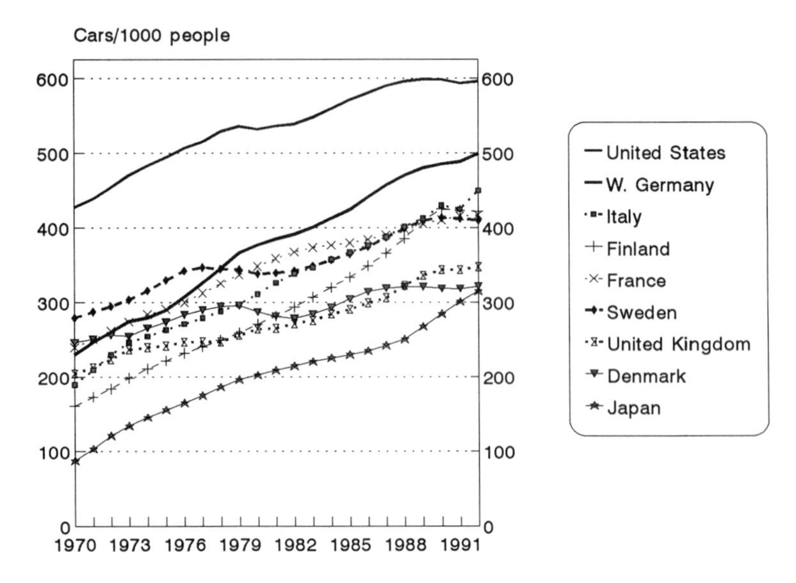

Figure 5 Growth in automobile ownership in the OECD since 1971.

16). In some cases (particularly the United States), registrations overstate the number of cars in use because of inaccuracies in keeping track of cars moving from one jurisdiction to another, and cars that have been retired or wrecked but not "de-registered." After extensive research we believe that the values shown in Figure 5 represent the best available estimates of cars and light trucks in use as passenger vehicles in the countries shown.

Car Use

Cars in the garage emit little pollution and consume no fuel (other than through evaporation of fuel). Nevertheless, it was established (17, 18) that the acquisition of a car tends to boost the mobility of an individual or a household by as much as a factor of five. It is the growth in numbers, not average use, that boosted both total automobile traffic and total mobility.

Figure 6 shows the yearly average distances cars are driven in the study countries. Arriving at these figures is tricky. In many countries or regions, authorities use fuel use to estimate total driving in one bureau but use fuel use and distance to estimate fuel economy elsewhere (19, 20). As Schipper et al (13) showed, this leads to errors in estimating both car use and fuel use, particularly if gasoline is not the only fuel used for automobiles. The figures presented here are based on top-down calculations (total driving divided by total number of cars in use), but are all adjusted to reflect information from

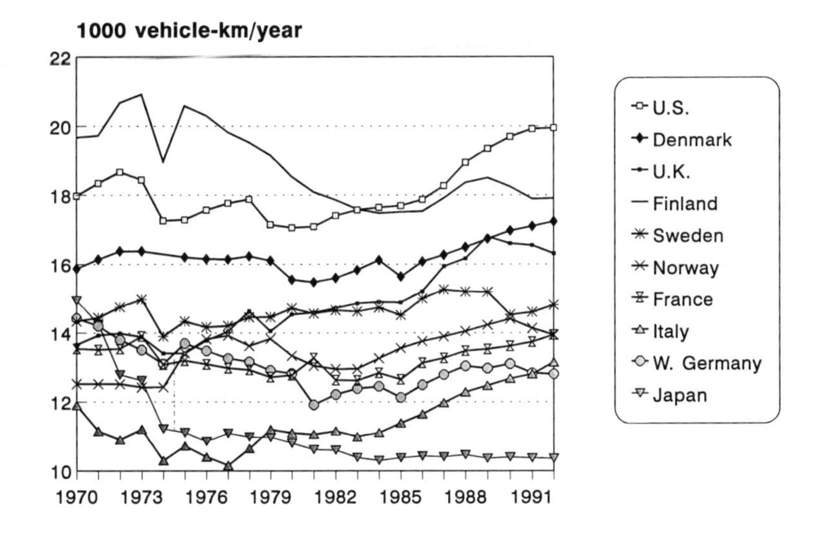

Figure 6 Distance traveled per automobile per year in the OECD.

bottom-up national car-use or travel surveys. Notable are the drop that occurred in every country after the first and second fuel crises, and the slow rebound that took place after the fuel price decrease in 1986. Equally interesting are the small declines that took place in some countries in the early 1990s as fuel taxes were raised. Overall, however, these fluctuations have been small; the overall trend is only very slow growth in use per car. In Denmark, Finland, and Britain, the European countries studied here with the fewest cars, annual driving distances are the greatest, bringing the per capita values there close to levels in countries with more cars, as people optimize use of vehicles. There are of course large differences within a country, according to the car user's income, place of residence, and age, and even the age of the car itself. In general, newer cars are driven more than older ones; the wealthier drive more than the less-wealthy; families with children or two workers drive more than those without; and those between the ages of 25 and 55 drive more than those younger or older.

Energy Use for Automobiles

It should be obvious that total automobile fuel use, total automobiles, total distance driven, and fuel use per unit of distance form an important set of circular variables (13). What is not obvious is that many national or local authorities derive distance driven from fuel use, while others derive fuel use from distance driven. In our work we have attempted to "square this circularity" by reconciling many national surveys of each parameter, for each kind of fuel used by automobiles (13, 14, 21, 22). Until recently, most statistical analyses of gasoline use assumed that "gasoline" could be used to represent the fuel consumed by automobiles (23–26). Schipper at al (13) showed clearly that this assumption is in error.

Table 1 shows automobile fuel in the study countries, with the share made up by diesel (and LPG for Denmark and Italy) indicated. A key finding is that automobile fuel use has grown in most countries far more rapidly than gasoline use alone, the quantity traditionally used to represent automobile fuel. This is even more striking if total gasoline is mistakenly used to represent the fuel for automobiles. Increasing numbers of motorists in Europe have switched to diesel fuel, while fewer and fewer trucks and buses use gasoline. This is one subtle but important aspect of energy use for transportation that has come to light in our research. The distortion is particularly large for Italy and France.

On-Road Fuel Economy

Fuel economy is a measure of how far a car can be driven on a given amount of fuel; fuel intensity is its inverse. Either is related to both the physical efficiencies of the components of the car, to the weight and features of the car, and to the behavior of drivers. Figure 7 shows the real on-the-road fuel intensity of the combined diesel and gasoline automobile fleets (including personal light trucks in Denmark and the United States) in OECD countries from 1970 to 1992. The data reflect all driving by domestically registered cars in the country indicated. Since a liter of diesel fuel contains more energy than one of gasoline, the actual quantities of fuel (in joules) were converted back to gasoline and expressed that way, at the calorific content of gasoline used in the country in question.

Real fuel economy means that the figures were derived for each fuel using measurements of fuel economy from surveys or good estimates consistent with the numbers of cars and driving distances shown previously, something that is difficult to do for an entire country (27–31). True fuel-economy surveys have only been carried out repeatedly in Canada (32), while occasional surveys have been carried out in France (30), the United States (33), and on a limited sample of vehicles in Sweden (29). For most other countries, authorities estimate fuel use and distance driven semi-independently from country-wide data and derive fuel economy from that figure (34–36).

The decline in fuel intensity in the United States is dramatic when compared with the slow changes in Europe. On the other hand, significant changes in the characteristics of cars have taken place. Such important determinants of fuel consumption as weight, engine displacement, and horsepower plummeted

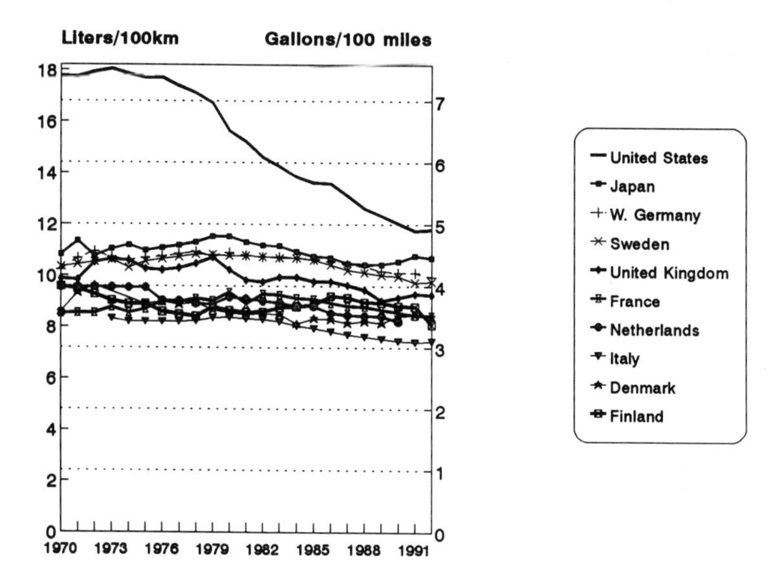

Figure 7 Automobile fuel intensities for the OECD; includes diesels and light trucks; liters per gasoline equivalent.

in the United States through the early 1980s, while these in general grew steadily in European countries. Figure 8, for example, shows the weight of new cars in many of the countries studied. Note how the United States figure is rising again. Trends in power and displacement are similar.

New-Car Fuel Economy

The fuel economy of the stock changes as the stock expands and new cars replace old ones. Yet this important parameter presents another important gap in our knowledge. Table 2 shows there are differences between the fuel intensity of new cars as measured in tests and what these cars appear to deliver on the road, in real traffic [(37), based largely on (27); (30), (38–41)]. Driver behavior, the influence of traffic, the difference between the test driving cycles and real driving cycles, and small differences in the make up of car models tested and those actually sold all distort the fuel economy drivers get on the road (30, 42). This gap can be substantial, except in Sweden (29). The reason for this exception is that the tabulation of test figures by authorities in Sweden gives more weight to urban driving than is the actual case for Swedish drivers, which roughly compensates for the inaccuracy of the tests. While the tests still serve an important function, namely informing potential car buyers of the relative fuel economy of different makes and models, they provide increasingly

Table 1 Parameters of Automobile Fuel Use in Study Countries

		1970	1972	1973	1975	1976	1977	1978	1979
UNITED STATES									
AUTO FUEL USE		9739	11088	11538	11352	11957	12191	12535	12031
Gasoline	PJ	9717	11066	11516	11296	11892	12123	12454	11930
Diesel	PJ	22	22	22	56	64	69	81	101
diesel %		0.2%	0.2%	0.2%	0.5%	0.5%	0.6%	0.6%	0.8%
VEHICLE USE	vkm/veh/yr	17978	18673	18445	17291	17571	17770	17883	17153
FUEL ECONOMY		17.8	17.9	18.0	17.7	17.7	17.4	17.1	16.7
Gasoline	l/100vkm	17.8	17.9	18.0	17.7	17.7	17.4	17.1	16.7
Diesel	l/100vkm								
New Auto Fuel Economy	l/100vkm	18.8	#N/A	18.1	15.4	14.1	13.3	12.7	12.6
VEHICLE STOCK									
Autos	10^3	89280	96860	101763	106713	110189	113696	116575	120248
Gasoline	10^3								
Diesel	10^3								
Light Trucks (Personal)	10^3	7143	8779	9804	11389	12613	13546	14807	15916
Source: Lawrence Berkeley and Oak Ridge National Laboratories.									
Energy Prices	(weight av.price)	0.22	0.22	0.22	0.27	0.26	0.25	0.29	0.35
Gasoline	USD/liter	0.08	0.09	0.09	0.13	0.14	0.14	0.18	0.24
Gasoline	85 PPP US$/L	0.22	0.22	0.22	0.27	0.26	0.25	0.29	0.35
Diesel	USD/liter	0.08	0.09	0.09	0.13	0.14	0.14	0.14	0.20
Diesel	85 PPP US$/L	0.22	0.22	0.22	0.27	0.26	0.25	0.24	0.30
JAPAN									
AUTO FUEL USE	PJ	518	627	713	745	788	871	935	1012
Gasoline	PJ	517	626	712	743	784	864	924	993
Diesel + LPG	PJ	1	1	1	2	4	6	12	20
diesel % +LPG%		0.2%	0.2%	0.2%	0.2%	0.5%	0.7%	1.2%	1.9%
VEHICLE USE	vkm/veh/yr	14948	12788	12623	11112	10858	11095	10986	10974
FUEL ECONOMY	l/100vkm	10.8	10.8	11.0	11.0	11.1	11.2	11.3	11.5
Gasoline	l/100vkm								
Diesel	l/100vkm								
New Auto Fuel Economy	l/100vkm			10.4	11.1	10.4	9.4	8.8	8.6
VEHICLE STOCK									
Autos	10^3								
Gasoline	10^3	17829	22481	24466	27794	29604	31525	33645	35670
Mini (gasoline)	10^3	9105	12964	14550	17373	18618	19940	21408	22750
Diesel	10^3	6575	9744	11363	14556	15923	17268	18794	20074
LPG	10^3	2328	2999	2954	2555	2412	2373	2224	2192
Source: Japan Institute for Energy Economics and Ministry of Transport.									
Energy Prices	(weigh ave.price)	0.55	0.53	0.54	0.69	0.64	0.67	0.58	0.68
Gasoline	JYen/liter	44.31	47.62	54.08	95.00	96.64	110.47	99.00	121.00
Gasoline	85 PPP US$/liter	0.55	0.53	0.54	0.69	0.64	0.68	0.58	0.69
Diesel	JYen/liter	27.44	27.51	30.04	46.84	56.05	59.76	57.00	73.00
Diesel	85 PPP US$/liter	0.34	0.31	0.30	0.34	0.37	0.37	0.34	0.42

1980	1981	1982	1983	1984	1985	1986	1987	1988	1989	1990	1991	1992
11255	11171	11088	11189	11246	11421	11808	11942	12139	12303	12345	12223	12690
11148	11042	10950	11038	11246	11262	11631	11744	11977	12140	12181	12056	12529
106	129	137	151	161	159	177	198	161	163	164	167	162
0.9%	1.2%	1.2%	1.3%	1.4%	1.4%	1.5%	1.7%	1.3%	1.3%	1.3%	1.4%	1.3%
17062	17091	17410	17570	17647	17695	17873	18270	18947	19354	19699	19922	20967
15.6	15.2	14.6	14.3	13.9	13.6	13.6	13.1	12.6	12.3	12.0	11.8	11.8
15.6	15.2	14.6	14.3	13.9	13.6	13.6	13.1	12.6	12.3	12.0	11.8	11.8
10.5	9.8	9.5	9.6	9.6	9.4	9.2	9.1	9.1	9.2	9.4	9.3	9.4
121724	123462	123746	126728	128271	132108	135431	137208	141252	143026	143453	142569	144213
16637	17491	18246	19361	20165	21534	22413	23350	24457	25295	26230	26672	27001
0.43	0.43	0.38	0.35	0.33	0.32	0.24	0.24	0.23	0.23	0.25	0.24	0.23
0.33	0.36	0.34	0.33	0.32	0.32	0.25	0.25	0.25	0.27	0.31	0.30	0.30
0.43	0.43	0.38	0.35	0.33	0.32	0.24	0.24	0.23	0.23	0.25	0.24	0.23
0.27	0.32	0.31	0.32	0.32	0.32	0.25	0.26	0.26	0.27	0.31	0.30	0.29
0.35	0.38	0.34	0.34	0.33	0.32	0.24	0.25	0.23	0.23	0.26	0.24	0.23
1036	1038	1060	1073	1071	1092	1121	1136	1180	1254	1357	1468	1526
1006	999	1010	1013	1003	1017	1043	1055	1090	1147	1227	1310	1357
30	39	50	60	68	74	78	81	90	107	130	158	169
2.9%	3.7%	4.7%	5.6%	6.3%	6.8%	6.9%	7.1%	7.6%	8.5%	9.6%	10.7%	11.1%
10803	10617	10602	10385	10302	10380	10428	10409	10467	10364	10401	10373	10360
11.5	11.3	11.2	11.2	10.9	10.8	10.7	10.5	10.4	10.4	10.6	10.8	10.7
8.3	8.1	7.7	7.8	7.8	8.1	8.3	8.5	8.6	8.8	8.7	7.6	
37172	38828	40445	42068	43626	45271	47022	49180	51440	54143	56543	59535	61231
23646	24578	25435	26320	27038	27790	28538	29600	30712	32937	35151	37311	39165
20916	21752	22425	23108	23638	24236	24853	25731	26596	28060	29140	31134	31980
2103	2064	2046	2038	2011	1943	1851	1776	1737	2056	2715	3360	3930
0.77	0.75	0.79	0.71	0.67	0.63	0.55	0.54	0.51	0.55	0.57	0.56	0.54
147.00	152.00	164.00	151.00	145.00	140.00	123.00	121.00	116.00	119.00	125.00	127.00	124.00
0.78	0.76	0.80	0.73	0.68	0.65	0.56	0.55	0.53	0.53	0.54	0.53	0.51
102.00	107.00	118.00	108.00	104.00	100.00	84.00	72.00	68.00	69.00	74.00	77.00	75.00
0.54	0.54	0.58	0.52	0.49	0.46	0.38	0.33	0.31	0.31	0.32	0.32	0.31

Table 1 *(Continued)*

		1970	1972	1973	1975	1976	1977	1978	1979
FRANCE									
AUTO FUEL USE	PJ	455	506	563	570	605	626	661	671
Gasoline	PJ	441	488	543	542	572	585	611	611
Diesel	PJ	13	19	20	28	33	41	51	60
diesel %		2.9%	3.7%	3.6%	5.0%	5.5%	6.5%	7.7%	8.9%
VEHICLE USE	vkm/veh/yr	13539	13540	13900	13200	13090	12980	12930	12720
FUEL ECONOMY		8.5	8.5	8.8	8.7	9.0	9.0	9.1	9.0
Gasoline	l/100vkm	8.5	8.5	8.8	8.6	8.9	8.9	9.0	8.9
Diesel	l/100vkm	9.3	9.3	8.3	9.1	9.1	9.3	9.5	9.7
New Auto Fuel Economy	l/100vkm				8.6	8.4	8.3	8.3	8.1
VEHICLE STOCK									
Autos	10^3								
Gasoline	10^3	12165	13528	14270	15350	15875	16610	17355	18080
Diesel	10^3	12041	13349	14055	15050	15513	16162	16799	17410
Source: Agence Francaise pour la Matrise d'Energie, now Agence d'Environment et Matrise de l'Energie.									
Energy Prices	(weight ave.price)	0.64	0.60	0.56	0.66	0.63	0.68	0.75	0.76
Gasoline	FF/l	1.06	1.11	1.12	1.69	1.76	2.09	2.55	2.88
Gasoline	85 PPP US$/l	0.65	0.61	0.57	0.68	0.64	0.70	0.78	0.79
Diesel	FF/l	0.72	0.78	0.78	1.16	1.25	1.34	1.54	1.86
Diesel	85 PPP US$/l	0.44	0.43	0.40	0.46	0.46	0.45	0.47	0.51
W. GERMANY									
AUTO FUEL USE	PJ	664	783	795	847	886	930	986	1013
Gasoline	PJ	629	743	752	797	835	875	925	941
Diesel	PJ	35	40	43	50	51	55	61	73
diesel %		5.2%	5.1%	5.4%	5.9%	5.7%	5.9%	6.2%	7.2%
VEHICLE USE	vkm/veh/yr	14422	13800	13504	13696	13480	13262	13165	12914
FUEL ECONOMY	l/100vkm	10.2	10.9	10.7	10.7	10.7	10.8	10.9	10.8
Gasoline	l/100vkm	10.2	11.0	10.7	10.7	10.8	10.9	11.0	10.8
Diesel	l/100vkm	9.2	9.3	9.5	9.5	9.5	9.5	9.5	9.6
New Auto Fuel Economy	l/100vkm			10.3				9.4	9.3
VEHICLE STOCK									
Autos	10^3	13941	16055	17023	17898	18920	20020	21212	22535
Gasoline	10^3	13506	15546	16464	17254	18230	19269	20373	21542
Diesel	10^3	435	509	559	644	690	751	839	993
Source: "Verkehr in Zahlen" ("Transportation in Figures", published by the Ministry of Transport) compiled by Deutsches Institut fuer Wirtschaft, Berlin.									
Energy Prices	(weight ave.price)	0.50	0.48	0.47	0.56	0.58	0.55	0.56	0.58
Gasoline	DM/l	0.55	0.58	0.61	0.82	0.89	0.88	0.93	1.00
Gasoline	85PPP US$/l	0.51	0.48	0.47	0.56	0.59	0.56	0.57	0.59
Diesel	DM/l	0.54	0.59	0.63	0.86	0.89	0.89	0.79	0.86
Diesel	85PPP US$/l	0.50	0.49	0.49	0.59	0.59	0.56	0.48	0.51

1980	1981	1982	1983	1984	1985	1986	1987	1988	1989	1990	1991	1992
726	743	760	773	784	780	809	823	852	863	877	888	911
662	670	677	682	685	672	690	694	707	695	686	673	668
64	73	82	91	99	108	118	129	145	168	191	215	242
8.9%	9.9%	10.9%	11.8%	12.6%	13.9%	14.7%	15.7%	17.0%	19.5%	21.8%	24.2%	26.6%
13020	13270	12630	12620	12840	12770	13130	13270	13480	13280	13597	13675	13945
9.4	8.9	9.3	9.3	9.1	9.1	8.9	8.8	8.8	8.7	8.6	8.5	8.4
9.2	9.2	9.3	9.3	9.2	9.1	9.1	9.0	9.0	8.9	8.9	8.9	8.9
8.5	8.3	8.1	8.1	7.8	7.6	7.4	7.2	7.1	7.1	6.9	6.8	6.7
7.8	7.4	7.1	7.0	6.9	6.7	6.7	6.6	6.6	6.5	6.5	6.5	7.5
18785	19440	20025	20450	20701	20945	21295	21735	22245	22765	23280	23680	23916
17975	18456	18842	19080	19144	19172	19290	19477	19650	19733	19760	19655	19340
0.80	0.80	0.82	0.80	0.81	0.82	0.67	0.66	0.63	0.64	0.64	0.61	0.57
3.38	3.87	4.43	4.77	5.16	5.59	4.72	4.85	4.82	5.18	5.35	5.36	5.26
0.82	0.83	0.85	0.84	0.84	0.86	0.71	0.71	0.68	0.71	0.71	0.69	0.66
2.35	2.83	3.26	3.52	3.76	3.96	3.10	3.05	2.90	3.00	3.09	3.12	2.92
0.57	0.61	0.63	0.62	0.61	0.61	0.47	0.45	0.41	0.41	0.41	0.40	0.37
1046	985	1021	1055	1085	1078	1149	1205	1261	1270	1315	1314	1323
962	891	909	928	948	929	976	1016	1057	1061	1096	1094	1092
84	94	112	128	138	150	173	190	204	210	219	220	230
8.0%	9.6%	11.0%	12.1%	12.7%	13.9%	15.0%	15.7%	16.2%	16.5%	16.6%	16.8%	17.4%
12825	11913	12212	12384	12450	12127	12491	12789	13039	12956	13088	12961	12804
10.9	10.8	10.7	10.7	10.7	10.6	10.6	10.4	10.4	10.2	10.1	10.0	10.0
10.9	10.9	10.9	10.9	10.9	10.9	10.9	10.8	10.7	10.5	10.4	10.3	10.2
9.7	9.4	9.1	9.1	9.0	8.8	8.5	8.3	8.3	8.3	8.4	8.3	8.4
8.8	8.4	8.1	8.1	7.5	7.4	7.2	7.4	7.7	7.9	7.9		
23192	23730	24105	24580	25218	25845	26917	27908	28878	29755	30685	31322	32007
22054	22413	22487	22742	23184	23504	23950	24407	25027	25752	27313	27641	27967
1138	1317	1618	1838	2034	2341	2967	3501	3851	4003	4122	4250	4430
0.64	0.73	0.68	0.65	0.65	0.65	0.48	0.46	0.44	0.51	0.52	0.56	0.59
1.17	1.42	1.39	1.37	1.40	1.44	1.08	1.05	1.02	1.23	1.28	1.44	1.53
0.66	0.75	0.70	0.67	0.67	0.67	0.50	0.49	0.47	0.55	0.56	0.60	0.64
1.00	1.13	1.16	1.12	1.15	1.18	0.87	0.82	0.77	0.83	0.87	0.95	0.94
0.56	0.60	0.58	0.54	0.55	0.55	0.40	0.38	0.35	0.37	0.38	0.40	0.39

Table 1 *(Continued)*

		1970	1972	1973	1975	1976	1977	1978	1979
ITALY									
AUTO FUEL USE	PJ	320	360	400	414	417	421	458	499
Gasoline, LPG, CNG	PJ	320	360	372	414	417	421	458	435
Diesel	PJ			27					64
diesel %		5.0%		6.8%	8.4%	9.5%	10.9%	11.8%	12.8%
VEHICLE USE	vkm/veh/yr	0	0	0	0	0	0	0	0
FUEL ECONOMY	l/100vkm	9.9	10.5	10.6	10.3	10.2	10.3	10.5	10.7
Gasoline	l/100vkm								
Diesel	l/100vkm								
New Auto Fuel Economy	l/100vkm								9.0
VEHICLE STOCK									
Autos	10^3	11328	12466	13231	13517	13792	13798	13801	14307
Gasoline	10^3								
Diesel	10^3								
Source: Data provided by Agip Petroli, 1990 (private communication) and Italstat, National Accounts of Transportation.									
ENERGY PRICES	(weight ave.price)	0.75	0.81	0.69	0.92	0.82	1.10	1.01	0.95
Gasoline	Lire/liter	130.00	152.00	152.00	287.00	300.00	480.00	500.00	541.00
Gasoline	'80 PPP US$/l	0.75	0.81	0.72	0.97	0.87	1.19	1.11	1.04
Diesel	Lire/liter	75.00	75.00	80.00	135.00	146.00	150.00	144.00	175.00
Diesel	'80 PPP US$/l	0.44	0.40	0.38	0.46	0.42	0.37	0.32	0.34
GREAT BRITAIN									
AUTO FUEL USE	PJ	527	634	675	643	670	689	731	744
Gasoline	PJ	524	630	671	639	667	685	726	740
Diesel	PJ	3	4	4	4	4	4	4	4
diesel %		0.6%	0.6%	0.5%	0.6%	0.5%	0.6%	0.5%	0.5%
VEHICLE USE	vkm/veh/yr	13658	13988	13884	13411	13779	14040	14635	14052
FUEL ECONOMY	l/100vkm	9.9	10.5	10.6	10.3	10.2	10.3	10.5	10.7
Gasoline	l/100vkm								
Diesel	l/100vkm								
New Auto Fuel Economy	l/100vkm								9.0
VEHICLE STOCK									
Autos	10^3	11328	12466	13231	13517	13792	13798	13801	14307
Gasoline	10^3								
Diesel	10^3								
Source: UK Digest of Transport Statistics, also Energy Technology Support Unit, D.Martin (priv.comm.).									
Energy Prices	(weight ave.price)	0.63	0.58	0.54	0.79	0.72	0.65	0.58	0.68
Gasoline	pence/lit	0.07	0.07	0.07	0.15	0.16	0.17	0.17	0.22
Gasoline	80 PPP US$/l	0.63	0.58	0.54	0.79	0.72	0.66	0.58	0.68
Diesel	pence/lit	0.07	0.07	0.08	0.12	0.14	0.17	0.17	0.21
Diesel	80 PPP US$/l	0.66	0.59	0.55	0.62	0.59	0.65	0.60	0.65

1980	1981	1982	1983	1984	1985	1986	1987	1988	1989	1990	1991	1992
520	540	566	568	581	606	627	655	691	685	710	736	818
445	445	446	435	429	444	441	442	458	455	482	516	593
75	96	120	133	152	162	186	212	232	230	228	220	224
14.4%	17.7%	21.2%	23.4%	26.2%	26.8%	29.7%	32.4%	33.6%	33.6%	32.1%	29.9%	27.4%
0	0	0	0	0	0	0	0	0	0	0	0	0
10.2	9.8	9.8	9.9	9.9	9.8	9.8	9.6	9.5	9.0	9.2	9.3	9.3
			10.0	9.7	9.5	9.8	9.5	9.3	9.1	9.3	9.4	9.3
			7.7	7.5	7.4	7.5	7.3	7.1	7.0	7.2	7.2	6.8
			7.9	7.6	7.5	7.5	7.4	7.4			7.3	
14772	14943	15303	15543	16055	16454	16981	17421	18432	19248	19742	19737	20116
13589		14231	15473		16285	16747	17109	18029	18733	19116	19001	19203
28		35	69		166	233	311	402	512	623	734	909
1.01	1.06	1.03	1.01	0.99	0.93	0.82	0.78	0.78	0.75	0.79	0.79	0.76
700.00	887.00	1030.00	1172.00	1290.00	1321.00	1280.00	1300.00	1360.00	1376.00	1480.00	1536.00	1527.00
1.11	1.20	1.19	1.18	1.18	1.10	1.01	0.98	0.98	0.93	0.94	0.92	0.87
275.00	332.00	427.00	505.00	555.00	608.00	517.00	553.00	603.00	676.00	830.00	954.00	951.00
0.44	0.45	0.50	0.51	0.51	0.51	0.41	0.42	0.43	0.46	0.53	0.57	0.54
758	744	765	793	837	845	892	946	996	1032	1060	1074	1069
754	740	761	789	825	832	876	926	970	984	1004	1007	993
4	4	4	4	11	12	16	20	25	48	56	67	77
0.5%	0.5%	0.5%	0.5%	1.4%	1.5%	1.8%	2.1%	2.6%	4.6%	5.3%	6.3%	7.2%
14537	14586	14723	14854	14897	14883	15207	15940	16169	17211	17013	16983	16644
10.2	9.8	9.8	9.9	9.9	9.8	9.8	9.6	9.5	9.0	9.2	9.3	9.3
			10.0	9.7	9.5	9.8	9.5	9.3	9.1	9.3	9.4	9.3
			7.7	7.5	7.4	7.5	7.3	7.1	7	7.2	7.2	6.8
			7.9	7.6	7.5	7.5	7.4	7.4			7.3	
14772	14943	15303	15542.7	16055	16454	16981	17421	18432	19248	19742	19737	20115.6
13589		14231	15473		16285	16747	17109	18029	18733	19116	19001	19203
28		35	69		166	233	311	402	512	623	734	909
0.74	0.79	0.78	0.81	0.79	0.79	0.67	0.65	0.61	0.61	0.61	0.63	0.62
0.28	0.34	0.37	0.39	0.41	0.43	0.37	0.38	0.37	0.40	0.45	0.49	0.50
0.74	0.79	0.79	0.81	0.80	0.80	0.67	0.65	0.61	0.61	0.62	0.63	0.63
0.25	0.29	0.31	0.33	0.33	0.37	0.31	0.30	0.30	0.32	0.35	0.38	0.39
0.65	0.68	0.67	0.67	0.65	0.67	0.55	0.52	0.48	0.48	0.49	0.49	0.49

Table 1 *(Continued)*

		1970	1972	1973	1975	1976	1977	1978	1979
NORWAY									
AUTO FUEL USE	PJ	31	36	38	43	47	51	52	54
Gasoline	PJ	31	36	38	42	46	51	51	53
Diesel	PJ	0	0	0	0	1	1	1	1
diesel %		0.8%	0.7%	0.8%	1.1%	1.1%	1.2%	1.4%	1.7%
VEHICLE USE	vkm/veh/yr	12522	12520	12423	13423	13821	13923	13625	13831
FUEL ECONOMY		10.3	10.3	10.3	10.2	10.2	10.2	10.2	10.1
Gasoline	l/100vkm	10.3	10.3	10.3	10.2	10.2	10.2	10.2	10.1
Diesel	l/100vkm	9.3	9.3	9.3	9.2	9.2	9.2	9.2	9.1
New Auto Fuel Economy	l/100vkm				9.5	9.3	9.2	9.1	9
VEHICLE STOCK									
Autos	10^3	747	854	913	954	1023	1107	1147	1190
Gasoline	10^3	742	849	907	945	1013	1095	1133	1172
Diesel	10^3	5	5	6	9	10	12	14	18
Source: Transport Oekonomisk Institut, various publications; Norsk Esso (priv.comm.).									
Energy Prices	(weight ave.price)	0.46	0.48	0.46	0.49	0.47	0.45	0.49	0.51
Gasoline	NOK/l	1.27	1.49	1.56	2.01	2.13	2.20	2.62	2.83
Gasoline	85 PPP US$/l	0.47	0.48	0.47	0.49	0.48	0.45	0.50	0.51
Diesel	NOK/l	0.37	0.44	0.51	0.78	0.92	0.94	0.88	1.03
Diesel	85 PPP US$/l	0.14	0.14	0.15	0.19	0.21	0.19	0.17	0.19
SWEDEN									
AUTO FUEL USE	PJ	105	117	123	128	133	137	140	140
Gasoline	PJ	100	112	118	123	128	132	134	134
Diesel	PJ	4	5	5	5	6	6	6	6
diesel %		4.0%	4.1%	4.2%	4.1%	4.3%	4.1%	4.3%	4.5%
VEHICLE USE	vkm/veh/yr	14339	14758	14980	14336	14168	14207	14457	14460
FUEL ECONOMY	l/100km average	10.3	10.5	10.6	10.6	10.6	10.7	10.8	10.8
Gasoline	l/100vkm	10.4	10.6	10.7	10.6	10.7	10.8	10.9	10.9
Diesel	l/100vkm	9.0	8.8	8.7	8.5	8.4	8.4	8.3	8.3
New Auto Fuel Economy	l/100vkm							9.3	9.2
VEHICLE STOCK									
Autos (3)	10^3	2298	2427	2513	2760	2872	2857	2856	2856
Gasoline	10^3	2229	2354	2435	2666	2770	2757	2755	2750
Diesel	10^3	69	73	77	94	102	100	101	106
Source: Transport Raadet (Transportation Council), National Board of Industry (Energy Board), and Ministry of Communications and Transport.									
Energy Prices	(weight ave.price)	0.40	0.41	0.39	0.43	0.45	0.41	0.43	0.48
Gasoline	SEK/Liter	0.87	1.01	1.01	1.36	1.58	1.58	1.86	2.20
Gasoline	USD 1985 PPP/l	0.41	0.42	0.40	0.44	0.47	0.42	0.45	0.49
Diesel	SEK/Liter	0.63	0.78	0.78	0.55	0.62	0.64	0.74	0.93
Diesel	USD 1985 PPP/l	0.30	0.33	0.31	0.18	0.18	0.17	0.18	0.21

1980	1981	1982	1983	1984	1985	1986	1987	1988	1989	1990	1991	1992
54	54	56	57	59	62	65	67	67	68	68	67	66
53	53	54	55	57	60	63	65	65	66	66	64	63
1	1	2	2	2	2	2	2	2	2	2	2	3
2.0%	2.6%	3.0%	3.2%	3.3%	3.2%	3.1%	2.9%	2.9%	3.0%	3.2%	3.6%	4.0%
13336	13046	12951	12955	13260	13558	13758	13897	14052	14232	14405	14137	13941
10.1	10.0	9.9	9.7	9.5	9.3	9.1	9.1	9.1	9.1	9.0	9.0	9.0
10.1	10.0	9.9	9.7	9.5	9.3	9.1	9.1	9.1	9.1	9.0	9.0	9.0
9.1	9.0	8.9	8.7	8.6	8.4	8.2	8.2	8.2	8.2	8.1	8.1	8.0
8.8	8.6	8.3	8.1	7.8	7.76	7.57	7.53	7.28	7.7	7.41	7.41	7.41
1234	1279	1338	1383	1430	1514	1592	1623	1622	1613	1613	1615	1619
1211	1249	1302	1343	1387	1470	1548	1579	1578	1568	1564	1562	1560
22	30	36	40	43	44	44	44	44	45	49	53	59
0.60	0.62	0.58	0.58	0.57	0.55	0.46	0.46	0.45	0.46	0.49	0.55	0.58
3.71	4.35	4.61	4.93	5.21	5.13	4.76	5.10	5.36	5.78	6.38	7.43	7.97
0.61	0.63	0.59	0.59	0.59	0.56	0.47	0.47	0.46	0.47	0.50	0.56	0.59
1.60	2.00	2.19	2.27	2.34	2.35	1.73	1.75	1.78	1.94	2.38	2.84	2.73
0.26	0.29	0.28	0.27	0.26	0.25	0.17	0.16	0.15	0.16	0.19	0.22	0.20
140	139	141	143	147	148	154	159	163	166	161	159	162
133	133	133	134	139	139	146	152	156	160	155	154	157
7	7	8	8	9	8	8	7	7	6	6	5	5
4.7%	5.0%	5.4%	5.8%	5.8%	5.6%	5.1%	4.6%	4.2%	3.7%	3.4%	3.3%	3.1%
14718	14565	14657	14622	14737	14515	14998	15259	15196	15188	14530	14605	14810
10.8	10.8	10.7	10.7	10.7	10.6	10.4	10.2	10.1	10.0	10.0	9.7	9.8
10.9	10.9	10.8	10.8	10.8	10.7	10.5	10.3	10.2	10.1	10.0	9.9	9.8
8.2	8.2	8.2	8.2	8.2	8.2	8.2	8.2	8.2	8.2	8.2	8.2	8.2
9	8.7	8.6	8.6	8.5	8.5	8.4	8.2	8.3	8.3	8.3	9.3	10.3
2883	2893	2935	3007	3081	3151	3254	3367	3483	3578	3600	3619	3587
2770	2774	2811	2878	2951	3023	3130	3247	3369	3469	3496	3520	3493
113	119	124	129	130	128	124	120	114	109	105	99	93
0.56	0.60	0.62	0.59	0.56	0.57	0.49	0.48	0.48	0.49	0.60	0.58	0.55
2.95	3.53	3.96	4.19	4.22	4.66	4.15	4.19	4.47	4.81	6.47	6.80	6.62
0.58	0.62	0.64	0.62	0.58	0.60	0.51	0.49	0.50	0.50	0.61	0.59	0.56
1.47	1.81	2.34	2.56	2.89	3.07	2.33	2.57	2.65	3.24	3.96	4.15	3.90
0.29	0.32	0.38	0.38	0.40	0.39	0.29	0.30	0.30	0.34	0.38	0.36	0.33

Table 1 *(Continued)*

		1970	1972	1973	1975	1976	1977	1978	1979
DENMARK									
AUTO FUEL USE	PJ	54	63	66	62	65	66	68	66
Gasoline	PJ	54	61	66	61	63	64	66	64
Diesel	PJ		2		2	2	2	2	2
diesel %			2.4%		2.7%	2.9%	3.1%	3.5%	3.1%
VEHICLE USE	vkm/veh/yr	15228	16493	15779	16188	15689	15708	16291	16105
FUEL ECONOMY		9.0	9.1	9.6	8.5	8.9	8.7	8.5	8.3
Gasoline	l/100vkm		9.7		8.5	8.6	8.5	8.5	8.2
Diesel	l/100vkm		8.6		7.8	7.7	7.9	8.0	7.9
New Auto Fuel Economy	l/100vkm				8.4		7.8		7.7
VEHICLE STOCK									
Autos (4)	10^3	1214	1278	1329	1384	1437	1474	1502	1514
Gasoline	10^3	1058	1182		1274	1316	1352	1382	1392
Diesel	10^3	8	8		10	11	12	15	18
LPG	10^3	5	6		5	4	4	4	6
Source: Ministry of Traffic (formerly Ministry of Public Works) data books and Energistyrelsen (Danish Energy Agency).									
Energy Prices	Weighted av.price)	0.53	0.47		0.57	0.53	0.58	0.54	0.61
Gasoline	DKK/l	1.29	1.37	1.44	2.16	2.18	2.65	2.76	3.38
Gasoline	85 PPP US$/l	0.53	0.47	0.48	0.57	0.53	0.58	0.55	0.61
Diesel	DKK/l	0.35	0.46	0.49	0.93	0.98	1.05	0.92	1.27
Diesel	85 PPP US$/l	0.14	0.16	0.16	0.25	0.24	0.23	0.18	0.23
FINLAND									
AUTO FUEL USE	PJ	42	49	52	57	57	57	58	62
Gasoline	PJ	40	46	49	54	53	53	54	57
Diesel	PJ	2	2	3	3	4	4	4	5
diesel %	%	4.5%	5.0%	5.3%	5.6%	6.3%	6.7%	7.5%	8.8%
VEHICLE USE	vkm/veh/yr	19680	20680	20918	20590	20304	19827	19538	19171
FUEL ECONOMY	l/100vkm	9.6	9.3	9.0	8.9	8.6	8.5	8.4	8.8
Gasoline	l/100vkm	9.6	9.2	9.0	8.8	8.5	8.4	8.3	8.7
Diesel	l/100vkm	8.9	8.8	8.8	8.8	8.8	8.7	8.7	8.7
New Auto Fuel Economy	l/100vkm								
VEHICLE STOCK									
Autos	10^3	1411	1619	1769	1966	2035	2115	2189	2288
Gasoline	10^3	712	818	894	996	1033	1075	1115	1169
Diesel	10^3	699	801	875	970	1002	1039	1073	1119
LPG	10^3								
Source: Energistatistik 1992 (Ministry of Trade and Industry).									
Energy Prices	'85 PPPUS$/literga	0.49	0.47	0.47	0.48	0.49	0.53	0.54	0.53
Gasoline	FIM/liter	0.74	0.80	0.89	1.25	1.47	1.83	1.99	2.15
Gasoline	'85 PPPUS$/liter	0.54	0.52	0.52	0.53	0.54	0.60	0.60	0.61
Diesel	FIM/liter	0.42	0.46	0.54	0.78	0.94	1.16	1.24	1.41
Diesel	'85 PPPUS$/liter	0.31	0.29	0.31	0.33	0.35	0.38	0.38	0.40

1980	1981	1982	1983	1984	1985	1986	1987	1988	1989	1990	1991	1992
61	59	59	61	61	63	67	68	70	70	73	74	75
59	57	56	58	58	59	61	61	62	62	64	65	65
2	2	3	3	4	4	6	7	8	8	9	10	10
3.6%	4.0%	4.3%	4.8%	6.2%	7.1%	8.6%	10.4%	11.5%	11.8%	11.8%	12.8%	12.7%
15543	15468	15592	15827	16117	15644	16069	16267	16493	16738	16973	17103	17239
8.2	8.1	8.1	8.1	7.8	8.0	7.9	7.8	7.9	7.8	8.0	8.1	8.0
8.1	8.0	8.0	8.0	7.7	7.8	7.8	7.6	7.6	7.5	7.8	7.7	7.7
8.3	8.3	8.2	8.0	8.3	8.3	8.3	8.8	9.2	8.7	8.7	9.5	9.6
	7.5		7.0		6.7		6.8		6.6	6.6	7.7	7.2
1471	1441	1426	1453	1498	1556	1609	1637	1645	1646	1636	1639	1659
1433	1389	1367	1389	1430	1483	1531	1554	1560	1557	1546	1550	1572
22	24	28	36	47	58	67	74	79	85	87	87	86
17	29	31	28	22	16	11	8	6	5	3	2	2
0.73	0.77	0.79	0.75	0.68	0.66	0.65	0.63	0.59	0.59	0.53	0.50	0.47
4.54	5.31	5.93	6.06	5.99	6.14	6.42	6.63	6.53	6.85	6.28	6.06	5.81
0.73	0.77	0.78	0.74	0.69	0.68	0.68	0.68	0.64	0.64	0.57	0.54	0.51
1.84	2.30	2.68	2.67	2.64	2.81	1.84	1.75	1.69	1.86	1.89	2.65	2.89
0.30	0.33	0.35	0.33	0.31	0.31	0.20	0.18	0.17	0.17	0.17	0.24	0.25
62	62	65	68	71	74	80	84	89	94	95	94	94
55	55	57	59	60	63	68	72	75	80	81	81	81
6	7	8	9	10	11	12	13	14	14	14	13	12
10.1%	11.3%	12.5%	13.7%	14.5%	14.9%	14.7%	15.2%	15.5%	15.0%	15.0%	13.8%	13.3%
18537	18090	17860	17596	17483	17516	17533	17913	18364	18506	18246	17896	17906
8.6	8.5	8.6	8.7	8.7	8.7	8.9	8.8	8.6	8.5	8.4	8.5	8.5
8.5	8.5	8.5	8.6	8.6	8.6	8.8	8.7	8.4	8.4	8.3	8.4	8.4
8.6	8.5	8.5	8.5	8.4	8.3	8.3	8.4	8.4	8.4	8.3	7.9	7.9
2389	2472	2608	2714	2831	2964	3102	3253	3441	3641	3698	3667	3700
1226	1273	1347	1408	1473	1545	1619	1699	1796	1897	1926	1910	1924
1164	1199	1260	1306	1358	1419	1483	1554	1646	1744	1771	1758	1777
0.63	0.64	0.60	0.58	0.55	0.53	0.42	0.41	0.40	0.40	0.45	0.46	0.47
2.89	3.33	3.47	3.68	3.80	3.86	3.15	3.24	3.35	3.53	4.17	4.42	4.55
0.73	0.75	0.71	0.70	0.67	0.65	0.51	0.51	0.50	0.49	0.55	0.56	0.56
1.96	2.32	2.43	2.60	2.71	2.79	2.21	2.26	2.38	2.62	3.09	3.13	2.99
0.49	0.52	0.50	0.49	0.48	0.47	0.36	0.35	0.35	0.37	0.41	0.40	0.37

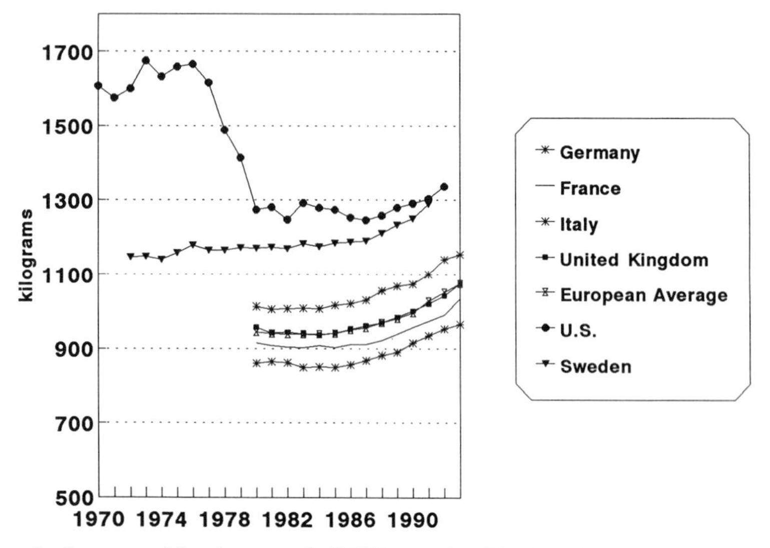

Figure 8 Average weight of new cars in OECD countries. (For the United States, cars excludes light trucks.)

unreliable information about actual fuel use for forecasting purposes or for estimating emissions in real driving conditions.

With these caveats, we present in Figure 9, the evolution of test fuel economy of new cars, calculated as sales-weighted averages of the fuel economy of individual models by authorities in each country. Except where noted, the fuel economy of diesels is included. In most countries, buyers of diesel cars choose somewhat more powerful motors, on average, than buyers of gasoline cars, so the importance of diesels to the figures presented is reduced. What is striking is the contrast between the rapid and dramatic declines in US (and it turns out Canadian) new car fuel economy and fuel economy in other countries. There is no question that automobiles became technically more energy-efficient in all countries. As Figure 8 suggests, however, new cars in Europe became heavier and more powerful while those sold in North America became smaller. As a result of these changes, fuel economy changed very little in Europe.

Given the international nature of both automobile manufacturers and parts suppliers, it makes little sense to argue whether new cars in any country are more efficient than those sold elsewhere. Instead, differences in weight, power, and features (such as air conditioning, automatic transmission, and power steering) distinguish what is bought in each country.

Table 2 Fuel Economy gap, test/actual, various countries (1/100 km)

Country	Year of Comparison	Test	Actual	Average Gap	% Gap	Comments	Ref.
Canada[a]	1988	8.0	10.0	2.0	20	Actual fuel efficiency from driver surveys. Test from laboratory test	32, 38
Individual car models[b]	1985	8.6	10.7	2.1	19.6		
France[c]	1988	6.5	8.4	1.9	23	Travel diaries compared to one third city, one third highway, one third road test values	30
Germany[d]	1987	7.7	9.8	2.1	21.4	DIN (test) vs DIW (actual)	22
Sweden[e]	1987	8.2	8.5	0.3	3.5	KOV compared with consumer reported survey data	29
U.S.[f]	1985						
Cars		9.7	11.9	2.2	18.5	RTECS survey vs EPA fleet average from dynanometer test	39
Trucks	11.6	14.5	2.9	20			
U.K.[g]	1989	7.2	9.3	2.1	22.6	Test value for registration-weighted average	27, 42

Sources:
[a] Statistics Canada 1990
[b] SOM, Inc. 1988; Energy, Mines, and Resources 1992
[c] Bosseboeuf 1988
[e] KOV 1987
[f] Mintz et al. 1993
[g] Sorrell 1992

Fuel Prices

Fuel prices are recognized as an important determinant of total automobile fuel demand (43–45). Table 1 showed fuel prices in each country studied, converted to real 1985 currency and then to US dollars using purchasing power parity for 1985 [(12), based on (1)]. Differences in taxation account for almost all of the differences in the prices shown.

Fuel prices shot up twice in the 1970s and fell back in the mid-1980s. During each period of high prices, driving fell slightly (as suggested by Figure 6). By 1988, however, the real cost of each fuel was close to its 1973 value in almost every country. Between 1988 and 1992, Germany, Sweden, and Norway had boosted fuel prices significantly with taxes, resulting in increases (after infla-

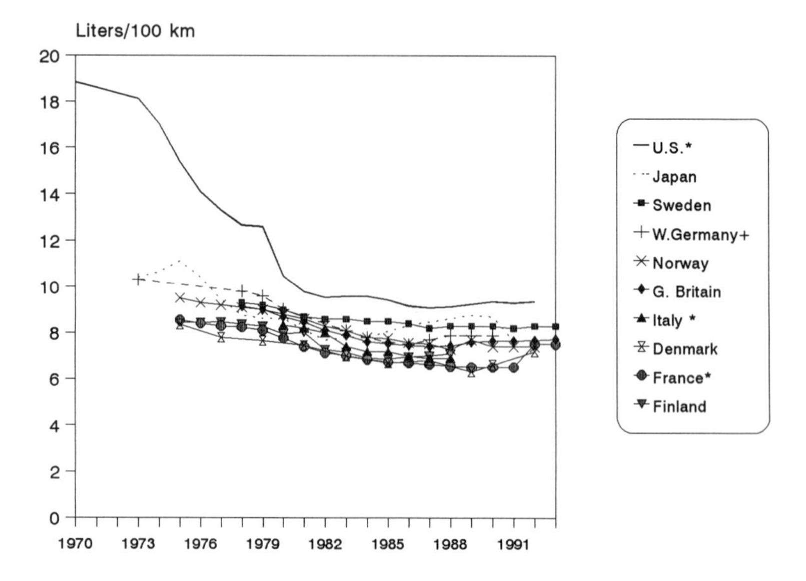

Figure 9 New car fuel intensities, using sales-weighted test values. US numbers include diesels and light trucks. Italy and France include diesels. West Germany excludes diesels.

tion) of 10–25%. Yet only in these countries were real prices in 1992 more than 10% above their 1973 levels. This means that for drivers in most countries, the "energy shock" of the 1970s was more or less gone by 1990.

There are two important reasons for this surprising result. First, the real price of crude oil fell significantly after 1985. Since as much as 75% of the price of fuel in Europe and Japan was made up of taxes, the price crash itself did not lead to a dramatic decline in prices, but this same buffer had also restrained the relative price increases that occurred in 1973 and 1979–1980.

The shift to lower-priced diesel mitigated some of the increase in fuel prices as well. Because both diesel and gasoline are used for automobiles, the prices of both fuels must be considered if effects of fuel-price changes are to be understood. Figure 10 shows the results when we average the prices of these two fuels (with taxes), using actual quantities of gasoline and diesel consumed as given in Table 1. The results are considerably different than what is obtained for gasoline alone in Germany, Italy, and France because of the important share of lower-cost diesel fuel, 25–30% of automobile fuel use in these countries in the early 1990s, up from less than 5% in 1973.

As a result of the decline in real fuel prices, the switch to diesel, and the improvements in fuel economy, the cost/vehicle-km for fuel was lower in 1992 than in 1973 in six of nine countries and only more than 15% higher in Sweden

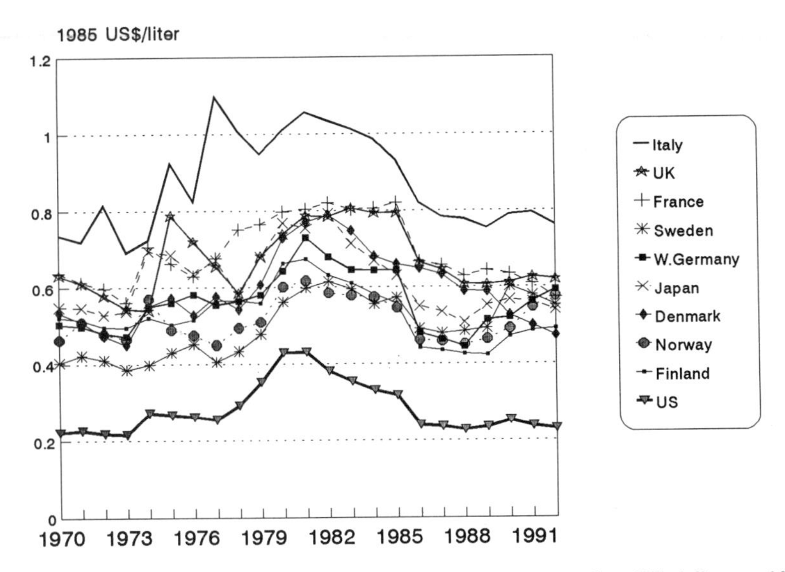

Figure 10 Automobile fuel prices, in real local currency converted to US dollars at 1985 purchasing power parity. Weighted by the actual mix of gasoline and diesel used by automobiles.

and Germany. This of course is not meant to imply that fuel prices are the only determinant of energy use for automobiles, but it does show that whatever the effect of fuel price increases during the entire period we studied, most consumers in 1992 felt less pressure to save fuel than they did in the early 1980s. Considering that their real incomes in 1992 were 20–40% higher than in 1973, it is not surprising European drivers owned and used considerably more cars that in turn were more powerful and only slightly less fuel-intensive than cars in 1973.

Other Modes of Travel

While the importance of the automobile to energy use for travel is clear, its role in Europe of providing travel in Europe is still less than in the United States because of Europe's significant (~15%) share of rail and bus travel. And the automobile now provides only slightly more than half of all domestic travel in Japan. The share of automobiles in total travel grew in all countries except in the United States and Scandinavia, where air travel grew even more rapidly, reaching 10% and 3% of travel, respectively. This section reviews briefly important trends in total travel. Data come from the same national references.

Travel in Figure 2 is measured in passenger-kilometers (pass-km). For automobiles, pass-km are derived from surveys of load factor (average occupancy or people/automobile) available from driving or travel surveys (listed in the Appendix to the Bibliography) and data on automobile use in veh-km. For other modes, domestic travel is reported by statistical agencies, trade associations representing local or intercity modes, and air travel authorities. We report automobile fuel intensities in Mj/vehicle-km but report the intensity of travel in automobiles in Mj/passenger-km for comparability with other modes.

Automobile travel did not grow as rapidly as automobile vehicle-km because load factors fell, from around 2 people per car in the early 1970s to around 1.5 in the 1990s in every country. This fall can be attributed to several factors. The most obvious is the increase in the number of families with two cars. Similarly the large rise in the number of single-person households with no others to share use caused car occupancy to fall. However, increased numbers of people in all countries used cars to go to work (increasing its share even in the United States between 1970 to 1990), for which single-occupancy vehicles dominate. A more subtle decline in average household size from over 3 to around 2.5 persons/household meant that even for family business and shopping or free time, the automobile has become less fully occupied.

The alternatives for local motorized travel are bus and rail. Most bus travel is in or around cities. Bus travel grew significantly in the United States, Scandinavia, and Europe. Growth in the United States was led by local bus travel, while intercity bus travel did increase elsewhere. Bus travel fell in Japan as automobiles gained share for local traffic. Rail traffic (including local rail transit, which is a small share of the total in most countries) increased 10–25% everywhere but in the United States, the increases reflecting mainly intercity rail. Domestic air travel grew by a factor of three outside of the U.S, but more than doubled in the United States. International air travel grew by an even greater amount, but is not easily counted country-by-country (46).

Energy intensities for travel in these modes (Table 3) behaved in differing ways. Because of the fall in the load factor, the energy intensity of automobile

Table 3 Energy intensities of travel modes

		USA (73)	USA (92)	JAP (73)	JAP (92)	EU-4 (73)[a]	EU-4 (91)	NO-4 (73)[b]	NO-4 (92)
Energy Intensity	Mj/pass-km	3.08	2.55	1.41	1.72	1.51	1.53	1.55	1.67
Car	Mj/pass-km	3.07	2.62	2.79	2.67	1.72	1.68	1.69	1.79
Bus	Mj/pass-km	0.79	0.88	0.54	0.66	0.58	0.74	0.82	0.88
Rail	Mj/pass-km	1.81	2.10	0.34	0.42	0.58	0.48	0.89	0.83
Air	Mj/pass-km	4.92	2.61	3.49	2.25	4.73	2.96	2.78	3.36

[a] EU-4: France, G. Britain, Italy, W. Germany
[b] NO-4: Denmark, Finland, Norway, Sweden

travel, in Mj/passenger-km, increased in most countries, the United States being an important exception. The intensities of bus travel increased in all regions, while those of rail travel fell in Scandinavia and Europe but increased in Japan and the United States The major component of change in all cases was utilization, i.e. the load factor. In Europe and Japan, buses and rail remain less energy-intensive than cars, but in the United States city buses and light rail (not shown separately in Table 2) were about as energy-intensive as average car travel by the 1990s, because the intensities of the former modes increased while those for cars fell.

The intensity of air travel fell everywhere, typically by 50%, although figures for European countries are uncertain because authorities do not keep records of fuel used in domestic travel. Schipper, Meyers, et al. (47) gathered data from many individual companies that confirmed the clear trend from published US and British data as well as the Market Outlook published each year by Boeing (48). Engines became considerably more efficient, the number of seats on an aircraft increased some 30–50%, and the share of seats filled increased from below 55% to 62–65%. Counteracting these trends somewhat was the increased popularity of business class, which led to removal of seats on long-haul aircraft in the 1980s, as well as increased congestion at airports and in air space, particularly in Europe; but the overall energy intensity air travel still fell dramatically.

Summary: Travel and Energy Use Over Twenty Years

When all the foregoing data are combined, the changes in automobile fuel use shown in Figure 4 are better understood. The growth in automobile ownership was the main component of increased travel, and thus the main component of increased energy use for travel. Because per capita auto ownership and use in the United States started from a much higher level in 1970 than in other countries, growth in the United States was considerably slower than in Europe and Japan. While the fuel intensity of the US fleet fell more than 30% from 1973 to 1992, the fleets in Europe and Japan fell at most 13% (Britain, Norway, and Denmark), and only 8% for Europe as a whole. Consequently, per capita fuel use for automobiles in 1992 was lower in the United States than in 1973, but considerably higher than in 1973 in all other countries. Needless to say, the gap between per capita automobile fuel use in the United States and Europe shrank considerably, from nearly five times greater in 1973 to "only" three times greater in 1992. Changes in the role of the automobile totally dominate the overall changes in energy use, the energy consumed by other modes being relatively small.

To illustrate the subtleties of these changes, the trends can be decomposed using Laspeyres or Divisia indices to isolate the changes in three components: per capita travel, modal structure, and the energy intensities of individual

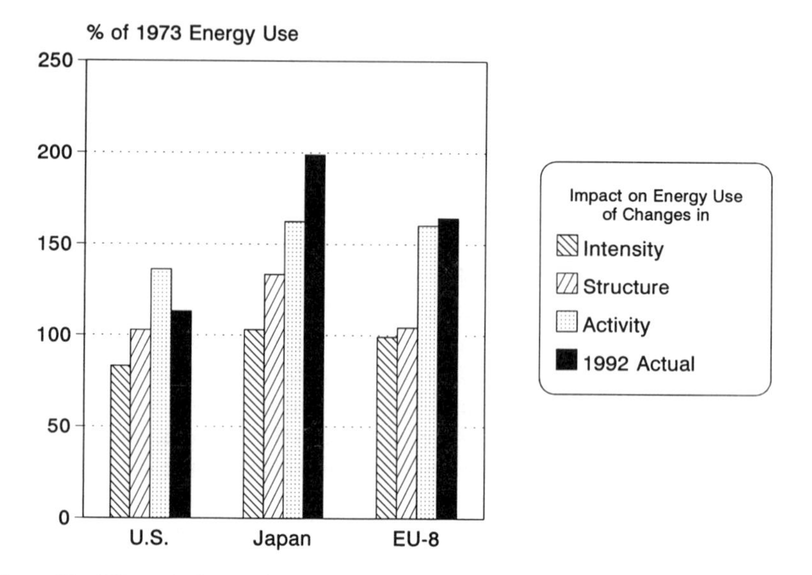

Figure 11 Effects of changes in energy intensity, modal structure, and total travel activity on energy use for travel in the OECD, 1973–1990

modes (49). Danelis (50) used Divisia indices for Italy, and Greene & Fan (51) performed a similar decomposition on the United States The factors are not necessarily independent. Reductions in fuel intensity of a given mode could lead to reductions in the cost of using that mode and thereby stimulate use of that mode. Greene (52) estimates this feedback for US automobile use and found that the elasticity of use with respect to fuel intensity to be less than –0.15, i.e., a 1% decline in fuel intensity led to less than 0.15% additional use. Holding two of these factors constant and letting the third vary gives the results shown in Figure 11: 1. Increased per capita travel itself (activity) boosted energy use 30% in the United States to as much as 60% (Japan). 2. Modal shifts (structure) boosted energy use significantly (33%) in Japan, by 3% in the United States, and by 5% in Europe, except in Denmark and Sweden, where a slight increase in the role of rail and bus actually reduced energy use slightly. 3. Changes in modal energy intensity (i.e. energy use per passenger kilometer) alone reduced energy use in the United States by 18%, but had only a marginal impact on other countries, or indeed increased energy use.

Thus, more energy was used to transport an average European in 1992 than in 1973 and about the same for a Japanese. Only in the United States, which started with energy intensities far higher than those in Europe, was there a clear decline in the overall energy intensity of transportation, led by the drop in the intensities of car and air travel. Yet the gap between per capita energy use in the United States and in Europe (or Japan) remained after 1992. Why?

ENERGY USE FOR TRAVEL IN EUROPE AND THE UNITED STATES COMPARED: WHY ARE THERE DIFFERENCES?

Three factors—motorization (or car ownership), mobility (or travel), and macho (or the characteristics and fuel economy of cars—share roughly equally in explaining the three-to-one gap between per capita fuel use in the United States and values in Europe in 1992. The influence of these factors is summarized in Figure 12. If Americans had driven the European fleet of vehicles in 1992, per capita fuel use would be about 25% lower than it was, as the second column (EU-8 intensity) shows. If Americans had driven their present cars the distances Europeans drive their cars, fuel use would have been about 30% lower, as the third column (EU Distance/car) shows. (Differences in driving and mobility are reviewed below.) The fourth column shows the impact on US fuel use of combining European car ownership and driving distance (EU Distance/capita), the fifth column, the combined effect of European distance,

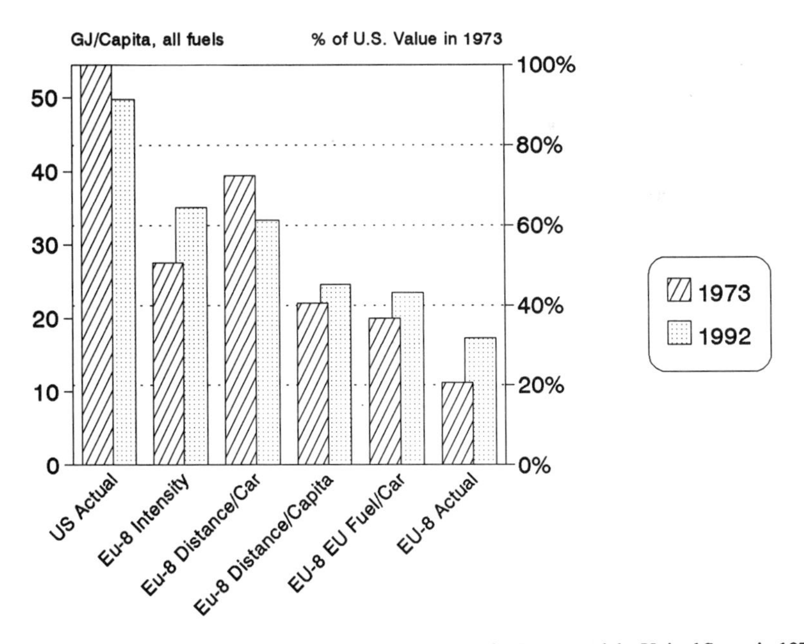

Figure 12 A comparison of per capita automobile fuel use in Europe and the United States in 1973 and 1992. EU-8 includes West Germany, Italy, France, Great Britain, Denmark, Sweden, Norway, and Finland. Fuels include gasoline, diesel, LPG, and CNG used in cars and personal light trucks.

fuel intensity, and US car ownership (EU Fuel/car), while the last column shows actual per capita fuel use in Europe in 1992. This breakdown shows that differences in distance and car ownership together make up a larger part of the US/Europe difference than do differences in fuel intensity alone.

The same comparison for 1973 shows a larger gap between the United States and Europe in per capita energy use, because the per capita car ownership gap was larger and because US cars used so much more fuel per kilometer than in 1992. The convergence occurred because the US automobile fleet became 35% less fuel-intensive, while per capita car travel in Europe increased by nearly 75%.

In comparisons between the United States and Japan, the components are different, because the energy intensity of car travel in Japan is almost as high as in the United States, but the overall level of travel is less than half of that in the United States and the share of rail and bus still over 45% of the total. The most important component of the difference in per capita fuel use for travel in the United States vs. Europe or Japan is thus the level of travel and not the fuel intensity of vehicles or modes.

Why is per capita fuel use for transportation in Europe so much lower than in the United States? That is, why is travel so much lower in Europe? The simplistic answer is that fuel prices in Europe are far higher than in the United States, and automobile prices are much higher in Europe (because of various purchase taxes), which boost the marginal cost of travel in Europe (or Japan), while disposable incomes are slightly lower than in the United States. Consequently, Americans can go farther on the same budget. An alternative hypothesis is that Americans have farther to go. As we shall see, there is more to the story than prices, incomes, or geography alone, but the first two factors have been demonstrated repeatedly to explain many of the differences in automobile fuel use (44, 45, 53). At the same time, cross-sectional studies of regions where incomes, prices, and socio-demographic variables are relatively uniform suggest geography, (i.e. city size and density, place of residence within a community, and so on) has a role, too. Here we do not repeat econometric studies, but instead offer some indicators of the role of various factors.

Pricing and Taxation

Fuel pricing has had an impact on both fuel use and distance traveled. The relationship among fuel use and its determinants is complex, since many other factors influence the use of automobile fuel. A preliminary cross-sectional econometric study shows that fuel prices primarily affect fuel use not only through the fuel economy of (new) cars, but also through the effect of fuel prices on car use. The mean value for the long-term price elasticity of fuel use is –0.85 from cross-sectional estimation (LJ Schipper & O Johansson, submitted for publication), with about –0.55 arising from differences in the fuel

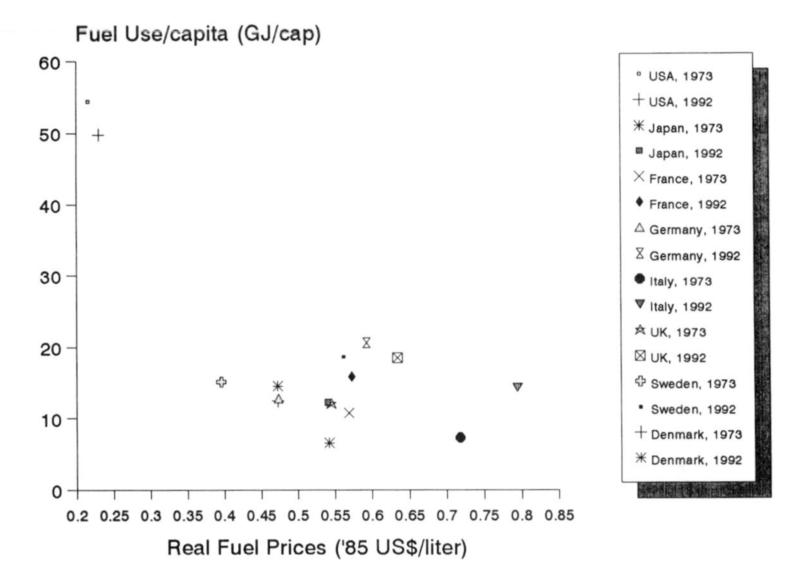

Figure 13 Automobile fuel use and fuel prices in the OECD in 1973 and 1988. The United States and Denmark include personal light trucks. Prices are a weighted average of diesel and gasoline.

intensity of the car stock, –0.2 from differences in distance traveled, and the remainder from differences in car ownership. If we include the population density of each country, the overall price elasticity falls very slightly, suggesting that national average population density plays a small but appreciable role. Income elasticities lie between values slightly below 1 to somewhat above 1, depending in part on the saturation of automobiles. These findings are consistent with a wealth of studies [reviewed in (44); see also (55) for a review of elasticities that look beyond fuel prices], although Schipper and Johansson was the first international, cross-sectional study to use national data for determining the size of the automobile fleet, auto usage, and the correct mix of fuels used by these automobiles.

Some simple plots show the cross-sectional relationships well: Figure 13 shows the inverse relationships between fuel prices and fuel consumption per capita; Figure 14, automobile distance driven per capita; and fuel use per kilometer (Figure 15), the three quantities portrayed in Figure 12. In these figures the United States is at the far left, Italy at the far right, and the other countries distributed in between. These relationships are stable throughout the 1973–1992 period. The one striking exception is the United States, for which 1992 fuel use per capita had fallen (in spite of rising incomes and car ownership) while fuel intensity was markedly lower, even though the 1992 real fuel price was close to its 1973 value. These correlations are just that, and do not

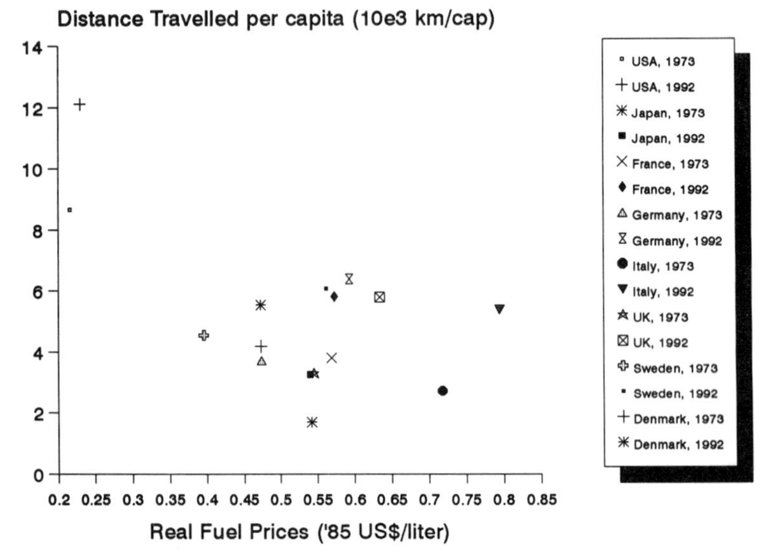

Figure 14 Car mobilitiy and fuel prices in the OECD in 1973 and 1992. The United States and Denmark include personal light trucks. Prices are the weighted average of diesel and gasoline.

by themselves prove causality. However, the fact that per capita automobile use is observed to decline when fuel prices increase in the short run does indicate a causal link, albeit with a low elasticity. Similarly the fact that both new-car and on-road fleet fuel intensity in the United States increased most rapidly during the period of highest fuel prices indicates a relationship between these variables as well. Gately (56) gives evidence for an asymmetrical result: Fuel use falls more when prices rise than it increases when prices fall, because of improvements in technology stimulated by higher prices that are not abandoned when prices fall.

Virtually all the cars in circulation in 1992 were bought after 1975; thus the points for 1992 (as seen in Figure 15) portray the impact on fuel intensity of car manufacturing-and-purchase decisions made after the first oil shock. In this light, the nearly inverse linear relation between fuel intensity and price is particularly revealing. Is there a universal law here, broken otherwise by the United States in 1973, which by 1992 lay close to the other countries, even though fuel prices had fallen back to their 1973 levels? Probably not, but something jolted the US stock, which was a clear outlier in 1973 but part of the pack by 1992. In Schipper et al (13) it was suggested that the imposition of the CAFE fuel economy standards in the United States was one reason that the United States stands out from the other countries, based on Greene's study

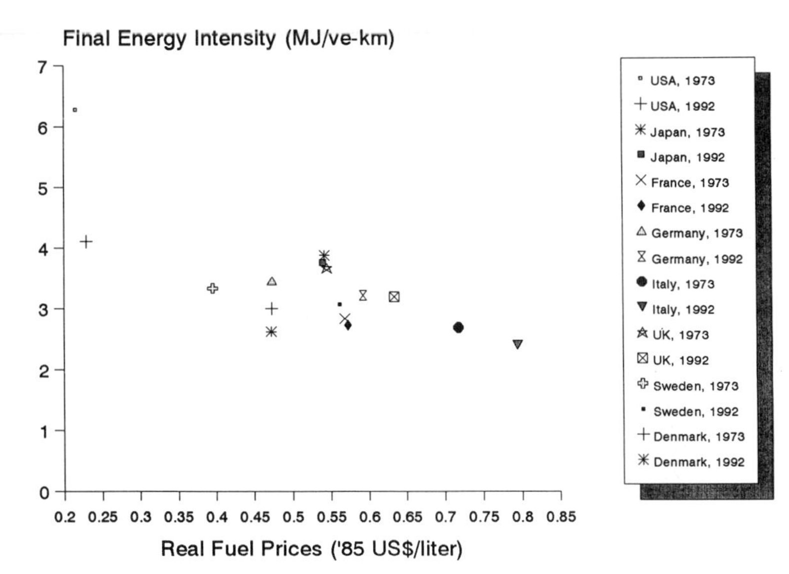

Figure 15 Car fuel intensity and fuel prices in the OECD in 1973 and 1992. The United States and Denmark include personal light trucks. Prices are the weighted average of diesel and gasoline.

(57). Another reason, suggested by Gately (1993 private communication) is that the ratio of peak to average fuel price for the United States during this period is much higher than for other countries.

Another important observation can be made based on Figure 15. By the 1980s the differences in the technologies used to power cars differed little from company to company or country to country. Instead, size, power, and features selected by buyers in each country, as well as driving conditions, determine test and actual fuel economy.

Although economic analyses based on fuel prices and incomes tend to explain most of the variation in automobile fuel use over time and between countries, such analyses have important shortcomings. For one thing, subtle differences in the way cars and other parameters of car use are taxed are also important to ownership and use, although the impact on fuel use is small because fuel taxation is the largest single effect in most countries. Schipper & Eriksson (58) calculated the potential impact of the tax schemes in seven countries (including six of the countries analyzed here) on the total costs for acquisition, yearly fees, and fuel for the mix of cars sold in the United States in 1990. Figure 16 shows both the untaxed US price of the car (and 10 years' worth of fuel at 14,000 km/year and 8 l/100 km), plus the taxes on acquisition, fuel, and yearly ownership over the same 10-year period (assuming constant costs or prices for each component from 1990). The enormous burden of *ad*

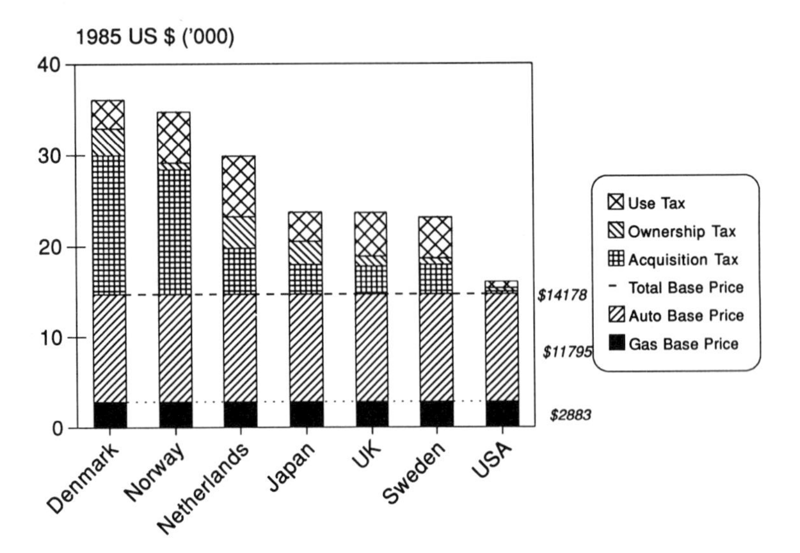

Figure 16 A comparison of 1990 automobile tax schemes and their effect on the cost of a new car.

valorem taxes in Denmark, Norway, and even Holland contrasts sharply with the near absence of taxation in the United States and the modest taxes levied by other countries with important car industries. These differences in car taxation help explain why car ownership is lower in Denmark than in other countries.

Some policies have increased the number of cars, the distance cars are driven, and the fuel used by cars. Company car policies (particularly in Holland, Great Britain, Sweden, Norway, and Germany) boost both car ownership and car size or power, which increases both use and fuel consumption (20, 58–62). This policy often extends to fuel use; those who get cars (and fuel) pay a flat sum as income tax on the benefit of having a company car, a sum unrelated to actual usage. Figure 17, derived by Schipper et al (13) [see also (64)] shows that new company cars (one third of new cars) in Sweden are heavier and more powerful than those bought by private individuals. Indeed, Jansson & Wall (59) found that for a given increase in fuel taxes in Sweden, roughly half is given back through a combination of policies and the indexing of some social benefits to inflation. A related measure is the policy in many European countries of permitting tax deductions for commuting expenses. Finally, employer-provided free parking for employees encourages the choice of the car over other modes for trips to work. Thus the real cost of driving, and often the marginal cost, in much of Europe is less than implied by the

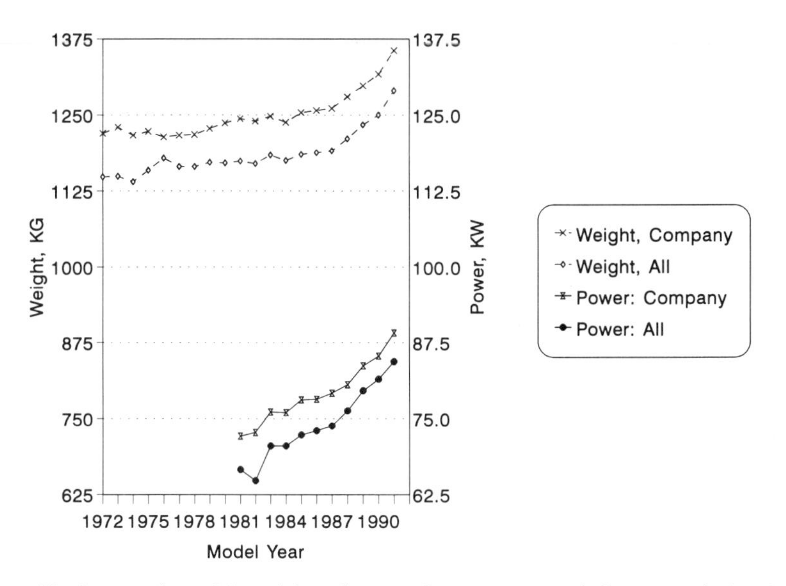

Figure 17 A comparison of the weight and power of company cars and all new cars, in Sweden, 1972–1991.

comparison of fuel prices in Figure 10 or the comparative taxation in Figure 16. This may in turn explain why the slope of Figure 15 is not steeper.

On the whole, car taxation (or tax relief) affects both the household's ability to afford a car as well as the value and characteristics of cars. Where heavily taxed by value and weight (Italy, Denmark, Britain, Holland), cars are somewhat smaller than where lightly taxed. Where taxes are light and company car privileges widely available (Sweden, Germany), cars are the most common and most powerful (or largest). Car taxation both raises the prices of cars and increases the differential in price between larger, more fuel-intensive cars and smaller ones. Needless to say, alternative forms of taxation are under intense scrutiny from authorities looking for ways to reduce environmental and other problems arising from car use (65, 66).

On balance, car taxation appears to restrain ownership somewhat. But in Britain and Denmark (the European countries closely examined that had the fewest cars), car usage is high, so apparently mobility is not restrained as much as is car ownership. This suggests we must examine more closely differences in the way cars are used.

People on the Move: Differences in Patterns of Mobility

While differences in the cost of driving a kilometer explain some of the difference in fuel economy between the United States and Europe, they explain

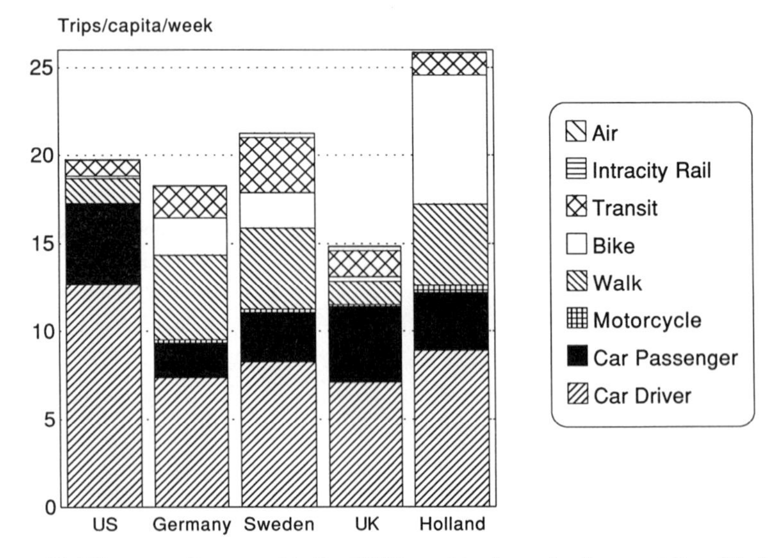

Figure 18 Trips per capita per week in five OECD countries, by mode of transport. Transit includes buses, subways, and other modes of public transport.

only a part of the differences in mobility. The relative distances between urban areas, and the pattern and density of settlements, offer some explanation. These factors also act to give such transit systems high enough ridership to keep costs low and service high. Although trends show the car slowly gaining in Europe, it is doubtful that cars will ever provide the same level of mobility per capita as they now do in the United States.

Travel surveys have been used for many years to analyze patterns of individual and household travel by mode and purpose (67–73). To understand more about these differences in transportation patterns and how they may explain differences in total travel between countries, we have begun to analyze how people travel, how often, and how far, using national personal-transportation surveys.

Figure 18 shows the overall pattern of trips by mode. The values shown in subsequent figures for per capita travel differ from those shown in Figure 2 because the coverage of the bottom-up household surveys reported here is different from that of the top-down data used as a basis for much of Figure 2. The car dominates everywhere and is used mostly for local trips. Air travel in the United States is as important for longer trips as are public ground modes in many of the European countries. Americans appear to make slightly fewer than the average number of trips, but this is because they make (and also report) fewer walking and cycling trips. Americans make more trips by car than

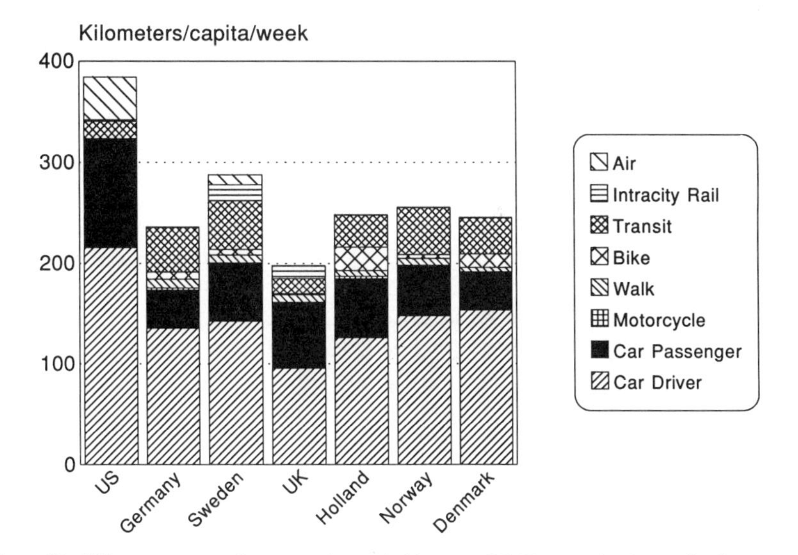

Figure 19 Kilometers per capita per week traveled in seven OECD countries, by mode of transport.

anyone else—13 trips per week as car drivers and 5 trips per week as car passengers—while Europeans make only 8–9 as drivers and 3–4 as passengers. Americans make fewer than 1 trip/week in some form of transit, Europeans 1.5 to 4. This extra share of car trips accounts for most of the difference in total distance traveled, with travel by air the other significant difference.

More than 90% of trips to work in the United States are by car as driver or passenger, vs 55% to 66% in all other countries. US work trips are 15% longer than those in Europe. The car also has a dominant share in the United States for trips for other than work, 86% vs 50%–55%. This difference in part may reflect the under-reporting of walking trips in the United States, but Americans take 5% of their trips on bus or local rail, vs 6–15% of trips that others take on these modes. Even if one discounts the non-motorized modes, it is clear that the automobile dominates far more in the United States than elsewhere.

If we weight trips by average length, we obtain total distance traveled by mode (Figure 19). The role of non-motorized travel falls sharply. The automobile is by far the most important mode in each country we analyzed, providing between 70–75% of travel (former West Germany and Holland) to as much as 85% of total travel (US). Motorized, collective ground modes, by contrast, provide only 5.5% of travel in the United States vs 15–20% in the other countries. But intercity rail forms an important component of travel in Europe. Since large distances in the United States boost the share of air over

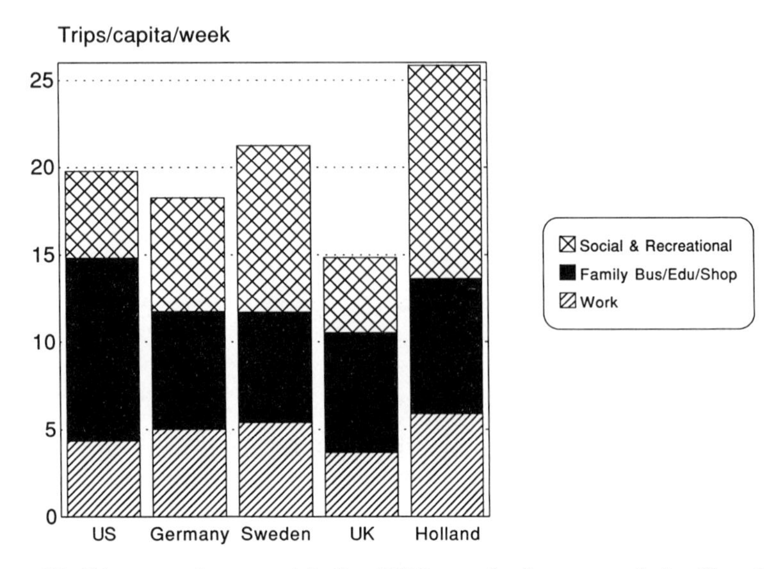

Figure 20 Trips per capita per week in five OECD countries, by purpose of trip. All modes of transport included.

rail, relative to Europe (and within Europe, boost the importance of air in Sweden relative to ground modes), comparison of travel excluding intercity rail is somewhat more justified in terms of comparing alternatives to the automobile. For the countries where this is possible, we still find that the share of local public transportation in the United States is far smaller than in other countries for all purposes.

Trip distance varies by mode. Air trips (where reported) are the longest. Intercity rail, where reported, are the second longest, which is not surprising. What is interesting is that in the United States, bus and local rail (including tram and metro) trips are longer than car trips, while in Europe the reverse is generally (but not always) true. This interesting contrast suggests, but does not prove, that bus and commuter rail within cities (i.e. for short distances) plays a significant role in Europe but an almost insignificant one in the United States This suspicion is further boosted by noting that practically all public transport trips in the United States are commuter trips, while in the other countries commuting trips account for only about one-third of bus and local rail travel.

Figure 20 shows trips per person per week by purpose. Americans make the average number of work trips, but make considerably more trips for family activities (including education, shopping, and civic purposes) or during free time. They tend to cover much larger distances for these latter purposes. It is these trip purposes for which distance and frequency, (i.e. total travel) vary

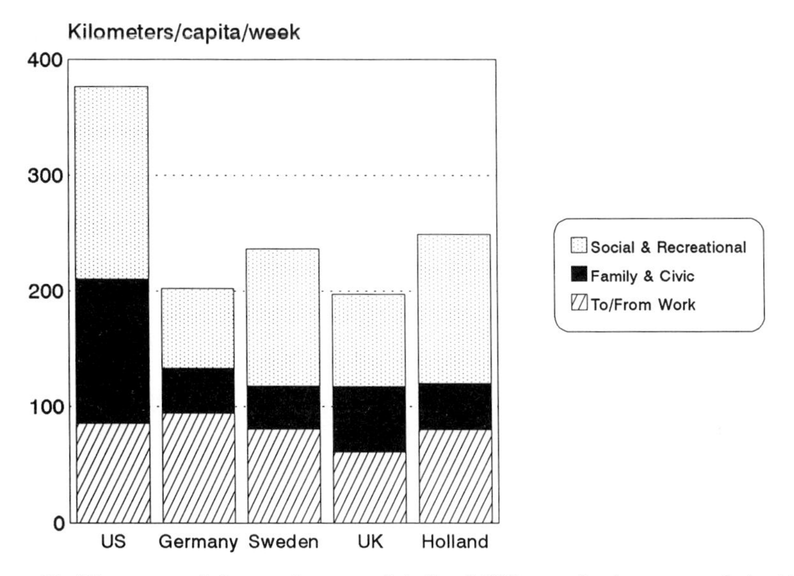

Figure 21 Distance traveled per capita per week in five OCED countries, by purpose of trip. All modes of transport are included.

the most between countries. And it is the frequency of this trip type that has been growing the most in the last 20 years. Moreover, these types of trips are not regular in either time or space, in contrast to regular trips to work. This means these trips are less likely to be suitable for collective modes, as the data indicate.

If we consider total travel (Figure 21), there are some striking similarities among the countries. First, roughly 25% of all travel is for work (30% of car kilometers driven), 30–35% for family activities (including civic purposes, shopping, and education), and about 35–40% for free time, including vacations. Americans show the highest per capita travel for these purposes. Trips for or within work tend to be long, but typically less than 5% of all trips. Roughly two thirds of all travel and 75% of trips are for other than work-related purposes, and this share has been growing slowly. This pattern is apparent in every country. The modal splits within trip types and overall differs considerably between countries.

Trip length by purpose varies somewhat. Trips to work vary from 11 kilometers (Sweden and the former West Germany) to 17 kilometers in the United States, with Holland and Britain lying around 13 kilometers. This is an interesting finding, since Sweden—the largest and least dense of the European countries we studied—shows the shortest trips to work, while those in the

United States—the largest, least dense country—lie at the higher end, and the three dense European countries lie in the middle range. The average length of a trip for other purposes is longer in the United States than in Europe, even if we exclude air travel. The most dramatic difference is that for shopping and family activities, where the average trip in the United States (for all modes) is 12 km vs 4–9 km for other countries. US car trips for these purposes (and for social and recreational purposes) are longer than trips in Europe for these purposes, too. Hence the average non-work trip in the United States is longer than that in Europe, even if we ignore non-motorized trips.

Finally, when we combine all the observations for cars, we find that a single journey in a car in the United States is only slightly longer than one in Europe. The range is surprisingly narrow, between 12 km (Norway, Germany, Denmark in the 1980s) to 15+ km for the United States, with Denmark (1992), Holland, and Sweden (1994) intermediate. It may be that in the United States a large number of very short car trips and a good number of very long ones give the same average trip as one in Europe. What explains the enormous gap between the United States and Europe in per capita distance traveled by car is thus the number of car trips per capita more than the distance per trip. We saw above that these trips are primarily for services and free time, not for work.

Does this mean that America's allegedly great distances are not important? The sprawl of America's suburbs certainly contributes to reducing walking and cycling trips to work, services, and leisure time. More subtly, however, it appears that the large number of trips Americans make by car both complement and substitute for shorter trips that Europeans make with their feet, their cycles, and urban transit—or simply don't make. These car trips are longer than the corresponding trips Europeans make on other modes, but short enough to reduce the average trip an American makes in a car to about the same distance as a car trip made in Europe. That is, Europeans have virtually the same access to travel destinations as Americans, but they do not travel as far to achieve this access. How much of the difference in actual distance traveled occurs because travel is more expensive (and in some cases time-consuming) in Europe, and how much because destinations in Europe are closer (i.e. population densities in urban areas are higher) is uncertain. But one reason why the trip distances appear to vary so little among countries is that most travel is local. Where the largest differences show up are in domestic air travel, almost negligible in Denmark, much more important in Sweden and Norway, and very large in the United States. Clearly this is a function of the size of the country.

Figure 2 revealed this difference in another light. We see that the distance Europeans travel by rail and bus (both intercity and within urban areas) is three to five times that covered by Americans. Europeans' travel by collective modes neither compensates for the American's greater travel by car, nor acts as a direct substitute. Instead, a complex set of factors related to land use, urban density, and fuel and transit prices both constrains Europeans to travel less and offers a more attractive framework for using collective modes. We examine the question of urban form or density next.

Other Determinants of Travel: Location, Income, and Demographic Factors

How does mobility, particularly automobile use, vary as a function of the nature and population density of the surroundings in which a person lives? The comparison above showed that Europeans take more trips with non-motorized modes than Americans, reducing car use accordingly. Naess (74) and Naess et al (75) find that most of the variation in average car use between Scandinavian cities is explained by city size and density, with other factors not important. Is this geographical explanation sufficient? Is this because of different ways in which towns are laid out, in particular, the density of settlements and proximity of housing to work? Or was the expansion of towns into suburbs in Europe limited because travel was expensive? These questions are being debated as planners seek the design of "transport efficient cities" (76, 77) or simply try to stem the flood of travelers from collective to individual modes (78, 79). To understand the role of location and demography, we examined national, multi-state regional, and local surveys.

Figure 22, based on the US Department of Energy's Residential Energy

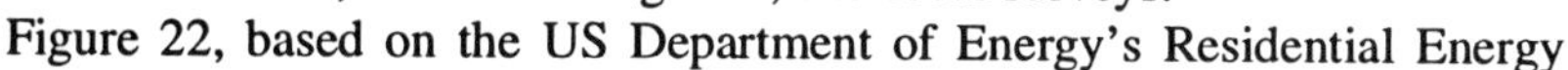

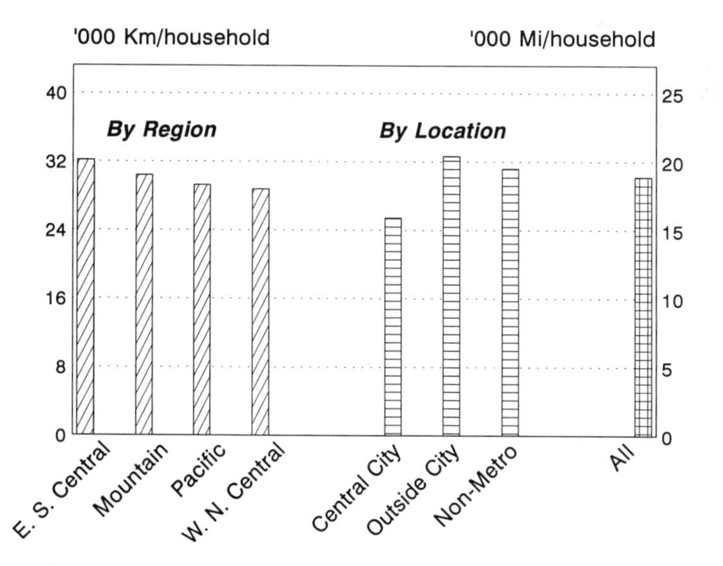

Figure 22 Distance traveled per household in the United States, by region of the country, and by location in relation to urban area.

Transportation Consumption Survey for 1991 (80), shows that household driving in the United States varies by the location of household (central city, suburban, or rural). Households in the urban areas have the lowest level of travel (measured in vehicle/kilometer of travel [vkt]/capita), those in the suburbs the highest vehicle use, and those in nonmetro areas have vehicle use close to but below that of those in suburban areas. Another national US survey, the National Personal Transportation Survey (73, 81) shows the same general results. Using the NPTS, Dunphy & Fisher (82) found a weak inverse relationship between population density where the survey respondent lived and total travel. Moreover, Dunphy and Fisher found that both total travel and the share of the automobile in the total rise as the population density falls. Similar variations between travel of those living in cities and those in suburbs or rural areas are apparent from surveys of Denmark (70), Sweden (71), and Britain (84), as well as from comparing the inner and outer parts of Copenhagen (85), Stockholm (86), and London (76). From these studies it is also clear that the average trip length is shortest for those living in the densest cities (i.e. Copenhagen, Stockholm, and London), but as with Dunphy & Fisher, Banister finds a relatively small difference in total per capita travel over a very wide variation in city size and density.

There is an important caveat underlying these comparisons. The incomes and socio-demographic characteristics of those living in less-dense parts of cities are different from those in rural areas or those in city centers. If we examine travel location as a function of both location and income as reported in the US EIA survey (Figure 22), the variation over location is significant and broader than the variation over region (we show two groups from the bottom and top of the distribution). Similarly, if we stratify by income and whether or not the household has children (Figure 23), we also obtain a significant difference in vehicle use, even if we control for income. Life-cycle characteristics are important considerations as well. Single adults in the United States over 60 years old travel 22,500 km/year, ranging from a low of 10,000 km/year for those earning under $15,000 to a high of 40,500 km/year for the wealthiest group. Those under 35 travel 50% more, within a given income band, than those over 60. This life-cycle effect was confirmed by Greening & Jeng (90) for a region of the United States, and by Vilhelmson (71) for Sweden. [See also (90) or (91).] The effect is apparent from inspection of the other national travel surveys, all of which display results as a function of life-cycle and family characteristics. The key point is that the variability of travel from these socio-demographic factors is much larger than the variability arising solely from differences in urban density, which is apparent from comparing Dunphy & Fisher's analysis with published cross tabs from the US NPTS survey. The same characteristics are also important determinants of travel in Europe (70, 76).

The difficulty of analyzing entire countries suggests we need to examine

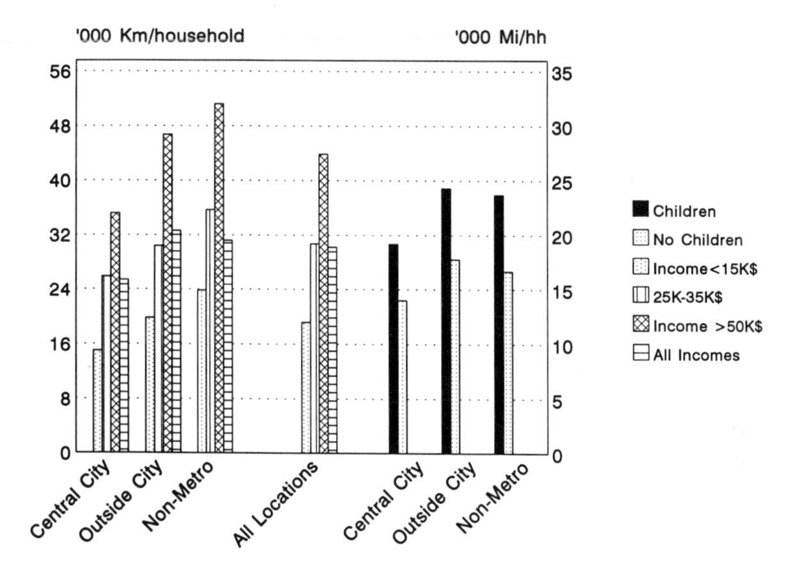

Figure 23 Distance traveled per year per household in the United States by income, location, and the presence of children.

household travel behavior in smaller regions if we wish to understand the relationship between location and travel. The San Francisco Bay Area region was studied in some detail by Deakin & Harvey, based on a household survey (93). Figure 24, showing automobile use (vkt) per household vs residential density (in residential units per hectare), reveals a sharp decline in travel as density increases. But the same survey shows that controlling for urban density, higher income households travel more, on average, than their less-affluent counterparts. Conversely, high-income families in higher density areas travel less than those in lower density areas. In addition, higher income families are more likely be larger and live in low-density areas, consistent with Dunphy & Fisher's observations. Therefore, the differences in driving shown in Figure 24 depend on more than just the location of the household.

As an international example, consider travel in the inner and outer parts of two Scandinavian cities, Copenhagen and Stockholm, two major cities with important car-free areas downtown and good transit service. According to the survey of Stockholm (86) and that for Denmark (84), those living in the inner city travel less, and use the car for a smaller share of their travel, than those in the suburbs, as Figure 25 shows. One reason those in central Stockholm do not move as far is that congestion is bad six days a week. Interestingly, weekend travel per capita is higher than weekday travel and is 75% by car vs

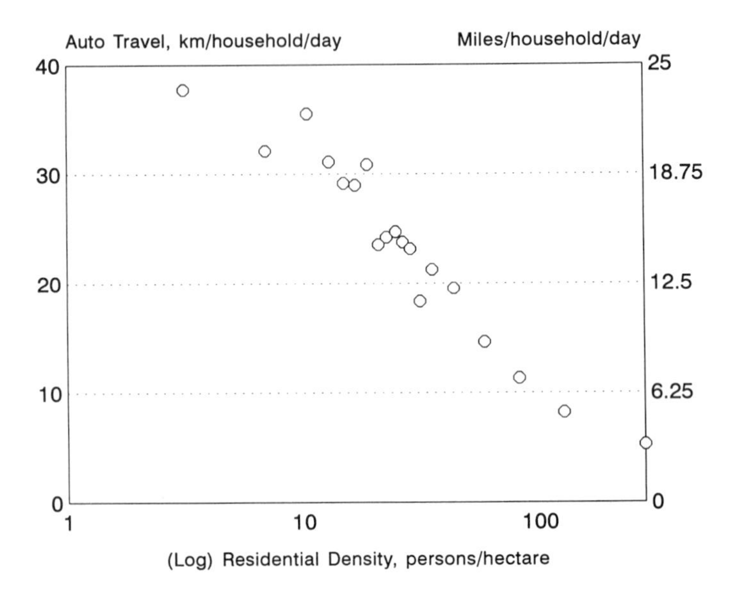

Figure 24 Residential density and average daily auto distance per household, San Francisco Bay area, 1991.

only 50% during the week. This points out the importance of measuring travel throughout the week and not simply on work days. During the week, travel within greater Stockholm makes up 80% of total travel, falling to 75% on weekends as more residents leave the city. This suggests that measuring the mobility of people living in a given place requires following their travel, not simply measuring traffic in a given region and then ascribing that to residents.

Stockholm reflects the importance of demographic and economic factors. Inner city residents own fewer cars (half as many per household), are older, have fewer children, and have smaller household incomes than those living in the suburbs. How much of the reduced travel (or lower share of car ownership) is caused by these factors, how much by the higher density and proximity to stores and work? Thirty-five percent of those in inner Stockholm do not use cars or public transport to get to work; presumably they walk or use their bikes (or are retired). In the rest of Stockholm, this share falls to well below 1%. The average distance for all trips for those in the inner city is 4 km (bus) and 5 km (car), vs well over 10 km for each mode for those living outside of the inner city. These relationships suggest that there is an important proximity effect.

Many factors contribute to these differences in car use between regions. Maintaining a car in downtown Stockholm—where virtually all on-street park-

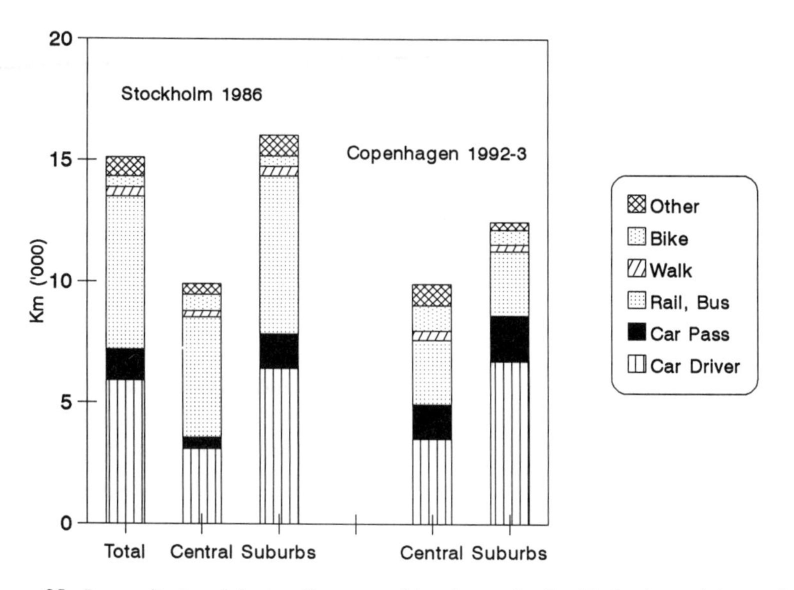

Figure 25 Per capita travel in two European cities, by mode. Stockholm is weekdays only; Copenhagen includes weekends.

ing is both paid or restricted to residents, or both—is far more expensive than in the suburbs, and many other costs are higher as well. The Stockholm survey shows that average speeds are far lower within the inner city than elsewhere. Thus, while urban form and location certainly affect travel, they affect many other aspects of daily life that also affect travel. (These observations also apply to Copenhagen.) And many factors besides urban form and population density affect travel. So while we have found relatively prosperous cities where the use of the car is low and use of other modes high, it is clear from examining national travel patterns that most people live outside of these regions and travel more than do those living within the cities. And all residents are more mobile today, with more cars, than ten or twenty years ago. Clearly, then, the size, density, and nature of the location where people live and work influences their travel, but so do many other factors.

At issue, then, is not simply whether people who live in large, dense cities travel less, but why they travel less. To what extent is less travel an expression of the constraints of congestion, higher car ownership and use costs, or the socio-demographic characteristics of the population, and to what extent a result of destinations being closer than in other locations? Income and demographic characteristics also vary from city to city, or from place to place within a city. We cannot tell the extent to which these variations themselves are causes of differences in travel behavior, or effects of both the locality where people live and the nature of the people themselves. Although some have proposed that increasing density would reduce travel, it is difficult to say how travel would vary if the densities of residential areas changed, and even more complicated to predict how other important aspects of life would change.

TRENDS IN TRAVEL: IMPLICATIONS FOR ENERGY AND ENVIRONMENT

Upward Mobility: Trends in Travel

The analysis of Section II showed that it is growth in mobility, mostly from the car, that has raised energy use for travel. The national travel surveys taken over time confirm that this is occurring both among individuals in a given location and as a result of growth of settlements outside of central cities. The city survey of Stockholm revealed this growth even within the regions of lowest mobility. Why has travel been increasing? Where are people going? Is there any saturation of the level of travel?

Certainly, rising income in many countries seems to encourage (or at least permit) people to spread out away from denser urban environments—those that could support frequent, and convenient, transit service and walking. Rising car ownership is the main route along which this change occurs (17, 18). Suburban development tends to undermine transit service (except for certain commuting corridors) and make people more auto-dependent. This change in the physical layout of society is not spontaneous. Powerful tax policies have influenced both public and private decisions affecting housing, services, and other development (94a).

Car ownership is "saturated" in the United States in the sense that there is nearly 1 car per driver's license. Moreover, 92% of males and 82% of females over driving age, and 68% of all resident Americans, had licenses in 1992 (34, Table DL20). Between the ages of 16 and 49, the share of females with licenses lies about 5 points below that of men; above 49 the difference jumps to 10% and higher. This difference between men's and women's licensing thus appears generational; one would expect that as the population ages, the increasing share of younger, licensed women pulls the share of women with licenses close to that for men. This means that in twenty years 93% of eligible Americans would have licenses instead of 87% in 1992, yielding a small increase in total car use.

By contrast, the share of the Swedish, German, or Norwegian population with drivers licenses is below 70% and the gap between men and women larger, particularly among those over 50 (these data come from national transport statistics and travel surveys). And in these countries there are only 0.6–0.7 cars/licensed driver. Thus even for these three countries that lie at the high end of European mobility, there is considerable room for growth in personal mobility. In Denmark, Britain, and Holland, and above all in Japan, the potential is even larger.

Other important forces have boosted travel demand. The structure of employment, with more women in the work force and more part-time or self-employment, has raised travel demand. So has decentralization of work away from city centers. And the increase in women in the workforce has led to shrinking of the gap between men's and women's driving. Figure 26 shows the distances men and women in different age groups in the United States drove in 1990 and previously. Since the number of women with access to cars grew so dramatically in the 1970s and 1980s in the United States (a phenomenon now apparent in Europe), this change drove important increases in total travel. Data from the United States and European countries show that the number of women having drivers' licenses at a given age is closing in on that figure for men, at least among populations under 55 years old. Soon women will be as mobile as men. And the surveys show workers with cars drive farther than nonworkers with cars or workers without cars. Finally, household size has been shrinking through a number of factors, including aging. This means more single drivers, particularly women, which increases travel.

The aging of populations adds a new dimension to travel demand, particularly as baby-boomers who grew up with full access to automobiles retire, bringing their cars and their mobility with them. As Figure 26 shows, distance driven also depends on age. Note that older people drive less than younger.

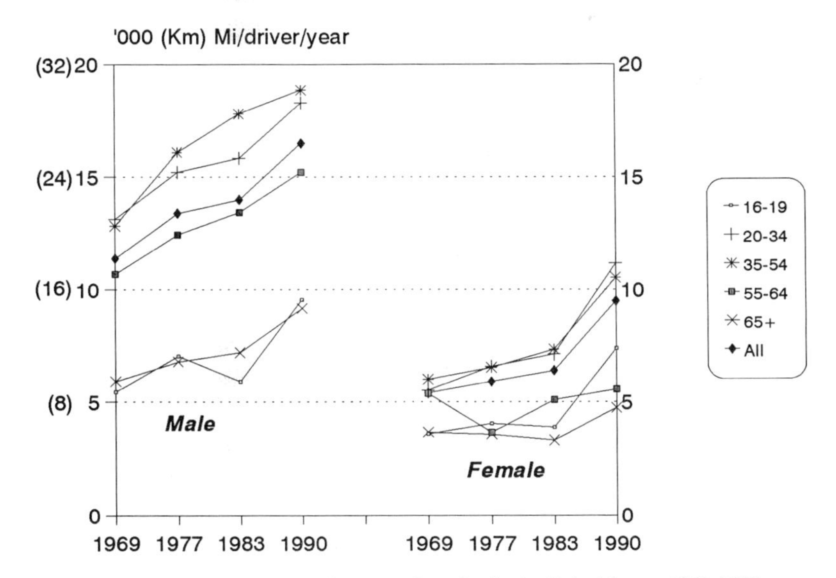

Figure 26 Distance driven per year, by age and gender, in the United States, 1969–1990.

Will these patterns persist as the present baby-boomers approach retirement? So far, the mobility of the oldest citizens in the United States and Europe has increased steadily. Although the mobility of every group has been increasing steadily, the swelling of the ranks of retirees is likely to put some damper on the growth in total mobility.

Finally, important but subtle social forces and traditions affect overall mobility. Take one example, the hours that businesses, stores, and shops are open (94b). In some European countries most shops must be closed by the early evening, and few are open on Sunday or even Saturday afternoons. In the United States, by contrast, and increasingly in Sweden and France, opening hours are less restrictive. And US-style malls or supermarkets are appearing now in all European countries, offering cheaper goods than smaller stores in cities provided the shopper has transportation to the location. But the present differences in opening hours may explain why use of the car in Europe for shopping and other personal services is so much less than in the United States.

Is total travel saturated? Probably not in the United States, and certainly not in Europe. While congestion limits speed in certain parts of the travel cycle, most growth in travel is occurring outside of congested cities or the worst hours of crowding. And Europeans are traveling more in the evenings and on weekends, as stores and other attractions are open for longer hours. In Europe, the growing network of relatively high-speed rail links and motorways, and the availability of inexpensive vacation charter flights and other inclusive packages augur for more travel between the congested cities. For the most part, this extra travel is car- or air-based, modes that are faster and more energy-intensive than buses or railways. While the hidden incentives for more travel and the socio-demographic forces stimulating travel may have begun to approach saturation, the full impact of a mobile society have yet to be felt in the industrialized countries.

Will Europeans ever travel as much as Americans? Since the automobile dominates travel in both cultures, consider the difference in the cost/km of using cars. Since the fuel cost of using cars in Europe is 50–125% more per km for fuel than in the United States, it is not surprising if Europeans only travel 40–55% as much as Americans by car. For their total travel to converge on the US levels, something approaching "supercars" (Lovins et al 1993) would have to drive down fuel costs to where they were irrelevant, and significant growth in travel would have to occur away from congested cities and suburbs. Since most travel and car use is short distance and presumably close to home, it is hard to imagine European travel growing to US levels. Could US travel decline? Schipper et al (1989) constructed a scenario of a "home-oriented" lifestyle that may be plausible, but no one knows how to reduce travel other than by raising its cost. They speculated that telecommuting and other uses of information might substitute for some travel; all that can be said in 1995 is that these important trends have simply not been observed yet in the aggregate.

Trends in CO_2 Emissions from Travel

Because of concern over rising levels of CO_2 in the atmosphere, emissions of (CO_2) have become a useful indicator of environmental problems arising from energy use. Other emissions from the use of fuels for transportation are far more important locally because of negative impacts on health and property, but these have fallen (or are falling) dramatically, both in absolute levels and measured per kilometer of traffic or travel.

Emissions of CO_2 have not declined. Figure 27 presents per capita emissions in 1973 and 1992 (96). The decline in per capita emissions for the United States was offset by increases in population. In general the changes in emissions resemble those for energy use, not surprising since almost all energy used in the transportation sector (counted here) is oil. (Note that emissions from freight traffic lay at roughly two thirds of those for travel in 1992, but had risen considerably more than those from travel since 1973. Energy and emissions from international air and sea traffic are also omitted; these represent roughly 15% as much energy and CO_2 as the combined domestic travel and freight sectors modeled here.) Finally, it should be clear from Figure 12 that it is differences in the volume of travel that account for most of the difference in per capita emissions; differences in energy intensity account for perhaps one quarter of the US/Europe difference; and differences in modal shares a

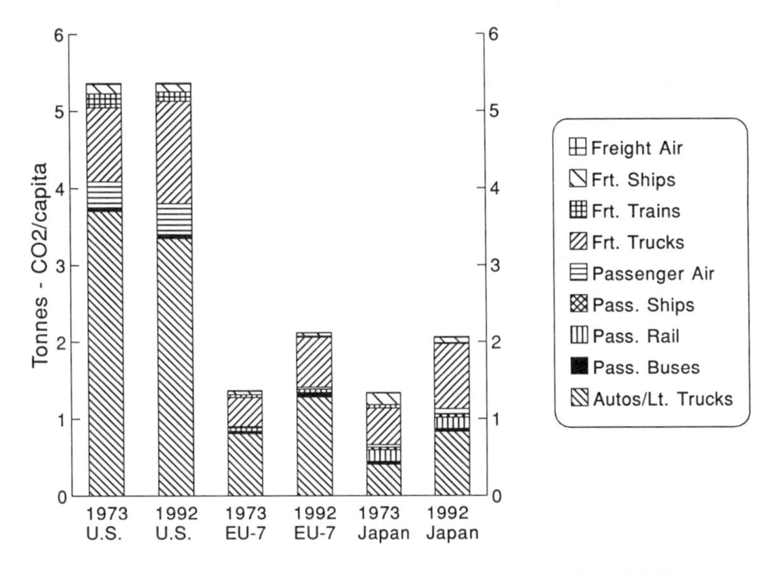

Figure 27 CO_2 emissions per capita from OECD transportation in 1973 and 1992, by mode.

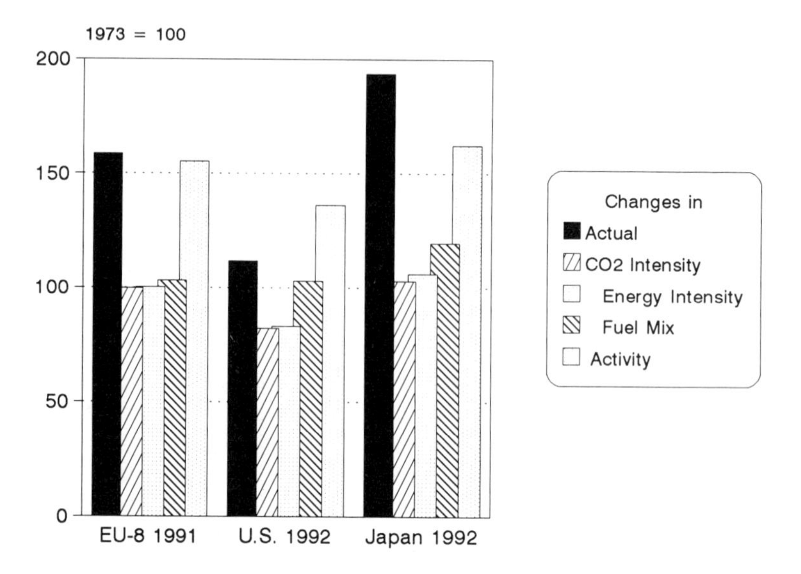

Figure 28 Decomposition of changes in CO_2 emissions: impacts of changes in activity, structure, energy intensity, and fuel mix, 1973 and 1992, by mode.

small amount. Differences in fuel mix (diesel vs gasoline) are insignificant compared with these other factors.

We can apply the methodology of disaggregating the components of changes in energy use to CO_2. In doing this for transportation, we pay particular attention to the small differences in CO_2 emissions per unit of energy released from gasoline, diesel fuel, LPG, jet fuel, and residual oil for inland shipping. We also calculate the average CO_2 emissions from electricity generation in each country.

Figure 28 gives the results, from Scholl, Schipper & Kiang (96): Rising activity levels and modal shifts drove up CO_2 emissions from travel. Fuel shifts had a very small impact. Only in the United States did overall changes in fuel intensity restrain CO_2 emissions significantly.

The calculations present a challenge to those concerned about CO_2 emissions. Activity levels and changes in modal mix boosted emissions in most countries; their combined effect boosted emissions everywhere, moderated to a significant degree only in the United States, where the energy intensity of passenger travel fell significantly. The fuel mix itself became slightly more energy intensive, diesel fuel emitting slightly more CO_2 per unit of energy than gasoline. To the extent that diesel fueled vehicles are less fuel intensive than their gasoline counterparts, however, this effect was minimal. But if the era of high fuel prices still saw such a significant increase in fuel use and CO_2, it is unclear how well future efforts to restrain the latter will succeed.

Recent trends do not point to restraint in CO_2 emissions. Energy intensities of car use are constant in the United States and declining only very slowly in Europe, while activity is increasing. Similarly, the decline of the intensity of air travel has slowed, while traffic is booming. Not surprisingly, this has led to concern from national authorities (6, 97) as well as from the Intergovernment Panel on Climate Change (98, 99), and the International Energy Agency (1).

Transportation and CO2 Emissions in a Broader Context

The factorial analysis itself suggests "solutions" for reducing CO_2 emissions from travel, all of which offer significant potential: switch fuels (99–102), reduce fuel intensities (28, 104, 105), change to less energy- and CO_2-intensive modes, or restrain travel through demand measures or planning (79, 105, 107). In the IPCC review (99) it is clear that all of these changes together could significantly alter the trends in CO_2 emissions (and energy use for travel as well). But will they? Current trends say no.

Carbon dioxide is not the only externality arising from travel; indeed, it may be one of the least serious (3). Given the wide range of costs that may be associated with these problems, it seems clear that many are more serious and immediate (on a local level) than CO_2. But most of these problems are associated with vehicle activity as well as with fuel use per se, which is related to activity. In this sense, it is likely that strong measures dealing with congestion or noise, whether fiscal or regulatory, would lead to somewhat reduced travel. Fuel taxes to reflect the hidden costs of pollution should reduce travel somewhat as well as stimulate development and purchase of less fuel-intensive cars. And a tax on the carbon content of fuels reinforces the previous effects and stimulates development and deployment of fuels that lead to lower CO_2 emissions than gasoline or diesel use does today. Taken together, such a package of taxes and other stimuli could markedly reduce CO_2 emissions from travel, affecting all of the determinants of energy use for travel.

Although they are often times politically unpopular, taxes on externalities are considered the most "efficient" way to deal with these problems (107). Since most of the technologies that reduce the impact of externalities require investments, price signals contained in these taxes convey information about appropriate levels of investment. Equally important, environmental taxes on transportation externalities would tend to restrain travel somewhat. This is important, because we saw that growth in CO_2 emissions was related both to behavior—the amount of travel—and technology, the amount of emissions (or externality per unit of travel). While prices for fuels or for other aspects of travel are not the only influence on the level of travel (and modal choice), they are always a critical factor. There seems no question that pricing these kinds of externalities into the private cost of transportation would help define a balance among response from technology and from behavior to travel-related externalities.

Exactly how much CO_2 or other externalities should be taxed is of course of great debate, as is the resulting change in travel demand or modal mix that could be expected (9, 66, 112–114). The response will differ within the population according to life-cycle and income (LA Greening, LJ Schipper, G Bell, submitted for publication). If the taxes chosen were small in comparison to present costs, then even a perfect pricing system might lead to only small changes in travel behavior and travel technology. This is the case so far for the Nordic countries, where the focus has been primarily on fuels (110). However, the Swedish CO_2 commission has begun to look at the problems of transportation in a more holistic way (97).

Finally, it should be borne in mind that present taxation schemes, which were designed to raise revenue, do not reflect the externalities of transportation. The issue of taxation of the marginal externalities is important. Webster et al (17, 18) identified acquisition of a car as the single most important step in boosting mobility. Yet owning a car per se leads to few environmental problems: A gas guzzler sitting in the garage pollutes and congests less than an efficient car driven several hours per day. What some authorities have begun to realize, then, is that taxes should be shifted away from fixed costs (such as those based on the value or weight of cars) towards variable costs to better reflect the fact that most externalities arise from use of transportation. But the taxes imposed on use should reflect as closely as possible real marginal externalities.

THE DETERMINANTS OF AUTOMOBILE ENERGY USE: CHALLENGE FOR MODELERS

Several kinds of forces can be considered determinants of travel that interact with the fuel-use properties of automobiles (and other vehicles) to yield fuel consumption. These determinants are economic, socio-demographic, and geographic. The economic determinants of travel are incomes and the cost of travel. Laced in this web are socio-demographic links between family size and structure, the age of the population, work status, and gender. That is, controlling for incomes and prices, these factors play a dominant role in explaining differences in travel, as has been demonstrated by the analysis of the various national surveys of car use and travel. Finally, geographical elements also shape travel through both population density—which affects the speed and cost of travel (and owning a car)—and distances between destinations.

These determinants interact. As incomes rise and more househlds acquire first (and second) cars, the fraction of the population living in suburbs and driving to work appears to rise in all countries. Families with two or more workers have higher incomes and use cars more, not simply for getting to work but also for other purposes.

It is nevertheless difficult to formulate an exact statistical relationship based on national aggregate means for measures of the above determinants. This is because the variation among the countries in a few of the quantities, namely prices, incomes and national population density is far greater than variations in share of the population employed, household size, and composition. But studies based on individual household data (71, 82, 90) tend to reveal significant explanatory power of the socio-demographic determinants of travel.

Fuel use is determined by the product of travel (by mode) and the fuel intensity of each mode. As we have seen, fuel use for automobiles depends primarily on technological factors, but the actual patterns of car use (trip length, congestion, driving behavior) have an impact as well. Car load factor links car use (in vehicle-km) to travel (in passenger-km). Fuel intensity affects total car use by its impact on the cost of driving. Conversely, driving distances affect the choice of fuel efficiency somewhat, since longer driving distances justify greater investments in fuel economy. For other modes, both technology and utilization, primarily as load factors, determine fuel use, but only for air travel is fuel use itself a significant portion of total operating costs.

The above summary overlooks an important determinant of car fuel use, however. Car size, power, and features exert a strong influence over car fuel economy. In the short term, car buyers can save fuel by choosing smaller, less powerful, or less equipped cars, something that occurred in the United States and Europe during brief periods of very high fuel prices (1974–1975 and 1979–1981). Over the longer term, technology can reduce fuel use markedly, but consumer preferences can erase these gains in favor of size, power, and features, which is clearly what happened to the mix of new cars bought in both Europe and Japan since 1973 and, since the late 1980s, in the United States as well. The debate over the costs and benefits of additional technological measures that could reduce fuel intensity (105) leaves many analysts puzzled: If technologies to save fuel are cost effective, why are manufacturers not employing them or consumers not demanding them? One reason may be that consumers derive greater surplus from investing extra money in car features than in spending the same funds to reduce fuel costs.

Thus there is a chain of uncertainties that challenge the energy modeler. First she must estimate the demand for travel by mode, then estimate the automobile share, then estimate fuel intensity, then total fuel use. Strictly technological models of fuel intensity ignore consumer tastes in car characteristics, while strictly econometric models of future fuel use based on past experience ignore both changes in the determinants of travel and real—or imagined—opportunities for saving fuel. Neither approach can measure the impact of changing policies.

Consider first travel. The key drivers of energy demand for transport are associated not simply with the costs of energy for transport (and incomes), but with geographical, demographic, and indeed lifestyle variables as well. While fuel prices have been stable in recent years, these other variables continue to change, and may be the most important sources of changes in mobility and automobile use: 1. the aging of the population and the changing shape of the labor market; 2. continued spreading of the population to the suburbs, as well as moves in the locations of jobs, services, and leisure activities; and 3. imposition of a variety of taxes, controls, or other policy tools designed to relieve some of the problems of transport not related to fuel use per se.

Some policy instruments, like a carbon tax on fuel, can be modeled (116), but more-radical changes, such as the mandates in California for zero-emission vehicles or the appearance of new low-pollution technologies cannot because there is so little historical experience with them. Sperling (28; see also 117) makes a strong case for alternative fuels, but it is hard to know how the ultimate fixed and variable costs associated with using these fuels will affect travel and fuel use. Similarly, policy changes that might stimulate alternative fuels are uncertain. For example, the Honda Motor Co. (118) announced a motor that would meet California's "Ultra Low Emission Vehicle" requirements. Or advanced diesel engines may be available that significantly reduce both fuel use and various emissions. If successful, these technologies may reduce the perceived importance of requiring electric drive or other fuels (besides gasoline or diesel) to combat air pollution or reduce CO_2 emissions. How can we model the impact of such a technological change and policy changes that might follow? And if a region (or country) agreed to internalize the costs of using cars within the prices of fuel and car use, there could be significant relocation of homes, workplaces, and other destinations over long periods of time.

A second theme in this paper is our argument for integrating the CO_2 problem into a broader menu of transportation problems. Our present modeling is often limited to either engineering estimates of how particular technologies should change car characteristics, car fuel use, and car emissions (110) or to economic estimates of how fuel price changes would change fuel use and emissions (116). But we need a new approach that broadens the variables used in energy modeling to those that reflect policy changes not directly associated with fuels, such as charging for using roads or entering congested areas, changes in parking policies or company car taxation rules. Hensher (120, 121) and his colleagues are trying to develop a model of future CO_2 emissions from cars that include traditional (i.e. economic) variables, technological variables, lifestyle variables, locational variables (using surveys of urban areas in Australia) and even the housing market itself. The effort is very data intensive, but could yield interesting results.

But in the end the problem is not about cars, but about people. As economist William Finger, formerly of Exxon USA, reminds us, "Cars do not drive cars; people drive cars." So it is that the "determinants" of energy use for travel demand an understanding of why people move around so much.

PRINCIPAL SOURCES OF COUNTRY DATA AND NATIONAL TRAVEL SURVEYS

In the course of this work we have relied regularly on certain sources for the data on transportation activity and energy use shown in Table 1. These sources are listed here by country.

Denmark

Trafikministeriet (Danish Ministry of Transport). 1990. *Transportstatistik 1984–1988, (Transport statistics 1984–1988).* Copenhagen, Denmark: Trafikministeriet. Now Published Yearly

Veidirektorat. 1994. *Transport statistics.* Copenhagen: Ministry of Transport

Trafik-og Kommunikationsministeriet (Danish Ministry of Transport and Communications). 1988. *Persontrafik i 1975, 1981 og 1986 (Personal travel in 1975, 1981, and 1986)* Copenhagen, Denmark: Trafik-og Kommunikationsministeriet

Tofte and Joergensen. 1992. *Befolknings Rejsevaner (The Travel Habits of the Population).* Copenhagen: Trafikministeriet

Trafikministeriet. 1993. *Transportstatistik, 1980–1991 (Transportation statistics 1980–1991).* Copenhagen: Trafikministeriet

Automobil-importorernes Sammenslutning (VIS). 1994. *Vejtransporten i tal og tekst (Road transportation statistics).* Hellerup: VIS. Editions from 1975 onward

Vejdirektoratet, 1994. *Tal om Vejtrafik (Data on road traffic).* Copenhagen: Veijdirektorat Sektorplanafdelingen

Finland

Bureau of Statistics (TK). *Annual Abstract of Transportation Statistics.* Helsinki: Central Bureau Statistics

France

INSEE and OEST (Institut National de la Statistique et des Etudes Economiques and Observatoire Economique et Statistique des Transports). 1987–

1994. *Les Comptes des Transports (Transport accounts)* Paris, France: INSEE. (Published Yearly)

Ministry of Industry. 1975–1994. *Tableaux de Consommation d'Energie en France (Tables of Energy Consumption in France)*. Paris: Ministry of Industry

Germany

Deutsches Institut fuer Wirtschaft (DIW) 1972–1994. *Verkehr in Zahlen 1994. (Traffic in Figures)*. Bonn, Germany: Bundesminister fuer Verkehr

Vergleichende Auswertungen von Haushaltsbefragungen zum Personennahverkehr (KONTIV 1976, 1982, 1989). Berlin, West Germany: Deutsches Institut fuer Wirtschaftsforschung (DIW). Original is Emnid-Institut GMBH & Co. 1990. KONTIV 1989. (Four Volumes.) Bielefeld, West Germany

Kloas J. and Kuhfeld H. 1987. *Verkehrsverhalten im Vergleich (Comparisons of traffic conditions)*. (KONTIV) DIW (Deutsches Institut fuer Wirtschaftsforschung). Berlin, West Germany: Duncker and Humboldt

Italy

Ministero Dei Trasporti, Direzione Generale Programmazione, Organizzazione e Coordinamento. 1993. *Conto nationale dei Trasporti (National Traffic Statistics)*. Rome: Istituto Poligrafico e Zecca dello Stato. (1989, 1991 and 1992 editions)

Unpublished calculations of AGIP, the State Oil Company. (Private communication)

Japan

Institute of Energy Economics (Japan). 1992 (undated). *Energy Data and Demand of Transportation Sector in Japan*. Tokyo, Japan: Institute of Energy Economics

Institute of Energy Economics (Japan). 1989. *Study on Transportation* (unpublished), Tokyo, Japan

Institute of Energy Economics Energy Data Modeling Center. *Annual Energy Statistics*

Ministry of Transportation (Japan). 1991. *Land Transportation Statistics*, Tokyo, Japan: Nihon Jidoshakaigisho

Netherlands

Centraal bureau voor de statistiek. 1991. De mobiliteit van de nederlandse bevolking 1990. (Mobility of the Dutch population in 1990.) The Netherlands: Voorburg/Heerlen

Norway

Central Bureau of Statistics (SSB), 1970–1994. *Samferdsel Statistikk (Transport Statistics)* Kongsviner: SSB

Norsk Esso (Private communication)

Transport Oekeonomisk Institutt, various years. *Transportarbeid paa Norsk Omraade (Transport in Norway)*. Oslo: Transport Economic Institute

Transport Oekeonomisk Institutt. 1993. Norsk reisevaner. *Dokumentasjonsrapport for den Landsomfattande Reisevaneundersoekelsen 1991-2 (National Survey of Travel Habits 1991-2)*. Report 183. Oslo: Transport Economic Institute

Sweden

A large number of smaller official and unofficial publications reviewed in Appendix 3 of Schipper LJ and Johnson F, with Howarth R, Andersson BE, Anderson BG, and Price LK. 1993. *Energy Use in Sweden: An International Perspective*. Lawrence Berkeley Laboratory Report LBL-33819. Berkeley, CA: Lawrence Berkeley Laboratory. Published as Schipper and Price 1994 in Nat. Res. Forum (May)

National Central Bureau of Statistics (Sweden). 1985 Resavanorundersoekning. *Statistiska Meddelanden (1985 Survey of Travel Habits)*. Stockholm, Sweden: Statistics Sweden

See also Vilhelmson B. 1990. *Vaar dagliga roerlighet (Our Daily Mobility)*. Stockholm, Sweden: Transportforskningsberedning, TFB Rapport 1990:16

United Kingdom

Department of Transport (DOT). 1970–1994. *Transport Statistics: Great Britain*. London, UK: Her Majesty's Stationery Office

Transport Department, various years. National Travel Survey. (1972/3, 1982/3, 1985/6, 1990/91) London, UK: Her Majesty's Stationery Office

United States

Davis SC. 1994. *Transportation Energy Data Book: Edition 14*. Oak Ridge, TN: Oak Ridge National Laboratory, ORNL-6710 (Edition 14 of ORNL-5198)

US Department of Transportation. 1992. *US Nationwide Personal Transportation Survey*. Washington, DC: US Dept. of Transportation

Klinger D, Kuzmyak R. 1986. *Personal Travel in the US: Nationwide Personal Transportation Study (NPTS)*. Washington, DC: US Department of Transportation, Federal Highway Administration. See also the 1990 Edition, released in preliminary form by the Federal Highway Administration in 1993 and 1994

US Department of Transportation. 1992. *US Nationwide Personal Transportation Survey*. Washington, DC: US Dept. of Transportation

ACKNOWLEDGMENTS

Work sponsored by the US Department of Energy, the Department of Transportation, the Environmental Protection Agency, the Swedish Transportation Research Board (now Communication Research Board), the Energy Foundation, Volvo AB, General Motors, Nissan North America, Shell Oil, Exxon USA, Conoco, and TOTAL SA. Other principal participants include Prof. Elizabeth Deakin of the Institute for Urban and Regional Development at UC Berkeley, and Dan Sperling of the Institute for Transportation Studies at UC Davis. Additional participants have included: Ruth Steiner, Lynn Scholl, Nancy Kiang, Roger Gorham, Wienke Tax, Nancy Kiang, and Maria Josefina Figueroa, all from UC Berkeley; Molly Espey from the UC Davis, Gunnar Eriksson from NORDPLAN, Stockholm; and Olof Johansson from the University of Gothenburg. Rob Socolow and Dennis Anderson provided helpful comments.

Opinions expressed herein are strictly those of the author, who gratefully acknowledges the hospitality of the Industry and Energy Department of the World Bank, Washington DC, where he was a Visiting Fellow during the writing of this report. The author acknowledges the help of Charles Campbell (LBL), Paul Wolman, and Mary Nash (World Bank), and Ted Gartner (LBL) in the preparation of this manuscript.

Literature Cited

1. International Energy Agency (IEA). 1994. *Energy Balances of OECD Countries 1980-92*. Organ. Econ. Country Dev., France, OECD/IEA. 449 pp.
2. Denmark Ministry of Transport. 1990. *Handlingsplan for Transport og Miljoe (Plan for Transport and the Environment)*. Copenhagen: Minist. Transp.
3. Barde JP, Button K. 1990. *Transport Policy and the Environment, Six Case Studies*. London, UK: Earthscan Publ.
4. International Energy Agency (IEA). 1991. *Energy Efficiency and the Environment*. Paris, France: IEA
5. Johnson E. 1993. *Avoiding the Collision of Cities and Cars: Urban Transportation Policy for the Twenty-First Century*. Chicago, IL: Am. Acad. Arts Sci. Aspen Inst.
6. Houghton J. 1994. *Royal Commission on Environmental Pollution: 18th Report, Transport and the Environment*. Sir John Houghton, Chairman. London: UK Dep. Environ.
7. World Bank. 1995. *Transport Sector Review*. Washington, DC: The World Bank
8. Deluchi M. 1995. *Original estimates of social costs of motor vehicle use*. Presented at Transp. Res. Board. Washington, DC
9. Kaageson P. 1993. *Getting the Prices Right: A European Scheme for Making Transport Pay its True Costs*. Stockholm: Eur. Fed. Transp. Environ.

10. Small KA, Kazimi C. 1995. On the costs of air pollution from motor vehicles. *J. Transp. Econ. Policy*
11a. Schipper LJ, Howarth R, Andersson B, Price LK. 1993. Energy use and efficiency in Denmark in an international perspective. *Nat. Res. Forum.* May:83–103
11b. Schipper LJ, Johnson F, with Howarth R, Andersson BE, Anderson BG, Price LK. 1993c. Energy use in Sweden: an international perspective. *Rep. LBL-33819,* Lawrence Berkeley Lab., Berkeley, CA
12. Kiang N, Schipper LJ. 1995. Energy use from transportation in Japan. *Transp. Policy.* In press
13. Schipper LJ, Figueroa MJ, Price LK, Espey L. 1993. Mind the gap: The vicious circle of measuring automobile fuel use. *Energy Policy* 21:1173–90
14. Schipper LJ, Steiner L, Figueroa MJ, Dolan K. 1993. Fuel prices and economy: Factors affecting land travel. *Transp. Policy* 1(1):6–20
15. Jansson JO, Cardebring P. 1986. *Avstaellda bilar och bilstatistiken. (Unregistered Cars and Car Statistics).* Linkoeping, Sweden: Vaeg-och Trafik-Inst., VTI. Rapport 445
16. Jansson JO, Cardebring P, Junghard O. 1986. Personbilsinnehavet i Sverige: 1950-2010. (Car ownership in Sweden: 1950-2010). *Linkoeping, Sweden: Vaeg-och Trafik-Inst. VTI. Rapport 301*
17. Webster FV, Bly PH, Johnson RH, Dasgupta M. 1986. Part 1: Urbanization, household travel, and car ownership. *Transp. Rev.* 6(1):49–86
18. Webster FV, Bly PH, Johnson RH, Dasgupta M. 1986. Part 2: Public transport and future patterns of travel. *Transp. Rev.* 6(2):129–72
19. Greene DL, Loebel A. (TERA, Inc.). 1979. Vehicle Miles of Travel Statistics. *Oak Ridge, TN: Oak Ridge Natl. Lab., ORNL/TM 6327*
20. Wall R. 1990. Bilanvaendningens bestaemningsfaktorer (Car usage determinants). Linkoeping, Sweden: Vaeg-och Trafik-Inst. VTI Meddelande 648
21. Schipper LJ, Steiner R, Meyers SP. 1993. Trends in transportation energy use, 1970-1988: An international perspective. *In Transportation and Global Climate Change,* ed. DL Greene, DJ Santini, pp. 51–89. Washington, DC: Am. Counc. Energy-Effic. Econ.
22. Deutsches Institut fuer Wirtschaftsforschung (DIW). (German Institute for Economic Research). 1987. *Gesamtfahrleistungen und Kraftstoffverbrauch im Strassenverkehr weiter deutlich gestiegen [Total road traffic and fuel use has increased significantly].* Berlin: DIW Wochenbericht 44/87
23. Wheaton WC. 1982. The long-run structure of transportation and gasoline demand. *Bell J. Econ.* 13(2):439–54
24. Pindyck RS. 1979. *The Structure of World Energy Demand.* Cambridge, MA: MIT Press
25. Griffin JM. 1979. *Energy Conservation in the OECD: 1980 to 2000.* Cambridge, MA: Ballinger
26. Kouris GJ. 1978. Price sensitivity of petrol consumption and some policy implications. *Energy Policy* 6(3):209–16
27. Sorrell S. 1992. Fuel efficiency in the UK vehicle stock. *Energy Policy* 20(8): 766–80
28. Martin DJ, Shock RAW. 1989. *Energy Use and Energy Efficiency in UK Transport Up to the Year 2010.* London: HMSO
29. Konsumentverket (KOV). (National Swedish Board of Consumer Affairs). 1989 (and previous years). *Bilunderhaall (Car maintenance).* Vaellingby, Sweden: KOV. Rapport 1988/89:11
30. Bosseboeuf D. 1988. *Les Consommations Unitaires des Voitures Particulieres [Unit Fuel Consumption in Private Automobiles].* Paris, France: Agence Francaise pour la Maitrise de L'Energie
31. Aastangen K, Oestmoe K, Aa B. 1987. *Trafikkarbeide, Kjoerlengder og Bensinforbruk for den Norske Personbilparken i Perioden 1975–1986, [Vehicle miles traveled, driving distances and gasoline use for the Norwegian automobile fleet during the period 1975–1986],* Oslo, Norway: Transportoekonomisk Inst. E-723 TOeI
32. Statistics Canada. 1989. *Fuel Consumption Survey, 1979–1988.* Ottawa: Stat. Can.
33. Energy Information Administration (EIA). 1987. *Residential Transportation Energy Consumption Survey.* Washington, DC: US Dep. Energy, Energy Inf. Admin. DOE/EIA-0464(85)
34. Federal Highway Administration (FHWA). 1993 (and previous years). *Highway Statistics 1992.* Washington, DC: US Dep. Transp., Fed. Highw. Admin., FHWA-PL-93-023
35. Department of Transport (DOT). 1970–1994. *Transport Statistics: Great Britain.* London: HMSO
36. Deutsches Institut fuer Wirtschaftsforschung (DIW). 1994 (and previous years from 1972). *Verkehr in Zahlen 1993*

[Traffic data for 1993]. Bonn, Germany: Bundesminist. Verk.

37. Schipper LJ, Tax W. 1994. New car test and actual fuel economy: Yet another gap? *Transp. Policy* 1(4):257–265
38. Strategy, Organization, and Method, Inc. (SOM). 1988. *Differences Between Laboratory-Tested and On-the-Road Fuel Consumption Rates*. Montreal: Rep. Transp. Can., Road Safety, April. MP-18–86
39. Mintz MM, Vyas ARD, Conley L. 1993. *Differences between EPA-test and in-use fuel economy: Are the correction factors correct?* Presented at Annu. Meet. Transp. Res. Board, 72nd, Washington, DC
40. Westbrook F, Patterson P. 1989. *Changing driving patterns and their effect on fuel economy*. Presented at 1989 SAE Gov./Ind. Meet., May 2, Washington, DC
41. Maples J. 1993. *The Light Duty Vehicle MPG Gap: Its Size Today and Potential Impact in the Future*. Draft Rep. Knoxville: Univ. Tenn., Transp. Cent.
42. Watson RL. 1989. *Car Fuel Consumption: Its Relationship to Official List Consumptions*. Crowthorne, Berkshire, UK: Transp. Road Res. Lab. Dig. Res., Rep. 155
43. Dahl CA. 1986. Gasoline demand survey. *Energy J.* 7(1)67–82
44. Sterner T. 1990. *The Pricing of and Demand for Gasoline*. Stockholm: Swedish Transp. Res. Board, TFB-Rapport 1990:9
45. Dahl C, Sterner T. 1991. Analyzing gasoline demand elasticities: A survey. *Energy Econ.* 13(3)203–10
46. International Civil Aviation Organization. 1993. *The World of Civil Aviation 1992–1995*. Circ. 244-AT/96. Montreal: ICAO
47. Schipper L, Myers S, Howarth R, Steiner R. 1992. Trends in transportation energy use 1970–1988. In *Energy Efficiency and Human Activity: Past Trends, Future Prospects*, Chapter 5. Cambridge: Cambridge Univ. Press
48. Boeing Commercial Airplane Group. 1991 (and successive years) *Current Market Outlook: World Market Demand and Airplane Supply Requirements*. Seattle, WA: Boeing Corp.
49. Schipper LJ, Steiner R, Duerr P, Feng A, Stroem S. 1992. Energy use in passenger transport in OECD countries: Changes since 1970. *Transportation* 19: 25–42
50. Danielis R. 1994. *PTRC Conf. Transport and Energy in Italy: Changes in the Period 1975–1991, Warwick, England*, pp. 257–68
51. Greene D, Fan Y. 1994. *Transportation Energy Efficiency Trends 1972–1992*. ORNL 6828. Oak Ridge, TN: Oak Ridge Natl. Lab.
52. Greene DL. 1992. Vehicle use and fuel economy: How big is the "rebound" effect? *Energy J.* 13(1):117
53. Gately DL. 1990. The U.S. demand for highway travel and motor fuel. *Energy J.* 11(3):59–73
54. Deleted in proof
55. Goodwin P. 1992. A review of new demand elasticities with special reference to short and long run effects of price changes. *J. Transp. Econ. Policy* 26(2):155–69
56. Gately D. 1992. Imperfect price reversibility of oil demand. Asymmetric responses of U.S. gasoline consumption to price increases and declines. *Energy J.* 13(4):179–207
57. Greene DL. 1990. CAFE or price?: An analysis of the effects of federal fuel economy regulations and gasoline price on new car MPG, 1978–89. *Energy J.* 11(3):37–57
58. Schipper LJ, Eriksson G. 1995. Taxation policies affecting automobile characteristics and use in Western Europe, Japan, and the United States. *Proc. Transp. Energy: A Conf. Exploring Energy Strategies Sustainable Transp. System. Workshop Inst. Transp. Studies, UC Davis, Asilomar, CA, Aug. 1993, pp. 217–242*
59. Jansson JO, Wall R. 1994. *Bensinskattefoeraendringarseffekter (Effects of Changes in Gasoline Taxes)*. Ds 1994:55. Stockholm: Minist. Finance
60. Schol E, Smokers R. 1993. Energy use of company cars. In *The Energy Efficiency Challenge for Europe*, ed. H Wilhite, R Ling. 1993 Summer Study. Oslo, Norway: Eur. Counc. Energy Effic. Econ.
61. National Economic Development Council (NEDC). 1991. *Company Cars: An International Perspective*. Road Res. Lab. Res. Rep. 61 TRRL. London: Natl. Econ. Dev. Counc. Traffic Manage. Syst. Work. Party
62. Fergeson M. 1990. *Subsidizes Pollution: Company Cars and the Greenhouse Effect*. London: Earth Resour. Res.
63. Deleted in proof
64. Schipper LJ, Price L. 1994. Efficient energy use and well being: The Swedish example after 20 years. *Nat. Res. Forum* 18(2)125–42
65. Transportoekonomisk Institutt (Norway). 1991. *Alternative Avgiftssyste-*

mers Effekt Paa Bilhold, Bilutskifting og Utslipp, [The effect of alternate policies on car ownership, stock turnover, and emissions], Oslo, Norway: Transportoekonomisk Inst. TOeRapport 00963/1991
66. COWI Consult. 1993. *Internalisation and the External Costs of Transport*. Lyngby, Denmark: Danish Minist. Energy, EFP-9
67. Jones PM, Dix MC, Clarke MI, Heggie IG. 1983. *Understanding Travel Behavior*. Oxford, UK: Oxford Stud. Transp.
68. Salomon I, Bovy P, Orfeuil JP, *eds. 1993. A Billion Trips Per Day*. Amsterdam: Kluwer Sci.
69. Kloas J, Kuhfeld H. 1987. *Verkehrsverhalten im Vergleich (KONTIV). (Comparative travel behavior)*. DIW. Berlin: Duncker & Humbolt
70. Christiansen L, Jensen D. 1994. *Preliminary Analysis of Transportation Surveys. PETRA Work. Pap. 2*. Copenhagen: Danmarks Miljoe Undersoegese (Natl. Environ. Res. Inst.)
71. Vilhelmson B. 1990. *Vaar dagliga roerlighet. (Our Daily Mobility)* Transporforskningsberedningen, TFB-Rapport 1990:16. Stockholm, Sweden: Allmaenna Forlaget
72. Wood C, Banister D, Banister C. 1994. *The Relationship between Energy Use in Transport and Urban Form*. London: Univ. College
73. Hu PS, Young J. 1992. *Nationwide Personal Transportation Survey (NPTS): Summary of Travel Trends*. Washington, DC: US Dep. Transp., Fed. Highw. Admin., FHWA-PL-92-018
74. Naess P. 1993. *Hvor bor de som Kjoere Mest? (Where Do Those Who Drive the Most Live?)* Oslo: Norsk Inst. by-og regionforskning
75. Naess P, Larsen SL, Roee PG. 1994. *Energibruk til transport I 22 nordiske byer (Energy Use for Transport in 12 Nordic Cities)*. Oslo, Norway: Norsk Inst. By-og regionforskning
76. Banister D. 1992. Energy use, transport, and settlement patterns. In *Sustainable Development and Urban Form*, ed. MJ Breheny, pp. 160–81. London: Pion
77. Owens S. 1991. *Energy Conscious Planning*. London: Counc. Prot. Rural England
78. Storstads Trafik Kommittee (STORK). 1989. *STORK Utredning (Report of the Commission on Traffic in Large Cities)*. SOU 1989:15. Stockholm: Liber Foerlag
79. Ecotec. 1993. *Reducing Transport Emissions through Planning*. UK Dep. Transp. London: HMSO
80. Energy Information Administration (EIA). 1993. *Household Vehicles Energy Consumption 1991*. Rep. DOE/EIA-0464(91). Washington, DC: US Dep. Energy
81. Department of Transportation. 1992. *U.S. Nationwide Personal Transportation Survey*. Washington, DC: US Dep. Transp.
82. Dunphy R, Fischer K. 1994. *Transportation, congestion, and density. New insights*. Presented at Transp. Res. Board, Washington, DC
83. Deleted in proof
84. Banister D, Watson S, Wood C. 1994. *The Relationship between Energy Use in Transport and Urban Form. Work. Pap. 10*. London: Plan. Dev. Res. Cent., Univ. College
85. Toft E, Joergensen J. 1994. *Befollkningens rejsevaner (RVU) 1992 (Peoples Travel Habits 1992)*. Copenhagen: Minist. Transp. See Ref. 70 for English interpretation
86. Tomth JE. 1992. *Saa Resor Vi I Stockholms Laen (Here Is How We Travel in Stockholm County)*. Stockholm, Sweden: Stockholms Laens Landsting, Regionplane och Trafikkontoret
87. Deleted in proof
88. Deleted in proof
89. Greening LA, Jeng HT. 1994. Life-cycle analysis of gasoline expenditure patterns. *Energy Econ.* 16(3):217–28
90. Solomon I, Ben-Akivajm. *1982. Lifestyle segmentation in travel-demand analysis. Transp. Res. Rec.* 879:37–45
91. Zimmerman CA. 1982. The lifecycle concept as a tool for travel research. *Transportation* 11:51–69
92. Charns H, Nelson B, Pitta N. 1991. *1990 Bay Area Travel Survey Final Report*. Prepared for Metrop. Transp. Comm., Oakland CA
93. Purvis C. 1994. *San Francisco Bay Area 1990 Regional Travel Characteristics*. Work. Pap. 4. Oakland, CA: Metrop. Transp. Comm.
94a. Office of Technology Assessment (OTA). 1994. *Saving Energy In U.S. Transportation*. OTA:ETI-589. Washington, DC: US Gov. Print. Off.
94b. Schipper L, Bartlett S, Hawk D, Vine E. 1989. Linking life-styles and energy use: A matter of time? *Annu. Rev. Energy* 14:273–320
95. Lovins A, Barnett JW, Lovins LH. 1993. Supercars: The coming light-vehicle revolution. In *The Energy Efficiency Challenge for Europe. Proc. 1993 ECEEE Summer Study, June 2–5, I:349. Rungstedgaard, Denmark: The Eur. Counc. Energy-Effic. Econ.*

96. Scholl P, Schipper LJ, Kiang N. 1995. CO_2 emissions from passenger transport: A comparison of international trends from 1973–1990. *Energy Policy.* In press
97. Trafik och CO2 Delegation. (Traffic and CO2 Delegation (TOK). 1994. *Trafiken och koldioxiden: Principer foer att minska trafikens kodioxidutslaepp (Traffic and carbon dioxide: principles for reducing emissions of carbon dioxide from transport).* SOU 1994:91. Stockholm: Dep. Environ.
98. Intergovernmental Panel on Climate Change (IPCC). 1990. *Climate Change: The Scientific Assessment 1990,* ed. JT Houghton, GJ Jenkins, JJ Ephraums. New York: Cambridge Univ. Press. 341 pp.
99. Michaelis L, et al. 1995. *Report of the Working Group II, Mitigation of CO_2. Emissions from Transportation.* London/Washington: Intergovernmental Panel on Climate Change. (Draft). In press
100. Sperling D. 1988. *New Transportation Fuels: A Strategic Approach to Technological Change.* Berkeley: Univ. Calif. Press
101. Sperling D. 1994. *Future Drive: Electric Vehicles and Sustainable Transportation.* Washington, DC: Island Press
102. Sperling D, DeLuchi MA. 1989. Transportation energy futures. *Annu. Rev. Energy* 14:375–424
103. Deleted in proof
104. National Research Council (NRC). 1993. *Automotive Fuel Economy, How Far Should We Go?* Washington, DC: Natl. Acad. Press. 259 pp.
105. Pischinger R, Hausberger S. 1993. *Measures to Reduce Greenhouse Gas Emissions in the Transport Sector. Study for the IPCC.* Graz: Inst. Intern. Combust. Engines Thermodyn., Graz Univ. Technol.
106. Deleted in proof
107. Steiner R. 1992. *Least-cost planning for transportation? What we can learn about transportation demand management from utility demand-side management.* Presented at Annu. Conf. Transp. Res. Board, Washington, DC. Abridged version in *Transp. Res. Rec.* 1346:14–17
108. National Research Council (NRC). 1994. *Curbing Gridlock: Peak Period Fees to Relieve Traffic Congestion. Special Rep. 242.* (2 Vols.) Washington, DC: Natl. Acad. Press
109. Baumol W, Oates W. 1988. *The Theory of Environmental Policy.* New York: Cambridge Univ. Press
110. Mangusson J, Brandel M. 1991. *Energi OchMiljoe i Norden (Energy and Environment in the Nordic Countries),* NORD 19991:23. Copenhagen: The Nordic Counc.
111. Deleted in proof
112. MacKenzie J, Dower R, Chen DDT. 1992. *The Going Rate: What It Really Costs to Drive.* Washington, DC: World Resourc. Inst.
113. Roelofs C, Komanoff C. 1994. *Subsidies for Traffic: How Taxpayer Dollars Underwrite Driving in New York State.* New York: Komanoff Energy Assoc.
114. California Energy Commission (CEC). 1994. *California Transportation Energy Analysis Report.* P300–94–002. Sacramento, CA: CEC
115. Deleted in proof
116. Sterner T, Dahl C, Franzen M. 1992. Gasoline tax policy: Carbon emissions and the global environment. *J. Transp. Econ. Policy* 26(2)
117. Sperling D, DeLuchi M. 1993. Alternative transportation energy. In *The Environment of Oil, Studies in Industrial Organization,* ed. RJ Gilbert, Chapter 4, pp. 85–141. Boston/Dordrecht/London: Kluwer Acad.
118. Honda Motor Co. 1995. *Honda First to Have Gasoline Engine Verified at ULEV Exhaust Technologies.* (press release) Torrance, CA: Honda Motor Co.
119. Deleted in proof
120. Hensher D. 1995. Opportunities to reduce greenhouse gas emissions in the urban passenger transport sector. *Econ. Pap.* In press
121. Henscher D. 1995. *An intergrated approach to modelling the impact on urban travel behavior of strategies to reduce enhanced greenhouse gas emissions.* Prepared for the 7th World Conf. Transp. Res., Sydney, Australia, Inst. Transp. Stud., Univ. Sydney

Annu. Rev. Energy Environ. 1995. 20:387–424

HISTORY OF, AND RECENT PROGRESS IN, WIND-ENERGY UTILIZATION

Bent Sørensen
Roskilde University, Institute 2, PO Box 260, DK-4000 Roskilde, Denmark

KEYWORDS: power systems, wind turbine history, environmental-impact assessment, life-cycle analysis, wind turbine technology, wind resources

CONTENTS

ABSTRACT

This review presents the current status of wind turbine technology and recent advances in understanding the long history of wind energy. Reasons for the convergence of technological solutions towards a horizontal axis concept with two or three blades are discussed, and the advances in materials science are identified as determinants of the change toward increasing optimum turbine size. The modest environmental impacts of wind turbines are illustrated by recent life-cycle analyses, and the economic incentive structure and power buy-back rates in different countries are invoked to explain the variation in wind technology penetration in countries with similar resource potentials. Finally, the possible future role of wind technology is discussed, based on resource estimates, competing land demands, government commitments and technological trends, including the recent offshore wind farm developments.

Status 1995

The new interest in wind energy over the past 20 years has led to successful development of commercial wind turbines of increasing unit size. The present

1056-3466/95/1022-0387$05.00

Table 1 Economy of current wind technology

Capital costs (per unit of rated power):[a]	
Turbine capital cost installed	6000 DKr/kW (about 800 ecu or 1000 US$)
Typical foundation costs	300 DKr/kW (about 40 ecu or 50 US$)
Typical grid connection costs	100 DKr/kW (about 33 ecu or 17 US$)
Running costs (totalling 2% of capital cost annually):[b]	
Insurance costs	40 DKr/kW/y (about 5 ecu or 7 US$)
Service contract	25 DKr/kW/y (about 3 ecu or 4 US$)
Other O&M costs, etc.	55 DKr/kW/y (about 7 ecu or 9 US$)
Assumed capital charge[c] 10%/y	
Assumed capacity factor[d] 35%	
Implied cost of power	0.25 DKr/kWh (about 0.034 ecu or 0.045 US$)

[a] Wind turbine price list, Danish Technological Institute, October 1994.
[b] Source: (76).
[c] Typical Danish 1994 cost of capital for an investor's loan financing with a 20-year back-payment period.
[d] Based on [a]: 3066 kWh/y per kW (rated power) or 146 W/m^2 swept, corresponding to a prime location in Denmark (and equal to the actual cost of the best placed turbines; S. Frandsen, background paper to the IPCC working group IIa).

generation ranges from 500 kW–1 MW. These wind turbines are a technical success in that they have a long working life, allowing long-term maintenance contracts and commercial insurance to be issued at favorable terms. They are an economic success in that new wind turbines at windy locations (such as the northwestern coastal regions of Europe) are competitive with the consumer cost of modern coal-power generation, which in several European countries includes substantial energy and environmental taxes. However, as discussed in the environmental-impact section below, these coal-power costs probably do not fully include the externalities associated with carbon-dioxide emissions. An example of current cost calculations, based on a 500-kW Vestas turbine, is given in Table 1. The kWh cost is substantially lower than it was 15–20 years ago, in real terms, as indicated in Figure 1a. Note that the capital cost scale on the left side is the fundamental one, whereas the kWh cost scale on the right side is strictly valid only for 1995 because of the 10% capital cost used across the board. In 1980, the actual cost of capital was considerably higher, implying that the kWh cost would also be higher.

The sensitivity analysis in Figure 1b indicates that the most important factors influencing the cost of wind power are wind resources and cost of capital. Wind resources are discussed in a later section. The capacity factor (average produced power over rated power) used in Table 1 corresponds to a good coastal site in Denmark, which is also one of the best European sites, although not as good as the western coasts of Scotland or Ireland. In the interior of Denmark, and in many locations worldwide, a typical capacity factor is 20–25% and the cost of power correspondingly higher. On the other hand, further cost reductions are expected for turbines produced over the next decade.

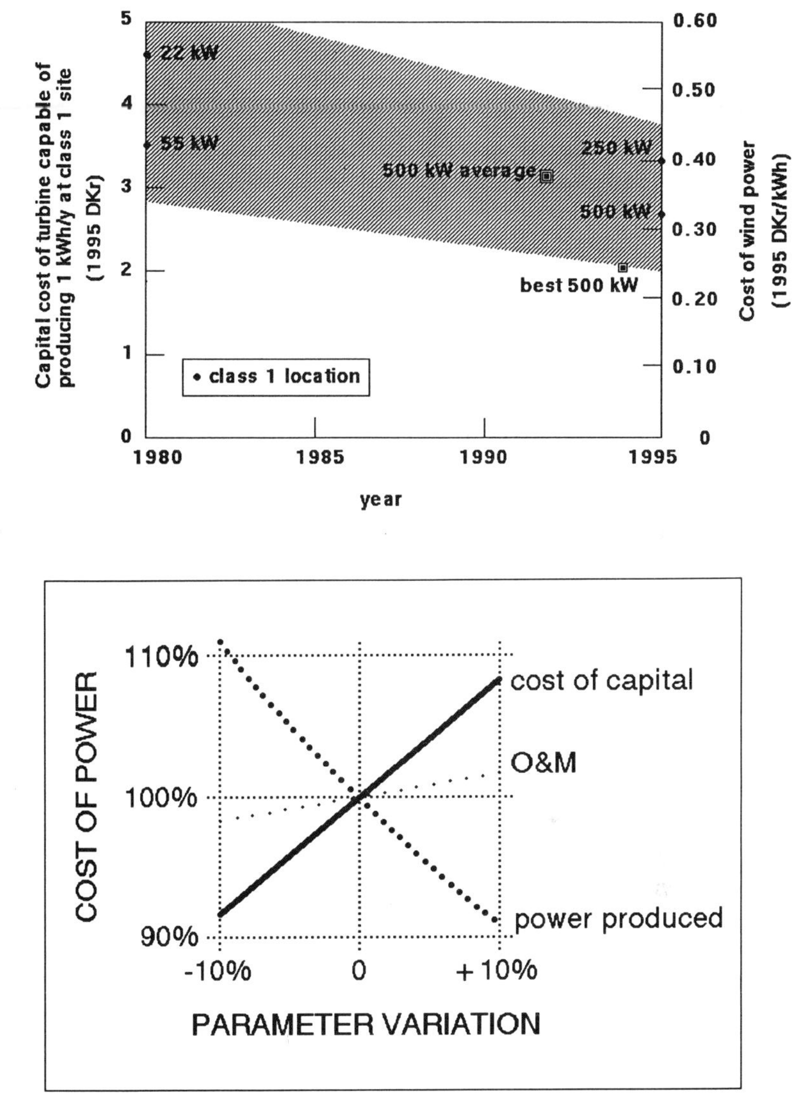

Figure 1 (*a*) Development in cost of wind power. The scale on the left side gives capital cost (including foundation and grid connection, and adjusted to 1995 prices by use of gross national inflation indices) for various turbines sold in Denmark, scaled down to an annual production of one kWh at a roughness class 1 site (best Danish inland location, e.g. extended plain with little vegetation or buildings). The scale on the right side gives an indication of power cost, using the values for turbine life and the 1995 costs of capital and operation and maintenance (O&M) given in Table 1. The early 1995 exchange rate was 5.5 DKr (Danish Kroner) = 1 US$. Also indicated are the 1992 average for 500 kW turbines at various sites and the power cost for the best-located 500-kW turbine in 1994 [based on official Danish wind turbine price lists and (48), (71), and (72)]. (*b*) Sensitivity analysis indicating the dependence of the cost of wind power on the most uncertain parameters.

The other important factor, the cost of capital, is determined in part by the general economic climate and in part by particular financing schemes pertaining to wind-power investments. The figure used in Table 1 is typical of the financing currently available to average citizens wishing to erect a wind turbine or buy a share in a shared installation. Most power utilities have access to financing at a lower interest rate. On the contrary, private investors just four years ago paid 17% on similar loans, reflecting the economic uncertainty prevailing at the time. Average interest rates over the twentieth century have been roughly 3% above inflation, which currently would mean a 5% total interest rate in Denmark. The interest value of around 7.8% assumed in Table 1 is thus a conservative estimate when viewed over the long term, but low when viewed over the short term.

The remaining parameters, such as operation and maintenance costs and turbine lifetime, are less uncertain, and the capital costs (for the same annual power production on a fixed location) fluctuate little between manufacturers, but are expected to continue to decline with time and turbine unit size. The decline is limited, however, by technology (which can only increase blade length as more sophisticated materials become available). Turbines that are exported used to be more expensive on average than those sold locally because of transportation costs and, in some cases, lack of a company base in the recipient country, but export prices have increasingly approached local prices. For example, the average Danish 1993 export cost was 7.5 DKr/W as compared with the 6 DKr/W quoted in Table 1, but exports included many smaller-sized turbines (1). An important factor in determining prices on a given market is government subsidies (see the section on Economic Issues and Trends). However, the trend is toward similar prices on all markets, including that of the United States (73).

The installed power in various parts of the world is summarized in Table 2, and the total stock's development over time is illustrated in Figure 2. Figure 3 gives the average unit size of turbines installed in Denmark in each year from 1980–1995 and the specific power production. The increase in unit size has clearly increased specific production, for reasons discussed below.

Just over 50% of the worldwide wind turbine stock has been manufactured in Denmark, the rest chiefly in the United States, Germany, England, Holland, Belgium, Japan, and Italy. However, the industry structure has been changing in recent years; for example, a considerable number of the Danish manufacturers are presently owned by international investment consortia.

With current government plans for wind power expansion, the installed stock will reach 9 GW by the year 2000 (1). About a decade ago, the strongest expansion was in California, financed through schemes that could be exploited for tax evasion purposes, but Europe and Asia currently have plans for the greatest expansion, justified mainly as a means of reducing greenhouse gas emissions.

Table 2 End-of-1994 installed wind power worldwide

World	3930 MW
Europe	1819 MW
Denmark	510
Germany	510
UK	300
Netherlands	200
Spain	90
Greece	50
Italy	40
Sweden	38
Portugal	20
France	14
Belgium	10
Ireland	10
Norway	9
Finland	8
Poland	5
Czech Republic	4
other	1
North America	1793 MW
USA	1755
Canada	38
Latin America	50 MW
Asia	211 MW
India	120
China	60
Israel	6
other	25
Africa	37 MW
Egypt	33
other	4
Australia & Pacific	20 MW

Sources: (77, with estimated corrections; 78)

Figure 3 shows the gradual increase in unit size and the increase in realized specific power production for Danish locations. Tower heights have increased less than rotor diameters, but higher winds at elevated hub heights are still responsible for part of the specific power increase. Selection of better wind-turbine sites in Denmark also contributes to this increase. The average potential production at the selected sites increased by 25% between 1980 and 1988, but has been constant from 1988 to 1993 (2). This increase is probably a result of the fact that early wind turbines were erected at the owner's farm, whereas now turbines that are shared (typically by 100–200 shareholders) or owned by utilities are placed more freely at the locations

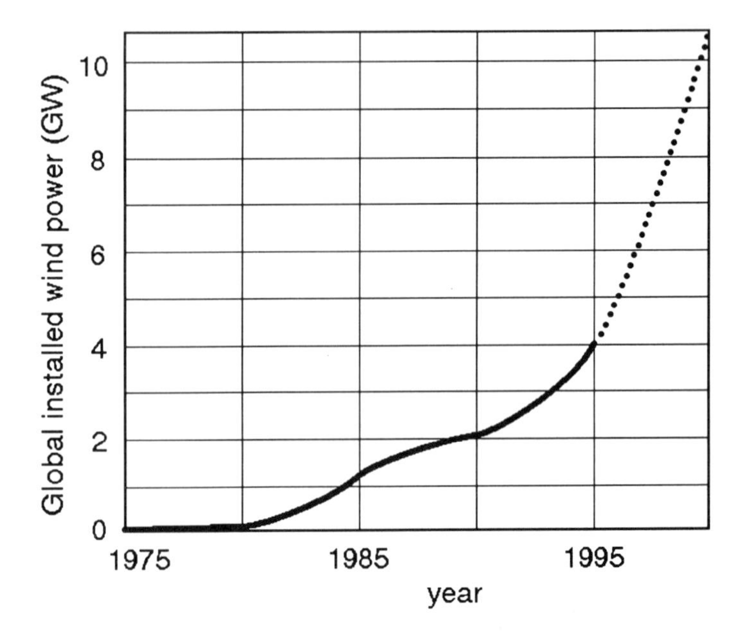

Figure 2 Worldwide penetration of wind power (GW installed) and government planned expansion to year 2000 [with use of (1)].

with the best wind conditions. The discontinuation of this trend after 1988 may mean that the best available sites have now been taken. Strict environmental siting criteria prevent the use of coastlines, but installation of offshore wind turbines is expected to increase, as is the number of large turbines replacing small ones at the same site. Present offshore wind farms (see Figure 8b) are situated 2–4 km from the coast and thus cause very little visual intrusion. The cost penalty for offshore foundation work and sea cables currently increases the turbine price by about 80%, which is not fully counterbalanced by the improved wind resource. However, when offshore siting becomes more routine the price is expected to drop considerably.

Brief History Of Wind-Energy Utilization

As a renewable energy source directly available in most parts of the world, wind energy has played an important role throughout human history. For example, wind energy provided the ventilation needed to sustain combustion of firewood and bring in new oxygen to maintain an acceptable level of air quality when cooking became an indoor activity. Persian chimneys, invented several hundred years ago, made use of wind flow to air-condition rooms by

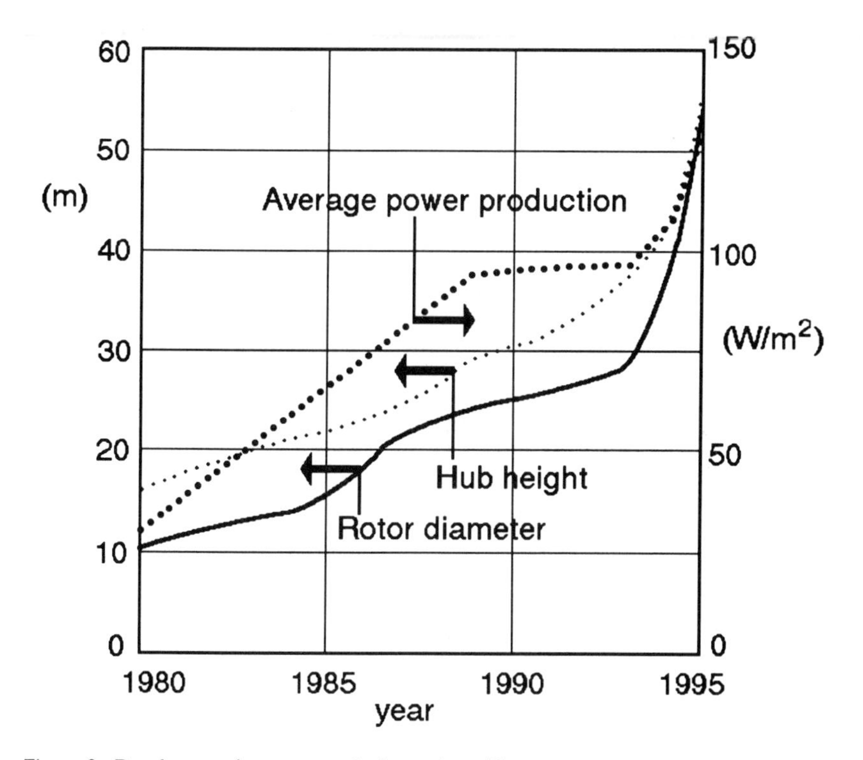

Figure 3 Development in average unit size and specific power output for turbines installed in Denmark [Based on (2) *Wind Turbine Price List,* Danish Technological Institute, October 1994; and *JOULE II Project Synopses,* Brussels: European Commission 1994].

pulling air over a body of water, which caused evaporative cooling. In most parts of the world today, the supply of fresh air in buildings is controlled by wind rather than mechanical systems. Wind is also important for dispersing and diluting pollutants, e.g. emissions released from stacks.

Active use of wind for energy supply purposes is believed to have started earlier than 5000 years ago in the Mediterranean region, when wind propulsion by sailing ships is first documented (3), cf Figure 4a. New types of merchant and war ships were developed by the Greeks about 550 BC and advanced by the Romans during the first centuries AD (4). Hydrodynamically-optimized boat shapes were used by the Vikings of northern Europe from 600 to 800 AD. The development of wind-driven ships continued to advance with the Flemish and Portuguese ships refined from the Roman types. The Europeans colonized several continents with such ships (Figure 4b) (4). The use of large sailing ships continued until about a hundred years ago when ships with diesel

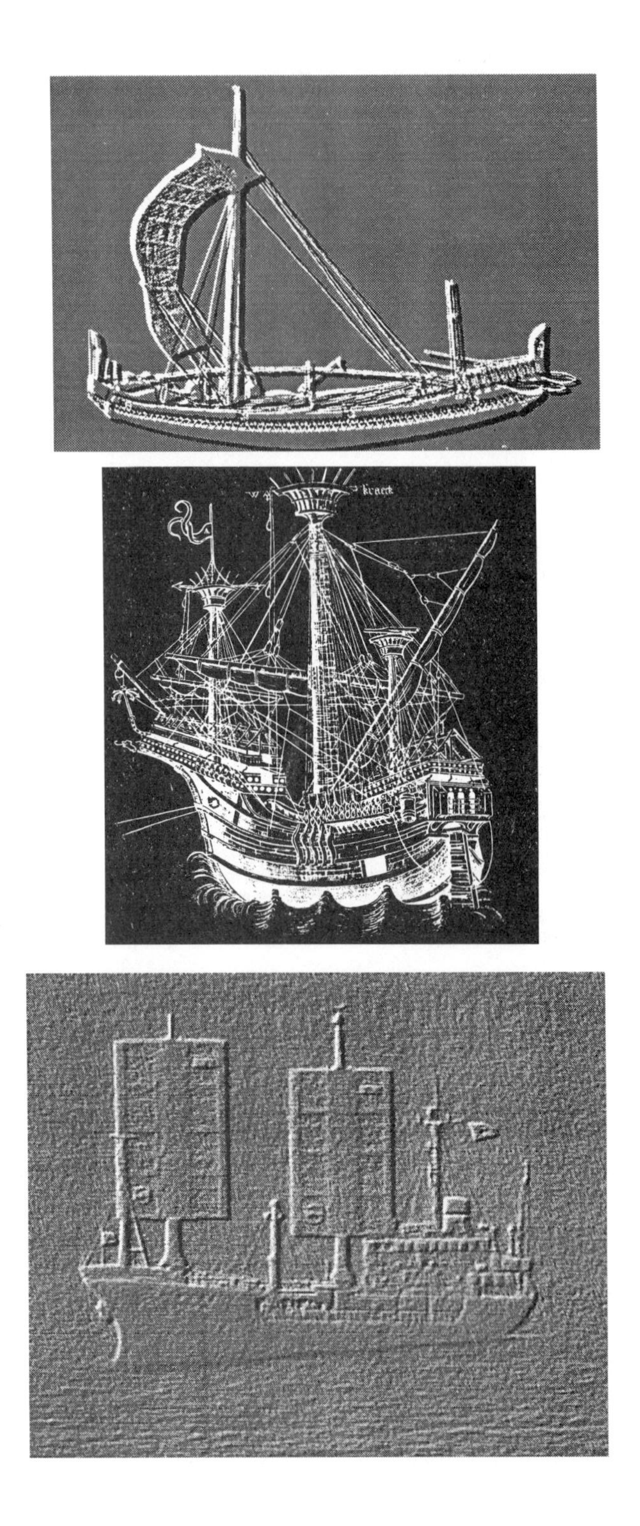

a.

b.

c.

engines became dominant. Sailboats are still used for leisure, and occasional attempts have been made to develop large fuel-saving ships that combine sails and engines (4c) (5).

Possibly the earliest mention of windmills dates back to about 400 BC: A Hindu book called *Arthasastra of Kantilya* suggests the use of windmills for pumping water (6), but no other evidence shows that they were used this way. Toy-like applications are the wind-powered organ (Figure 5a) described in the first century AD by Heron of Alexandria (7, 8), and the wind-driven Buddhist prayer-wheels mentioned around 400 AD (9). Heron replaced the blowing source for a pipe-organ invented by Ktesibios a few hundred years earlier with what appears to be a horizontal-axis wind turbine. Heron's idea to reverse the action of already well-known fans did not lead to wind turbine applications right away, but it may have influenced the transition from vertical-axis to horizontal-axis machines in Europe several hundred years later (10). The suggestion often made in the literature [based on (9)] that the drawings in Heron's book were added later by someone familiar with wind turbines is clearly false, because many existing copies of Heron's book were produced several hundred years before either eastern- or western-type windmills became known (11).

Evidence for actual development of vertical-axis windmills for grinding (but also for irrigation and rice pounding) comes from Persia during the late part of the first millenium. The application of wind to power a water fountain was hinted at in 800 AD (12). Two sources from the early 10th century (13, 14) suggest that windmills were first proposed by Abu Lulua in 644 and developed sometime during the two following centuries as a direct continuation of the technique for hand-milling and milling by waterwheels (10, 15). The millstones were even placed above the rotor blades, which were made of cloth, as evidenced by the first known drawing of a Persian windmill (Figure 5b), which dates from the late 13th century (16). This arrangement was convenient for a low-head waterwheel, but later it was realized that placing the rotor above the millstones was more efficient with wind and more convenient for the miller [cf the Persian windmills found in the Sistan region early in this century (15)].

Persian-style mills were found on Sumatra in 1154 (17) and in India in the 14th century, suggesting a gradual eastward dissemination. Lewis (10) speculates that the abundance of mills in the region that was derived from the state of Bactria may indicate a common denominator for the development of wind energy. Bactria was created by Alexander the Great, populated by many Greek emigrants, and existed independently of Greece and Rome from about 200

Figure 4 Wind-powered ships: (*a*) Egyptian ship ca. 2500 BC (3); (*b*) Flemish merchant ship ca. 1480 (4); (*c*) Japanese wind-powered ship, 1981 (5).

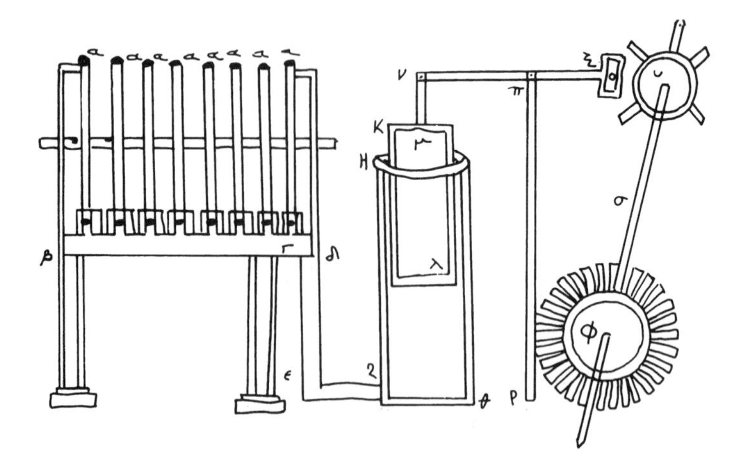

a.

b.

Figure 5 Early windmills: (*a*) 516 AD drawing of Heron's wind-powered organ (7, 10); (*b*) earliest drawing of Persian windmill, ca. 1300 (16); (*c*) earliest drawing of European postmill, 1270 [Canterbury psalter, see (18)]; (*d*) vertical-axis mill on Crete [Grünemberg, 1486, see (10)].

a.

b.

BC–200 AD. This suggestion implies that the ideas of Heron may have been further developed by the Greek emigrants and first put to practical use in the Sistan region of Persia, known for its steady and strong winds during three months of the year.

The horizontal-axis windmill first appears in Europe, where it was used primarily for cereal grinding, but also for pumping water and (in the Netherlands) for drainage. This type of windmill is considerably more efficient than the vertical-axis machines of Persia. The first description of the concept is in

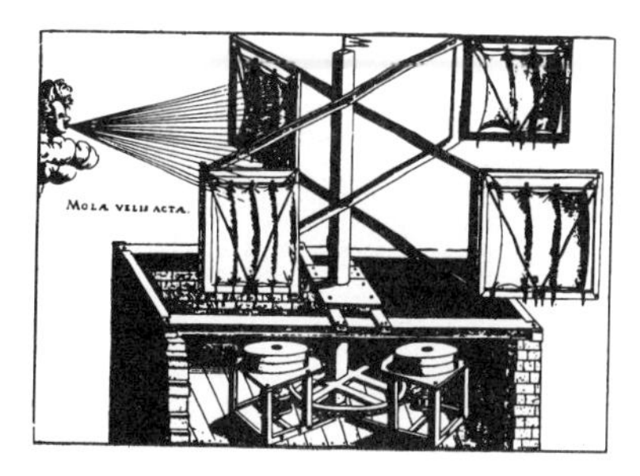

c.

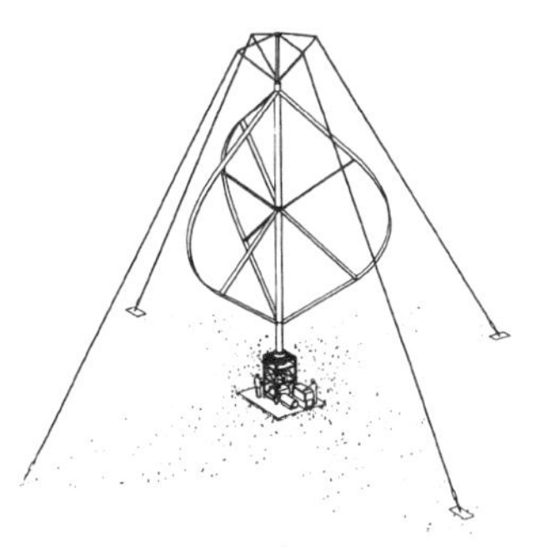

d.

Figure 6 Vertical axis turbines: (*a*) from Veranzio 1595 [see (18)]; (*b*) from Besson, 1578 (23a); (c) also from Veranzio 1595; (d) Darrieus turbine 1990 (24).

a French tax announcement from 1105, and the earliest picture of this type of windmill is in the 1180 Canterbury windmill psalter (Figure 5c) (18). The Persian concept may have been brought to Europe by the technicians accompanying the crusaders to the Middle East. But European developers likely added the idea of a horizontal-to-vertical shaft transmission, which would have been long familiar to them, e.g. from the gear drive of the Vitruvian waterwheel introduced during the first century BC (19). However, stimulus diffusion is perhaps even more likely; i.e. the Jerusalem travelers told those back home that it could be done and European entrepreneurs then made the detailed design of the windmill, incorporating their knowledge of gear mechanisms.

Attempts to find a path of gradual dissemination from Persia to Europe have revealed at least two possible routes. One leads through Constantinople to the Greek islands and on to western parts of the Mediterranean including Portugal. Along this route one finds some vertical-axis turbines similar to the Persian one (on Karpathos) and a further development of it (1486 in Kandia on Crete, see Figure 5d), but also an abundance of tower mills; i.e. horizontal-axis

a.

b.

turbines mounted on a tower, sometimes with the capacity to yaw the top. The other route goes through Russian provinces and comprises horizontal-axis mills. Some of the early ones had the millstones above the rotor, and the entire construction in later models could be turned by a lever, albeit with considerable effort. Similar models (sometimes called postmills) found further along the suggested route, in Flanders and the surrounding countries, are more advanced in that they use a pivoting design that required much less force. Figure 5c shows a postmill and Figure 7a a tower mill.

The horizontal-axis windmills soon came into widespread use, starting in the Normandy-England-Belgium triangle and quickly spreading to Holland, Germany, and Denmark (20). Figure 7b shows a Dutch windmill as developed during the seventeenth century. These windmills, which are often quite large,

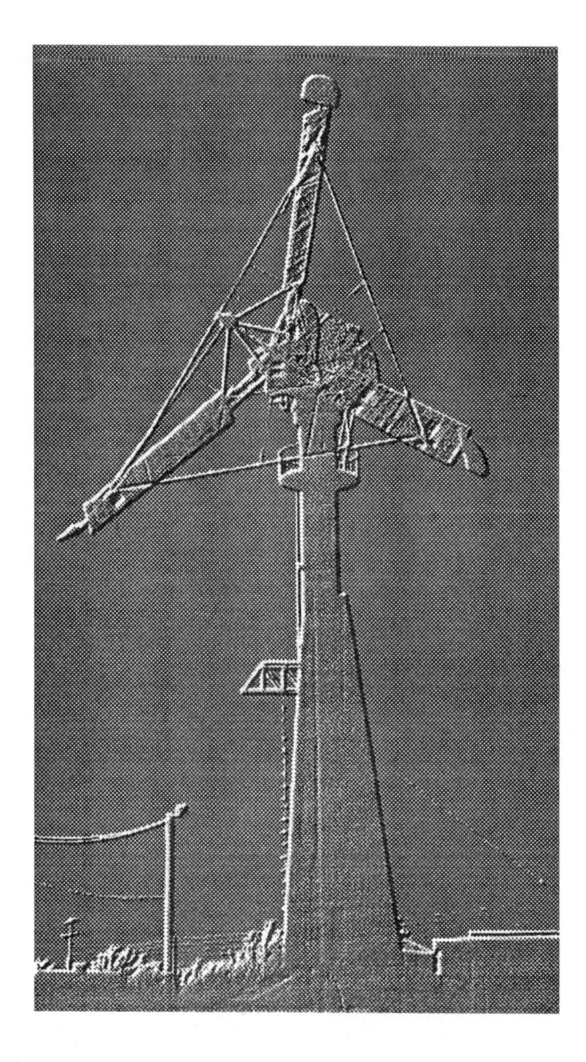

c.

Figure 7 Horizontal axis turbines: (*a*) Mediterannean sailtype; (*b*) Dutch top-yawing mill; (*c*) Danish AC-producing turbine at Gedser, 1959.

became abundant in most of Europe. They were frequently placed around cities and were used for grinding and irrigation as well as drainage of low countries. By the end of the nineteenth century, the efficiency of these mills had increased from below 1% to nearly 5% (cf Figure 12b). In addition to the dominant type, one finds early examples of nearly all alternative designs known today including the vertical-axis tower (Figure 6a), the self-starting Savonius (Figure 6b),

a.

Figure 8 Modern windmills: (*a*) Danish turbines, 1979; (*b*) Danish offshore windpark at Vindeby, 1993 (Bonns turbines).

hinged vertical blades [Figure 6c; recently reconsidered in the UK (21)] and tornado type [reconsidered from time to time (22), but never successful]. Rotors built into walls, as in the Sistan mills, have been reconsidered (23), but they would not be very efficient. For pumping and irrigation, the American multiblade design came into widespread use during the nineteenth century. This type is a low-efficiency mill, but one that will start at very low winds (1–2 m s^{-1}). Designs developed in the twentieth century include Darrieus machines (24) (Figure 6d) and blades with tip-vanes (25). They could match or surpass the efficiency of horizontal-axis machines, but they have never sufficiently attracted manufacturers.

The development of wind turbines for power generation started about 100 years ago in countries such as Denmark, England, Germany, and the United States (26). Serious work on wind-turbine efficiency and the aerodynamics of blade performance began in England in the eighteenth century, continued through the nineteenth century in France (27, 27a), and brought the technology to its present stage by the 1920s, owing mainly to German work (42, 39). Systematic airfoil studies began in Denmark in 1892 with la Cour's wind tunnel and his experimental windmills at Askov high school (28). la Cour's turbines, along with those of US pioneers (26), were the first to produce electricity. His goal was rural electrification, and the electric power produced by one of his experimental turbines was used for electrolysis, which generated hydrogen that was piped to lamps in an adjacent building. The design of aerodynamically-efficient rotor blades soon advanced with the aid of theoretical principles developed for the infant airplane industry. During the following two decades, wind- and coal-based electricity competed fiercely for the Danish market, and not until the 1920s did fuel-based power become cheap enough to force wind power out of the market, except for use on small islands. Isolated mountain settlements in countries such as Switzerland used windmills in the early twentieth century, as did farmers in many countries in areas far from the grid. Wind-diesel combinations regained popularity toward the end of the twentieth century, again in areas far from the grid (29).

During World War II several wind turbines were produced and erected in countries that were cut off from their traditional sources of fuel supply, such as Denmark. This new generation of wind turbines had concrete tower constructions, but most of them were still DC producers feeding into DC grids. Only after another fuel supply problem, the Suez crisis in 1957, did the Danish engineer Juul combine modern turbines with AC production based on wind-driven induction generators (30). His demonstration project, the 200-kW Gedser turbine shown in Figure 7c, became the prototype for the new generation of wind turbines put into production following the 1973–1974 oil crisis. While several other countries experimented with megawatt-size wind turbines (projects that all failed), the Danish firms began to manufacture models of modest rating (15–30 kW, see Figure 8a) in 1975 and soon had a booming business and a fairly good record of reliability and power production (31). The industry then gradually increased unit size, a trend that continued through the development of present models rated at about 1 megawatt (cf Figure 3). Advanced glass fibre and composite materials are now used for the blades, and towers today are often made of steel tubes.

Wind Resources

Winds are caused by differential heating of the atmosphere, in combination with the Earth's rotation and friction both within the atmosphere and between the atmosphere and the Earth's surface. Lorenz has estimated (32) that about 1200 TW (terawatts) are used on average to sustain the general circulation, which is under 1% of the incoming solar radiation. The kinetic energy in the circulation is 750 EJ, so the turnover time is approximately eight days (33, 34). About half the energy dissipates in the upper atmosphere and half near the surface. Only an estimated 3 TW are used to create waves (35), so land surfaces are responsible for most of the friction. Placing wind turbines near the ground helps this process. The placement of wind turbines at what are considered good wind sites amounts to increasing the roughness of the surface, whereas placing them at other locations might instead replace other features that create roughness in the surface.

One may ask what the effect would be of extracting power corresponding to a large fraction of the energy influx that maintains the circulation. These effects are currently under study, but if wind turbines cause the average friction to increase substantially at the ground, then the low-altitude wind speeds would likely become lower, whereas the gradient farther up in the atmosphere might become stronger because the stratospheric jet streams are unlikely to change. Ultimately, more of the atmosphere might follow the Earth's rotation. Such an effect could have a significant impact on global circulation, which in turn could have an influence on climate.

Using wind power to supply the entire world electricity demand would currently require only about one TW on average, so any significant effect on global circulation would be highly unlikely. In other words, the general resource availability is huge. The real question is how much of this resource could be harvested conveniently and economically, and how the resources are geographically distributed compared with power demand.

The winds at a particular location on the Earth's surface are determined by the geostrophic wind aloft and the roughness of the terrain (34). The geostrophic winds follow regular patterns determined by the midlatitude jet streams and the equatorial trade winds (see Figure 9), modified only by the effects of short-term weather systems. The roughness is generally high for uneven to-

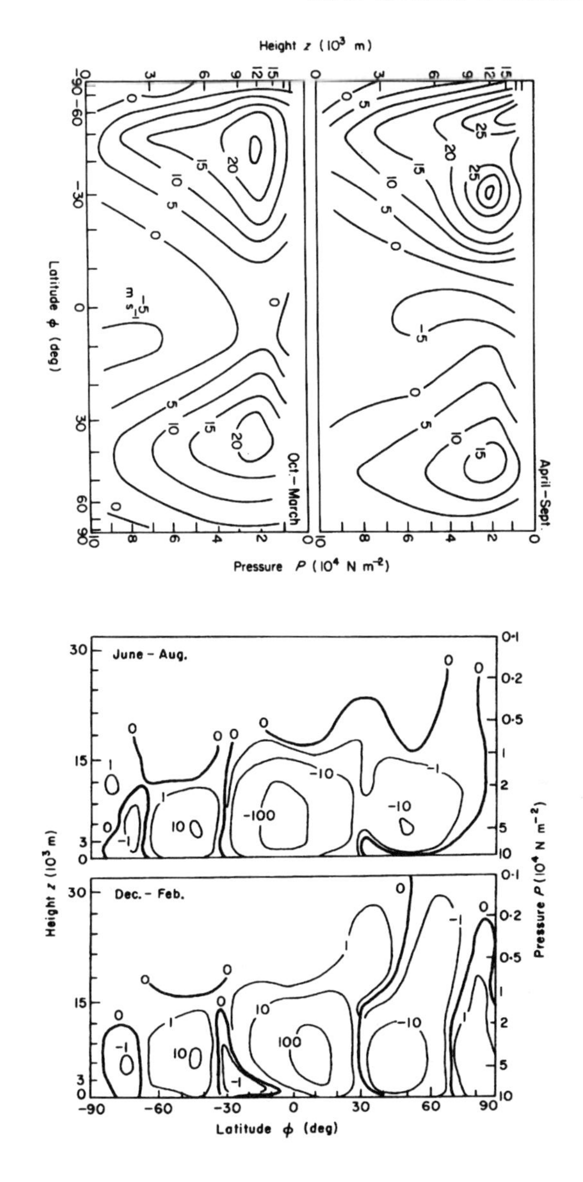

Figure 9 (*a*) Summer and winter longitudinal winds, i.e. winds following lines of constant latitude and height and (*b*) streamlines of latitudinal wind patterns; mass transport is given in 10^9 kg s^{-1} (a positive value means northward motion aloft) (34).

Table 3 Estimate of wind power potential, including technical and environmental constraints (average possible production in GW)

World	6050
Africa	1200
Australia	330
North America	1600
Latin America	610
Western Europe	550
Eastern Europe and former SU	1200
Rest of Asia	560

Sources: (79, 80).

pography, including the built environment, and low for flat plains and open sea (34).

Based on cartological information on topography and measured geostrophic winds, wind maps have been constructed, e.g. for Europe and the United States (36, 37). These maps divide the land into areas with average wind speeds in given intervals. Wind maps reliably predict wind-turbine production for fairly flat and uniform areas (in Denmark, manufacturers even give a production guarantee based on wind-map estimates), but are not very accurate for more complex terrain such as mountain passes and valleys. Furthermore, because wind-turbine power production depends on the third power of wind speed, and because averages may mask underlying distribution differences, using average wind speeds can lead to erroneous estimates.

Table 3 gives estimates of exploitable wind resources on land areas. These estimates are based on wind maps as well as consideration of other land uses (to avoid use conflicts). Only sites with an average wind speed exceeding 5.1 m s^{-1} a height of 10 m have been included, and a fairly arbitrary measure of environmental concerns is considered (chiefly visual intrusion). The exploitable potential is over five times the current rate of electricity use, but with a distribution substantially different from that of demand.

The average height profile of wind speeds at a particular location depends on the stability of the atmosphere. A simple eddy diffusion model predicts a logarithmic height profile, with height scaled by the roughness parameter and velocity by an atmospheric stability parameter (34). The variations in wind speed depend on passing weather–front systems, heating from the sun, and on a short time scale on stability (gustiness). Figure 10 exhibits clear peaks in the variance spectrum (which is designed to reveal periodicity) corresponding to the mentioned periods. Seasonal variations of the power in the wind vary with location, as illustrated in Figure 11, and in countries such as Denmark are anticorrelated with variations in solar radiation (38).

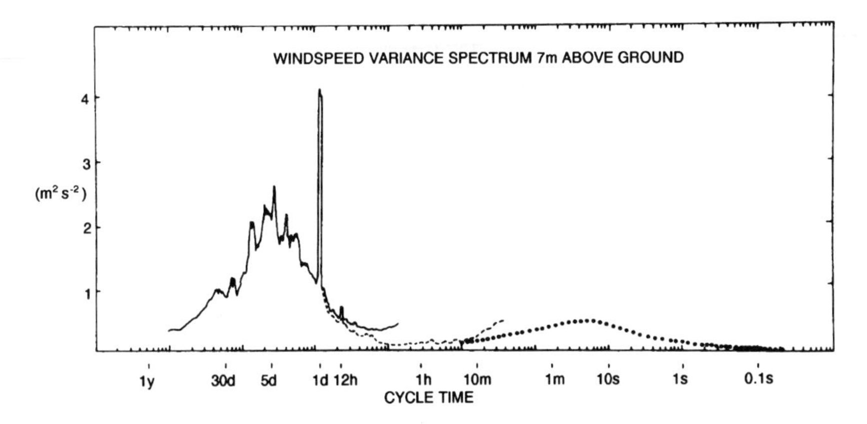

Figure 10 Frequency spectrum of winds near the ground, showing synoptic periods (5d), solar radiation effects (1d), and turbulent motion (10s). Based on (38a) and (34).

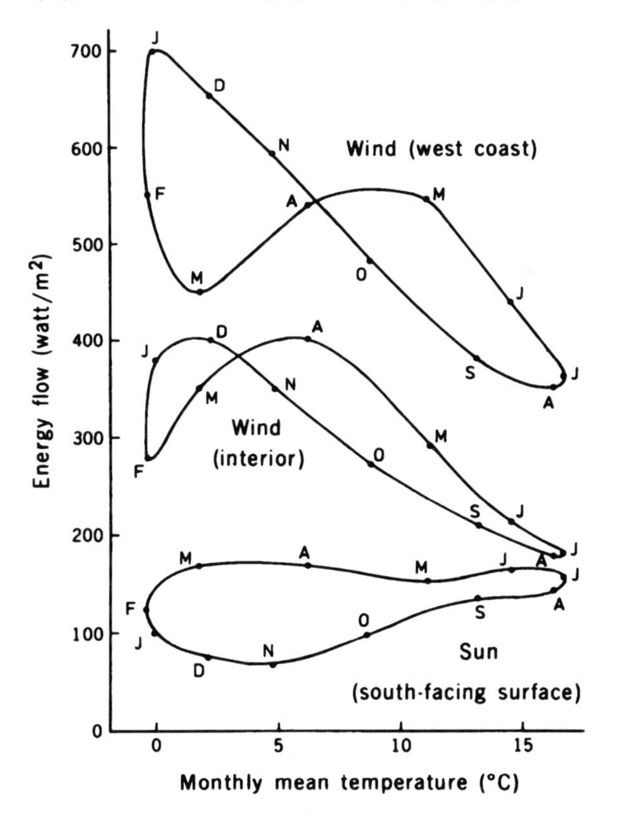

Figure 11 Comparison of inland and coastal site seasonal variation in wind and solar radiation, in Denmark (38).

Wind-Energy Technology

The present generation of wind turbines is characterized by advanced, aerodynamic blade shapes, use of highly advanced materials in blade construction, and state-of-the-art conditioning and control equipment. The manufacturing process has become fairly specialized, with some companies concentrating on blades or controls, others on tower and gear mechanisms. Only the more mundane parts of the process, such as foundation and site work, are left to local contractors.

Most wind turbines are currently made with three aerodynamic, glass- or carbon-fiber blades 20–30 meters long, and a tower 40–60 meters high (for large turbines, a steel tube tower is most common). The blades are mounted on a drive shaft, which usually incorporates a gear-box transmission and the electric generator within a nacelle, which can be yawed to make the blades face the wind at any time. The blade profile is optimized for a design wind speed using three-dimensional modeling (34) and verified in wind-tunnel tests. For each fixed profile, a compromise is made between maximum power at the design wind speed and high performance at other wind speeds.

Although blade shape cannot be differentially adjusted, the entire blade can be rotated on some turbines. Such turbines are still optimized at one particular wind speed, but the penalty at other speeds can be reduced by rotating the existing profile to the angle that gives the best average performance (pitch regulation). Most manufacturers do not make turbines with this feature, claiming that the cost of adjustable blades is not recovered by the modest increase in output. In the case of a fixed blade profile, increasing wind speed will eventually create turbulent eddies near the blade and a corresponding loss of lift force (34). This mechanism is called stall regulation, reflecting the fact that overall power production as a function of wind speed will reach a maximum and then stay constant or decline (Figure 12a). If maximum rotational speed is properly chosen, the danger of "running away" can be minimized. Stall regulation also means that if the connection to the grid is lost the wind turbine will not increase its rotational speed uncontrollably, but will stay at a constant speed, from which it can be stopped by conventional brake technology. Combinations of stall and pitch regulation are expected to be incorporated into future designs.

The choice of three blades is a compromise between high power output and the stability and lifetime of the blades. Conventional blade theory, ignoring interference phenomena [as first proposed by Betz in 1926 (39)], predicts a slight gain from more blades, but two blades appear economically optimal. However, adding more blades to rotors with few blades can help eliminate problems caused by each blade experiencing a different wind regime (if they are separated by 25 m, for example). Wind flows around blades also interact

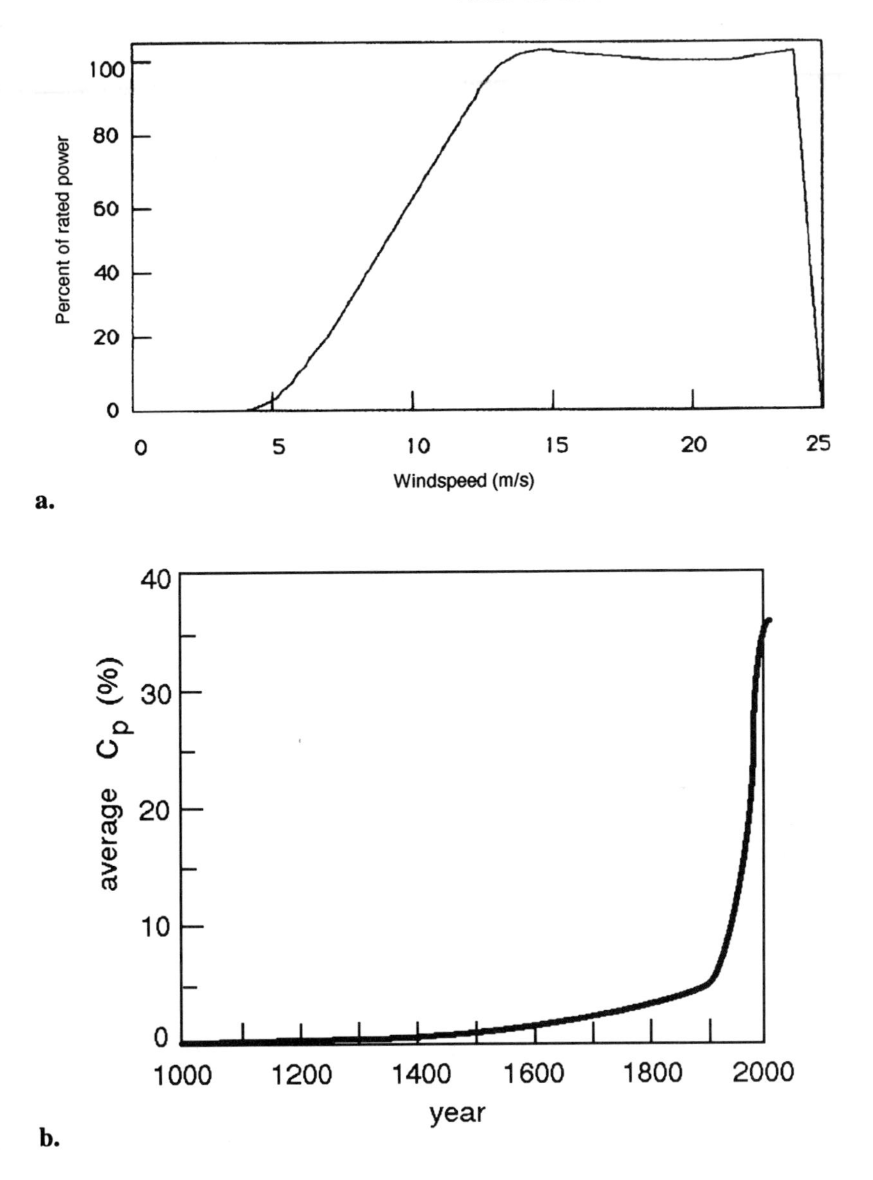

Figure 12 (*a*) Typical power output from modern glass-fiber rotor blades. B Sorensen, *The Status of Wind Generators in Europe,* Energy Authority of NSW report 86/18, Sydney; (*b*) Average power coefficient C_p (power produced by the turbine over power of the undisturbed wind passing through an area equal to that swept by the rotor) for typical wind turbines through time.

with the tower, a problem that appears to be largest for one- and two-bladed solutions (40). This lesson has been learned in practice from two-bladed constructions such as the US MOD-2 vis-a-vis three-bladed constructions of the same size (41, 41a). However, a few modern wind turbines have reintroduced the two-bladed concept, but it is too early to tell whether they will be successful.

Overall efficiencies for leading technologies over time are depicted in Figure 12b, which also explains the efficiency parameter C_p. Simple axial flow calculations first made by Betz (42) indicate a maximum C_p of 16/27 or about 59%. This may be exceeded by turbines generating a rotational or axial flow, such as ducted rotors or the tip-vane concept mentioned earlier (cf 34).

The electricity generator favored for current machines is the induction generator, which operates at near-constant rotational speeds and allows the AC grid to maintain the correct frequency without active controls [this concept was already in use with the first AC grid–connected wind turbines around 1920 (43)]. However, a few manufacturers presently use synchronous generators, which allow varying speeds and thus improve average efficiency. Most manufacturers are expected to shift to synchronous generators with advanced computer control over the next 5–15 years because the best variable-speed turbines have already increased power output (71) and because these generators are expected to reduce blade fatigue, contributing to longer life.

The world-market success of Danish wind generators is largely a result of the excellent quality control offered by the Danish Wind-Turbine Test Station. This station was started by four unemployed engineers in the late 1970s who received support from a government unemployment program designed to further innovative ideas. Once the wind test–station proved successful, it was incorporated into the Risø National Laboratory. The test station not only offers tests and certification, but will collaborate closely with manufacturers to improve designs. This collaboration has been surprisingly smooth, benefitting the industry without causing competitive friction.

Environmental Impacts And Life-Cycle Costs

The direct and indirect impacts of an energy technology come from each stage of development: equipment manufacture and installation, the productive operation of the plant, and the decommissioning process. Wind energy, like many renewable energy technologies, is expected to have many, if not most, of its impacts during the manufacturing stage, in which the bulk of financial expenditure, environmental effects of materials procurement, and hazards of industrial work environments is concentrated. On the other hand, these impacts are well-known from other industries, so they can be fairly accurately assessed. The same is true for the installation and operational phase, for which impacts are also similar to those of familiar enterprises. For final equipment disposal,

Table 4 Energy input to and emissions from the manufacture and operation of wind turbines

	Turbine manufacture			O&M	Total
	a	b	c	d	e
Energy inputs (GJ/kW rated)	—	—	6.6	2.2	8.8
Energy ratio (20y outputs/inputs)	—	—	—	—	18
Energy payback time (years)	—	—	0.8	0.3	1.1
Carbon dioxide (g/kWh)	9.1	11	12.1	3.8	15.9
Sulphur dioxide (g/kWh)	0.087	<0.05	0.05	0.01	0.06
Nitrogen oxides (g/kWh)	0.036	<0.05	0.04	0.02	0.06

[a] Source: (58).
[b] Source: (82).
[c] The energy ratio and payback are calculated on the basis of (83), from which the emissions (c) are directly taken.
[d] Total O&M energy input is given for the assumed 20-year operation period, based on data from (83).
[e] Sum of (c) and (d), i.e. total life-cycle estimates except for energy used for decommissioning, which is not estimated.

only the dismantling and the disposal or recycling of materials must be considered, because long-term effects (such as radioactivity in the case of nuclear power plants) are not an issue.

Detailed environmental analyses of wind energy were first made around 1980 (44). These analyses focused on safety hazards during construction and maintenance, the risk of parts (such as blades) being expelled in case of failure, the impacts of noise and electromagnetic interference, risk to birds in the area, visual intrusion, and climatic impacts caused by the redistribution of energy flows near the turbine.

Recent estimates of the environmental impacts from wind-turbine construction are summarized in Table 4. The underlying calculations are top-down estimates, based on energy imbedded in materials and used in manufacture, per unit of capital outlay or per unit of material weight. Emissions are derived from existing data on average emissions from the production of the various types of energy assumed to be used in wind-turbine manufacture. The emissions are distributed over the turbine's power production, assuming a 20-year lifetime. The slight differences between the emissions shown in Table 4 are consistent with differences in fuel mix in the three countries considered (the United Kingdom, Denmark, and Germany).

Other potential impacts during manufacture come from toxic materials, particularly chemicals used during the production of glass-fiber blades (epoxy resins, acids, etc). In modern production facilities, such chemicals are used in sealed environments with restricted access, and residuals undergo maximum recycling (44). Thus, occupational exposure as well as environmental emis-

Table 5 Accident risks associated with wind-turbine manufacture and construction, for a power production of 10^{12} kWh[a]

	Materials & manufacture	Construction	Operation
Deaths	5	20	b
Injuries with hospitalization	400	540	
Minor injuries	2800	2600	

[a] Source: (58).
[b] In addition to the injuries quoted, Eyre estimates road accidents for construction and maintenance workers, which would roughly double the values given. However, the estimates are admitted to be too high, based e.g. on four independent maintenance visits a day to a 100–wind turbine farm from a base 50 km away.

sions are primarily associated with accidents. No statistics on accidents in the blade-manufacturing industry are available, but data from similar industries suggest that the risks are very small.

The occupational hazards of installing and repairing turbines are similar to those of constructing and maintaining any tall structure such as a high-rise building or a bridge. As turbine sizes have increased, concrete towers have been phased out in favor of steel-tube towers, which usually consist of large prefabricated parts that need only simple on-site assembly. Therefore, inside stairs can be used for most access to the nacelle, during construction as well as maintenance, and stepping out of the nacelle or using a crane would only be necessary for some blade work.

In a Swiss study, Fritzsche examined accidental deaths during manufacture and construction of fairly small wind turbines and in public road accidents involving the transport of wind-turbine equipment (45). More recent average statistical data from the United Kingdom on occupational injuries in the mechanical and electrical engineering industries, and in the construction sector, show accidental death rates some four times lower than the Swiss estimate when applied to wind-turbine manufacture on a value-added basis. The new data on death and injury are summarized in Table 5. To this should be added any economic loss. Statistics on outage and component failure indicate that some 10% of operating turbines experience a component failure in any given month and that it is usually corrected in a day or two. Only 4% of the failures result in the component being replaced, and around 0.2% may exhibit a safety risk such as a detached blade (31). The risk to the public arising from expelled blades is estimated to be negligible (46).

Land use by wind turbines is estimated as 10 m^2 per kW rated power (47). This estimate takes into account the fact that wind-turbine siting is only allowed in nonurban zones and that activities such as farming may be going on very close to the turbine. The bulk of the land needed for a turbine is for access roads for maintenance crews.

The degree of visual intrusion, which has been studied by several authors, depends on the technology used (e.g. steel truss towers vs tubular towers) and on the density and layout of turbines (wind farms vs individual, dispersed units). Ownership is also a factor; e.g. the share-financed wind turbines in Denmark (for which legal requirements include that owners live near the turbines and that share size correspond to electricity use) are considered a positive contribution to the landscape, whereas utility-owned wind turbines erected without participation of the local population have sometimes induced protests (although mainly from individuals employed in the fossil-fuel sector). Consumer surveys reveal very positive attitudes to both present and planned levels of wind turbine penetration: 83% of the population wants an increased effort to expand wind power in Denmark, whereas only 8% are against it (48). Efforts by architects to design wind farms that have an attractive layout and are integrated into their surroundings have helped wind farms gain acceptance (49). In other locations (e.g. California), wind farms have been installed without much consideration for how best to integrate them visually.

Noise from wind turbines is of two kinds. One is the noise from the gear-box and transmission, which can be reduced to any small, prescribed value by sound insulation of the nacelle. The other kind is the aerodynamical noise from the wind coming off the wing blades. This cannot be arbitrarily reduced; it is the same type of environmental noise created by trees or buildings. This noise will necessarily increase with increasing wind speed and can only be regulated by changing the blade profiles or the rotational speeds of the rotor, which may not be compatible with optimal energy collection or the control strategy. For example, the stall regulation described in the preceding section creates eddies and therefore noise around the blades in high winds, and to allow either increased rotation or a minimal–intrusion incident angle may create other problems. Noise from air flow around the tower or from blade/tower interaction can be minimized by proper choice of tower construction and rotor separation from the tower. Smooth tubular towers generally create much less noise than steel truss towers. Thus, one must expect a noise level that will increase with wind speed for speeds below the break-in speed, remain fairly constant during operation near the design wind-speed, and increase in the high-wind stall region (44).

The basic frequency of rotor noise is linked to the rotational frequency (typically around 0.3 Hz) and its harmonics. Transmission noise has a higher base frequency, but may be dealt with by insulation, as mentioned. Harmonics of the rotor frequency, as well as new frequencies created by interaction with objects in the environment, may have components in the audible frequency region that significantly exceed the background level at distances within a few rotor diameters from the turbine. Substantial temporal variations in the directional pattern do occur, and the turbine may occasionally be heard up to about

25 rotor diameters away. Measurements show an increase in noise level of about 10 dB(A) at rated turbine output and measured two rotor diameters away from the turbine, due to aerodynamical noise from wind near the blades (50, 51). dB(A) is a logarithmic scale using a frequency averaging believed to approximate average human noise perception. In many countries, noise profiles are routinely measured at each new, significant wind installation.

Whether or not people are annoyed by noise depends on several qualities of the noise itself such as tonality, intermittency, and background level, as well as what people are doing (e.g. sleeping) and their individual sensitivity. A study by Keast (52) finds that the percentage of people annoyed by levels between 40–50 dB(A) ranges from 0–3%. The percentage of people annoyed by noise in the 50–60 dB(A) interval ranges from 0–15%, by 60–70 dB(A) from 3–33%, by 70–80 dB(A) from 20–60%, and by 80–90 dB(A) from 35–90%. Regulatory limits for the general public range from 70 dB(A) in industrial areas to 35 dB(A) at night in residential and recreational areas (53). For infrasound (0.1–5.0 Hz), a few countries have regulatory limits of 70 dB(A) (51, 54).

Several models are available to calculate the propagation of sound. The models furnish noise levels at specified locations as a function of sound-source locations and strengths (55, 56). Ground-absorption must be considered in such models in order for the predictions to agree with measurements (57). In complex terrain, the models tend to be fairly inaccurate, but for typical wind-turbine sites modeling can provide a fair estimate of the combined noise impact at inhabited locations in the neighborhood of the turbine(s) (58). For a 103 × 250-kW turbine wind farm in Wales, the noise level increments reach occasional maxima of 3.4 dB(A) at a nearby house, are below 2.0 dB(A) at any location 2 km from the wind farm, and are below 0.05 dB(A) at the nearest village, which is 3.5 km away. These are outdoor levels and would be considerably reduced inside any building.

Telecommunications interference has been thoroughly discussed (59). As with other structures, the reflection of electromagnetic radiation from a wind turbine depends on the materials that compose it (large scattering crosssection from metal parts, medium from glass fiber, and small from wooden parts). The disturbance depends on the amplitude of the scattered signal relative to that of the direct one, and the disturbance typically increases with signal frequency. Interference with television signals is typically below a modulation threshold of about 2 dB at distances less than 10 rotor diameters from the turbine. The cost of providing cables instead of aerial antennas to any building located within this radius is normally negligible and is routinely considered part of the wind-turbine installation cost. For frequencies used by navigational and military intelligence systems, the interference is less than that from a building of the same height as the wind turbine (44).

Impacts on ecosystems would come mainly from mining and manufacturing of construction materials. An operating wind turbine is a smaller hazard for birds and flying insects than a similar static structure (51). A few bird kills have been observed among turbines on major songbird migration paths. Larger birds usually migrate at heights above 500 m. The blades of modern wind turbines turn slowly, so birds can generally avoid them, as several studies have shown. The hazard is thus greatest in misty weather with winds below the turbine minimum (44).

Decreasing the speed of the wind, which is necessary for wind-energy extraction, has microclimatic impacts. Slowing the wind can lead to altered evapotranspiration patterns and may have positive impacts on soil erosion. A slowing of the wind is felt over roughly twice the area swept by the turbine and the impact has disappeared ten rotor diameters downwind (60), or sooner, depending on the stability of the natural wind direction. Effects on rainfall and carbon-dioxide concentrations at ground level are limited to the area immediately below the turbine (44).

A common unit of measure, such as monetary value, would be useful for comparing the impacts described above and for comparing different energy technologies with different types of impacts. Systematic approaches to this monetizing approach have proven difficult. Certain damages have identifiable costs, but others (e.g. visual impacts) do not. One method that has been used is to calculate avoidance cost, i.e. find the cheapest technology that does not have the impact in question. This approach is unsatisfying because the replacement technology will always have impacts that differ from those of the original technology. An alternative methodology is contingency valuation, e.g. asking people how much they are willing to pay to avoid a certain type of damage. This approach is clearly influenced by day-to-day changes in the information people receive and by competing priorities. Hedonic costing, which is based on preferences revealed through e.g. property values near and far from a given energy facility, and the value of life, defined as the amount of life insurance people take out, are also clearly loaded with problems and uncertainties. Some studies further propose to distinguish between internalized and external costs, i.e. to identify which costs are included in market prices (e.g. 60a).

The technique of life-cycle analysis may be defined in such a way that identifying a common unit is not essential (61). Those who have to assess life-cycle analysis and choose among alternative solutions to a given problem can be given impact profiles in which each item has its own unit, but the same unit is used for that item for all alternative technologies. Monetizing may be done in cases in which it does not introduce new levels of uncertainty, but the impact-profile technique ensures that monetized and otherwise quantified impacts are presented in the same way. Even nonquantifiable impacts can be presented on equal footing as long as the qualitative impacts of different

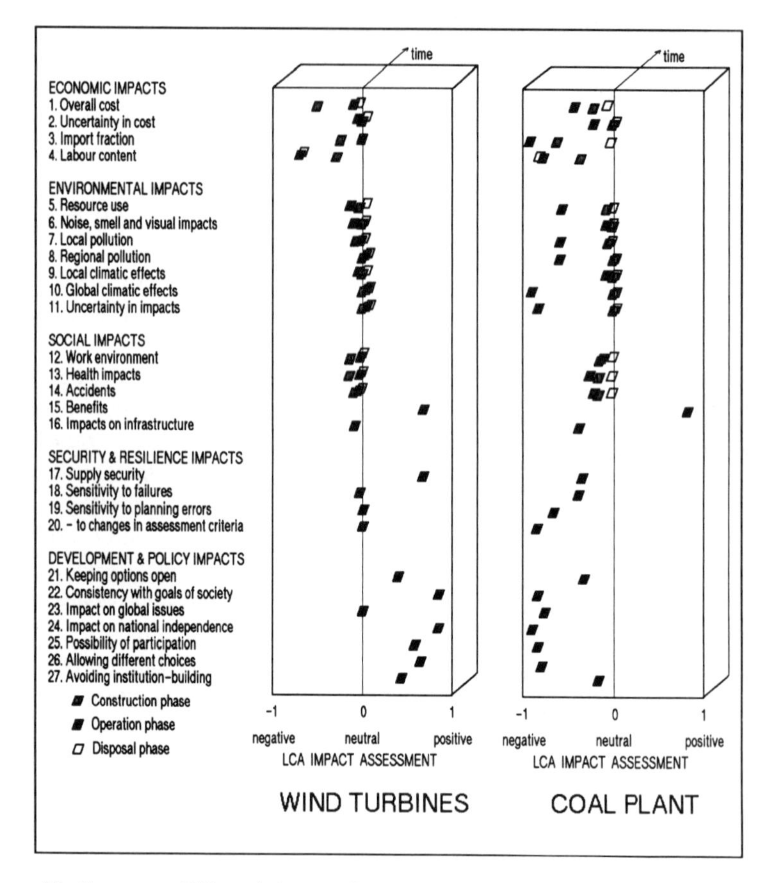

Figure 13 Summary of life-cycle impacts for wind and coal power plants. Some of the impacts are quantified (see text), others are not (62).

technology solutions can be distinguished. Figure 13 gives an example of such impact profiles for wind-based and coal-based electricity production (47, 62). An assumption in these profiles for wind-based production is that no specific energy storage facilities are required, e.g. if wind-based production accounts for less than some 30% of electricity production in a large, internationally-connected utility system (63).

Systems Integration Issues

Issues of integration arise for both wind-energy systems themselves and for wind in combination with other power systems. The advantage of having wind turbines scattered in a decentralized fashion for systems of wind-energy converters is rather evident: The wind regime will change over length- and time-scales in unison, according to what is termed the ergodic hypothesis (34). Therefore, differences in power production over time may to some extent be counteracted by having wind turbines at separate locations feed into a common grid. Small spatial separation will take care of short-term fluctuations, whereas continental dispersal is needed to average out synoptic differences (i.e. the effects of passing front–systems). This type of integration may reduce the need for energy storage in case of a large penetration of wind power in the electricity system, just as a short-term storage with a capacity equal to 10–20 hours of demand makes the loss of load probability (LOLP) of a single wind turbine similar to that of a typical nuclear power plant (64). To reduce the LOLP to the same value as ordinary fossil power plants, about 1000 hours (a little over a month) of storage is needed (Figure 14) (65). Alternatively, the integration of turbines over a distance of 240 km will increase the availability of half the average power output from 50% to 70% of time (65). In order to average out synoptic differences, clearly a larger area is needed. Even in cases of very large dispersal, storage will be useful in a pure wind-energy system because with storage minimum production need not match demand and thus the installed capacity can be reduced.

Ease of construction, as well as maintenance conditions similar to those of conventional power stations, has led to the clustering of wind turbines into wind farms, particularly those owned by power utilities. Wind turbines are similarly clustered in the recently-added offshore wind farms (Figure 8b), which are of interest in regions where many interests are competing for land use. Dense siting makes interference between wind flows in wind farms a potential problem and consequently raises the issue of how to optimize the balance between density of wind turbines and power loss resulting from reduced winds. Calculations of wake effects indicate that the power losses behave approximately as indicated in Figure 15. In order to avoid significant power losses in a large wind farm, a spacing of 20 rotor diameters is needed (34, 66, 67, 74). This distance would be somewhat reduced in regions with incoming winds from shifting directions. In bi-directional flow regions, such as valleys or passes, the turbine separation perpendicular to the wind direction can be just a few rotor diameters. Current wind farms often accept a fairly significant energy loss.

Electricity systems involving a combination of wind turbines with other forms of generating equipment raise interesting layout and control questions. Which energy systems integrate easily with wind, and if they can be regulated, how is this best done? Considering first combinations of wind with other renewable energy systems, there is an advantage as regards diurnal and annual variations to combining wind with solar (photovoltaic) and with hydro power, at least in some regions of the world (68). For hydro power systems with

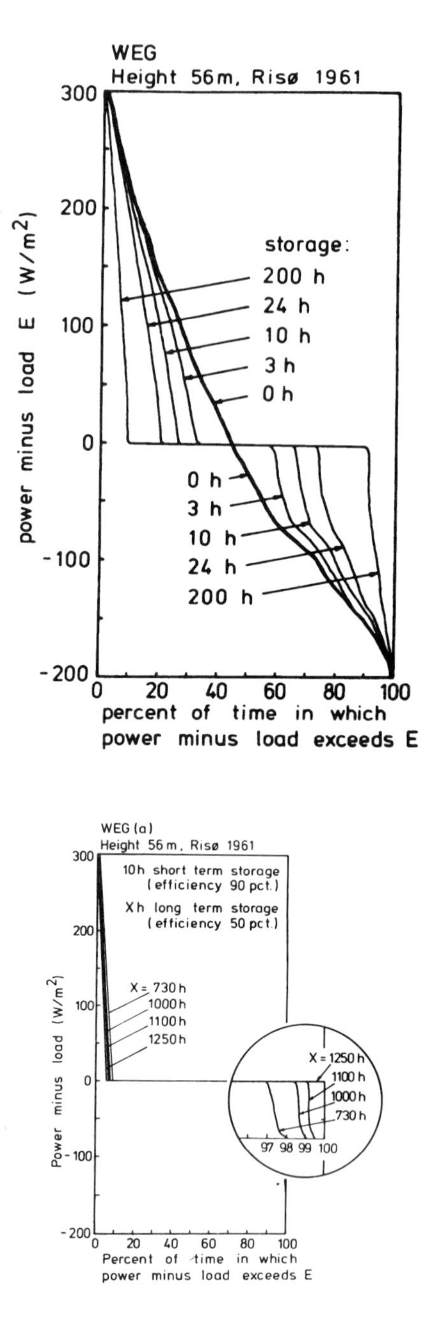

a.

Figure 14 Model investigations of the effect of short-term (*a*) and long-term (*b*) storage on the power duration curve of a typical wind turbine (65).

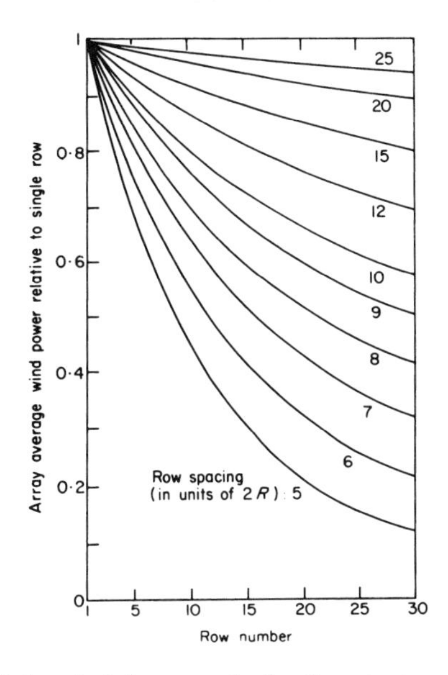

Figure 15 Model calculation of wind-power reduction through a large wind farm with a regular spacing of turbines. Based on (66) and (34).

annual reservoirs, found, for example, in Norway, no drawback derives from using this system as a backup for even a fairly large wind system because, as mentioned above, periods with low wind-power production will last only a few weeks, and therefore, the borrowing of hydro power that is subsequently paid back with wind power will only slightly influence the water level in the seasonal hydro reservoirs.

Combinations of wind power with biomass used in power plants (either directly by combustion or indirectly after biogas production) will behave like fossil-fuel combinations, as biomass can be stored in much the same way as, for example, coal. Various control strategies can be used to regulate such a combined system. A system can be equipped with either enough quick-start gas turbines to take care of sudden changes in wind production or several coal- or gas-fired power plants that function as spinning reserves to avoid problems with the warm-up period of conventional plants (although this is getting shorter with modern injection technology). The control may be based on wind-power forecasts taken from general meteorological forecasts or on extrapolations from previous levels of wind production. Simulations have shown that the latter method is probably sufficient for all practical purposes (63).

Economic Issues And Trends

The very uneven penetration of wind turbines in countries with similar wind resources is the result of widely different support in terms of utility interfacing, compensation for externalities, and financing schemes. Connection to the grid involves an installation cost: In some cases, the distribution line is too weak and must be reinforced, a measure which, however, may have been needed anyway as a result of increasing customer load. A fair arrangement for sharing this cost between the power company and the wind turbine owners must be found.

The buy-back of surplus wind power is also an issue. Is the utility required to buy back such power, and if so at what price? Do they pay only the avoided-fuel cost, or is wind given some capacity credit, and a premium, for polluting much less than the average power plant? If the utility cost structure does not consider externalities, should society give tax credits to less-polluting energy systems?

Another issue is investment financing. In some countries, utilities are given concessions that allow them to accumulate capital for future investments, whereas in other countries this is not the case (leaving the utilities to normal market financing of new investments). Wind-turbine owners are confined for the most part to private loans, sometimes on terms less attractive than those offered a large utility.

The United States was the first country to establish fixed power buy-back rates [through the Public Utilities Regulatory Policy Act of 1978 (PURPA)], but some countries still do not allow independent producers to hook up to the grid and sell power. Table 6 shows buy-back rates in some European countries, along with information on whether investment subsidies are given. Denmark

Table 6 Wind power buy-back rates and investment subsidies[a]

Country	Small consumer's electricity price (ecu/kWh)[b]	Wind power buy-back rate (ecu/kWh)	Investment subsidy (%)
Denmark	0.112	0.083	0
The Netherlands	0.084	0.035–0.069	up to 35
Germany[c]	0.133	0.118 (0.086)	up to 60 or 46 kecu
United Kingdom	0.109	0.136 (0.092)[d]	0
Italy	0.113	0.087	20–40
Spain	0.130	0.077	0
Sweden	0.078	0.034	25

[a] Source: (81).
[b] 1 ecu is about 1.25 US$.
[c] The high buy-back price is available only to those not receiving an extra investment subsidy.
[d] The lower price applies after 1998.

is the only country that has a fixed procedure for calculating grid-connection costs: Wind-turbine owners pay the connection installation cost and the power company pays for any grid reinforcement that is needed because of the turbine. Subsidy and buy-back rules have changed over time in the countries listed. Denmark, for example, initially provided a 30% capital subsidy to wind-turbine buyers in order to boost the wind business during the late 1970s (69).

In Denmark, wind-turbine shareholders are governed by very strict rules designed to prevent use of wind power as a tax evasion scheme. Each shareholder must live close to the installation and cannot hold shares corresponding to more than 150% of his own electricity consumption. If the shareholders use a particular company form called IS, profits up to 10% are not taxed. Wind power is also exempt from energy and carbon taxes. Discussions are under way on whether the nearness condition should be relaxed so that people living in big cities can also support wind energy.

In the Netherlands, most wind turbines are owned by power companies. They can use the carbon taxes paid by consumers for wind-energy investments. In Germany and the UK, project-financing is common, and in Germany 50% can be financed by low-interest loans through the subsidized Ausgleichbank (DAB).

The ambitious plans for wind-energy expansion in several European countries is likely to raise two issues: capacity credit and energy storage, or how to eliminate the need for storage. The discussion in the previous chapters shows that intelligent dispatch (which may involve collaboration among several countries) and relatively short-term storage facilities will make it possible for wind energy to cover a very high fraction of the demand. The resources to do so are available, but the extra expense of offshore siting may be necessary in some cases to avoid land-use conflicts. In countries that already have a considerable share of wind power, substituting large turbines for smaller ones at the best wind sites will minimize environmental intrusion. In Denmark, which has about 4000 turbines, this substitution could increase the electricity provided by wind power from 3% to nearly 100% of current electricity use (70).

Literature Cited

1. Madsen BT. 1994. Consultancy report to the Association of Danish Wind Turbine Manufacturers, quoted in *Naturlig Energi*, Sept:6, 1994, and *Ingeniøren*, Aug. 5:2
2. Nielsen P. 1994. 500MW windpower in Denmark. *Naturlig Energi* Sept:14–17 (In Danish)
3. Digby S. 1954. Boats and ships. See Ref. 75, p. 735
4. Lethbridge TC. 1954. Shipbuilding. See Ref. 75, pp. 564–86

5. Azarin B. 1981. In the wake of the flying cloud. *Science* 81:81–85
6. Nath N. 1957. Preface. In *Windmills and Mill Wrighting*, by F Freese, p. 1. Cambridge, UK
7. Heron of Alexandria. ca. 80. *Pneumatics*. Transl. ed. B Woodcroft, 1951. London: Taylor Walton & Maberly (From Greek)
8. de Camp LS. 1960. *Ancient Engineers*. London: Tandem
9. Forbes RJ. 1954. Power. See Ref. 75, p. 615
10. Lewis MJT. 1993. The Greeks and the early windmill. In *History of Technology*, ed. GH Short, FA James, 15:141–89. London: Mansell
11. Drachmann AG. 1961. Heron's windmill. *Centaurus* 7:145–51
12. Banu Musa bin Shakir. ca. 800. *The Book of Ingenious Devices*. Transl. D Hill, 1979, p. 222. Dordrecht (From Arabic)
13. Al-Tabari. ca. 900. *Croniques*. Transl. ed. H Zotenberg, 1958, 3:628–29. Paris
14. Al-Mas'udi AH. ca. 956. *Meadows of Gold and Mines of Gems*, a Historical Encyclopedia. Transl. ed. A Sprenger, 1841, London; *Les Prairies d'or*. Transl. ed. B Meynard, P Courteille, C Pellat, 1971, 3:607. Paris (From Arabic)
15. Wulff H. 1966. *The Traditional Crafts of Persia*. Cambridge, MA: MIT Press
16. Al-Dimashqi MAT. ca. 1300. *Manuel de la Cosmographie du Moyen Age*. Transl. ed. Mehren, 1874. Paris (From Arabic)
17. Al-Idrisi. 1154. Quoted in *Relations de voyages et textes géographiques arabes, persans et turcs relatif à l'Extrême-Orient du 8e au 18e siècles*, 1913. p. 194. Paris
18. Wailes R. 1954. A note on windmills. See Ref. 75, p. 623
19. Vitruvius P. ca. 60 BC. *De Architectura*, ed. Gwilt, 1860. (From Latin)
20. Forbes RJ. 1955. *Studies in Ancient Technology*, 2:116–17. Leiden
21. Musgrove P, Mays I. 1978. *Proc. Int. Symp. Wind Energy Syst., 2nd, Pap. E4*. Cranfield: Br. Hydromechanical Assoc.
22. Yen J. 1976. *Proc. Int. Symp. Wind Energy Syst., Pap. E4*. Cranfield: Br. Hydromechanical Assoc.
23. Hyypia J. 1980. Incredible power fence. *Sci. Mech.*, Winter:41–44
23a. Besson J. 1578. *Theatre des Instruments*. France: Vincent Lyons
24. Darrieus GJM. 1931. *US Patent 1835018*. Recent experiments reported in *Initial structural response measurement and model validation for the Sandia 34m VAWT test bed*, by TD Ashwill, 1990. Albuquerque: SAND88-0633
25. van Bussel GJW, van Holten T, van Kuik GAM. 1978. Flow visualization study of the boundary layer on rotating tipvanes. *Rep. M-302*. Univ. Tech: Delft
26. Hills R. 1994. *Power from Wind*. Cambridge, MA: Univ. Press
27. Smeaton J. 1796. *Experimental Inquiry Concerning the Natural Powers of Wind and Water to turn Mills and Other Machines Depending on a Circular Motion*. London: I & J Taylor
27a. Drzewiecki S. 1892. Method for determining the mechanical elements of rotor screws. *Bull. l'Assoc. Techniques Maritime* 3:11–31 (In French)
28. la Cour P. 1900–1903. *The Experimental Mill*, Vols. 1–4. Copenhagen: Det nordiske forlag (In Danish)
29. Sørensen B. 1986. *A Study of Wind-Diesel/Gas Combination Systems*. Sydney: Energy Auth. New South Wales
30. Juul J. 1964. *Proc. UN Conf. New Sources Energy, Rome 1961*, 7:229–39. New York: UN
31. Windstats Newsletter. 1987–1994. *A quarterly survey of Wind Turbine performance*. Knebel, Denmark: T. Møller
32. Lorenz N. 1967. The nature and theory of the general circulation of the atmosphere. *Publ. 218TP115*. Geneve: World Meteorol. Organ.
33. Newell R, Vincent D, Dopplick T, Ferruzza D, Kidson J. 1969. In *The Global Circulation of the Atmosphere*, ed. G Corby, pp. 42–90. Englewood Cliffs, NJ: Prentice Hall
34. Sørensen B. 1979. *Renewable Energy*. London: Academic
35. Gregg MC. 1973. Microstructure of ocean. *Sci. Am.* 228:65–77
36. Troen I, Petersen EL. 1989. *European Wind Atlas*. Roskilde: Risø Natl. Lab.
37. Elliott DL, Holladay CG, Barchett WR, Foote HP, Sandusky WF. 1986. *Rep. DoE/CH 100934*. Dep. Energy, Washington, DC
38. Sørensen B. 1975. Energy and resources. *Science* 189:255–60
38a. Petersen EL. 1975. On the kinetic energy spectrum of atmospheric motions in the planetary boundary layer. *Rep. 285, Risø Natl. Lab.*, Risø, Denmark
39. Betz A. 1926. *Wind Energy and Its Exploitation*. Göttingen (In German)
40. Kottapalli S, Friedman P, Rosen A. 1978. *Proc. Symp. Wind Energy Syst., 2nd*, pp. C4.49–66. Cranfield: Br. Hydrodynamic Res. Assoc.
41. Bonneville Power Administration. 1981. *Mod-2*, pamphlet, 4 pp.

41a. Flying in the face of the wind. 1976. *Consulting Engin.*, Nov:2–3
42. Betz A. 1920. Maximum theoretically possible use of wind through wind turbines. *Z. Gesamte Turbinenwesen,* Vol. 17:20 (In German). The main assumption was developed in Betz A, Prandtl L. 1919. Rotors with minimum energy loss. *Nach. Ges. Wiss. Göttingen Math. Phys. Kl.,* pp. 193–217 (In German)
43. Adler E. 1920. Using windpower to produce electricity. *Ingeniøren* 78:595–98 (In Danish). Similar practical work performed 1918–1924 by P Vinding and RJ Jensen, who describes it in 1924. *Ingeniøren* 50:585-87
44. Sørensen B. 1981. Wind energy. In *Renewable Sources of Energy and the Environment,* ed. EE El-Hinnawi, AK Biswas, pp. 98–116. Dublin: Tycooli Intl.
45. Fritzsche E. 1989. The health risks of energy production. *Risk Anal.* 9:565–77
46. Macqueen J, et al. 1983. Risks associated with wind turbine blade failures. *IEE Proc. A 130,* pp. 574–86
47. Sørensen B. 1993. Environmental impacts of photovoltaic and wind-based electricity production evaluated on a life-cycle basis. *Proc. Int. Conf. Heat and Mass Transfer, Cancun,* pp. 516–21. Morelos, Mexico: Universidad Nacional Autonoma de Mexico
48. Møller T. 1994. AIM Survey 1993. *Vindstyrke* Jan:1 (In Danish)
49. Nielsen FB. 1980. Placing windmills. *Rep.* Aarhus: School Architecture (In Danish)
50. Danish Department of Energy. 1978. *Newsletter* 7. Copenhagen: Wind Energy Program 12 pp. (In Danish)
51. Rogers S, Cornaby B, Rodman C, Sticksel P, Tolle D. 1977. Environmental studies related to the operation of wind energy conversion systems. *Rep. TID-28828.* Dep. Energy, Washington, DC
52. Keast D. 1978. Noise-control needs in the developing energy technologies, *Rep. COO-4389-1.* Dep. Energy, Washington, DC
53. Danish Department of the Environment. 1974. *Recommendation* 2. Copenhagen: Environmental Agency (In Danish)
54. Holley W, Bell B. 1990. Low frequency sound from wind turbine arrays. In *European Community Wind Energy Conference, Madrid,* pp. 743–47. Bedford: Stephens
55. Ljunggren S, Gustafsson A. 1988. Acoustic measurements of noise emissions from wind turbines. *Rep. IEA Programme R&D Wind Energy Conversion Syst.,* 20 pp. Stockholm: Natl. Defense Res. Inst.
56. Manning CJ. 1981. The propagation of noise from petroleum and petrochemical complexes to neighbouring communities. *CONCAWE Rep. 4/81*
57. Bass J. 1992. Noise assessment at Carland Cross Wind Farm Site. In *Proc. DTU/BWEA Workshop on Wind Turbine Noise,* Harwell: ETSU N-123
58. Eyre N. 1995. The environmental costs of wind energy. See Ref. 60a, Vol. 6, pp. 3–118
59. Sengupta D, Senior T. 1978. Electromagnetic interference by wind energy turbine generators. *TID-28828.* Dep. Energy, Washington, DC
60. Alfredsson P, Bark F, Dahlberg J. 1980. *Proc. Symp. Wind Energy Syst., 3rd,* pp. 469–84. Cranfield: Br. Hydrodynamical Res. Assoc.
60a. European Commission, DGXII, Science, Research, and Development, JOULE. 1995. *Externalities of Fuel Cycles "ExternE" Project* (In press)
61. Sørensen B. 1993. What is life-cycle analysis? In *Life-Cycle Analysis of Energy Systems,* pp. 21–53. Paris: OECD
62. Sørensen B. 1994. Life-cycle analysis of renewable energy systems. *Renewable Energy* 5:1270–77
63. Sørensen B. 1978. The regulation of an electricity supply system including wind energy generators. In *Proc. Int. Symp. Wind Energy Systems, 2nd,* pp. G1–G8. Cranfield: Br. Hydrodynamical Res. Assoc.
64. Sørensen B. 1976. Dependability of wind energy generators with short-term energy storage. *Science* 194:935–37
65. Sørensen B. 1978. On the fluctuating power generation of large wind energy converters, with and without storage facilities. *Solar Energy* 20:321–31
66. Crafoord C. 1975. An estimate of the interaction of a limited array of windmills. *Rep. DM-16,* Univ. Dep. Meteorol. Stockholm
67. Builtjes PJH, Smit J. 1978. Calculation of wake effects in wind turbine parks. *Wind Eng.* 2:135–45
68. Sørensen B. 1981. A combined wind and hydro power system. *Energy Policy* 9:51–55
69. Danish Academy of Technical Sciences. 1975–1976. *Wind Power 1-2. Proposed Action Plan.* Lyngby: Dan. Acad. Tech. Sci. (In Danish, English transl. available)
70. Sørensen B, Nielsen LH, Pedersen SL, Illum K, Morthorst PE. 1994. Renewable energy system of the future. *Danish Tech. Counc. Rep. 1994/3,* Copenhagen: Dan. Technol. Counc. (In Danish)

71. Godtfredsen F. 1993. Economic comparison of Danish and foreign wind turbines. *Risø Rep. R-662.* Risø: Risø Natl. Lab. (In Danish)
72. Sørensen B. 1981. Turning to the wind. *Am. Sci.* 89:500–8
73. Cavallo A, Hock S, Smith D. 1993. Wind energy: technology and economics. In *Renewable Energy Sources for Fuels and Electricity,* ed. T Johansson, H Kelly, A Reddy, R Williams, pp. 121–56. Washington, DC: Island
74. Milborrow DJ, Surman PI. 1987. CEGB wind energy strategy and future research. In *Proc. BWEA Conf., 9th,* ed. J Gault. Glasgow: Br. Wind Energy Assoc.
75. Singer C, Holmyard EJ, Hall AR, eds. 1954, 1956. *A History of Technology,* Vols. 1, 2. London: Clarendon. 912 pp. 798 pp.
76. Skriver S. 1994. Economic analysis of windmills. *Naturlig Energi* Nov/Dec: 12–14 (In Danish)
77. Møller T. 1994. Reality surpasses 'super-optimism'. *Naturlig Energi* Sep:6 (In Danish)
78. Nielsen P. 1993. Wind energy. *Energi- og Miljødata* 3:1–2 (In Danish)
79. van Wijk AJM, Coelingh JP, Turkenburg, WC. 1993. Wind energy. In *Renewable Energy Resources.* London: World Energy Conf.
80. Grubb M, Meyer N. 1993. Wind energy: resources, systems, and regional strategies. In *Renewable Energy Sources for Fuel and Electricity,* ed. T Johansson, H Kelly, A Reddy, R Williams, pp. 157–212. Washington, DC: Island
81. van Zuylen EJ, van Wijk AJM, Mitchell C. 1993. *Comparison of the Financing Arrangements and Tariff Structures for Wind Energy in European Community Countries.* Utrecht: Ecofys consult
82. Fritsche U. 1993. TEMIS. In *Life-Cycle Analysis of Energy Systems,* pp. 103–11. Paris: OECD
83. Meyer H, Morthorst P, Schleisner L, Meyer N, Nielsen P, Nielsen V. 1994. Omkostningsopgørelse fer miljøeksternalitates i forbindelse med energi produktion. *Risø Rep. R-770 (DA),* Risø Natl. Lab., Risø

Annu. Rev. Energy Environ. 1995. 20:425–61

THE EMERGING CLIMATE CHANGE REGIME[1,2]

Daniel M. Bodansky
School of Law, University of Washington, Seattle, Washington, 98105

KEY WORDS: global warming, greenhouse effect, international law, United Nations

CONTENTS

[1]Abbreviations used: AOSIS, Alliance of Small Island States; COP, Conference of the Parties; EC, European Community; FCCC, United Nations Framework Convention on Climate Change; GEF, Global Environment Facility; GHG, greenhouse gas; ICSU, International Council of Scientific Unions; IEA, International Energy Agency; IGBP, International Geosphere-Biosphere Program; INC, Intergovernmental Negotiating Committee for a Framework Convention on Climate Change; IPCC, Intergovernmental Panel on Climate Change; JI, joint implementation; NGO, nongovernmental organization; OECD, Organization for Economic Co-operation and Development; SBI, Subsidiary Body for Implementation; SBSTA, Subsidiary Body for Scientific and Technological Advice; UNCED, United Nations Conference on Environment and Development; UNDP, United Nations Development Programme; UNEP, United Nations Environment Programme; WMO, World Meteorological Organization.

[2]The present article draws on the author's previous work—in particular, References 17 and 120.

1056-3466/95/1022-0425$05.00

ABSTRACT

The emerging climate change regime—with the UN Framework Convention on Climate Change (FCCC) at its core—reflects the substantial uncertainties, high stakes and complicated politics of the greenhouse warming issue. The regime represents a hedging strategy. On the one hand, it treats climate change as a potentially serious problem and, in response, creates a long-term, evolutionary process to encourage further research, promote national planning, increase public awareness, and help create a sense of community among states. But it requires very little by way of substantive— and potentially costly—mitigation or adaptation measures. Although the FCCC parties have agreed to negotiate additional commitments, substantial progress is unlikely without further developments in science, technology, and public opinion. The FCCC encourages such developments, and is capable of evolution and growth, should the political will to take stronger international action emerge.

INTRODUCTION

On March 21, 1994, the UN Framework Convention on Climate Change (FCCC) took effect as international law, following its ratification by 50 states. Adopted two years earlier by a UN negotiating committee, and opened for signature at the June 1992 UN Conference on Environment and Development (UNCED),[3] the FCCC serves as the constitution for the emerging international climate change regime.[4] This article examines the principal features of that regime, its development since UNCED, and its prospects for the future.

The climate change problem is by now familiar. It arises from the buildup of carbon dioxide and other greenhouse gases (GHGs), which trap heat in the atmosphere. Although enormous uncertainties persist about virtually every aspect of the problem, scientists estimate that if current emissions patterns continue unchecked, the earth's surface will on average warm by 1.5 to 4.5°C by the end of the next century (1; see also 2), [5] with potentially severe effects

[3]Multilateral conventions such as the FCCC become international law through a multistage process: (*a*) adoption of the final text by a negotiating committee (such as the INC) or a diplomatic conference; (*b*) signature by states, indicating their preliminary support; (*c*) formal acceptance of the treaty through ratification or accession, in many countries requiring parliamentary approval; and (*d*) entry into force, which generally depends on ratification or accession by a certain minimum number of states (in the case of the FCCC, 50).

[4]As of April 4, 1995, 165 states and the European Community had signed the FCCC and 128 states had ratified or acceded to it, including nearly all OECD countries, Russia, and many leading developing countries (e.g. Brazil, China, India, Indonesia, and South Korea).

[5]This 1992 IPCC estimate does not take into account several factors that tend to reduce the net rate of warming, in particular, the forcing from sulfate aerosols and stratospheric ozone depletion. Recent computer modelling experiments that include sulfate aerosols project a global mean rate of change of about 0.2°C per decade through the early part of the next century (3).

on coastal areas, agriculture, forests, biological diversity, and human health (4, 5).

Although the general theory of greenhouse warming has been understood since the end of the nineteenth century, an international regime to address the problem of anthropogenic climate change began to develop only in the late 1980s, due to improved scientific knowledge, promotion of the issue by a relatively small group of scientists and environmentalists, and an upswing in public concern about global environmental issues generally (6, 7) (see Table 1). The first intergovernmental response came in 1988, when WMO and UNEP created the Intergovernmental Panel on Climate Change (IPCC) to provide a scientific and policy assessment of the problem (2, 4, 8). Two years later, the UN General Assembly established an Intergovernmental Negotiating Committee (INC) to develop a climate change convention. The INC negotiated the FCCC in little more than 15 months, finalizing the text on May 9, 1992, shortly before UNCED (9).

The climate change regime is the network of rules, institutions, programs, and decision-making procedures that, internationally, shape expectations and structure and guide activities relating to climate change. This article focuses primarily on the FCCC, which serves as the focal point of the emerging climate change regime. (For general discussions of the FCCC, see References 17–20.) But the regime has other important elements. Most climate research, for example, is conducted outside the Convention framework, through international and national research programs (10). The activities of industry and environmental groups—as well as of international institutions such as the World Bank—will also play crucial roles in addressing the climate change problem. One challenge for the climate change regime will be to coordinate these extra-Convention activities and ensure that they further the FCCC's objectives (11, 12).

Several general features of the emerging climate change regime are noteworthy. First, the regime aims at the widest possible participation, due to the global nature of the greenhouse problem and the recognition that anthropogenic climate change is the "common concern of humankind" (FCCC preamble, para. 1). Currently, the United States contributes nearly one fifth of global CO_2 emissions, and all OECD countries contribute roughly two fifths (13). But emissions from developing countries are growing rapidly and are projected to surpass industrialized-country emissions in the first half of the next century (14). Although some have argued that a regime could potentially be effective comprising only the dozen largest industrialized and developing countries (which together account for more than three fourths of global GHG emissions), such a regime is unlikely, given fears that it would provide a competitive advantage to nonparticipants, and would lack the international legitimacy of a regime representing a true global consensus (15, 16).

Table 1 Landmarks of the Emerging Climate Change Regime

	Date	Organizer	Status	Outcomes, conclusions, recommendations
Villach Conference	1985	WMO & UNEP	scientific	• Significant climate change highly probable. • States should initiate consideration of a global climate convention.
Toronto Conference	1988	Canada	nongovernmental	• States should cut global CO_2 emissions by 20% by 2005. • States should develop comprehensive framework convention on the law of the atmosphere.
UN General Assembly	1988	UN	intergovernmental	• Climate change a "common concern of mankind."
Hague Summit	1989	Netherlands, France, Norway	summit	• Signatories will promote the development of new institutional authority within the UN involving non-unanimous decision-making.
Noordwijk Conference	1989	Netherlands	ministerial	• Industrialized countries should stabilize GHG emissions as soon as possible. • "Many" countries support stabilization of emissions by 2000.
IPCC First Assessment Report	1990	WMO & UNEP	scientific	• Global mean temperature likely to increase by 0.3°C/decade under business-as-usual emissions scenario.
Second World Climate Conference	1990	WMO & UNEP	ministerial	• Countries need to stabilize GHG emissions. • Developed states should establish targets and/or national programs or strategies to limit emissions.
UN General Assembly	1990	UN	intergovernmental	• Establishment of INC.
INC 5	1992	UN	intergovernmental	• Adoption of FCCC.
FCCC's entry into force	1994			
Berlin Conference	1995	FCCC	Conference of Parties	• Berlin Mandate for negotiations to strengthen FCCC commitments.

Second, as its name indicates, the FCCC is a framework instrument. It puts in place a long-term, evolutionary process to address the climate change problem, which (*a*) enunciates the regime's ultimate objective and guiding principles, (*b*) establishes an infrastructure of institutions and decision-making mechanisms, (*c*) promotes the systematic collection and review of data, and (*d*) encourages national action. It does not, however, impose strong substantive commitments on states—for example, a clear quantitative limitation on GHG emissions (targets and timetables), or common response measures such as carbon taxes or energy-efficiency standards (17–20). Although the lack of strict controls on GHGs was decried by many environmentalists (21), the climate change problem has a long time horizon, and a delay of 10 or even 20 years should "make little difference in atmospheric concentrations at the end of the next century" (16; see also 22).

Third, the FCCC reflects a carefully balanced compromise among the often conflicting positions of states. Many of its provisions are "constructively" ambiguous, to allow for further negotiation of contentious issues. From this perspective, the FCCC represents not an end point, but rather a punctuation mark in an ongoing process of negotiation (17). This process has continued uninterrupted since UNCED, without even pausing to await the FCCC's entry into force (23). Throughout 1993 and 1994, the INC continued to work on several unfinished issues—notably the regime's reporting and review procedure and financial mechanism. In March 1995, this work was assumed by the FCCC's Conference of the Parties, which adopted the Berlin Mandate, authorizing a new round of negotiations to develop more stringent commitments (24).

Fourth, the FCCC—at least potentially—has an exceptionally broad scope, encompassing not simply environmental protection as traditionally conceived (i.e. limiting emissions of pollutants), but economic and development policies more generally. Virtually the entire range of human activities contributes to GHG emissions (25, 26). GHG emission scenarios, for example, are highly sensitive to population and economic growth assumptions (1). [6] Thus, population and economic policies could conceivably play a prominent role in the climate change regime.

Fifth, as a framework instrument, the FCCC is neutral regarding policy options. Although climate change is primarily a problem of carbon dioxide (which accounts for roughly two thirds of the predicted radiative forcing through 2050 for business-as-usual scenarios) (28), fossil fuels (which account for 65–90% of CO_2 emissions and 30% of methane emissions) (8),[7] and coal

[6]According to one scenario, population growth will account for half of the increase in global CO_2 emissions from fossil fuels over the next 60 years (27).

[7]The initial reports by industrialized countries indicate that fossil fuels account for 97% of their CO_2 emissions and that CO_2 accounts for 75% of their overall GHG emissions (61).

(which represents more than 90% of the carbon in estimated fossil fuel reservoirs) (29), the FCCC does not single out any particular greenhouse gas or economic sector for special attention. Indeed, although attention has tended to focus thus far on mitigation, the FCCC addresses adaptation as well. As a result, states have maximum flexibility in designing response strategies.

ESTABLISHING A FRAMEWORK OF GOVERNANCE

An international regime is a network of rules, institutions, programs, and decision-making procedures governing a given area of international relations (30–32). A widely shared assumption is that legal regimes consist of rules that constrain behavior, with effective mechanisms to resolve disputes and sanction violations. But international environmental regimes can serve a variety of other functions. They can:

1. Provide a general system of governance for a given subject area, including basic goals and principles, institutions, and decision-making procedures. This is the core function of so-called "framework" conventions.
2. Impose procedural duties on states—for example, to report on national emissions and response measures—and create international monitoring and review procedures.
3. Improve technical understanding of a problem, by coordinating scientific research and promoting the systematic collection and review of data.
4. Facilitate state action—for example, through the provision of financial and technical assistance.

Often, adopting binding rules is impossible, due to lack of agreement among states. Enacting stringent environmental laws is hard enough within national legal systems, with well-established institutions, majoritarian decision-making procedures, and shared cultural traditions and values. It is much more difficult internationally, where these elements are generally lacking. For this reason, international environmental regimes have tended to proceed incrementally; they first establish a framework of institutions and mechanisms that foster the development of consensus, and only later attempt to elaborate more substantive obligations (33–35). This was the pattern, for example, in both the acid rain and ozone regimes (36–38).

In the climate change negotiations, most states were willing from the outset to accept a framework convention. The main question has been whether to go further—in particular, by adopting more specific commitments to limit greenhouse gas emissions. For several reasons, such commitments have proved difficult:

1. *High uncertainties*—On the one hand, the costs of anthropogenic climate change are uncertain (39). Although many scientists believe that the consequences of greenhouse warming will be severe, substantial uncertainties persist about virtually every aspect of the issue: future emissions of GHGs, the effects of these emissions on atmospheric concentrations, the sensitivity of the climate system to increased concentrations, regional variations in climate change, and the impacts of global warming on natural ecosystems and human activities (1, 2, 4). Meanwhile, economists tend to argue that the costs of climate change would be low, given the adaptability of human societies (40, 41; but see 42). Relatively few states appear genuinely frightened by the prospect of climate change, and most of these are small island countries vulnerable to sea-level rise, who do not contribute much to the problem and have comparatively little power internationally. At least some countries might actually reap benefits from climate change and have no direct incentive to undertake mitigation measures (although no country seems to base its negotiating positions on this assumption). Moreover, because of the climate system's inertia, the benefits from reduced emissions of GHGs would be realized only in the long term—a time horizon beyond the consideration of most governments.

2. *High stakes*—In contrast, the costs of limiting GHG emissions—both economic and political—would be felt more immediately. Although estimates vary markedly (43–45), the economic costs of abatement are potentially very high. Moreover, the activities that produce GHG emissions are deeply entrenched in national economies and are politically difficult to change, even when doing so would yield net savings (26). (Consider, for example, the political obstacles to eliminating market-distorting energy subsidies.)

3. *Divergent state interests*—States have divergent interests concerning climate change, which must be reconciled. The potential costs of abatement and adaptation differ substantially from state to state, depending on each state's energy resources, starting point, and vulnerability to climate change (46). Some states have special grounds for opposing strong mitigation measures: in particular, countries with large potential sources of emissions—such as oil or coal reserves or forests—whose use might be restricted, or who have already cut emissions (for example, through nuclear power or energy efficiency) and have little potential for further reductions. In contrast, other states have low- or negative-cost mitigation options.

4. *Complicated political dynamic*—Finally, the climate change regime has involved negotiations not merely within the North, between the North and the South, or within the South, but of all three types simultaneously (9, 46). Industrialized countries have differed on the targets and timetables issues (47). Although the United States was the only OECD country in the INC to oppose targets and timetables altogether, the others have disagreed about the types of

gases that should be covered (CO_2 or all GHGs not controlled by the Montreal Protocol), whether to use a historical or per capita baseline, whether to control net or gross emissions, and whether to allow states to implement their obligations jointly. Meanwhile, industrialized and developing countries have differed on the terms of financial and technological transfers, and more fundamentally on whether all states should undertake commitments to help prevent climate change, or whether the problem is mainly the responsibility of the North, due to its historically high emissions of GHGs (48).[8] Finally, developing countries have tended to separate into three groups: the large, semi-industrialized states such as India, China, and Brazil, who seek to ensure that the climate change regime does not hinder their ability to grow economically; the Alliance of Small Island States (AOSIS), whose members fear sea-level rise and support strong international action;[9] and the Gulf oil-producers, who stress the scientific uncertainties and argue for a "go slow" approach.

Given these obstacles, criticisms of the climate change regime for lack of progress are unrealistic. Like politics generally, building a climate change regime is the art of the possible. Rather than impose strong substantive obligations, which some states might not have accepted, the FCCC promotes consensus-building and cooperation. Because it requires relatively little, most states have been willing to become parties.[10] But once adopted, it may create a dynamic that eventually produces more substantive commitments. It sets forth basic goals and principles to guide the development of the regime. It encourages further research and data gathering, which may lead to a better understanding of the climate change problem and thereby raise mutual concern. It establishes a process for the adoption of additional commitments, should they eventually be deemed necessary. It creates a regular forum for discussions, thereby helping to focus international attention and pressure. And it creates common expectations which, as the process builds momentum, may draw along even reluctant states. This framework goes well beyond earlier framework conventions on acid rain and ozone depletion. Although it by no means guarantees action, it helps generate the scientific and normative consensus necessary to adopt more substantive obligations in the future (33, 50).

[8]Industrialized countries, for example, are responsible for 85% of cumulative past emissions of CO_2 (49).

[9]At COP 1 in Berlin, India and many other developing countries joined forces with AOSIS (forming the so-called "Green Group") in pushing for stronger emissions commitments by industrialized countries.

[10]In little more than two years, more than 120 states have ratified the FCCC—a rapid acceptance rate compared to most international environmental agreements (49a).

OVERVIEW OF THE FCCC[11]

Basic Goals and Principles

OBJECTIVE The FCCC defines the regime's ultimate objective as the stabilization of atmospheric concentrations of greenhouse gases at safe levels (i.e. levels that would "prevent dangerous anthropogenic interference with the climate system"). Stabilization should be achieved within a time frame that (*a*) allows ecosystems to adapt naturally, (*b*) ensures that food production is not threatened, and (*c*) enables sustainable economic development (FCCC art. 2).

Two features of this objective are noteworthy. First, it focuses on atmospheric concentrations of GHGs rather than anthropogenic emissions. Second, it addresses not only concentration levels, but also rates of change. In large part, the regime's evolution will involve spelling out what this objective means. In particular: (*a*) What concentration levels and rates of change are "safe"? (*b*) What emission reductions are necessary to achieve these levels and in what time frame? Although science can provide guidance on these questions (for example, about the effects of concentration levels on the climate system and on ecosystems), ultimately they will require political answers (51–53).

PRINCIPLES In addition to defining the regime's ultimate objective, the FCCC enunciates several guiding principles (FCCC art. 3). These include:

1. *Equity*—States should protect the climate system for the benefit of present and future generations, in accordance with their common but differentiated responsibilities and respective capabilities. Developed countries should take the lead in combating climate change and its adverse effects.
2. *Special needs*—The special needs of developing countries, especially those that are particularly vulnerable or that would bear a disproportionate burden under the FCCC, should be given full consideration.
3. *Precaution*—Lack of full scientific certainty should not be used as a reason for postponing action.
4. *Cost-effectiveness and comprehensiveness*—Measures should ensure global benefits at the lowest cost; cover all relevant sources, sinks, and reservoirs of GHGs; and comprise all economic sectors (54). Efforts to address climate change may be carried out cooperatively.
5. *Sustainable development*—States have a right to sustainable development. Policies and measures to protect the climate system should be appropriate

[11]See Table 2

Table 2 Key Provisions of the FCCC

Objective	Stabilize atmospheric greenhouse gas concentrations at a level that would prevent dangerous anthropogenic interference with the climate system, within a time-frame sufficient to (*a*) allow ecosystems to adapt naturally, (*b*) protect food production, and (*c*) allow sustainable economic development (art. 2).
Principles	Intra- and intergenerational equity; differentiated responsibilities and respective capabilities; special needs of developing-country parties; right to sustainable development; precaution; cost-effectiveness; comprehensiveness; supportive and open international economic system (art. 3).
Commitments	All countries General commitments to develop national GHG inventories; formulate national mitigation and adaptation programs; promote and cooperate in scientific research, education, training, and public awareness (arts. 4(1), 5, 6).
	Developed countries (listed in Annex 1) Recognize that a return to earlier emission levels of CO_2 and other GHGs by the end of decade would contribute to modifying long-term emission trends, and aim to return to 1990 emission levels (art. 4(2)).
	OECD countries (listed in Annex 2) Commitments to fully fund developing-country inventories and reports; fund the incremental costs of agreed mitigation measures; provide assistance for adaptation; facilitate, promote, and finance technology transfer (art. 4(3)–(5)).
Institutions	Conference of the parties (art. 7), secretariat (art. 8), subsidiary body for scientific and technological advice (art. 9), subsidiary body for implementation (art. 10), financial mechanism (art. 11).
Reporting ("communication of) information"	All countries National GHG inventories; steps taken to implement the Convention (art. 12(1)). Developed countries Detailed description of policies and measures to limit GHG emissions and enhance sinks, and a specific estimate of their effects on emissions (art. 12(2)).
	OECD countries Details of financial and technological assistance measures (art. 12(3)).
Adjustment procedure	Reassessments of the adequacy of commitments every three years, based on the best available scientific information (art. 4(2)(d)).

for the specific conditions of each party and should be integrated with national development programs.

6. *International economic system*—States should promote a supportive and open international economic system, and should not take measures that constitute arbitrary, unjustifiable, or disguised barriers to trade.

Most of these principles reflect more general principles of international law—for example, the principles of common but differentiated responsibilities (55), intra- and intergenerational equity (56, 57), and precaution (58). The principle of common but differentiated responsibility, together with the right to sustainable development, addresses the concern of developing countries that the greenhouse issue not be used as a basis to limit their economic development. The FCCC specifically acknowledges that the GHG emissions of developing countries will need to grow (FCCC preamble, para. 3). Similarly, precautionary measures are appropriate given the time lags in the climate system (by the time we detect climate change unambiguously, we may have already committed ourselves to substantial warming) and the possibility of nonlinear responses and catastrophic harm.

The agreed principles in the FCCC establish the general framework for the development of the climate change regime. They provide benchmarks against which to evaluate proposals such as carbon taxes, tradeable emission entitlements, or targets and timetables. Significantly, the Berlin Mandate for the next round of negotiations refers to the FCCC's principles, and in particular the principle of equity (24).

But while the principles set the terms of debate for future discussions and negotiations, they do not determine directly what measures should or should not be taken. The precautionary principle, for example, does not specify the appropriate level of precaution, or how much certainty is needed before taking action (59). Nor does the principle of common but differentiated responsibilities specify the basis or degree of differentiation. In practice, these difficult issues will need to be resolved through a process of further negotiation.

Commitments

The FCCC imposes relatively few obligations on states. These obligations fall into three categories, applicable to different categories of countries, in accordance with the principle of common but differentiated responsibilities.

GENERAL COMMITMENTS The FCCC's general commitments apply to all parties, both developed and developing, and are intended to promote long-term national planning and international review. Each country must develop a national inventory of its anthropogenic emissions by sources and removals by sinks of GHGs; formulate and implement national programs to mitigate and adapt to climate change; report to the FCCC's Conference of the Parties on its national inventories and implementation steps; and promote and cooperate in scientific research, exchange of information, education, training, and public awareness related to climate change (FCCC arts. 4(1), 5, 6, 12(1)).

These commitments are general not only in their applicability to all states but also in their content. They do not compel particular actions; rather, the FCCC takes a "bottom-up" approach, encouraging states to develop and implement their own national programs to address climate change. The requirement of filing periodic reports, in particular, encourages more comprehensive and systematic review of existing policies, and may act as a catalyst for better coordination and planning among the various parts of national governments.

As a component of their bilateral assistance programs, several OECD countries have supported country study programs in developing and eastern European states, to help prepare GHG inventories, assess vulnerability to climate change, and evaluate possible response strategies. The US program has been the biggest of these, involving 55 countries as of March 1995.

TARGETS AND TIMETABLES In addition to these general commitments, the FCCC sets forth more specific commitments for industrialized countries (listed in Annex I to the FCCC),[12] including (*a*) a quasi-target and -timetable to return GHG emissions to 1990 levels by the year 2000; and (*b*) more stringent reporting requirements (FCCC arts. 4(2), 12(2)). In recent years, targets and timetables have become the preferred international method for controlling atmospheric pollution. While they constrain states, they allow each country to choose how it will meet its national target, whether by means of direct regulation, market mechanisms, or taxes, making them more politically acceptable than specific regulatory requirements.

In the INC, targets and timetables were perhaps the most contentious—and certainly the most publicized—issue. The European Community, joined to varying degrees by other Western states, supported a commitment to stabilize CO_2 emissions at 1990 levels by the year 2000, while the United States, under the Bush Administration, adamantly opposed targets and timetables (47). Ultimately, these countries compromised on a nebulous provision recognizing the desirability of "returning to" 1990 emission levels, apparently by the year

[12]Annex I lists 1. thirty five states—including (*a*) all members of the OECD except Mexico (which had not yet joined the OECD when the FCCC was adopted) and (*b*) countries with economies in transition (i.e. the former members of the Soviet bloc)—and 2. the European Community. Newly industrialized countries such as China, South Korea, and Malaysia are not included in Annex I and hence are not covered by the FCCC's quasi-target and -timetable.

2000 (FCCC art. 4(2)).[13] Among its many ambiguities, this quasi-target and -timetable does not clearly indicate (*a*) whether it applies to each GHG individually or to GHGs as a group, (*b*) whether it applies to gross or net emissions (i.e. emissions from sources minus removals by sinks), and (*c*) whether states must implement it within their own territory or may do so through cooperative activities with other countries (joint implementation).[14]

Even if interpreted as a binding target, Article 4(2) constitutes only a modest first step in combating greenhouse warming. First, it falls short of a long-term emissions stabilization target, since it does not address the period after the year 2000. Second, it applies only to industrialized countries, and does not address the projected increase in emissions by developing countries. According to preliminary estimates by the IPCC, even if all Annex I countries stabilized their emissions of CO_2 at 1990 levels through 2100, this would reduce annual emissions in 2100 by less than 15% and cumulative emissions by less than 10%, compared to the IPCC's mid-range business-as-usual scenario (IS92a) (B Bolin, report to INC 10, Aug. 22, 1994). Finally, even if the target were extended to all countries, global stabilization of CO_2 emissions at 1990 levels would not stabilize atmospheric concentrations until they had more than doubled, several hundred years from now. To stabilize concentrations of CO_2 at lower levels, global anthropogenic emissions would need to be substantially reduced, not merely capped (60).[15]

Nevertheless, while modest, the target and timetable set forth in Article 4(2) is proving difficult to meet. Most OECD countries are projected to have higher emissions in 2000 than in 1990 (61–63). The European Community has been unable to adopt a key part of its strategy for stabilizing emissions—a CO_2/energy tax—due primarily to opposition by the United Kingdom (64). As a result, unless it adopts additional measures, the EC is likely to fall several percent short of its collective stabilization target (23). Similarly, the US Climate Change Action Plan contains mostly voluntary measures, which industry may or may not implement, and rests on optimistic economic assumptions. In March

[13]For a detailed analysis of Article 4(2), see RA Reinstein. 1994. *The Framework Convention on Climate Change: Developed Country Commitments on Greenhouse Emissions.* Jan. (unpublished paper).

[14]In addition to the FCCC's quasi-target and -timetable, virtually all OECD countries have adopted national (or, in the case of several EC countries, regional) emissions targets. These were generally adopted prior to the FCCC, either to provide international leadership or in response to pressures from domestic groups or other states. These national targets, however, differ widely in approach, and in some cases are largely symbolic, consisting of general plans or proposals that have not been implemented through specific measures. Moreover, several are tied to reciprocal action by other countries (62, 66).

[15]According to the IPCC's latest figures for methane, stabilization of global emissions would stabilize atmospheric concentrations within 50 years, at 11% higher levels. A 10% reduction in methane emissions would stabilize concentrations at current levels (60).

1995, the United States conceded that its current climate plan is unlikely to meet the FCCC target. The basic problem is that few countries have been willing to make difficult political decisions to limit emissions. Currently, only five OECD countries have taxes on CO_2 (Denmark, Finland, the Netherlands, Norway, and Sweden), and even these countries are not assured of stabilizing emissions, since their taxes are not directly related to their emissions targets (62, 65). Little effort has been made to address the problem of emissions from the transportation sector, which account for the largest predicted increases in most OECD countries. Indeed, to the extent countries are likely to meet their targets, it may be for reasons unrelated to the climate change problem—for example, the collapse of eastern Germany's economy has more than offset emissions increases in western Germany (23). In short, the record to date illustrates the political and institutional—if not the technical—difficulties in limiting GHG emissions in most industrialized countries (66). These difficulties make the adoption of additional international regulatory requirements quite problematic, at least in the near term.

FINANCIAL AND TECHNOLOGY TRANSFER Even if industrialized countries were to succeed in meeting their collective and/or national goals of limiting GHG emissions, "the effect on total world emissions ... is likely to be dwarfed by growth in emissions in the large, rapidly growing developing countries" (62). A long-term solution to the climate change problem requires limiting these countries' emissions. For this reason, the FCCC includes commitments by OECD countries (listed in Annex II) to provide financial and technical assistance to developing countries (FCCC art. 4(3)–(5)) and explicitly acknowledges that developing-country implementation will depend on the adequacy of this assistance (FCCC art. 4(7)). Financial assistance may be provided either through the FCCC's financial mechanism, i.e. the Global Environment Facility (GEF), or bilaterally.

Unlike ordinary development assistance, the financial and technological assistance mandated by the FCCC is a form of North-South partnership; its fundamental purpose is to benefit the global environment by averting climate change, not to benefit developing countries themselves. Nonetheless, the amount of assistance provided thus far has been relatively modest—particularly compared to the expectations (and, in some cases, demands) of developing countries. In the FCCC, OECD countries agreed to provide full funding only for the costs of developing-country inventories and reports. Financial assistance for other mitigation measures depends on approval by the GEF and covers only a project's "incremental" costs (i.e. the additional costs relating to climate change mitigation) (FCCC art. 4(3)). Meanwhile, assistance for adaptation is limited to particularly vulnerable developing countries and covers unspecified types of costs (FCCC art. 4(4)).

ADEQUACY OF COMMITMENTS AND ADJUSTMENT PROCEDURE The FCCC acknowledges that its limited obligations may be inadequate, and that the regime will need to evolve in response to new scientific information. Accordingly, it establishes a process to reassess its commitments, modeled on that of the ozone regime, which has led to progressively stricter international regulation of ozone-depleting substances (67). The first reassessment occurred in April 1995 at COP 1, which concluded that the commitments of developed countries set forth in Article 4(2)(a) and (b) of the FCCC are inadequate, and agreed to begin negotiations on appropriate actions for the post-2000 period, including the strengthening of these commitments. The next reassessment is scheduled to take place not later than the end of 1998. By requiring periodic reassessments of the FCCC's commitments on the basis of the best available scientific information, the FCCC creates channels for scientific developments to influence the policy process. To maximize their impact, future IPCC assessments should be timed so that they can feed into this review cycle.

Institutions

When the climate change problem first emerged as a policy issue, some leaders felt that it required the development of supranational institutions, with authority to adopt and enforce regulatory standards (68). At the 1989 Hague Conference, 17 heads of state called for the establishment of "new institutional authority" to address climate change, with non-unanimous decision-making powers. This radical proposal was never pursued in the FCCC negotiations, and the FCCC instead relies on more traditional types of international institutions, which play a coordinating and facilitative role (see Table 3).

CONFERENCE OF THE PARTIES The annual Conference of the Parties (COP) serves as the FCCC's "supreme body," with authority to examine the Convention's obligations and institutional arrangements, to supervise its implementation, and to develop amendments and protocols (FCCC art. 7). Among its functions, it provides a permanent forum for discussion and negotiation, keeps states' attention focused on the climate change problem, and helps build a sense of community (69). Moreover, by allowing environmental and industry groups to attend as observers, it gives them a forum for making inputs and exerting pressure.

Although the COP has no explicit regulatory powers (unlike its analog in the ozone regime, which can tighten existing control measures on ozone-depleting substances by a two-thirds vote), its other decision-making authority is broad. COP 1, for example, initiated a new round of negotiations to strengthen the FCCC's commitments, established a pilot phase of joint implementation, adopted reporting and review procedures, designated a permanent secretariat, and defined the roles of its subsidiary bodies. The COP's decision-

Table 3 Climate Change Institutions

Name	Abbreviation	Description
Intergovernmental Negotiating Committee	INC	Established December 1990 by UN General Assembly. Negotiated the FCCC. Now replaced by the FCCC Conference of the Parties.
Conference of the Parties	COP	Established by FCCC art. 7. "Supreme body" of FCCC. Functions: regular review of FCCC implementation; decisions necessary to promote effective implementation; adoption of amendments and protocols. Meets yearly.
Secretariat		Established by FCCC art. 8. Administrative functions in support of COP and other Convention institutions.
Subsidiary Body for Scientific and Technological Advice	SBSTA	Established by FCCC art. 9. Composed of government experts. Provides assessments of scientific knowledge; reviews scientific/technical aspects of national reports.
Subsidiary Body for Implementation	SBI	Established by FCCC art. 10. Composed of government experts. Reviews policy aspects of national reports; assists COP in assessing aggregated effect of implementation measures.
Financial mechanism		"Defined" by FCCC art. 11. Operation entrusted to GEF on interim basis.
Intergovernmental Panel on Climate Change	IPCC	Established in 1988 by WMO and UNEP to provide assessments of the science, impacts, and policy aspects of climate change. First Assessment Report in August 1990; Second Report scheduled for December 1995.
Global Environment Facility	GEF	Established by World Bank, UNDP, and UNEP in 1991. Restructured in 1994.

making authority makes its voting rules vital, and states have not yet been able to reach agreement on this question. Some states insist that consensus should be required for important decisions such as the adoption of protocols, while others prefer a two-thirds or three-fourths voting rule for all substantive matters, so that a small minority of countries cannot block agreement. Until the COP is able to resolve this question, consensus decision-making will be the default rule, given the FCCC's requirement that the COP's rules of procedure themselves be adopted by consensus.

SUBSIDIARY BODIES The FCCC creates two subsidiary bodies to assist the COP—the Subsidiary Body for Scientific and Technological Advice (SBSTA) (FCCC art. 9) and the Subsidiary Body for Implementation (SBI) (FCCC art. 10). In theory, the SBSTA is intended to have a technical/analytic function, and the SBI a policy function. But because both bodies are open to participation by all parties and consist of government rather than independent experts, both could develop a political character. In practice, the exact roles of these bodies—and their relationship to the COP and to outside groups such as the IPCC and the GEF—will emerge only over time.

SECRETARIAT At COP 1, the parties decided that the interim secretariat initially established for the INC should become the FCCC secretariat, providing general administrative and policy support to the COP and its subsidiary bodies (FCCC art. 8). In the past several years, the secretariat has grown substantially in size. During the negotiations, it served a primarily administrative function. But since the FCCC's adoption, it has played an increasingly important role; for example, by proposing policy options that seek to bridge differences among states, organizing the FCCC's review processes, and serving as an information clearinghouse (70).

Over the past several years, the Secretariat has developed a program to promote technical cooperation activities (CC:COPE). Components include (*a*) a computerized information network (CC:INFO), in conjunction with UNEP, to disseminate information about the increasingly large volume of climate change activities and to match potential donors and recipients and (*b*) a training program (CC:TRAIN), undertaken in collaboration with the UN Institute for Training and Research (UNITAR), to facilitate the development of national implementation strategies in developing countries. COP 1 bolstered the Secretariat's technical cooperation function, directing it to prepare an inventory and assessment of potentially transferable technologies to mitigate and adapt to climate change.

FINANCIAL MECHANISM Rather than create a new climate fund, the FCCC entrusted the GEF with the operation of its financial mechanism on an interim basis (FCCC art. 11). The GEF was established in 1991 by the World Bank, UNDP, and UNEP to help developing countries address global environmental issues (71). In the FCCC negotiations, the financial mechanism was one of the most contentious issues. The large donor countries insisted on using the GEF, an institution that they had initially proposed and that they dominated, while developing countries favored creating a new institution under the control of the COP. The FCCC represents a compromise between these positions. It gives the COP authority over the financial mechanism's policies, program priorities, and eligibility criteria, while leaving the GEF with authority over individual funding decisions.

In 1994, representatives of 73 countries participating in the GEF agreed to restructure it for the 1994–1997 triennium (72), in response both to developing-country demands for greater transparency and democracy and to criticisms by environmentalists and outside reviewers (73). The restructured GEF is functionally autonomous from the World Bank and is governed by a 32-member Council, evenly split between developing and developed countries. The GEF's rules require the concurrence of developing and donor countries for all substantive decisions. Despite this restructuring, however, the nature of the operational linkages between the COP and GEF remains a source of contention, and at its first meeting, the COP renewed the GEF's role as the Convention's financial mechanism only on an interim basis, reserving the final decision for later. In practice, since the same countries are generally represented in both the GEF and the COP, and both institutions are likely to continue their traditions of consensus decision-making, the precise allocation of authority may be more symbolic than real.

The GEF's mandate permits funding only of the "incremental" costs of projects that produce global environmental benefits (and hence are ineligible for ordinary World Bank lending, which focuses on the local benefits of projects) (71). This limitation has raised the question: Which costs should be considered "incremental"? The World Bank has favored limiting GEF assistance to "net incremental costs"—that is, the difference between the total costs of a project and its local benefits. Developing countries and environmental NGOs, in contrast, argue that the GEF should provide assistance for the full costs of projects to implement the FCCC; this would permit funding of "no regrets" strategies, which have a negative net cost. In practice, determining which costs produce local vs. global benefits is nearly impossible, so the issue will need to be worked out flexibly and pragmatically, on a project-by-project basis, through negotiations between the GEF and the country concerned (74).

In June 1994, the GEF Council agreed to pursue a two-track programming approach, developing a long-term operational strategy while undertaking some project activities to ensure a smooth transition from the pilot phase to the restructured GEF. Initially, the GEF plans to give priority to building the capacity of developing countries, through country studies, research, training, institutional strengthening, and other "enabling activities." In its pilot phase, which ended in 1994, the GEF financed 41 climate change projects with costs totalling $259.6 million (75). In March 1994, donor countries agreed to a $2 billion replenishment of the GEF Trust Fund for the 1994–1997 period. If the proportion of GEF funds devoted to climate change activities remains the same as during the last triennium (approximately 38%),[16] then GEF will have $760 million for climate change activities through 1997.

INTERGOVERNMENTAL PANEL ON CLIMATE CHANGE Until now, the IPCC has served the crucial function of providing collective appraisals of scientific knowledge. Although questions continue to be raised occasionally about the IPCC's role (76), the IPCC has, in general, walked the tightrope between governmental ownership and professional autonomy, thereby maintaining both political and scientific legitimacy.[17] Its intergovernmental character has given governments a sense of ownership and stake in its work, leading them to accept its assessments as authoritative. The restructuring of the IPCC in 1992 to ensure greater participation by developing-country scientists will likely strengthen its acceptance by the one group of states that had previously criticized it. On the other hand, the IPCC—and in particular, Working Group I on the science of climate change—has managed to maintain its autonomy as a scientific body, and thereby its scientific credibility.[18]

After first playing a crucial role in legitimizing the climate change issue and laying the foundations for the FCCC negotiations, but then falling into eclipse during the negotiations themselves, the IPCC has had renewed influence during the current implementation phase of the FCCC. In contrast to negotiations, which are a political not a scientific process, implementation raises technical questions that the IPCC is better equipped to answer. The development of inventory methodologies, in conjunction with the OECD and the International Energy Agency (IEA), provides one illustration. Another may be the IPCC's effort to work out the implications of the FCCC's objective (namely to stabilize atmospheric concentrations of GHGs at a safe level), by developing inverse emissions scenarios, which start with a given concentration level and work backwards to calculate the emissions scenarios likely to achieve stabilization at that level (51).[19]

[16]The remaining 62% was spent on the three other GEF priority areas: preservation of biodiversity, limiting pollution of international waters, and protection of the stratospheric ozone layer.

[17]I am indebted to David Victor for this point.

[18]The reorganized IPCC now has a working group on economic and social issues, which will provide assessments of the economic costs of mitigation measures. It remains to be seen whether these economic assessments will achieve the same political and professional legitimacy as the IPCC's scientific assessments.

[19]Although the ultimate choice of concentration level is a political decision, the question of what emissions scenarios will achieve that concentration level is technical.

The creation of a new body (the SBSTA) to provide scientific and technological advice to the COP leaves the IPCC's role somewhat uncertain. The SBSTA's composition of government representatives, however, makes it unlikely to displace the IPCC from its role of providing authoritative technical assessments. Instead, the SBSTA will likely play a complementary function, translating the work of the IPCC into policy-relevant terms for the COP.

Implementation Mechanisms

The development of a strong reporting and review procedure for industrialized countries has been one of the most important achievements of the climate change regime since UNCED.[20] Reporting and review serve several functions (77, 78). Although they do not guarantee meaningful action, they put pressure on states by holding them up to domestic and international scrutiny. States may be less likely to make extravagant claims about their national programs if they know their reports will be reviewed internationally. Second, by improving transparency, the review process helps build confidence among parties that others cannot free ride without being caught. Third, reporting and review serve an educational function: By sharing information, states can benefit from each others' experiences. Finally, reporting and review produce useful information for assessing the effectiveness of the FCCC and the need for further commitments.

NATIONAL REPORTING Under the FCCC, each Annex I party must submit an initial national report ("communication of information") within six months of becoming a party to the Convention (e.g. September 21, 1994 for the charter members of the Convention) (FCCC art. 12(5)), and a second national report in April 1997. National inventories by Annex I parties will be required every year. Developing countries have three years in which to file their initial reports. As of March 1995, 22 Annex I parties had submitted their reports in a timely manner and only one had not.[21] Compared to other international environmental agreements, for which compliance with reporting requirements has been low (79), this is an excellent record.

Industrialized-country reports must contain three types of information: (*a*) a national inventory of anthropogenic emissions and removals of GHGs; (*b*) a description of national policies and measures to implement the Convention,

[20]A reporting and review procedure for developing countries has not yet been elaborated, since developing-country reports are not due until 1997.

[21]The remaining Annex I countries, as of March 1995, either were not yet parties to the FCCC or had joined the FCCC less than six months before.

and (*c*) projections of the effects of these policies and measures on future GHG emissions (FCCC art. 12(2)).[22]

Inventories Systematic national inventories serve two purposes. First, they improve our understanding of the sources and sinks of GHGs. Currently, reliable inventories exist only for CO_2 emissions from fossil fuels and cement manufacturing—and even these use different methodologies for gathering, integrating, and reporting data (80). Second, they provide a baseline for evaluating the FCCC's implementation and effectiveness. To promote the reliability and comparability of inventories, the OECD and IPCC have taken the lead in developing common methodologies for estimating, reporting, and verifying data, which were adopted by INC 9 in February 1994. The IPCC/OECD guidelines provide a standard table format for common source/sink categories, documentation standards, and a default methodology containing step-by-step instructions for preparing inventories of the principal sources and sinks of the key GHGs. If a country chooses to use its own methodology rather than this default methodology, it must include sufficient documentation to allow the reconstruction of the inventory from national activity data and emissions factors.

National policies and measures and projections of their effects Comparable data across countries, GHGs, and time is also essential for reporting on national policies and measures, both to allow meaningful review of the performance of individual countries and to evaluate the parties' overall progress in responding to climate change. Currently, comparisons among countries are hampered by the use of varying assumptions and methodologies, which can make substantial differences in projections of future emissions, and by variations in data quality; meanwhile, comparisons among GHGs are sensitive to our relative understanding of various sources and sinks and to the choice of time horizon for calculating global warming potentials (GWPs) (81). The reporting procedure developed by the INC and COP attempts to address these problems by requiring the use of common methodologies if practical and, if not, by requiring states (*a*) to make available as much information as possible about their assumptions and methodologies, and (*b*) to disaggregate data on a gas-by-gas—and, in some cases, a sector-by-sector—basis.

INTERNATIONAL REVIEW National reports by industrialized countries are subject to international review by teams of experts nominated by the FCCC parties (and certain international organizations) and selected by the FCCC Secretariat.

[22]In addition, OECD country reports must contain information about their implementation of the FCCC's financial and technological assistance commitments (FCCC art. 12(3)).

Both the SBSTA and the SBI will help oversee the review process, the SBSTA focusing on the scientific and technical aspects of the reports and the SBI on the policy aspects. The review mechanism is intended to be nonconfrontational and facilitative in nature, and has two components:

1. In-depth reviews of each national report, to promote individual accountability and enhance comparability. These in-depth reviews are like outside audits; they examine the reliability, consistency, accuracy, and relevancy of national reports by reviewing key data points, verifying methodologies, and comparing assumptions across countries and with international sources.[23] At a minimum, they involve paper reviews; but in addition, an international review team may visit a country, with the country's approval, to allow a greater dialogue between the review team and national authorities.

2. A synthesis report, which compiles and aggregates the data in the various Annex I country reports, to determine their overall progress in implementing the Convention. The first synthesis report of Annex I country reports was completed prior to COP 1 (61).

In addition to the official, intergovernmentally-sponsored reviews, nongovernmental groups have established their own informal review process, releasing two reviews in 1994 and another in 1995 (63, 82). The NGO reports evaluate each country's policies and measures to combat climate change, its implementation record, and the quality and reliability of data and projections contained in its national report.

DISPUTE RESOLUTION As a cooperative, forward-looking instrument, which attempts to encourage and facilitate rather than coerce national action, the FCCC does not include a robust mechanism to resolve disputes. It includes a boilerplate dispute settlement provision, calling for the settlement of disputes by negotiation, conciliation and, if both sides agree, arbitration or the International Court of Justice (FCCC art. 14). But this type of procedure, found in virtually every international environmental agreement, is seldom if ever used—in part because global environmental disputes do not have the bilateral character of more traditional types of international disputes (e.g. boundary issues).

In the FCCC negotiations, states could not agree on an alternative procedure to address implementation questions—such as, for example, the noncompliance procedures found in the Montreal Protocol and the recent Sulphur Dioxide Protocol (83). Instead, the FCCC merely calls on the parties to "consider"

[23]International data sources include both intergovernmental sources such as the EC's Coordination of Information on the Air (CORINAIR), the Food and Agriculture Organization (FAO), IEA, OECD, and the UN Energy Statistics Yearbook, and nongovernmental sources such as the World Resources Institute, World Watch, and the World Energy Council.

establishing a "multilateral consultative process" for the "resolution of questions" concerning implementation (FCCC art. 13).

FUTURE DIRECTIONS

The FCCC establishes a framework to address the climate change problem. The development of the climate change regime will involve fleshing out that framework through decisions of the parties and the adoption of additional legal instruments. This development could proceed in two, potentially complementary, directions: (*a*) elaborating the regime's institutions and implementation machinery, and (*b*) adopting more stringent substantive commitments. In the period following UNCED, the former process got under way, with the elaboration of detailed reporting and review procedures. Some analysts have recommended that the parties continue along this procedural path, by strengthening the reporting and review process, building the FCCC's institutions, and developing an effective noncompliance procedure, rather than attempting immediately to negotiate additional targets and timetables (65; but see 84). At COP 1, however, the parties eschewed a purely procedural approach, concluding that the specific commitments of the industrialized countries are inadequate and agreeing on a mandate to negotiate additional commitments for the period after 2000, when the current quasi-target and -timetable expires. The so-called Berlin Mandate creates a negotiating group to develop a legal instrument (most likely a protocol), with a target completion date of 1997, in time for COP 3. The negotiating group will be open to all parties, both developed and developing, although significantly the Mandate specifies that new commitments may be introduced only for industrialized countries (24).

Driving Forces

Although most countries now politically acknowledge the need for additional commitments to limit GHG emissions and enhance sinks, the new negotiations will face the same obstacles as before: potentially high abatement costs, uncertain benefits, divergent state interests, and a complicated negotiating dynamic within the North, between the North and the South, and within the South (85). Given the difficulties that industrialized countries have had in meeting their current, modest aim of returning to 1990 emission levels by the year 2000, it is hard to see how these countries will be able to assume more ambitious commitments soon.

What might change this situation? The experience of other international environmental regimes suggests that science, public opinion, and technology will be the main driving forces (86, 87). Between 1985 and 1990, for example, the ozone regime developed from a skeletal framework convention into an agreement requiring the phaseout of most ozone-depleting substances by the year 2000. This rapid progress resulted from (*a*) scientific assessments, which painted an increasingly dire picture of the problem, leading to successively stricter regulations; (*b*) the discovery of the Antarctic ozone hole in 1987, which catalyzed public concern; and (*c*) advances in technology, which eventually convinced industry that acceptable substitutes could be developed for most ozone-depleting substances (37, 88, 89).

SCIENTIFIC DEVELOPMENTS Science played a major role in putting the climate change issue on the international agenda and prompting the FCCC negotiations (90). If scientists were to find a clear signal that the earth has begun a process of greenhouse warming, or substantially reduced the uncertainties about the timing, magnitude, and regional effects of climate change, then these developments could spur the political process forward. Recognizing the importance of science, the FCCC requires the systematic reporting of data on sources and sinks, promotes cooperation in scientific research and systematic observation, establishes a body to provide scientific advice, and calls for periodic review of the Convention on the basis of the best available scientific information.

Currently, major international and national research efforts are under way outside the formal FCCC process, under the auspices of WMO and UNEP's World Climate Research Program and ICSU's International Geosphere-Biosphere Program (91, 92). Important areas of research include ocean uptake of GHGs, cloud properties, ocean transport of heat, and improvements in global circulation models through better resolution and coupling.

The IPCC's second assessment report, scheduled for completion in December 1995, will reportedly contain no major new conclusions, however. Instead, the new report reflects incremental progress in the science of climate change. Developments include a better understanding of the carbon cycle (in particular, uptake by terrestrial sinks) and better quantification of the radiative forcing by aerosols. But uncertainties persist about the bottom-line issues, and there still is no clear empirical evidence that global warming has begun.

Scientific certainty is likely to remain elusive for quite some time. Scientists may not be able to say with high confidence that they have detected anthropogenic climate change until well into the 21st century (93; see also 94). Without greater certainty, many states are likely to remain reluctant to take costly actions to mitigate climate change (the FCCC's invocation of the precautionary principle notwithstanding).

PUBLIC CONCERN Public opinion also helped propel climate change onto the international agenda in the late 1980s. Although since UNCED public concern about climate change has waned, a series of extreme events over an extended period of time—for example, serious floods, droughts, heat waves, epidemics, or changes in ocean circulation patterns—could have an effect similar to the discovery of the ozone hole: Such events could increase public concern and make it politically possible to takc unpopular measures such as imposing a carbon tax or increasing fuel- and energy-efficiency standards. The problem is that if climate change occurs relatively slowly and tranquilly, it may not raise public alarm. Moreover, even high public concern may not lead to effective action unless people better understand the causes of climate change and the appropriate policy responses. Surveys, for example, indicate that most people do not understand the connection between energy efficiency and climate change (95). In this regard, the FCCC correctly emphasizes the need for educational and public awareness programs on climate change and its effects (96).

TECHNOLOGICAL DEVELOPMENT If technological developments lowered the costs of mitigation significantly, it is conceivable that the climate change problem could be resolved through market forces or become politically easy to address. Technological options include noncarbon energy sources such as nuclear, solar, or wind power; improved supply and end-use efficiency; and climate engineering. By providing a clear signal that there will be a market for alternative technologies, commitments to limit GHG emissions could help stimulate the process of technological development towards a less carbon-intensive future.

Future Commitments to Limit Emissions and Enhance Sinks

GENERAL ISSUES In negotiating additional commitments to limit GHG emissions and enhance sinks, the parties to the FCCC will need to address several fundamental questions, either explicitly or implicitly (97, 98):

1. What is the desired level of abatement? The targets and timetables suggested thus far—such as stabilizing emissions at 1990 levels by the year 2000, or reducing emissions by 20% by the year 2005—are essentially arbitrary; they are convenient focal points, but have no scientific or economic basis. Ultimately, the FCCC's objective—namely to stabilize atmospheric concentrations at a safe level—should determine the appropriate level of emissions abatement. Economists tend to define safety in cost-benefit terms: the optimal level of abatement is that which minimizes the total expected costs of abatement and adaptation. The FCCC's objective, however, suggests another possible interpretation—that safety be defined in an absolute manner, in terms of the rate of change compatible with natural ecosystem adaptation, protection of food supplies, and sustainable development (FCCC art. 2) (51).

2. How should the burdens of abatement be distributed? Who should abate, by how much, and at whose expense? Burden-sharing involves issues of both efficiency and equity. The efficiency question is the easier of the two: abate-

ment should be undertaken by whoever can do so most cheaply. The equity question is more difficult, since there are many different conceptions of equity. In the longer term, the challenge will be to reach agreement on more nuanced criteria of equity than the simple categorizations of countries in the FCCC. Criteria for distributing the burdens of abatement include historical responsibility, current emissions levels per capita or per unit gross domestic product (GDP), or present ability to pay (99–101).

3. Should additional legal instruments (such as protocols or amendments) be comprehensive in scope, like the FCCC itself, or should they focus on particular sectors or GHGs?[24] A comprehensive approach would have the benefit of flexibility, and might thus be more acceptable to states than protocols that focus on particular GHGs or economic sectors. But it would require political (if not scientific) agreement on the global warming potentials of the various GHGs, and might be difficult to verify, since our understanding and monitoring of many sources and sinks are still quite poor. A sector-by-sector or gas-by-gas approach, in contrast, would preclude trade-offs across sectors or gases to equalize the burdens on different countries, but would have the advantage of permitting more specific, and possibly more verifiable, commitments.

4. Should separate protocols be developed for different categories of countries (OECD, economies in transition, newly industrializing, developing, etc), or should protocols be general in application?

OPTIONS FOR INTERNATIONAL POLICY INSTRUMENTS During the INC negotiations, most states focused on uniform national targets and timetables based on historical emissions as the principal policy instrument for limiting GHG emissions. But since UNCED, several alternatives have received increased attention, including common policies and measures and joint implementation. The choice among policy instruments will depend on a variety of factors, including political acceptability, efficiency, equity, the feasibility of domestic implementation, and verifiability.

Nonbinding benchmarks Given the difficulties of adopting binding commitments, one modest beginning would be to adopt additional nonbinding benchmarks for future control of GHG emissions. In essence, the quasi-target and -timetable set forth in Article 4(2) serves as an initial benchmark. Because benchmarks are merely aspirational, they are less threatening to states, and thus easier to adopt, than legal commitments. If they prove difficult to meet, a state can always argue that the benchmarks merely establish desirable goals,

[24]For example, the protocols to the Long-Range Transboundary Air Pollution Convention each address a particular precursor of acid rain: SO_2, NO_x, and volatile organic compounds.

not legal obligations. Nonetheless, agreed benchmarks can exert considerable political pressure on states—witness the current efforts by Annex I countries to demonstrate that they will return their GHG emissions to 1990 levels by the year 2000.

Targets and timetables Substantially reducing GHG emissions will eventually require moving from a voluntary regime of nonbinding benchmarks to a mandatory regime, since states will want stronger assurance that they are not proceeding alone in adopting costly and potentially uncompetitive measures. Thus far, most states seeking strong international action have focused on targets and timetables based on historical emissions as the preferred policy instrument. Prior to COP 1, AOSIS introduced a draft protocol containing a target and timetable for industrialized countries to reduce their CO_2 emissions by 20% by the year 2005 (the so-called "Toronto target"). Five OECD states have already adopted this as their national target (62). At COP 1, while the parties could not agree on any particular target and timetable, they agreed to negotiate a legal instrument setting "quantified limitation and reduction objectives within specified time-frames"—apparently a reference to targets and timetables, not merely nonbinding benchmarks.

Although historically-based targets and timetables were used successfully in the ozone regime, policy analysts tend to criticize them as both inefficient and inequitable (102). Many developed countries argue that such targets are unfair, given differences between countries in starting points, economic structures, resource bases, and other national circumstances. In response, both Article 4(2) of the FCCC and the Berlin Mandate acknowledge that these differences should be "taken into account." The problem will be to reach agreement on how to do so. As targets and timetables become increasingly stringent, these equity issues will continue to gain in prominence.[25]

Negotiating stricter targets thus poses the following dilemma. On the one hand, a uniform target serves as a focal point and is generally the easiest to negotiate (see Reference 103 on the concept of focal points). Once the Pandora's Box of differential targets is opened up, negotiations generally bog down in distributional issues. But because uniform targets impose greater burdens on some countries than others, the countries with the bigger burdens will be reluctant to accept strict targets. To be acceptable, stricter targets will thus require a complex package of incentives and exemptions.

Moreover, even were industrialized countries to agree on a target and time-

[25]Alternative allocation formulas based on such factors as population or GDP, however, appear no more promising. Although such formulas are arguably more equitable than historically based targets, they would require massive reallocations of emissions among countries. As a result, they would be even more difficult to negotiate.

table to reduce their emissions, such agreement would still fall far short of achieving stabilization of atmospheric concentrations. Ultimately, to stabilize atmospheric concentrations, targets and timetables would need to encompass at least the large developing countries such as China, India, and Brazil. But convincing developing countries to accept an emissions target will be extremely difficult; it will depend, at a minimum, on the provision of substantial financial and technological assistance by the North, which has not been forthcoming thus far. Accordingly, the Berlin Mandate provides that the next round of negotiations will not introduce any new commitments for developing countries.

Common policies and measures An alternative to uniform national targets are common policies and measures to combat climate change—for example, efficiency standards or a carbon tax. Either specific policies and measures could be mandated for all or particular groups of countries, or a menu of options could be developed from which countries could choose. The menu approach appears more realistic, since it permits national differences in regulatory and institutional approaches to be taken into account. The inability of the EC thus far to adopt a carbon tax illustrates the difficulties of adopting specific common measures, even within a comparatively homogenous group of states (64).

Given the difficulties in allocating emissions targets, the common policies and measures approach appears to be gaining in attractiveness. The Berlin Mandate provides that the upcoming negotiations should aim to develop common policies and measures as well as targets and timetables. In the early stage of this process, the parties will undertake an analysis and assessment to identify possible policies and measures.

Joint implementation Allowing states to implement their targets jointly could help circumvent some of the obstacles to stricter emissions targets. Under a joint implementation (JI) scheme, countries could meet their commitments by undertaking projects in other countries that limit GHG emissions or enhance sinks (104). Ultimately (although still a very long way off), JI could evolve into a full-blown system of tradeable emission entitlements (101, 105). The FCCC recognizes the concept of JI by noting that industrialized countries may implement their policies and measures to limit GHG emissions and enhance sinks "jointly with other Parties" (art. 4(2)(a)).

Joint implementation exploits the fact that, in mitigating global climate change, it does not matter where emissions reductions or sink enhancements occur. Since GHGs remain in the atmosphere for a long time and mix globally, a reduction or enhancement anywhere in the world has the same effect on atmospheric concentrations.

Supporters of JI argue that it has three benefits. First, it is cost-effective, since it allows response measures to be undertaken wherever they are cheapest (106). Lower costs make it more likely that states will be willing to take action. Second, since most analysts believe that mitigation costs are generally lower in developing and eastern European countries than in OECD countries, joint implementation would involve net transfers of money and technology from OECD to non-OECD countries—transfers that could eventually dwarf those by the GEF. To the extent that the private sector has incentives to engage in JI (for example, by receiving credits towards their domestic regulatory requirements), JI would tap a major new source of funding for developing countries. Finally, JI could make it politically easier to adopt stricter emissions targets, by helping to equalize the burden of regulation. Assuming a global market in joint implementation projects, the marginal costs of abatement would be the same for all countries.

Despite these putative advantages, developing countries (and environmental NGOs) have generally opposed JI, fearing that it would allow industrialized countries to avoid taking action at home to limit emissions. Moreover, if developing countries eventually become subject to emission targets, then unless they receive credit for past JI projects, they would start from a lower emissions baseline, with the lowest-cost mitigation options already exploited by JI projects. Finally, determining the net abatement effects of JI projects presents serious practical problems, both in establishing the baseline from which to calculate how much a JI project has limited emissions or enhanced sinks and then in monitoring the project's effects (107).

Given these differing views, COP 1 decided to establish JI initially on a pilot basis (24). The pilot phase will allow states to gain practical experience with JI, before making final decisions about sensitive matters such as crediting. To allay the concerns of developing countries that JI would allow industrialized countries to avoid taking domestic measures, activities undertaken during the pilot phase may not be credited toward industrialized countries' existing aim of returning emissions to 1990 levels by the year 2000. On the other hand, developing as well as industrialized countries may participate in the pilot phase—an important point sought by OECD countries, who wish to undertake JI projects in the South. Although during the pilot phase no emissions credits will be allowed for JI activities, if JI is eventually adopted on a permanent basis, it will require some system of crediting, since crediting is what distinguishes JI from ordinary financial and technological assistance. JI without credits would be like *Hamlet* without the prince.

Even before COP 1 approved the pilot phase, several countries supported JI-like projects, in order to gain experience and demonstrate JI's feasibility. For example, in 1993, Norway contributed to GEF projects in Poland on fuel substitution and in Mexico on energy efficiency (109). In 1994, the United States initiated its own pilot JI program and, as of February 1995, had approved seven projects. In addition, several private businesses—particularly in the energy sector—have sponsored JI-like projects, apparently in the hope that these projects will eventually count as offsets against domestic constraints on GHG emissions.

Technology Diffusion and Transfer

Limiting the growth of GHG emissions in developing and former Communist countries will require the transfer of technologies that meet these countries' economic needs in an environmentally sound manner. According to model results, small differences in the rate of technology diffusion could make a large difference in the total costs of mitigation (16).

Intellectual property rights are often seen as the major barrier to technology transfer; but in practice, most technologies to reduce GHG emissions are not difficult to obtain through purchase or licensing, and many are already in the public domain (110). The real issues are whether a given technology is appropriate to the circumstances of developing countries (for example, do these countries have the infrastructure and human resources necessary to use and maintain the technology effectively), and whether developing countries can afford the technology. For technology transfer to be successful, it must be sensitive to the needs of the recipient country; it must involve soft as well as hard technology; the local population must understand the need for the new technology; and in the long run, the technology must be economically viable, rather than dependent on ongoing subsidies (110, 111).

The climate change regime could play several roles in helping overcome the current obstacles to technology development—including inadequate information and starting capital. These roles include (*a*) identifying appropriate technologies, (*b*) disseminating information about them worldwide, and (*c*) financing their transfer. Thus far, the OECD and IEA have taken the lead in facilitating the diffusion and exchange of information about GHG technologies, through Greentie (the Greenhouse Gas Technology Information Exchange), a joint initiative begun in October 1993 (112). But the private sector could also play a significant role in assessing the technical and market potentials of particular technologies, as it did so successfully in the ozone regime. Other potential mechanisms to facilitate the dissemination of information include the IPCC's Technology Characterization Inventory (TCI) and the FCCC Secretariat's CC:COPE program. Financing of technology transfer is primarily the GEF's responsibility, although bilateral and regional assistance may also play a significant role (113).

Adaptation

If the costs of climate change are low, as many economists argue, adaptation could prove to be a more cost-effective response strategy than mitigation (41).

Moreover, according to the projections of global circulation models, some global warming will occur even if relatively drastic measures are taken to abate GHG emissions (2); indeed, past emissions may already have committed us to substantial warming. As a result, measures to adapt to a warmer climate will likely be needed—including measures both to avoid the adverse effects of climate change (increasing the resilience of natural systems, developing heat- and drought-resistant crops, shielding areas from the impacts of climate change, and so forth) and to respond to harms once they have occurred.

In contrast to mitigation, governments and individuals have individual incentives to undertake adaptation, since the benefits tend to be local not global.[26] But making a state bear the costs of adaptation is inequitable if the state did not contribute substantially to the climate change problem—as is the case for most developing countries and, in particular, small island states, which are especially vulnerable to climate change. The function of an international regime here is not to overcome the tragedy of the commons, but rather to make the polluter pay—for example, through a liability or insurance scheme.

During the negotiations of the FCCC, small island developing states lobbied hard to make the climate change regime address the problem of adaptation as well as mitigation. In particular, they proposed establishing an insurance fund to compensate countries that suffer damage as a result of sea-level rise. Although this initiative proved unsuccessful, the island states did succeed both in including a reference to insurance as one possible option (FCCC art. 4(8)), and in adding a provision requiring Annex II parties to assist developing states with adaptation costs (FCCC art. 4(4)). Most of this assistance will likely be bilateral; but, at the insistence of developing countries, COP 1 authorized the FCCC's financial mechanism (the GEF) to provide funding for planning and reporting activities relating to adaptation. In addition, the IPCC and UNEP have supported assessments of adaptation options.

Currently, establishing the existence of anthropogenic climate change, let alone its adverse effects, is controversial; so assistance for specific adaptation measures may be premature. Instead, the parties have focused on preliminary work, such as assessing a country's vulnerability to climate change, identifying possible adaptation measures, and building local capacity. Although the geographical distribution and impacts of global warming are uncertain, states can begin now to prepare for climate change, through research, contingency planning, and the development of more robust and flexible ecological and social systems, which are less vulnerable to climate change (114). Indeed, many of these adaptation policies may be "no regrets," increasing the resilience of

[26]This is the principal reason why the GEF has resisted providing assistance for adaptation measures, since its mandate is limited to projects with global benefits.

systems to environmental and anthropogenic stresses other than global warming.

Compliance

The FCCC relies primarily on self-reporting to gather information about implementation; while the in-depth reviews of national reports check national data against available international sources, they do not at present involve full-scale verification. (On monitoring and verification of international environmental agreements, see Reference 116.) Due to the potentially high costs of abatement, some observers contend that states will undertake strong commitments to limit emissions only if the climate change regime contains mechanisms to verify that others are complying with their obligations (115). Although some environmental agreements have been successful without strong verification procedures—the Montreal Protocol on Substances that Deplete the Ozone Layer contains stringent control measures, but relies primarily on self-reporting rather than international verification—the incentives to defect are likely to be stronger in the climate case, making verification more important.

Thus far, the parties to the FCCC have not been able to decide whether to establish a multilateral consultative process (MCP) for the resolution of questions regarding implementation. At COP 1, they agreed to create an open-ended working group of technical and legal experts to study the relevant issues. These issues include (*a*) whether the MCP should focus on "noncompliance" or take a less judgmental approach, aimed at building confidence and encouraging mutual learning by the parties, and (*b*) what the relationship of the MCP should be to the existing review process for national reports (117).[27]

CONCLUSION

The climate change question raises complicated issues of decision-making under uncertainty. Experts disagree substantially over the degree of risk and the appropriate response (118). Waiting may lead to reduced scientific uncertainties and cheaper response options; but it could also dig us into a deeper hole, making it more difficult to get out. The enormous uncertainties suggest the need to buy time and flexibility through hedging strategies that include a mix of scientific research, technological development, and abatement measures (119).

[27]One model for the FCCC is the noncompliance procedure adopted under the Montreal Protocol on Substances that Deplete the Ozone Layer, which (*a*) is forward-looking, aimed at bringing a party into compliance rather than adjudicating guilt; and (*b*) involves decision-making by the parties themselves—through the Protocol's Implementation Committee and Conference of the Parties— rather than by a third-party arbitrator or judge (33).

The climate change regime represents one such hedging strategy. It views climate change as a potentially serious problem and, in response, creates a long-term evolutionary process that encourages scientific research and cooperation; induces states to begin a process of national planning and to explore possible response strategies; promotes transparency through its reporting and review procedures; helps build capacity in poorer countries; keeps attention focused on the issue through its institutions and meetings; and, more intangibly, helps create a sense of community among states. But while it sends a political signal that past emission trends will not be allowed to continue indefinitely, it requires very little by way of specific mitigation or adaptation measures. It does not require states to make significant—and potentially costly—changes in their economies and life-styles.

If the skeptics prove correct, then the climate change regime will represent a misguided, but relatively harmless, endeavor. Indeed, some of its elements represent "no regrets" strategies: they are useful whether or not climate change turns out to be a serious problem.

But if climate change proves as big a problem as many scientists believe, then planning and marginal adjustments will no longer be sufficient; states will need to take substantive—and possibly painful—measures to decrease net flows of GHGs into the atmosphere, or adapt to the adverse effects of climate change as best they can. The FCCC does not prejudge the appropriate response strategies; it does not indicate, for example, the proper mix of mitigation and adaptation. But while it does not ensure that states will reach agreement on the necessary actions, it at least puts them in a better position to do so.

With a complex, controversial, and continuing issue such as climate change, what is important initially is to establish a meaningful international process, to build a basis for future action, and to draw in skeptical countries. The international regime established by the FCCC is still in its formative stages. But it should keep attention focused on the climate change issue, both internationally and nationally, and is capable of evolution and growth, should the political will to take stronger international action emerge (120).

ACKNOWLEDGMENTS

The author would like to thank Salvano Briceno, Richard Kinley, Richard Moss, Jonathan Pershing, David Victor, and several anonymous reviewers for providing many helpful comments and corrections on this article. They are, of course, not responsible for any remaining mistakes or misinterpretations.

Literature Cited

1. IPCC. 1992. *Climate Change 1992: The Supplementary Report to the IPCC Scientific Assessment,* ed. JT Houghton, BA Callander, SK Varney. Cambridge: Cambridge Univ. Press
2. IPCC. 1990. *Climate Change: The IPCC Scientific Assessment,* ed. JT Houghton, GJ Jenkins, JJ Ephraums. Cambridge: Cambridge Univ. Press
3. Hadley Cent. Clim. Predict. Res. 1995. *Modelling Climate Change: 1860–2050.* London: UK Meteorol. Off.
4. IPCC. 1990. *Climate Change: The IPCC Impacts Assessment,* ed. WJ McG Tegart, GW Sheldon, DC Griffiths. Canberra: Aust. Gov. Publ. Serv.
5. Strzepek KM, Smith JB, eds. 1995. *As Climate Changes: International Impacts and Implications.* Cambridge: Cambridge Univ. Press
6. Bodansky D. 1994. Prologue to the Climate Change Convention. See Ref. 9, pp. 45–74
7. Pomerance R. 1989. The dangers from climate warming: a public awakening. In *The Challenge of Global Warming,* ed. DE Abrahamson, pp. 259–69. Washington, DC: Island
8. IPCC. 1991. *Climate Change: The IPCC Response Strategies.* Washington, DC: Island
9. Mintzer IM, Leonard JA, eds. 1994. *Negotiating Climate Change: The Inside Story of the Rio Convention.* Cambridge: Cambridge Univ. Press
10. Corell RW, Anderson PA, eds. 1991. *Global Environmental Change.* Berlin: Springer-Verlag
11. FCCC Interim Secretariat. 1994. Relevant Activities Outside the Financial Mechanism: Consistency with Convention Policies. *UN Doc. A/AC.237/71*
12. World Bank. 1995. *The World Bank and the UN Framework Convention on Climate Change.* World Bank Environ. Dept. Pap., Climate Change Ser. No. 008. Washington, DC: World Bank
13. Subak S, Raskin P, Von Hippel D. 1993. National greenhouse gas accounts: current anthropogenic sources and sinks. *Clim. Change* 25(1):15–58
14. World Energy Council Commission. 1993. *Energy for Tomorrow's World—The Realities, the Real Options and the Agenda for Achievement.* London: Kogan Page
15. Grubb M, Rose A. 1992. Introduction: nature of the issue and policy options. See Ref. 101, pp. 1–10
16. Off. Technol. Assess., US Congress. 1994. *Climate Treaties and Models: Issues in the International Management of Climate Change.* OTA-BP-ENV-128. Washington, DC: US Gov. Print. Off.
17. Bodansky DM. 1993. The United Nations Framework Convention on Climate Change: a commentary. *Yale J. Int. Law* 18(2):451–558
18. Barratt-Brown EP, Hajost SA, Sterne JH. 1993. A forum for action on global warming: the UN Framework Convention on Climate Change. *Colo. J. Int. Environ. Law Policy* 4(1):103–18
19. Grubb M. 1992. The Climate Change Convention: an assessment. *Int. Environ. Rep.* 15:540–43
20. Sands P. 1992. The United Nations Framework Convention on Climate Change. *Rev. Eur. Commun. Int. Environ. Law* 1(3):270–77
21. Kelly M, Granich S. 1992. A step in the right direction? *Tiempo* 6:1–2
22. Hammitt JK, Lempert RJ, Schlesinger ME. 1992. A sequential-decision strategy for abating climate change. *Nature* 357:315–18
23. Victor DG, Salt JE. 1994. From Rio to Berlin: managing climate change. *Environment* 36(10):6–15, 25–32
24. FCCC Conf. Parties. 1995. Report of First Conference of the Parties. *UN Doc. FCCC/CP/1995/7/Add. 1*
25. Sebenius JK. 1991. Designing negotiations toward a new regime: the case of global warming. *Int. Secur.* 15(4):110–48
26. Skolnikoff EB. 1990. The policy gridlock on global warming. *Foreign Policy* 79:77–93
27. DeCanio SJ. 1992. International cooperation to avert global warming: economic growth, carbon pricing, and energy efficiency. *J. Environ. Dev.* 1(1): 41–62
28. Shine KP, Derwent RG, Wuebbles DJ, Morcrette JJ. 1990. Radiative forcing of climate. See Ref. 2, pp. 41–68
29. Panel Policy Implications Greenhouse Warming, US Natl. Acad. Sci. 1992. *Policy Implications of Greenhouse Warming: Mitigation, Adaptation, and the Science Base.* Washington, DC: Natl. Acad. Press
30. Krasner SD. 1983. Structural causes and regime consequences: regimes as intervening variables. In *International Regimes,* ed. SD Krasner, pp. 1–21. Ithaca: Cornell Univ. Press
31. Levy MA, Young OR, Zurn M. 1994. *Working Paper: The Study of Interna-*

tional Regimes. WP-94–113. Laxenburg, Austria: Int. Inst. Appl. Syst. Anal.

32. Young OR. 1994. *International Governance: Protecting the Environment in a Stateless Society.* Ithaca: Cornell Univ. Press
33. Gehring T. 1994. *Dynamic International Regimes: Institutions for International Environmental Governance.* Frankfurt: Lang
34. Sand PH. 1990. *Lessons Learned in Global Environmental Governance.* Washington, DC: World Resour. Inst.
35. Victor DG, Chayes A, Skolnikoff EB. 1993. Pragmatic approaches to regime building for complex international problems. In *Global Accord: Environmental Challenges and International Responses,* ed. N Choucri, pp. 453–74. Cambridge, MA: MIT Press
36. Levy MA. 1993. European acid rain: the power of tote-board diplomacy. See Ref. 50, pp. 75–132
37. Parson EA. 1993. Protecting the ozone layer. See Ref. 50, pp. 27–73
38. Benedick RE. 1991. *Ozone Diplomacy: New Directions in Safeguarding the Planet.* Cambridge, MA: Harvard Univ. Press
39. Fankhauser S. 1994. The economic costs of global warming damage: a survey. *Glob. Environ. Change* 4(4):301–9
40. Nordhaus WD. 1993. Reflections on the economics of climate change. *J. Econ. Persp.* 7(4):11–25
41. Ausubel JH. 1991. Does climate still matter? *Nature* 350:649–52
42. Cline WR. 1992. *Global Warming: The Economic Stakes.* Washington, DC: Inst. Int. Econ.
43. Grubb M, Edmonds J, Ten Brink P, Morrison M. 1993. The costs of limiting fossil-fuel CO_2 emissions: a survey and analysis. *Annu. Rev. Energy Environ.* 18:397–478
44. Weyant JP. 1993. Costs of reducing global carbon emissions. *J. Econ. Persp.* 7(4):27–46
45. Wilson D, Swisher J. 1993. Exploring the gap: top-down versus bottom-up analyses of the cost of mitigating global warming. *Energy Policy* 21(3):249–63
46. Paterson M, Grubb M. 1992. The international politics of climate change. *Int. Aff.* 68(2):293–310
47. Kjellen B. 1994. A personal assessment. See Ref. 9, pp. 149–74
48. Dasgupta C. 1994. The climate change negotiations. See Ref. 9, pp. 129–48
49. Grubler A, Fujii Y. 1991. Inter-generational and spatial equity issues of carbon accounts. *Energy—The Int. J.* 16:1397–416

49a. Spector BI, Korula AR. 1993. Problems of ratifying international environmental agreements: overcoming obstacles in the post-agreement negotiation process. *Glob. Environ. Change* 3(4):369–81

50. Haas PM, Keohane RO, Levy MA, eds. 1993. *Institutions for the Earth: Sources of Effective International Environmental Protection.* Cambridge, MA: MIT Press
51. IPCC. 1994. *IPCC Special Workshop: Article 2 of the United Nations Framework Convention on Climate Change, Fortaleza, Brazil.* Geneva: IPCC
52. Swart RJ, Vellinga P. 1994. The "ultimate objective" of the Framework Convention on Climate Change requires a new approach in climate change research. *Clim. Change* 26(4):343–49
53. Moss RH. 1995. Avoiding "dangerous" interference in the climate system: the roles of values, science, and policy. *Glob. Environ. Change* 5(1):3–6
54. Stewart RB, Wiener JB. 1992. The comprehensive approach to global climate policy: issues of design and practicality. *Ariz. J. Int. Comp. Law* 9(1): 83–112
55. Magraw DB. 1990. Legal treatment of developing countries: differential, contextual and absolute norms. *Colo. J. Int. Environ. Law Policy* 1:69–99
56. Yamin F. 1994. *Principles of Equity in International Environmental Agreements with Special Reference to the Climate Change Convention.* London: Found. Int. Environ. Law Dev.
57. Weiss EB. 1989. *In Fairness to Future Generations: International Law, Common Patrimony and Intergenerational Equity.* Dobbs Ferry, NY: Transnational
58. O'Riordan T, Cameron J, eds. 1994. *Interpreting the Precautionary Principle.* London: Cameron & May
59. Bodansky DM. 1991. Scientific uncertainty and the precautionary principle. *Environment* 33(7):4–5, 43–44
60. IPCC. 1994. *IPCC Special Report: Summaries for Policymakers and Other Summaries.* Geneva: IPCC
61. FCCC Interim Secretariat. 1994. Compilation and Synthesis of National Communications from Annex I Parties. *UN Doc. A/AC.237/81*
62. IEA. 1994. *Climate Change Policy Initiatives—1994 Update.* Vol. 1: *OECD Countries.* Paris: OECD/IEA
63. Climate Action Network. 1995. *Independent NGO Evaluations of National Plans for Climate Change Mitigation: OECD Countries, Third Review, Jan. 1995.* Brussels: Climate Network Europe
64. Vellinga P, Grubb M, eds. 1993. *Cli-*

mate Change Policy in the European Community. London: R. Inst. Int. Aff.
65. Victor DG, Salt JE. 1995. Keeping the climate treaty relevant. *Nature* 373:280–82
66. Fish AL, South DW. 1993. *Global Climate Change: Actions and Policies of National Governments.* Argonne: IL: Argonne Natl. Lab.
67. Parson EA, Greene O. 1995. The complex chemistry of the international ozone agreements. *Environment* 37(2):16–20, 35–43
68. Palmer G. 1992. The implications of climate change for international law and institutions. *Transnatl. Law Contemp. Probl.* 2(1):205–57
69. Nitze WA, Miller AS, Sand PH. 1992. Shaping institutions to build new partnerships: lessons from the past and a vision for the future. In *Confronting Climate Change: Risks, Implications and Responses,* ed. IM Mintzer, pp. 337–50. Cambridge: Cambridge Univ. Press
70. Sandford R. 1994. International environmental treaty secretariats: stagehands or actors? In *Green Globe Yearbook 1994,* pp. 17–29. Oxford: Oxford Univ. Press
71. Jordan A. 1994. Paying the incremental costs of global environmental protection: the evolving role of GEF. *Environment* 36(6):12–20, 31–36
72. GEF. 1994. *Instrument for the Establishment of the Restructured Global Environment Facility.* Washington, DC: GEF
73. *Report of the Independent Evaluation of the GEF Pilot Phase.* 1993. Washington, DC: UNEP/UNDP/World Bank
74. Jordan A, Werksman J. 1994. Additional funds, incremental costs and the global environment. *Rev. Eur. Commun. Int. Environ. Law* 3(2/3):81–87
75. GEF. 1995. Report by the GEF to the First Conference of the Parties of the Framework Convention on Climate Change. *UN Doc. FCCC/CP/1995/4*
76. Berlin and global warming policy. 1995. *Nature* 374:199–200
77. Sachariew K. 1992. Promoting compliance with international environmental legal standards: reflections on monitoring and reporting mechanisms. *Yearbook Int. Environ. Law—1991* 2:31–52. London: Graham & Trotman
78. Chayes A, Chayes AH. 1993. On compliance. *Int. Organ.* 47(2):175–205
79. Gen. Account. Off., US Congress. 1992. *International Environment: International Agreements Are Not Well Monitored.* GAO/RCED-92-43. Washington, DC: US Gov. Print. Off.
80. Von Hippel D, Raskin P, Subak S, Stavisky D. 1993. Estimating greenhouse gas emissions from fossil fuel consumption: two approaches compared. *Energy Policy* 21(6):691–702
81. Harvey LDD. 1993. A guide to global warming potentials (GWPs). *Energy Policy* 21(1):24–34
82. Climate Action Network. 1995. *Independent NGO Evaluations of National Plans for Climate Change Mitigation: Central and Eastern Europe, First Review, Jan. 1995.* Brussels: Climate Network Europe
83. FCCC Interim Secretariat. 1995. A Review of Selected Non-Compliance, Dispute Resolution and Implementation Review Procedures. *UN Doc. FCCC/CP/1995/Misc.2*
84. Lashof DA, Subak SE. 1995. Reducing greenhouse gases. *Nature* 374:300 (letters responding to Victor & Salt, Ref. 65)
85. Sebenius JK. 1994. Towards a winning climate coalition. See Ref. 9, pp. 277–320
86. Hahn RW, Richards KR. 1989. The internationalization of environmental regulation. *Harvard Int. Law J.* 30(2):421–46
87. Morrisette PM, Darmstadter J, Plantinga AJ, Toman MA. 1991. Prospects for a global greenhouse gas accord: lessons from other agreements. *Glob. Environ. Change* 1(3):209–23
88. Morrisette PM. 1989. The evolution of policy responses to stratospheric ozone depletion. *Nat. Resour. J.* 29:793–820
89. Isaksen ISA. 1993. The role of scientific assessments on climate change and ozone depletion for negotiations of international agreements. *Int. Chall.* 13(2):76–84
90. Lanchberry J, Victor DG. 1995. The role of science in the global climate negotiations. In *Green Globe Yearbook 1995,* pp. 29–40. Oxford: Oxford Univ. Press
91. WMO. 1992. *World Climate Programme 1992–2001.* WMO No. 762. Geneva: WMO
92. IGBP. 1994. *Global Change—IGBP in Action: Work Plan 1994–1998.* Rep. No. 28. Stockholm: Swed. Acad. Sci.
93. Wigley TM, Barnett TP. 1990. Detection of the greenhouse effect in the observations. See Ref. 2, pp. 239–55
94. Schneider SH. 1994. Detecting climatic change signals: are there any "fingerprints"? *Science* 263:341–47
95. Kempton W. 1993. Will public environ-

mental concern lead to action on global warming? *Annu. Rev. Energy Environ.* 18:217–45
96. Rudig W. 1995. *Public Opinion and Global Warming: A Comparative Analysis.* Strathclyde Pap. Gov. Politics No. 101. Glasgow: Univ. Strathclyde
97. Andresen S, Wettestad J. 1992. International resource cooperation and the greenhouse problem. *Glob. Environ. Change* 2(4):277–91
98. Barrett S. 1991. Economic analysis of international environmental agreements: lessons for a global warming treaty. In *Responding to Climate Change: Selected Economic Issues,* pp. 109–49. Paris: OECD
99. Grubler A, Nakicenovic N. 1992. *International Burden Sharing in Greenhouse Gas Reduction.* World Bank Environ. Working Pap. No. 55. Washington, DC: World Bank
100. Grubb M, Sebenius J, Magalhaes A, Subak S. 1992. Sharing the burden. See Ref. 69, pp. 305–22
101. UN Conf. Trade Dev. 1992. *Combating Global Warming: Study on a Global System of Tradeable Carbon Emission Entitlements.* UNCTAD/RDP/DFP/1. Geneva: United Nations
102. Grubb M. 1990. *Energy Policies and the Greenhouse Effect.* Vol. 1: *Policy Appraisal.* London: R. Inst. Int. Aff.
103. Schelling TC. 1960. *The Strategy of Conflict.* Cambridge, MA: Harvard Univ. Press
104. Kuik O, Peters P, Schrijver N. 1994. *Joint Implementation to Curb Climate Change: Legal and Economic Aspects.* Dordrecht: Kluwer Academic
105. UN Conf. Trade Dev. 1994. *Combating Global Warming: Possible Rules, Regulations and Administrative Arrangements for a Global Market in* CO_2 *Emission Entitlements.* UNCTAD/GID/8. Geneva: United Nations
106. Barrett S. 1994. *The Strategy of Joint Implementation in the Framework Convention on Climate Change.* UNCTAD/GID/10. Geneva: United Nations
107. *Joint Implementation from a European NGO Perspective.* 1994. Brussels: Climate Network Europe
108. Deleted in proof
109. Anderson RJ Jr. 1995. *Joint Implementation of Climate Change Measures.* World Bank Environ. Dept. Pap., Climate Change Ser. No. 005. Washington, DC: World Bank
110. MacDonald GJ. 1992. Technology transfer: the climate change challenge. *J. Environ. Dev.* 1(1):1–39
111. Climate Network Africa. 1994. Climate change and technology: is technology the answer? *Impact* 12:1–10
112. Greentie. 1994. *Annual Report 1993/94.* Sittard, Netherlands: Greentie
113. Baldwin SF, Burke S, Dunkerley S, Komor P. 1992. Energy technologies for developing countries: US policies and programs for trade and investment. *Annu. Rev. Energy Environ.* 17:327–58
114. Off. Technol. Assess., US Congress. 1993. *Preparing for an Uncertain Climate.* OTA-0–567/568. Washington, DC: US Gov. Print. Off.
115. Wettestad J. 1991. Verification of international greenhouse agreements: a mismatch between technical and political feasibility? *Int. Chall.* 11(1):41–47
116. Ausubel JH, Victor DG. 1992. Verification of international environmental agreements. *Annu. Rev. Energy Environ.* 17:1–43
117. Victor DG. 1994. *Design Options for Article 13 of the Framework Convention on Climate Change.* Laxenburg, Austria: Int. Inst. Appl. Syst. Anal.
118. Nordhaus WD. 1994. Expert opinion on climatic change. *Am. Sci.* 82:45–51
119. Manne AS, Richels RG. 1992. *Buying Greenhouse Insurance: The Economic Costs of Carbon Dioxide Emission Limits.* Cambridge, MA: MIT Press
120. Bodansky D. 1993. Managing climate change. *Yearbook Int. Environ. Law—1992.* 3:60–74

Annu. Rev. Energy Environ. 1995. 20:463–92

NATIONAL MATERIALS FLOWS AND THE ENVIRONMENT

Iddo K. Wernick and Jesse H. Ausubel

Program for the Human Environment, The Rockefeller University, New York, NY 10021

KEY WORDS: environmental indicators, industrial ecology, dematerialization

---

CONTENTS

ABSTRACT

The functioning of modern societies requires large flows of materials to satisfy human wants both directly and indirectly; for example, 50 kg per day per American. The nature of these flows determines their impact on the natural environment. We develop and test a comprehensive framework to order materials flows in the US economy. We assess and quantify inputs to the national economy, outputs, foreign trade, and wastes from resource extraction, using mass measures of these flow components. The bulk of materials inputs satisfies demand for energy, construction, and food. Atmospheric emissions and materials embedded in long-lived structures dominate outputs, with smaller contributions from solid wastes and dissipated materials. Trade, accounting for approximately 10% of US materials flows, is dominated by bulk commodities such as fuel, food, and chemicals. Extractive wastes from fuel and nonfuel minerals account for more than double the amount of inputs and mostly remain at the site of generation. Metrics based on a consistent, periodic accounting of

1056-3466/95/1022-0463$05.00

physical materials flows can provide a powerful means to assess environmental performance at the national level. Improvements in the collection and organization of the data supporting national material accounts will further their utility.

Introduction and Method

Modern societies mine and metabolize large quantities of materials. Individuals directly consume food, clothing, and other goods ranging from accordions to zoom lenses. Less directly, people consume materials for shelter and power, for travel and communication, and in agriculture and industry. In 1990 each American mobilized on average about 20 metric tons of materials, over 50 kg day^{-1}, a mass equal to an average person's body weight every day or two. At this rate a person consumes about 1600 metric tons over the course of an 80-year life. This mass would occupy a cube almost 12 meters on a side with the density of water. With the typical density of a bag of household trash, the sides of the cube would exceed 18 meters. Such heavy use of materials necessarily raises environmental concerns.

This paper reviews materials flow at the national level from an environmental viewpoint. The framework we use gauges the flow at four stages: inputs to the economy, outputs, trade with other nations, and wastes from extractive industries. In assessing inputs we divide materials into six classes: energy, construction minerals, industrial minerals, metals, forestry products, and agriculture. We class outputs in five groups: domestic stock, atmospheric emissions, dissipation, other wastes, and recycled materials. We arrange foreign trade by individual commodities and classes of commodities. Extractive wastes include residues from the mining as well as the oil and gas industries. We provide brief comments on the flow of selected materials and materials classes at each of the four stages, with a focus on the relative size and the interdependence of the materials. Then, we propose a set of measures to evaluate national environmental performance with respect to materials. Finally, we consider applications of materials flow accounts and obstacles to their improvement.

We choose the United States as the subject of our study because it operates by far the world's largest economy and uses the most material by most measures (1). Moreover, it has well developed data that are familiar to us. The framework developed here should serve for other nations as well. In fact, contrasting national case studies would improve the framework and enhance interpretation of data. We select 1990 as the main reference year to allow for publication and revision of data yet remain relevant to current issues.

Various concerns have inspired comprehensive materials studies. In the 1950s in the United States, prospects of shortages of strategic materials in the event of conflict led to *Resources for Freedom* (*The Paley Report*), a congressional study which stressed questions of resource access (2). By the early

1970s, the adequacy and depletion of the natural resource base (3–6) in light of global economic development joined strategic fears as a major cause for assessment. During the 1980s, the environmental consequences of resource consumption motivated more studies (7, 8).

Since the Brundtland Commission report in 1987 (9), the need has grown for rigorous frameworks and measures to give substance to the debate over sustainable development (10, 11). Sustainability inevitably involves choices between the well-being and security of current and future generations. Such choice is arbitrary without periodic measures that monitor changes in national, as well as local and global, environmental performance.

Most recently, the emerging discipline of industrial ecology has inspired improved materials accounting (12–15). Industrial ecology is the study of the totality of the relationships among different industrial activities, their products, and the environment. Applications of industrial ecology should prevent pollution, reduce waste, and encourage reuse and recycling of materials. Understanding the relative scales and the relationships of the materials that flow through the national economy should expose areas of opportunity to better the performance of industrial ecosystems.

Analysts have examined anthropogenic materials flows from diverse viewpoints. Ayres has examined flows using the "materials balance principle" (16) and also introduced the biological metaphor of "industrial metabolism" (17). He has also followed global and national flows of specific environmentally sensitive heavy metals such as lead, chromium, and cadmium (18). Impressed by the heterogeneity of materials, other researchers have tried to translate them into common units of ecological impact (19). Other studies compared the scale of human activities globally with background or natural fluxes (20, 21). Researchers have also used monetary input-output models and insights from structural economics to describe the dynamics of materials stocks and flows in the economy as they affect environment (22–24).

Looking at secular trends in materials consumption, some authors emphasize the diminishing importance of basic materials in industrial economies relative to increasingly refined and complex ones (25–27). Correspondingly, several studies have suggested that after industrialization societies begin to "dematerialize:" Their economies grow while relative and even absolute demand for materials declines (28– 30).

All such studies would benefit from reliable, periodic information about national materials flows. The reasons such accounts do not already exist will become apparent in the course of this paper. First we mention three major difficulties with studying materials flows and how we resolve them, albeit tentatively.

The first difficulty is choosing a common currency to compare quantities of not only apples and oranges, but apples and aluminum. Quite independent units of environmental import describe materials: volume, energy content, and toxicity, for example. In the postindustrial age, the information content of materials becomes a more salient parameter, and future studies may use bits as well as kilograms as the unit of choice (31).

We choose weight as our initial standard. Although an incomplete indicator, weight conveys the sheer quantities of materials mobilized and consumed and is easily compared. Moreover, weight data are the most widely available. Because the boundaries of the system we examine are large, the natural unit of measure is million metric tons (MMT). We examine only physical and not monetary quantities. One should be able to prepare a monetary map corresponding to our physical map, indicating where value is added and lost.

The second difficulty is scope. We seek completeness by considering the reported weight of the great majority of all inputs, outputs, trades, and extractive wastes. We omit some materials used below an annual threshold of one MMT and unreported materials migrations, such as firewood consumed by individuals or black market trade.

Unless otherwise noted, we exclude water and give only the dry weight of materials. We do not explicitly treat water consumption because the mass of this ubiquitous and precious resource would obscure other materials. In 1990, consumptive use of fresh water (defined as water that has been evaporated, transpired, or incorporated into products, plant or animal tissue, and is therefore unavailable for immediate reuse) in the United States exceeded 34 trillion gallons or about 130 billion metric tons, some 25 times other inputs (32). For similar reasons, we do not consider consumption of atmospheric oxygen for biological respiration and in industrial processes. We do include atmospheric nitrogen fixed into NO_x emissions as well as that used for ammonia production.

Our framework also does not explicitly treat manufactured chemical products and by-products, many of which fall between our categories of inputs and outputs. Our framework could be further segmented to account more fully for intermediate products such as chemicals. Organic chemical production in 1990 was about 90 MMT (33). Inorganic chemical production was at a similar level (32). Our inputs category does account for the natural gas, petroleum, nitrogen, sulfur, phosphorus, sodium, chlorine, and metals used in making organic and inorganic chemicals. Initially benign starting materials used in chemical manufacture can acquire problematic (as well as useful) characteristics subsequent to thermal, chemical, and pressurized processing.The chemical industry generated about 350 MMT of hazardous waste (wet basis), more than half the US total in 1986 (34).

The third difficulty is the availability of data. The data presented here are gathered from various sources, primarily agencies of the federal government or published literature containing data from US government sources. The accuracy of the data depends on the accuracy of the reporting agency or author. Much of the data is self-reported, a method that has been criticized for underestimating actual waste values (35). Neither definitions of materials nor their end-uses are uniform across the data sources. We have labored to remain consistent in our own definitions to avoid redundancies and omissions in our account of the flows.

Both the accuracy and consistency of current data sources need to be improved. In the data notes and the table notes and comments we discuss our sources, their methods, and a sampling of inconsistencies between different sources and within individual sources. Looking forward, we seek a comprehensive framework that contains definitions sufficiently precise that data collectors can measure and report the same quantity or entity at different times with confidence.

Together, the considerations of common currency, scope, and quality of data recommend caution in drawing conclusions. Nevertheless, we do occasionally sum totals within and across categories and point out contrasts and resemblances. We venture to indicate the potential value of the accounts and to stimulate better analyses.

Inputs

Demand for energy, construction, and food largely forms the US menu of materials inputs (Table 1). The materials mobilized are primarily commodities such as coal, oil, sand, clay, steel, and grain, sold in bulk. The input menu contrasts with goods, i.e. consumer products, which weigh relatively little and sell more on the basis of value added during processing and manufacturing. Finished goods require heavy materials inputs, often not included in end products, in the form of facilities, equipment, and auxiliary production materials (e.g. coke and lime for steel, sodium and sulfur chemicals for paper). Thus, much of the bulk accounted in Table 1 constitutes the hidden consumption needed to support society.

Energy materials, adding to almost two billion metric tons, comprise just under 40% (by weight) of materials input to the US economy.[1] The residues of these materials present severe environmental problems, and handling and transporting this vast quantity of material requires significant energy use. Of all energy materials, coal consumption is highest. This solid fuel has the lowest energy density (i.e. BTU per kilogram) of the major fuels and is the hardest

[1]About 6% of energy materials are used for petroleum and natural gas products, such as road asphalt and plastics, not energy. Apparent US consumption of uranium concentrate (U_3O_8) in 1990, an energy material not listed in Table 1, was under 15 metric tons (61). See Table 4 for mining wastes from uranium production.

Table 1 Materials flows: US 1990—inputs

Material group	Apparent consumption[a]	(MMT)	Total US	Per capita per day (kg)	Total pcpd (kg)	Reference	Comments
Energy	Coal	843.2		9.26		32	Crude oil data are based on American Petroleum Institute (API) conversion values of 6,998 bbl./MT for foreign and 7.463 bbl./MT for domestic crude. We use an average value of 8 bbl./MT for petroleum products. Imports accounted for about 45% of crude oil and 10% of natural gas consumption. The value for petroleum products shows net U.S. import reliance for 1990.
	Crude oil	667.1		7.31			
	Natural gas	377.6		4.14			
	Petroleum products	62.8	1960.7	0.69	21.5		
Construction Minerals	Crushed stone	1092.8		11.98		74	
	Sand & Gravel	827.5		9.07			
	Dimension stone	1.1	1921.4	0.01	21.07		
Industrial Minerals	Salt	40.6		0.45		74	Anhydrous ammonia is the primary feed for nitrogen compounds, including ammonium nitrate, ammonium sulfate, urea, ammonium phosphates, and other fertilizer materials. The constituents of concrete and cement are accounted for separately in the table. What is included under the category Cement is net imports of hydraulic and clinker cement. Apparent consumption of cement in 1990 was 90.4 MMT (portland 75.6, hydraulic and clinker 11.5, and masonry 3.3). The value for lime is subtracted from crushed stone.
	Phosphate rock	39.9		0.44			
	Clays	38.8		0.43			
	Industrial sand & gravel	24.8		0.27			
	Gypsum	22.9		0.25			
	Nitrogen compounds	16.6		0.18			
	Lime	16.0		0.17			
	Sulfur	13.1		0.14			
	Cement	11.5		0.13			
	Soda ash	6.9		0.08			
	Other	17.7	248.6	0.19	2.73		

Metals	Iron & Steel	99.9		1.09		42	For 1990, U.S. apparent consumption of iron ore, agglomerates, and pellets was 81.7 MMT, iron and steel scrap consumption was 45.5 MMT and net imports of steel mill products was 11.7 MMT. Apparent consumption of bauxite, the starting mineral for aluminum production, in 1990 was 12.6 MMT, all imported (74).
	Aluminum	5.3		0.06			
	Copper	2.2		0.02			
	Other	4.2	111.6	0.58	1.23		
Forestry Products	Saw timber	122.9		1.35		75	Data for 1991. The category saw timber includes saw logs and veneer logs. We assume a specific gravity of 0.6 g/cc for hardwoods and 0.45 g/cc for softwoods on a dry weight basis, green volume.
	Pulpwood	72.8		0.80			
	Fuelwood	51.5		0.56			
	Other	12.6	259.7	0.14	2.85		
Agriculture	Grains	219.7		2.41		76	The category "Other" includes cotton, tobacco, hides, flaxseed, wool, and fishery products. Also see notes[b]
	Hay	133.2		1.46			
	Fruit & vegetables	70.5		0.77			
	Milk & milkfat	63.2		0.69			
	Sugar crops	50.6		0.55			
	Oilseeds	44.7		0.49			
	Meat & poultry (Qty, live weight)	42.3		0.46			
	Other	4.9	629.1	0.05	6.90		

[a] Apparent consumption is the consumption of the commodity at the feedstock stage (i.e. refined metal, ammonia, crushed stone). This number is arrived at by adding the domestic production and imports, and subtracting exports. Apart from metals we do not account for changes in inventory.

[b] The equivalent weight for data given in bushels, gallons, or other volume and unit measures is based on USDA conversion values. The grains category represents total disappearance from domestic use of wheat, rye, rice, corn, oats, barley, and sorghum grain. Estimates for total pasture and harvested roughage consumption in 1990 summed to over 233 MMT and are not included here. The category of fruits and vegetables includes potatoes, sweet potatoes, treenuts, cacao, coffee, and tea in addition to the general fruit and vegetable categories. Sugar crops are defined as sugar beets and sugarcane; the category includes apparent consumption of maple syrup and honey. Oilseed crops are defined here as soybeans, peanuts, and all varieties of sunflower. Eggs are included in the meat and poultry category.

to move.[2] In addition, coal has the highest carbon intensity (i.e. kilogram of carbon per BTU) of the fossil fuels. Reducing absolute energy consumption as well as changing the mix of fuels used for power generation and transportation would have substantial consequences in the materials sphere. Exploiting the properties of other materials can reduce energy use. Construction and other materials can substitute for energy materials. For instance, superior insulation can reduce energy demand.

Annual consumption of construction minerals occurs on the same scale as consumption of energy materials. These materials do not cause significant environmental impacts, with some exceptions. For example, the excavation of sand from stream and river beds alters the ecosystems from which they are retrieved. End-uses for these materials also have a potential environmental downside, as they provide the means for covering land with human artifacts. Today, these artifacts cover only 1 or 2% of global land (36, 37), yet the extent and radiative properties of construction materials affect Earth's albedo and other surface characteristics important to local and global climate (38, 39). Managing the albedo of the built environment may become important in a world economy 5–10 times as large as today 50–100 years hence. Because the quantities used are massive, transporting construction materials is energy intensive. Thus, with the exception of specialty building materials such as Italian marble, they are typically retrieved from within 50 kilometers of their final destination (40).

In 1990, Americans consumed about 250 MMT of industrial minerals, approximately 5% of all materials inputs, listed here. These minerals, valued for their chemical and physical properties, have sundry uses including construction (lime and gypsum for cement, soda ash for glass, clay for bricks), agriculture (nitrogen compounds, sulfur, phosphate rock for fertilizers), and manufacturing processes (lime and soda ash to control alkalinity and for flotation).

Minerals used for agriculture alone comprise over 60 MMT, equal to more than half the total apparent consumption of iron and steel. These materials help generate abundant food and large agricultural surpluses for the United States. They also create several environmental problems arising from fertilizer use. Water pollution problems stem from agricultural runoff from land saturated with elements intended for crop uptake. Enhancing the crop uptake of mineral nutrients from fertilizers could reduce this materials stream. Fertilizers also serve to distribute harmful trace elements, such as cadmium, to agricultural soils (41). Reducing consumption of agricultural minerals without a compen-

[2]To illustrate, 97 MMT or 12% of US coal production in 1984 was captive coal, defined as coal consumed by mining companies internally (35a). In 1990, coal accounted for 40% of US freight rail traffic by weight (32).

satory increase in productivity would cause the total land area cultivated for crops to rise and lessen the area that remains, or can revert back to its natural condition, another environmental concern (37).

In accounting for metals we consider the apparent consumption of finished metals. Future studies might integrate the consumption of auxiliary materials in metals processing as well as mining and mineral processing wastes, here treated in the section on extractive wastes. Metals constitute a trifle of material inputs, about 2% in mass terms. Nonetheless, the durability and formability of this material group in addition to other desirable physical characteristics (e.g. tensile strength, toughness, thermal and electrical conductivity) have made metals essential to technical and social progress throughout human development. Iron and steel dominate the metals group and, historically, have provided the backbone of industrial society. Though unlisted in Table 1, alloying metals such as molybdenum, manganese, and cobalt, sometimes referred to as metallic vitamins, are significant components of metal flows. Consumption of these metals is relatively small, yet their ability to improve properties in bulk materials makes them essential ingredients in modern metallurgy. These elements also constitute impurities that pose problems for recycling and use in the secondary metals industry.

Recycled metal accounts for over half the metals consumed in the United States (42). However, economic, physical, and regulatory factors keep recovery below 10% for arsenic, barium, chromium, and some of the other most biologically harmful metals in the *Toxics Release Inventory,* an annual listing of US toxic wastes emissions (43).

US consumption of forestry products exceeded 250 MMT in 1990, well over twice the weight of metals. Although forestry products comprise only 5% of total materials inputs, they affect environmental quality in numerous ways and have become highly symbolic in the public debate on the environment. Forests provide an important sink for greenhouse gases and, unless balanced, excessive timber harvesting upsets the global carbon cycle. Improper logging practices can also disturb forest ecosystems, adversely affecting the habitat for both plant and animal life. Saw timber used for lumber, plywood, and other structural applications accounts for almost half the forestry products total. Pulpwood, used for manufacturing paper, accounts for almost 30%. (Wood chips from sources other than pulpwood account for the 40% of the wood input to the paper industry that is not pulpwood, an efficient use of material.) Pulp and paper industries have historically polluted water bodies. Paper comprises over 30% of municipal solid waste (44). Reuse and incineration of paper waste are the chosen strategies for recovering cellulose and energy, respectively, and for reducing the paper component in the solid-waste stream.

Combined inputs for human food consumption in the United States in 1990 sum to approximately 630 MMT, more than double industrial mineral consumption. The composition of food inputs changes with the national diet, leading to important environmental impacts, particularly regarding land use. For instance, reduced meat consumption, accompanied by a rise in fruit, grain, and vegetable consumption, alters the balance of agricultural land devoted to grazing and feed as opposed to food crops. Cultivation of legumes and rice affect the atmospheric concentration of nitrogen and methane respectively. Fertilizer and pesticide use rates are tailored to specific crops. (We have already remarked on the quantity of minerals—mainly nitrogen, phosphates, potash, and, indirectly, sulfur—consumed in food production.) We also note that energy must be expended to bring lettuce and grapefruits to cold New York from warm Florida in February.

Outputs

Materials pass through the economy on different time scales. Dams and bridges embody materials that may remain untouched for centuries. Chewing gum wrappers do not last as long. Beverage cans can have many lives. Some materials wear down through normal use, whereas others undergo phase changes and volatilize into the air. Our national account of materials outputs (Table 2) encompasses all of these fates.

Data on materials outputs are generally scarcer and harder to interpret than input data. Two important factors are the absence of uniform definitions for material end-uses and the lack of weight data for most manufactured goods. Waste outputs are often given on a wet weight basis, and in many cases the solid fraction is swamped by water. In addition, the waste's absent or unrecognized value makes waste accounting of secondary importance in many industries, though high disposal costs and regulatory requirements influence this state of affairs by forcing firms to pay attention. Dissipation data are entirely based on estimated rather than direct empirical information, as explained in the data notes.

Forty percent of US material outputs contribute to the domestic stock of materials that are incorporated into the built environment and industrial infrastructure. Atmospheric emissions, primarily produced by fossil-fuel combustion, account for a similar fraction of outputs. Approximately 3% of materials dissipates directly into the environment. About 5% of outputs recycles directly into the economy. Other wastes account for the remaining share of outputs.

Domestic stock comprises objects not consumed during normal use and designed to last for a period greater than one year. Typically, increasing the mass of domestic material stocks links with economic development. Indeed, accelerated turnover of the goods that augment this stock generally serves economic interests under the present US tax and accounting systems (45) in part because the economic transactions leading to the production and distribution of durable goods do not cover the full social costs of disposal.

Almost 90% of the 1990 material contribution to US domestic stock, 1677 MMT, was in construction. Yet, the Environmental Protection Agency (EPA) estimates that US construction and demolition generate only about 29 MMT of waste annually (46). One explanation for this difference is that the physical capital in construction amasses at an astounding rate. Another possibility is that the data do not accurately reflect all the wastes associated with construction and depreciation. These explanations are not mutually exclusive.

The materials that inhabit the built environment, and those that support the industrial infrastructure, constitute the remaining 10% of domestic stock. For example, US materials consumption for land-based transportation was estimated at about 22 MMT in 1990 (47). Materials for machinery (mostly steel) and electrical uses (mostly copper and aluminum) account for over 11 MMT, and refractory uses account for over 7 MMT (48). An example of how domestic stock accumulates is that, as of 1987, over 680 MMT of recoverable iron resided in scrap piles across the United States, an amount equal to almost seven times the 1990 iron and steel input (49).

The variety of other products in the domestic stock category precludes precise statements about their environmental effects. This variety complicates efforts to retrieve primary materials from waste streams and masks the release of potentially harmful substances that occurs in the use and final disposal of consumer products.

The quantity of materials emitted into the atmosphere rivals all the sand and stone used for building. The carbon in CO_2 from fossil fuel combustion makes up the vast majority of emissions, followed by water vapor. Table 2 also reminds us that through emissions other trace elements are liberated from their geological formations and distributed in the atmosphere, reemphasizing the important role of energy strictly from materials considerations. Energy costs and other factors often make impractical the recovery of emissions (with the notable exception of sulfur), and scrubbers designed to clean smokestack emissions generate sludges and other spent scrubber materials. Source reduction through improved efficiency continues to be the main strategy for addressing this sizable material flow.

Although dissipated materials such as pesticides constitute a relatively small portion (~3%) of materials outputs, they have a considerable impact on the environment. In fact, these materials are a major focus of industrial ecology. As with atmospheric emissions, reconcentrating dissipated materials is frequently infeasible. Their dispersion changes the balance of elements biologically available to the plants and the other animals with which humans have evolved. Metal-loaded soil is inhospitable to plant life in general. In some

Table 2 Materials flows: US—outputs

Destination	Amount (MMT)	Per capita per day (kg)	Reference	Comments
DOMESTIC STOCK	1880.3	20.61		
Construction	1677.1	18.39	48, 75, 77	This figure is obtained by subtracting the amount of construction minerals recycled, dissipated, discarded and going to other uses from apparent consumption of construction minerals. Also included in the total are clay, asphalt, imported cement, gypsum, lime, steel, aluminum, copper, manganese, nickel, silicon, zinc, and lumber used for construction. These materials are used for residential and non-residential construction, in cement and concrete products, for construction fill, road base and cover, railroad ballast, and other permanent uses.
Other	203.2	2.23	48, 75, 77	Obtained by summing apparent consumption of industrial minerals, metals, forestry products, and energy materials used for industrial and consumer products and subtracting the amount recycled, dissipated, and discarded.
ATMOSPHERIC EMISSIONS[a]	1734.7	19.02		
CO_2 (carbon fraction only)	1367.0	14.99	77	Most (98%) CO_2 emissions are from energy sources.
Hydrogen	254.6	2.79		Calculated hydrogen fraction from water vapor emitted during fossil fuel combustion. Total water vapor emitted is 2290 MMT or about 600 billion gallons.
Methane	29.1	0.32	77	Sources of methane emissions include solid waste, coal mining, oil and gas production, leakage during transmission, and agriculture.
CO (carbon fraction only)	29.0	0.32	77	Most (78%) CO emissions are energy related.
NO_x	19.4	0.21	77	Most (95%) NO_x emissions are energy related.
VOC	17.6	0.19	77	
SO_2 (sulfur fraction only)	10.4	0.11	77	
Particulate matter	5.5	0.06	77	Includes PM-10 emissions defined as particles with a diameter of less than 10 microns. Does not include PM-10 fugitive dust.
DISSIPATION[b]	144.5	1.58	42	Data do not include the dissipated amounts of dimension stone, fluorspar, mica, perlite, steel, asphalt & road oil, petrochemicals, and natural rubber

OTHER WASTES	555.2	6.09		
Processing waste[c]	136.2	1.49	42	Data do not include processing wastes for barite, calcium, cement, diatomite, gypsum, industrial sand and gravel, magnesium, phosphate, aluminum, chromium, nickel, and refinery products.
Post consumer waste[d]	276.4	3.03	42	Data do not include post consumer wastes for cement, clays, diatomite, fluorspar, gypsum, industrial sand and gravel, perlite, talc, nickel, tin, and petroleum waxes.
Coal ash	85.0	0.93	60	Data are for 1984. This figure includes fly ash, bottom ash, boiler slag, and flue gas desulphurization [FGD] sludge. About 10% of coal used for energy recovery is ash. 20% of coal ash wastes find other uses in the economy, primarily in road construction (51).
Yard waste	35.0	0.38	44	Yard waste is organic matter and has no input accounted for in Table 1.
Food waste	13.2	0.14	44	This value is for food wastes entering the municipal solid waste steam and does not reflect waste generated in the food processing industry. Food processing wastes are overwhelmingly water.
Water and wastewater sludge	9.4	0.10	46	Based on an annual estimate from USEPA Office of Water Regulation and Standards.
RECYCLED	243.8	2.67	42	

[a] In seeking to satisfy a materials balance for the total mass of fossil fuels listed in Table 1, we note that by summing the fossil fuels emissions data in Table 2 with coal ash, we find a total of 1707 MMT. This is short of fossil fuel inputs by over 250 MMT. Several factors are responsible for the discrepancy. Crude oil and natural gas are not all combusted for energy; the fraction consumed as nonrenewable organic products (e.g. asphalt, plastic, lubricants) totaled 112.9 MMT in 1990 (74). The moisture content of coals can range up to 20% of total weight, overestimating reported tonnages. Additionally, trace elements such as fuel bound nitrogen and sulfur account for some of the discrepancy. US coals averaged 10.9% ash content and 1.6% sulfur content, in 1983 (59).

[b] Dissipation refers to the estimated annual quantity of materials released directly into the environment, where no attempt to recover the material is practical; examples are the application of fertilizers, pesticides and road salt. In this category we include TiO_2 used for paint and pigments. However, we do not account for materials dissipation resulting from normal wear of items such as bridges, brake pads, and coatings.

[c] Processing waste comes from the processing and manufacture of materials into a finished product following extraction from the original resource. Only the dry weight of the material is included, and any liquid or ancillary materials associated with mineral concentrator tailings, such as waste from nonmineralized rock, are not included. Because of the difficulty in determining the portion of waste released directly into the environment, all the estimated waste and losses from the processing phase are assumed to be disposed of in a controlled manner. We note that manufacturing wastes include varying amounts of water (frequently >90%). Deriving their dry weight directly from existing data sources is difficult.

[d] Postconsumer waste includes Municipal Solid Waste (MSW) and is generally defined as the estimated annual quantity of materials disposed of in a controlled fashion following use in product form.

cases, even minute contaminant concentrations (10s of ppm) damage plant growth and overpower their genetic adaptability (50).

Other (solid) waste estimates span a wide range. Our figures are comparatively low and are for dry weight only. Both process and postconsumer wastes contribute to the total, reflecting the implicit as well as the explicit consequences of economic activity. The substantial contribution of coal ash to wastes again illustrates the role of energy in materials flows. Of the few solid waste disposal options available, each is fraught with environmental risks. Both industry and government have begun to pursue more aggressively the opportunities for recovering the value, utility, and stored energy in this accumulating fraction of waste materials (51–54). This quest is one of the principal objectives of industrial ecology.

Household hazardous wastes include used products such as adhesives, cleaners, cosmetics, and batteries. Estimates of the size of this vexing waste stream range from .0015%–.4% of municipal solid waste, or 0.002–0.589 MMT using 1990 data, which is too low to appear explicitly in Table 2 (46).

Dry weight data for agricultural wastes are not available, but the EPA estimates 1 billion gallons (3.7 MMT) per day as an upper limit on daily waste generation from the US agricultural sector (46).

Recycled materials totaled approximately 244 MMT in 1990, about 5% of our estimate for materials outputs (42). Environmental considerations for recycling are material specific and, in all cases, impact energy consumption for processing and transportation. Fully restoring the function and value of organics, such as paper and plastic, may not make sense if it means increasing energy use and therefore consumption of energy materials. For metals, recycling usually requires less energy than production from virgin materials. For example, secondary aluminum production consumes only 5–10% of the energy needed for the electrolytic refining of primary alumina (55). Reprocessing steel and iron scrap can save over 60% in energy consumption (56).

Our list of outputs equals approximately 88% of the input total, an encouraging but perhaps deceptively good match. The bulk of the 600 MMT discrepancy is due to human and animal food consumption. Human fecal matter decomposes during treatment and is not fully accounted for in material terms. Neither is manure from livestock, which the US EPA estimates at about 2 billion metric tons annually on a wet basis (46), of which approximately 20% is dry matter. Additionally, we have not accounted for residuals in the timber and food-processing industries, frequently used for energy recovery.

Trade

Approximately 90% of US materials flows (by weight) appear confined to America when materials imports and exports are measured against our basic categories of inputs and outputs. Still, Americans import approximately 2.4

Table 3 Selected materials flows: US 1990—foreign trade

Category	Exports (MMT)	Imports (MMT)	Net annual per capita (kg)	Reference
Agricultural products	135.5	14.9	(482.6)	32, Tbl. 1123
Coal	96.0	2.4	(374.5)	32, Tbl. 945
Minerals	47.8	54.2	25.6	74
Metals and ores	27.0	76.4	197.8	74
Chemical and allied products	41.3	14.4	(107.6)	32, Tbl. 1079
Petroleum products	34.1	96.9	251.3	32, Tbl. 945
Timber products	16.4	18.4	22.8	32, Tbl. 1165
Paper & board	6.2	11.9	22.8	32, Tbl. 1165
Oil (crude)	5.6	307.4	1207.6	32, Tbl. 945
Natural gas	1.7	31.0	117.2	32, Tbl. 945
Automobiles[a]	1.2	5.9	18.8	32, Tbl. 1019
TOTAL	412.7	633.8	884.4	
Air transport	1.5	1.7	—	32, Tbl. 1076
Waterborne transport	406.9	524.9	512.4	32, Tbl. 1079
Trucks	151,000 (units)	766,000 (units)		32, Tbl. 1019
Other industrial & consumer products	?	?		

[a] Based on an estimated average vehicle weight of 1.5 metric tons.

kg net of material per capita per day (Table 3). The US monetary foreign trade imbalance corresponds to the mass imbalance presented here. In contrast, Japan is a heavy materials importer but maintains a positive monetary trade balance.

In mass, agricultural products, coal, and chemicals dominate US exports, whereas oil, oil products, and metals and ores dominate imports. Data on the mass, as distinct from the economic, value of US trade in manufactured products are not directly available. Because information content, rather than bulk, largely determines the value of manufactured products such as microchips, drugs, and clothing, the total weight of traded products would probably alter little the picture offered by Table 3. Even automobiles, among the most mass-intensive manufactured imports, amount to only about 1% of all imports by weight.

Measured by weight, most imports to the United States are transported by water. A minor fraction is shipped overland. For example, about 70 MMT (280 kgs per capita annually) of crude oil flowed into the United States from its North American neighbors in 1990, presumably through pipelines. Light, valuable items fly. In 1990, air freight formed less than 0.5% of US foreign trade by weight but over 22% by value (32). The disparity in Table 3 between import-export totals and waterborne transport totals must partly result from the mass of goods shipped by land and pipe as well as finished goods traded, for which mass values are not given. Further work in this area could account for flows more fully and would be timely in light of environmental concerns about the North American Free Trade Agreement and the General Agreement on Tariffs and Trade.

Examining the material flow across the borders of other countries is instructive. For Japan, the mass ratio of raw materials imports to product imports is greater than 10 to 1. Japanese exports consist of products exclusively (57). Data from Sub-Saharan Africa and Latin America for the late 1980s show that manufactured exports were commonly under 5% of total exports by value (presumably less by weight), and imports of machinery and transportation equipment averaged well over 25% (58).

Extractive Wastes

The retrieval and preparation of crude minerals and ores for human consumption create large amounts of wastes (Table 4). Although these materials seldom enter the economy directly, their generation is essential to its operation. Resource extraction activities require extensive land use, infrastructure, and auxiliary materials, such as barite ($BaSO_4$) for making muds to seal oil wells and nitrogen-based explosives for clearing rock.

Generally speaking, wastes from the extractive industries remain in the mines and wells where they are generated. Apart from oil and gas wastes, rock mobilized to access desired minerals and ores constitutes the majority of the extractive wastes. Waste rock may be harmless. However, displacing dirt and rock and exposing raw earth to wind and water affects local acidity levels and transports trace elements to water sources and neighboring biota. Chemical leaching and other operations for retrieving metals from ores can also create damage, spreading hazardous chemicals that seep into the earth.

To access coal seams, both surface and underground coal mining must remove overburden, gob in the underground case. The amounts mobilized vary from seam to seam, ranging from under 3 cubic yards per short ton for surface-mined coal in Wyoming to 48 cubic yards per short ton in Oklahoma. The average for coal surface mining in 1983 was 9 cubic yards of overburden per short ton of coal mined (35a). Coal mined for processing consists of 70% or more combustible material on average. The properties of coals from different sites vary considerably, even within the same seam. Variables include the content of sulfur, ash, pyrites, mercury, and sodium, as well as moisture content. Coal cleaning wastes include the 30% or less shale and clay in coal seam partings, as well as impurities such as sulfur and ash. The variability means waste generation from coal cleaning is site specific, and relevant data are scarce.

Table 4 Major materials flows: US—extractive wastes

Category	Amount (MMT)	Reference	Comments
Surface coal mining wastes	>10042.4	59	Data for 1983. Figure is for wastes from surface mining only which accounted for 60.7% of all coal mined. Based on DOE estimates of an average 6.88 cubic meters overburden per short ton of coal mined and using the density of granite (2.7 MT/m') for overburden.
Coal cleaning wastes	>84.1	59	Data for 1983. Represents the refuse from mechanically cleaned coal (32.5% of all coal mined). This number is based on an estimate of 32 tons of refuse for every 100 tons of coal cleaned mechanically. For the remaining 68.5% the data were not collected. Residues from coal mining are classified as subtitle D wastes and not normally reported.
Oil & Gas Produced Waters	(20.87 billion bbl.) 3318.2	60	Data are for 1985 and do not include some states. Produced waters are mixtures of naturally occurring water in geological formations, naturally derived constitutents such as benzene and radionuclides, and added chemicals. We assume water density for our mass value. Over 90% of this waste is reinjected into the ground.
Oil & Gas Drilling Fluids	(361.4 million bbl.) 57.2	60	Data are for 1985. Drilling fluids include drill cuttings removed during well boring, drilling muds pumped into wells to facilitate extraction, protect various geological layers, and remove drill cuttings. We assume an average density of .996kg/l.
Metals overburden	755.0	60	Data are for 1987.
Metals tailings	409.1	60	Data are for 1987. Mine tailings are calculated as the differences between the amount of crude ore and the amount of marketable product.
Phosphate overburden	262.3	60	
Phosphate tailings	108.1	60	
Uranium tailings	188.0	61	
Minerals Processing	93.8	78	Data are for the mid to late 1980s. The amount shown represents the total weight of the 20 waste streams considered by EPA to be 'high volume low-hazard' wastes from the mineral processing. One estimate for hazardous wastes from the mineral processing industry is 6.7 MMT including water.

Water encountered in oil and gas drilling operations combined with small amounts of minerals and other chemicals, derived naturally and added for drilling, to form so-called produced waters. These waters account for 96–98% of US oil and gas wastes (60). Producers reinject over 90% of the waters into impermeable geological formations. Although these waters are mixed with chemicals to enhance recovery, their treatment is designed to make them mostly inert. Drilling fluids, which include the rock removed during drilling and muds used to provide well back-pressure, lubrication, and sealing, make up the remaining 2–4%. These wastes are currently considered innocuous and generally remain on site.

For political and technical reasons, the US EPA ruled in 1986 that mining wastes should not be considered hazardous. Waste data are not collected for mined materials such as clay, stone, sand, and gravel because their environmental risk is judged negligible. Some data are withheld by companies. Of the overburden and tailings for which data were collected, in 1987, most were from copper (43%), phosphate rock (24%), gold (18%), and iron ore (10%) mining and processing. Uranium mill tailings, regulated separately from other metals by the Nuclear Regulatory Commission with assistance from EPA, amounted to 188 MMT in 1990 (61), comparable to gold mining wastes. Among nonmetallic nonfuel minerals (asbestos, gypsum, lime, sulfur, phosphate rock) we list only phosphate rock because waste generation from this sector is prominent. Thus, the data are incomplete.

Nonhazardous wastes from mineral processing, a further step in the refining of metals and minerals, amounted to approximately 94 MMT, according to EPA estimates from the late 1980s. Slightly more than half of this waste is phosphogypsum (primarily $CaSO_4$) from phosphoric acid production, which is unsuitable for other uses because it contains radionuclides and other contaminants. Iron slag accounts for a little more than a third. Iron and steel slags and some of the other so-called wastes in this category are in fact marketable and useful in the economy.

Environmental Metrics

We now turn to the use of the national materials flow accounts to improve environmental performance. We believe the accounts can form the basis of a set of environmental metrics that indicate the changing and comparative performance of a national economy. Devising metrics that incorporate materials flow data and provide information of consequence to environmental quality will require discussion, experimentation, and refinement (62). Table 5 presents our initial attempt at a set of metrics, most of which address either the productivity or efficiency of resource use. The list is not exhaustive, and some of the metrics are already in use. The metrics require economic as well as weight data. Below we comment briefly on selected proposed metrics:

1. Absolute national and per capita inputs aggregated by class and by individual material. We began this paper by mentioning that on average each American consumed about 50 kg of materials daily in 1990. This figure results from the addition of all inputs divided by population. Although the components of the aggregate differ in qualities and accuracy, we believe the total is meaningful and would be useful to track over time and to compare across nations. Similarly, an analysis of materials use over time and across nations would be useful for classes of materials, such as energy, and individual materials, such as lead.

2. Composition of the national materials (input) basket. With economic development and technical change, the demand for materials evolves. Knowing whether and why the shares of major materials classes change would be interesting.

Within the energy sector, the evolution of the ratio of coal:oil:gas, or in more elemental terms the hydrogen (H) to carbon (C) ratio in fossil fuels has environmental import (5, 63). Determining the balance between high-energy materials such as metals and organics and low-energy materials such as sand could also be useful. Shifting the balance in favor of low energy materials could bring environmental benefits, and great potential may lie in old materials, such as stones, that can be upgraded to glass and other forms.

The ratio of inputs to the economy of various nonenergy materials might also indicate trends relevant to national environmental performance. Ashby, for example, related the physical properties found in metals, ceramics, glasses, and polymers to those most often sought in material goods (e.g. Young's modulus, yield strength, and toughness) (64). He concluded that advances in metal alloys may be near an end and that many functions formerly fulfilled by metals will be provided by impact-resistant ceramics and polymers, stiffened by increasing the density of carbon-carbon bonds in the direction of loading, or filled with a material, such as sand or glass, of higher modulus. We suspect that the environmental and materials science communities have much to learn from each other about trends, needs, and capacities.

3. Intensity of use indicators. Intensity of use metrics quantify materials consumed against physical or monetary outputs. Often, materials such as steel, copper, and tin have been indexed to the Gross Domestic Product (GDP) in constant dollars (65). A historic example of declining intensity of use is the decarbonization of the economy, or decline in carbon inputs per unit GDP over the past century (66). Intensity of use measures may help gauge developmental status and define realistic goals that integrate economic growth and improved environmental quality. The concept could be applied more widely, assessing, for example, ratios of agricultural minerals to crops. A host of materials other than water and land now flow quite directly into agriculture. Measuring the

Table 5 National materials flows: sample metrics

Metric		Dimensions	Formula	Environmental significance
Total inputs		MMT and MMT/Capita	Aggregate total consumption of all materials classes and individual material classes, on absolute and per capita basis	Benchmarking national resource use
	Fuel mix (H/C)	Dimensionless	Consumption ratio of Btu's from Natural gas : Petroleum : Coal	CO_2 emissions
Input composition	Processed metal : ceramic : glass : polymer ratio in finished products	Dimensionless	Consumption ratio of said materials in finished products and structures	Gross shifts in materials use, materials efficiency and cyclicity, mining and processing waste, energy use
Input intensities	Intensity of use	MMT of inputs/$\$10^6$ GDP	Material consumption quantity for selected input materials/GDP in constant dollars	Relationship of resource use to economic activity
	Agricultural intensity	Dimensionless	Agricultural materials consumption/Total crop production	Materials efficiency, eutrophication of water bodies, topsoil erosion, ecosystem disruption, energy use
	Decarbonization	MMT of Carbon inputs/$\$10^6$ GDP	Carbon inputs/GDP in constant dollars	Relationship of carbon emissions to economic activity
	"Virginity" index	%	Quantity of all virgin materials/Total material inputs	Materials efficiency and cyclicity, mining and processing waste, energy use

Recycling indices	Metals recycling rate	%	Quantity of recycled and secondary metals production/Primary production from ores	Materials efficiency and cyclicity
	Renewable net carbon balance	%	Forest growth/Forest products harvested	Global carbon balance of sources and sinks, land use, ecosystem disruption
Output intensities	Green productivity	%	Quantity of solid wastes/Quantity of total solid physical outputs	Materials efficiency and cyclicity
	Intensity of use for residues	MMT/10^6 GDP	Generation quantity for selected materials waste streams/GDP in constant dollars	Relationship of waste generation to economic activity
Leak index		%	Quantity of materials dissipated into the environment/Total material outputs	Materials efficiency and cyclicity, media contamination
Conversion efficiency	Industrial conversion efficiency	%	Total output for an industrial sector/Total inputs	Materials efficiency
	Process to post-consumer wastes ratio	%	Process wastes from industry/Post consumer waste	Relating generally unseen to seen consequences of industrial production
Physical trade index		%	Mass value of net trade in manufactured products/Mass value of net trade in raw resources	Domestic resource consumption, domestic environmental burden caused by exported goods
Mining efficiency	Mining wastes	Dimensionless	Quantity of wastes generated/Ton of finished product	Solid wastes, acid mine drainage
	By-product recovery	Dimensionless	Total by-product recovery/Total output	Materials efficiency, solid wastes

amounts of these materials used for food and fiber production indexed to production might provide important information about performance.

4. Virginity and recycling indices. A virginity or raw materials index could indicate, nationally and per capita, absolute amounts of raw materials, and ratios of raw materials to national materials inputs. The indices would monitor the distance to a society that has largely stopped extracting materials from the earth and sustains itself through its materials endowment and recycling. As society capitalizes on the "mines above ground" or scrap piles, traditional mining and thus mining wastes grow dispensable and the materials loop closes. Of course, the demand for materials with highly specific properties also alters the pool of resources that can be used as inputs (67).

Among specific materials of interest for recycling are metals and wood. The high ratio of secondary to primary metals consumption indicates both the efficiency of metals use and success in overcoming the recycling problems caused by contaminants. For forestry products a simple environmental measure would be the net carbon balance from forest growth and harvesting.

5. Waste (or emission) intensities. Comparable to intensity of use, these metrics focus on residuals and emissions per unit of output measured in physical or economic terms. Corporate practice increasingly evaluates the ratio of wastes to total output, including products and salable byproducts (68). Similar national indicators would assess "green" productivity by evaluating the amount of materials considered as waste against various outputs.

6. Leak indices. Measuring the fraction of outputs dissipated to total outputs as a leak index would quantify the proportion of materials lost to further productive use and dispersed into the environment. Applying this measure would allow for easier identification and isolation of holes in the system and focus efforts to plug them. Environmental monitoring activities would have to be modified to support a genuine mass-balance account by identifying, for example, input-output discrepancies at production sites.

7. End-product materials efficiency. The relation between primary or total materials consumption and end-products is important. A parallel in the energy field is the relation between primary energy and end-use as a measure of system efficiency. Implementing a comparable materials measure would be more difficult, but might be possible within sectors, where products are measured against total inputs and waste.

8. Ratios of raw and manufactured materials traded. Exporting raw materials consumes national resources and scars landscapes. In contrast, using domestic industry to convert materials into export products can damage the environment in other ways. The ratios may indicate whether nations are displacing or exporting pollution and, if so, what kinds.

9. Extractive waste ratios. Geological characteristics primarily determine overburden and tailings, but judgments also affect mine wastes. A measure,

subject to some physical constraints, of mine wastes per ton of ore mined or primary metal produced could be explored. Some companies already use measures such as water and energy use per ton of finished product. Measuring the efficiency of by-product recovery, such as methane deposits trapped in coal seams, sulfuric acid from smelter emissions, and metals captured from flue dusts provides further opportunity. Mining concerns are increasingly alert to these dimensions of their operations (69). In fact, in numerous sectors firms are experimenting with new ways for managerial accounting to track environmentally significant materials flows (70). Several of these approaches might be evaluated for scale-up to monitor national materials performance.

Discussion

Having laid out a framework, supported it with data, and proposed how it might be used, we return to questions of feasibility and value.

The main obstacle to development is data. Relevant data are collected for one purpose or another. For our purposes, collection is patchy and sporadic. Synchronizing data collection among various federal departments and agencies to build more complete data sets for selected years could amplify the benefits of existing efforts. Equally important would be the development of consistent classifications of material commodities and end uses. Erroneous assumptions regarding materials classification lead to omitting and double counting of material components.

Procedural changes could ease the development of national materials accounts. However, weight data are simply not collected in important areas, because some companies fear disclosure of proprietary information, or because the perceived value or direct environmental threat of a materials flow is considered too small to justify collection efforts. As a result, high levels of uncertainty are associated with many materials streams (e.g. mining and industrial wastes). For most manufactured products economic considerations dominate, and weight data are neglected.

Moreover, weight data do not provide the complete picture. The environmental impact associated with materials flows differs considerably among and within materials classes. Total US dioxin and furan emissions, which annually amount to less than one metric ton, provide a vivid example (71). National material accounts would need to include these flows as well.

To realize their value, national materials accounts would have to be calculated periodically. A frequency of once every three to five years might balance the labor of collection with the utility of the product, and still allow for identification of trends and understanding of the primary and secondary effects of changes in the economy, industrial practice, and government regulation.

A fair question is whether national boundaries make sense for materials analysis. The answer is partly opportunistic: the data are collected at the

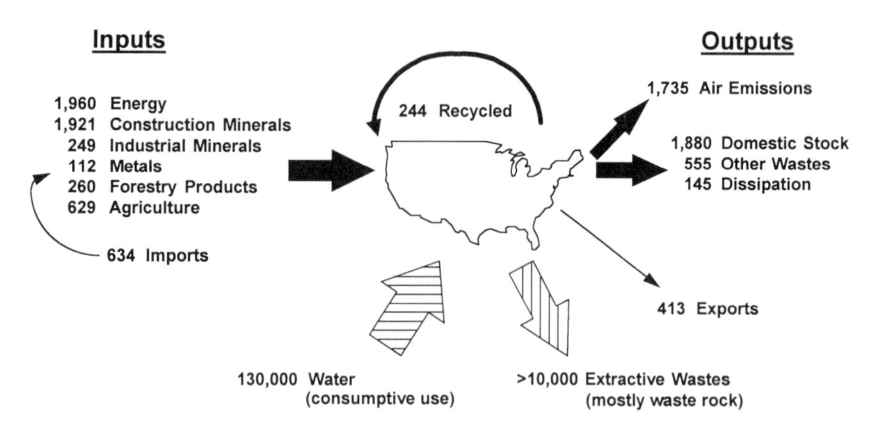

Figure 1 US Material flows, circa 1990. All values in million metric tons (MMT).

national level. Does enlarging system boundaries to encompass a nation forfeit critical environmental information? Clearly national materials accounts do not obviate the need for monitoring environmental variables at the level of localities and firms. Rather, measures based on comprehensive national accounts complement these smaller scale metrics in two ways. They help to identify macroscale trends important to the environment in much the same way that national economic indicators, such as GDP, are useful. They also capture the physical data lost in the patchwork of the current regulatory structure.

Moreover, important decisions relating to materials policies are made at the national level. We believe few if any nations have the means at present to assess meaningfully their environmental performance with respect to materials or to set priorities. If at the systems level nations genuinely hope to shift from a linear to a more circular industrial economy, the dimensions of the shift must be better understood (72). Figure 1 presents our rough schematic account of US 1990 national materials flows. Figure 2 presents the identical data at the level of the individual, which may have additional educational and political impact. More refined frameworks and more certain numbers are needed.

In addition to providing improvement benchmarks for domestic performance, national materials accounts would allow international comparisons. The type and quantity of materials used by a society describe its economic activity, level of industrial development, and environment. Japan provides an interesting example. According to a rough estimate, Japanese materials use in 1990 summed to 52 kg per capita per day, a figure similar to our estimate for the United States (57). The amount of building materials going to domestic stock came to 7.7 metric tons per capita per year in Japan as compared to 6.7 metric

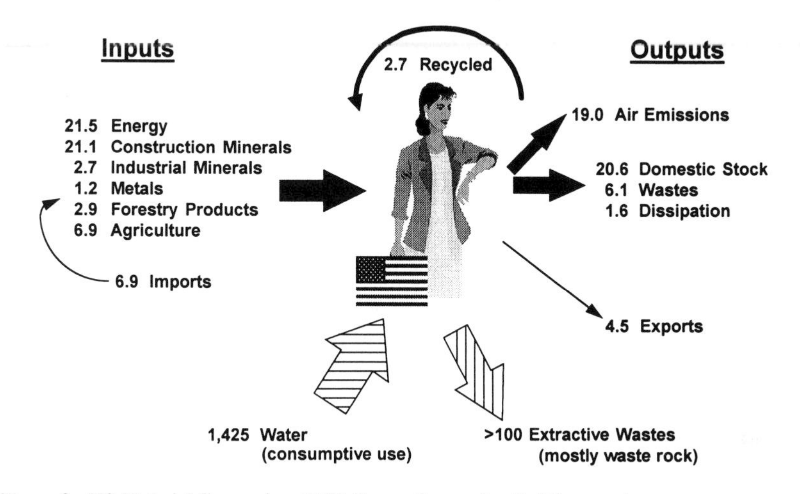

Figure 2 US Material flows, circa 1990. Per capita per day (in kilograms).

tons per capita per year in the United States, perhaps reflecting Japan's higher fraction of capital investment and recent building boom.

The dynamic pattern of materials flows of various nations could serve as references for other nations exploring paths of development. An unanswered question of great interest mentioned at the outset of this paper is whether industrialized nations are beginning overall to dematerialize. Are they reducing materials use, with the help of a broad array of measures, in the same way societies have begun to advance consistently in efficient energy use (30)? Comprehensive materials accounts could address this question.

As a tool, national materials accounts could be extended and applied in various ways: for example, on different geographical scales, in individual economic sectors, on the basis of toxicity or other environmental and health concerns, and in contrasting anthropogenic fluxes with natural or background reservoirs and fluxes (21). They would also provide the context for detailed horizontal studies of individual elements and comprehensive materials-balance studies.

National materials accounts could also help set the environmental research agenda for materials science and engineering (73). Three corrective strategies recur: reducing inputs, increasing the fraction of outputs that reenter the economy, and identifying alternative materials that satisfy human wants while lessening environmental damage. To our knowledge, the materials research community has not carefully evaluated its work from this perspective.

With materials consumption at 50 kg per day per American, even the rough profile developed here demonstrates the urgency of meshing environmental and materials research. We need to begin to consider our materials legacy as a dowry to future generations, rich in valuable ore.

In fact, future materials fluxes may be much larger than today, even with limited inputs and more looping. To make the fluxes environmentally compatible, we need a clearer picture than we have today. We can imagine an industrial ecosystem in which emissions, including carbon and water vapor, would be captured, solid wastes used productively, and waste streams separated and valuable materials recovered. The discipline of creating national materials accounts could be extremely useful in creating a consistent, realistic long-range technical vision.

Data Notes

Mineral consumption data from the US Bureau of Mines (USBM) (42) are consistent with their Mineral Commodity Summaries and Minerals Yearbook publications. The classification for materials is not universal. Furthermore, for some materials no data are given for apparent consumption, dissipative use, postconsumer waste, processing waste, and recycled amounts. Dissipation data are entirely based on estimates from mineral commodity specialists at the USBM. Original sources for the Rogich et al data include the Departments of Energy, Agriculture, and Commerce, the US International Trade Commission, the US Forest Service, the Environmental Protection Agency, Franklin Associates, and Modern Plastics magazine.

Data reported in separate tables of the *Statistical Abstract of the United States (SAUS)* are at times inconsistent, giving different mass values when reporting on the same commodity due to differences in definition or methods of data collection. Examples of such discrepancies include the quantity of petroleum imports and exports (*SAUS* 1993, tables 945 & 1194) and the total mass of waterborne commerce (*SAUS* 1993, tables 1079 & 1080).

Agricultural data are obtained from the US Department of Agriculture. They refer statistical methodology questions to the agency responsible for compiling the data, usually the USDA.

Waste data are in most cases traceable to the US Environmental Protection Agency (EPA). Data collection often begins with sampling and industrial surveys, and the results are statistically calculated to produce final estimates. For the case of hazardous wastes EPA reports a 95% confidence interval. For municipal solid waste data, the reporting group (44) uses a "materials flow methodology" based on a formulation by EPA's Office of Solid Waste and its predecessor in the US Public Health Service. Data sources for *EPA Solid Wastes Report to Congress* (46) include state and federal program offices, published and unpublished literature, the regulated community, and technical research including surveys and fieldwork done by EPA at selected landfills.

Coal wastes data were collected using the *EIA-7A Coal Production Report* from companies owning mining operations that produced, processed, or prepared 10,000 or more short tons of coal in 1984. These mining operations accounted for 99.4% of total US coal production in that year.

For oil and gas wastes, EPA and American Petroleum Institute (API) estimates differ significantly. Some states are not included in the reported data.

Metals mining waste data do not include information from beryllium, magnesium, manganiferrous, molybdenum, nickel, and tungsten segments. Tailings are calculated as the difference between the amount of crude ore and marketable product. Some wastes are not reported for reasons of confidentiality.

ACKNOWLEDGMENT

We are indebted to Donald Rogich, Jim Lemons, and Grecia Matos at the US Bureau of Mines for data and ideas on materials taxonomy. Many other individuals have helped in this effort by offering their comments on the text or by supplying data. They include: Richard Bonskowski at the Department of Energy (DOE), Harvey Brooks of Harvard University, Tom Bruulsema of the Potash and Phosphate Institute, Tasin Chung at DOE, Peter Ince at the US Forest Service, Arnulf Gruebler at the International Institute for Applied Systems Analysis (IIASA), Perrin Meyer at The Rockefeller University, and Paul Waggoner at the Connecticut Agricultural Experiment Station.

Literature Cited

1. World Resources Institute. 1992. *World Resources 1992–93: A Guide to the Global Environment*. New York: Oxford Univ. Press
2. United States Congress. 1952. Resources for Freedom: Report of the President's Materials Policy Commission. *US House of Representatives Doc. 527*, US Gov. Print. Off., Washington, DC
3. National Academy of Science/National Academy of Engineering. 1973. *Man, Materials, and Environment. MIT Press Environmental Studies Series*. Cambridge, MA: MIT Press
4. Barnett HJ, Van Muiswinkel GM, Shechter M. 1981. Are minerals costing more? *Int. Inst. Appl. Syst. Anal., Work. Pap. No. WP-81-20*. IIASA, Laxenburg, Austria
5. Goeller HE, Weinberg AM. 1976. The age of sustainability. *Science* 191 (4228):683–89
6. Grenon M, Lapillonne B. 1976. The WELMM approach to energy strategies and options. *Int. Inst. Appl. Syst. Anal., Res. Rep. RR-76-19*, IIASA, Laxenburg, Austria
7. Royal Society of Canada. 1990. *Planet Under Stress*, ed. C Mungall, DJ McLaren. Don Mills, Ontario: Oxford Univ. Press
8. Bolin B, Cook RB, eds. 1983. *The Major Biogeochemical Cycles and Their Interactions*. New York: Wiley
9. World Commission on Environment and Development. 1987. *Our Common Future*. New York: Oxford Univ. Press
10. Toman MA. 1992. The difficulty in defining sustainability. In *Global Development and the Environment: Perspectives on Sustainability*, ed. J

Darmstadter, pp. 15–23. Washington, DC: Resources for the Future
11. Taylor J. 1994. The challenge of sustainable development. *Regulation* 1:35–50
12. Frosch RA. 1994. Industrial ecology: minimizing the impact of industrial waste. *Phys. Today* 47(11):63–68
13. Socolow R, Andrews C, Berkhout F, Thomas V, eds. 1994. *Industrial Ecology and Global Change*. Cambridge: Cambridge Univ. Press
14. Allenby BR, Graedel TE. 1995. *Industrial Ecology*. Englewood Cliffs, NJ: Prentice Hall
15. Ausubel JH. 1992. Industrial ecology: reflections on a colloquium. *Proc. Natl. Acad. Sci. USA* 89(3):879–84
16. Ayres RU. 1978. *Resources, Environment and Economics: Applications of the Materials/Energy Balance Principle*. New York: Wiley
17. Ayres RU. 1989. Industrial metabolism. See Ref. 79, pp. 23–49
18. Ayres RU. 1992. Toxic heavy metals: materials cycle optimization. *Proc. Natl. Acad. Sci. USA* 89(3):815–20
19. Schmidt-Bleek F, Lehman H, Bringezu S, Hinterberger F, Welfens MJ, et al. 1993. *Fresenius Environ. Bull.* 2(8): 407–90
20. Vitousek PM, Ehrlich PR, Ehrlich AH. 1986. Human appropriation of the products of photosynthesis. *BioScience* 36: 368–73
21. National Research Council. 1994. *Material Fluxes on the Surface of the Earth*. Washington, DC: Natl. Acad.
22. Duchin F. 1992. Industrial input-output analysis: implications for industrial ecology. See Ref. 15, pp. 851–55
23. Jänicke M, Mönch H, Binder M. 1993. Ecological aspects of structural change. *Intereconomics* 28(4):159–69
24. Leontief W, Koo JCM, Naser S, Sohn I. 1982. *The Future of Non-Fuel Minerals in the US and the World Economy*. Lexington, MA: DC Heath
25. Williams RH, Larson ED, Ross MH. 1987. Materials, affluence and industrial energy use. *Annu. Rev. Energy* 12:99–144
26. Waddell LM, Labys WC. 1988. Transmaterialization: Technology and materials demand cycles. *Mater. Soc.* 12(1): 59–86
27. Curzio AQ, Fortis M, Zoboli R, eds. 1994. *Innovation, Resources, and Economic Growth*. New York: Springer-Verlag
28. Herman R, Ardekani SA, Ausubel JH. 1989. Dematerialization. See Ref. 79, pp. 50–69
29. Bernardini O, Galli R. 1993. Dematerialization: Long term trends in the intensity of use of materials and energy. *Futures* 25:431–48
30. Wernick IK, Herman R, Govind S, Ausubel JH. 1995. Materialization and dematerialization: measures and trends. See Ref. 80
31. Ayres RU. 1994. *Information, Evolution, and Economics: A New Evolutionary Paradigm*. Waterbury, NY: Am. Inst. Phys.
32. United States Bureau of the Census. 1993. *Statistical Abstract of the United States: 1993*. Washington, DC: US Gov. Print. Off. 113 ed.
33. Chemical Manufacturers Association 1991. *United States Chemical Industry Statistical Handbook*. Washington, DC: Chem. Manuf. Assoc.
34. Baker RD, Warren JL. 1992. Generation of hazardous waste in the United States. *Hazard. Waste Hazard. Mater.* 9(1): 19–35
35. Russell CS. 1990. Monitoring and enforcement. In *Public Policies for Environmental Protection*, ed. PR Portney, pp. 243–73. Washington, DC: Resources for the Future
35a. United States Energy Information Administration. 1985. *Coal Production 1984*. DOE/EIA—00118 (84). Washington, DC: US Dep. Energy
36. Gruebler A. 1994. Technology. In *Changes in Land Use: A Global Perspective*, ed. WB Meyer, BL Turner, pp. 287–328. Cambridge: Cambridge Univ. Press
37. Waggoner PE. 1994. How Much Land Can Ten Billion People Spare for Nature? *Counc. Agric. Sci. Technol. Task Force Rep. 121*, CAST, Ames, IA
38. Oke TR. 1989. The micrometeorology of the urban forest. *Philos. Trans. R. Soc. London Ser. B* 324:335–49
39. Mylne MF, Rowntree PR. 1992. Modelling the effects of albedo change associated with tropical deforestation. *Clim. Chang.* 21:317–43
40. Barsotti AF. 1994. *Industrial Minerals and Sustainable Development*. Washington, DC: US Bur. Mines
41. Thornton I. 1992. Sources and pathways of cadmium in the environment. In *Cadmium in the Human Environment: Toxicity and Carcinogenicity*, ed. GF Nordberg, RFM Herber, L Alessio, pp. 149–62. Lyon, France: Iarc Sci. Publ.
42. Rogich DG and staff. 1993. *Materials Use, Economic Growth, and the Environment*. Presented at Inter. Recycling Congr. REC'93 Trade Fair, Geneva, Switzerland

43. Allen DT, Behamanesh N. 1994. Wastes as raw materials. See Ref. 81, pp. 68–96
44. Franklin Associates, Ltd. 1992. Characterization of Municipal Solid Waste in the United States: 1992 Update, Final Report. *EPA Rep. 530-R-92-019*, Franklin Assoc., Prairie Village, Kans.
45. Stahel WR. 1993. *Product Design and Utilization*. Geneva, Switzerland: Prod. Life Inst.
46. US Environmental Protection Agency. 1988. Report to Congress: Solid Waste Disposal in the Unites States, Vols. 1, 2. *EPA/530-SW-89-033A*, US Gov. Print. Off., Washington, DC
47. Ginley DM. 1994. Materials flows in the transportation industry: an example of industrial metabolism. *Resour. Policy 20(3):169–81*
48. United States Bureau of Mines. 1991. *Mineral Commodity Summaries*. Washington, DC: US Bur. Mines
49. Frosch RA, Gallopoulos NE. 1989. Strategies for manufacturing. *Sci. Am.* 260:144–152
50. Bradshaw AD. 1984. Adaptation of plants to soils containing toxic metals—a test for conceit. In *Origins and Development of Adaptation*, ed. D Evered, GM Collins, pp. 4–19. London: Pitman
51. Ahmed I. 1993. *Use of Waste Materials in Highway Construction*. Park Ridge, NJ: Noyes Data Corp.
52. 3M Corporation. 1982. *Low- or Non-Pollution Technology Through Pollution Prevention: An Overview*. St. Paul, MN: 3M Corp.
53. Wald ML. 1992. Turning a stew of old tires into energy. *New York Times*, Dec. 27, p. F8
54. Edwards GH. 1993. *Consumption of glass furnace demolition waste as glass raw material*. Presented at Natl. Acad. Eng. Workshop Corp. Environ. Stewardship, Woods Hole, MA
55. Butterwick L, Smith GDW. 1986. Aluminum recovery from consumer waste: technology review. *Conserv. Recycl.* 9(3):281–92
56. Org. Econ. Coop. Dev. 1994. *Summary Report of the Workshop on Life-Cycle Management and Trade*. Paris: OECD
57. Gotoh S. 1994. *The potential and limits of using life-cycle approach for improved environmental decisions*. Presented at Intl. Conf. Ind. Ecol., Irvine, CA
58. The Economist Books, Ltd. 1990. *The Economist Book of Vital World Statistics*. New York: Times
59. Deleted in proof
60. US Congress, Off. Technol. Assess. 1992. Managing Industrial Solid Wastes from Manufacturing, Mining, Oil and Gas Production, and Utility Coal Combustion. *OTA Rep. OTA-BP-O-82*, US Gov. Print. Off., Washington, DC
61. United States Department of Energy. 1994. *Integrated Database for 1993: U.S. Spent Fuel and Radioactive Waste Inventories, Projections, and Characteristics*. Oak Ridge, TN: USDE
62. World Bank. 1992. *World Development Report 1992: Development and the Environment*. New York: Oxford Univ. Press
63. Marchetti C. 1989. How to solve the CO_2 problem without tears. *Intl. J. Hydrog. Energy* 14(8):493–506
64. Ashby MF. 1979. The science of engineering materials. In *Science and Future Choice*, ed. PW Hemily, MN Özdas, pp. 19–48. Oxford, UK: Clarendon
65. Malenbaum W. 1978. *World Demand for Raw Materials in 1985 and 2000*. New York: McGraw Hill
66. Nakicenovic N. 1993. Decarbonization: doing more with less. See Ref. 80
67. Wernick IK. 1994. Dematerialization and secondary materials recovery: a long-run perspective. *J. Miner. Met. Mater. Soc.* 46(4):39–42
68. 3M Corp. 1993. *3M Waste Minimization Guidelines*. St. Paul, MN: 3M Corp.
69. Chiaro P, Joklik F. 1994. *Industrial ecology in extractive industries*. Presented at Intl. Conf. Ind. Ecol., Irvine, CA
70. Todd R. 1994. Zero-loss environmental accounting systems. See Ref. 81, pp. 191–200
71. Thomas VM, Spiro TJ. 1994. An estimation of dioxin emissions in the United States. PU/CEES Rep. 285. Princeton, NJ
72. Allenby BR. 1992. Industrial ecology: The materials scientist in an environmentally constrained world. *Mater. Res. Bull.* (17)3:46–51
73. National Research Council. 1989. *Materials Science and Engineering for the 1990s*. Washington, DC: Natl. Acad.
74. United States Bureau of Mines. 1991. *Minerals Yearbook 1991*. Washington, DC: US Gov. Print. Off.
75. Smith WB, Faulkner JL, Powell DS. 1992. Forest Statistics of the US. *USFS Gen. Tech. Rep. NC168*, US For. Serv. North Cent. For. Exp. Stn., St. Paul, MN
76. United States Department of Agriculture. 1992. *Agricultural Statistics 1992*. Washington, DC: US Gov. Print. Off.
77. United States Bureau of the Census. 1994. *Statistical Abstract of the United*

States: 1994, Washington, DC: US Gov. Print. Off. 114 ed.

78. US Environmental Protection Agency. 1990. Report to Congress on Special Wastes from Mineral Processing: Summary and Findings, Methods and Analyses, Appendices. *EPA/530-SW-90-070C,* US Gov. Print. Off., Washington, DC
79. Ausubel JH, Sladovich HD, eds. 1989. *Technology and Environment.* Washington, DC: Natl. Acad.
80. Ausubel JH, Langford HD, eds. 1995. *Technological Trajectories and the Human Environment.* Washington, DC: Natl. Acad. In press
81. Allenby BR, Richards DJ, eds. 1994. *The Greening of Industrial Ecosystems.* Washington, DC: Natl. Acad.

Annu. Rev. Energy Environ. 1995. 20:493–573

ROUNDTABLE ON ENERGY EFFICIENCY AND THE ECONOMISTS—SIX PERSPECTIVES AND AN ASSESSMENT

CONTENTS

AIMS OF THE ROUNDTABLE

Dennis Anderson

What is the best way to improve energy efficiency? The desirability of improving energy efficiency raises little dispute. Almost all ways of producing, converting, and using energy have seen immense improvements in energy efficiency over the past two centuries, and the potential for further gains is appreciable. Energy policies, however, are still in dispute. How much should policies leave to market forces and how much to regulation? What is the role of public management programs that may actually erode the market for electricity services? What level of responsibility lies with the manufacturers of

1056-3466/95/1022-0493$05.00

electrical equipment and vehicles with architects and builders and so forth to demonstrate and market new, more efficient ways of using energy in their products? Will the institutional reforms now sweeping the industry in many countries, and the emergence of private suppliers of electricity supply, improve energy efficiency? Will these reforms lead to a rethinking about policies?

Such questions (and many could be added) are much debated in the industry, and viewpoints differ widely. The following papers provide six perspectives and an assessment, with special reference to the electricity industry. In contrast to the reviews usually requested by this *Annual Review*, the authors were each asked to write a short paper on their perspectives. Taken together, the papers are meant to provide a broad perspective on the debate, and a basis for people to take stock once again of where we stand on policies and on the question of how energy efficiency can best be improved.

ENERGY EFFICIENCY AND THE ECONOMISTS: The Case for a Policy Based on Economic Principles

Dennis Anderson
The World Bank, 1818 H Street NW, Washington, DC 20433

CONTENTS

INTRODUCTION

Despite the widespread debate over integrated resource planning, discussions of energy efficiency and how best to achieve it are often very narrow. Often, discussions focus on one particular element of policy, such as utility demand-side management programs, to the virtual exclusion of others, such as the roles of private investment and more commercial pricing and regulatory policies. The move in many countries toward greater commercialization and the private provision of electricity supplies has indeed raised concerns among environmentalists (in particular) that the goal of improving energy efficiency will be undermined, as will programs to reduce the environmental costs of energy production and use. Yet many people in the industry (including the present writer) believe that more commercial policies and greater openness to private investment, coupled with essentially noncontroversial environmental policies focused directly on pollution, not only will improve energy efficiency substantially but will enable people in all regions of the world to enjoy high levels of energy consumption *and* an improved environment. If so, this conclusion is especially relevant for developing countries, where per capita energy consumption is still low relative to that of the industrial countries, populations are large, and energy demands are set to grow to very large levels (perhaps four or five times the aggregate level of the industrial countries today).

This paper suggests that the key to reconciling high levels of energy consumption with greatly reduced pollution lies in achieving a policy based on economic principles. In particular, a desirable policy contains eight elements, not one or two. The elements are: (*1*) a stable and growing economy, something that the high-income industrial countries have long taken for granted, but which numerous other countries alas have still not achieved; (*2*) price efficiency, by which is meant prices that reflect the incremental costs of supply, including the incremental costs of compliance with environmental policies; (*3*) corporatization and independent regulation of the energy industry; (*4*) openness to private investment; (*5*) demand management programs emphasizing customer information services (including efficiency standards and codes for equipment, buildings, and materials) and the development of energy service companies willing to work not only with the utilities but, especially important, with the manufacturers of electrical machines and equipment, the designers of buildings, and so forth; (*6*) environmental taxes and regulations to reduce external costs; (*7*) an effective approach to rural energy supply problems in developing countries; and (*8*) research and development into new energy production and end-use technologies.

The paper appeals to all parties to see efficiency in this broader light. Disagreements (there are many) are not about the desirability of efficiency but about how best to achieve it. The argument has been developed mainly in relation to electricity supply, but it can be applied without much difficulty to the nonelectric markets. It is also concerned with the situations of developing countries.

ENERGY DEMAND GROWTH AND THE ENVIRONMENT

Even in an energy-efficient scenario, world demands seem set to grow to very large levels on account of the growth of demands from developing regions. Per capita consumption of commercial energy in developing countries is exceedingly low compared with the industrialized countries. In the United States, per capita consumption of commercial energy is 8 tonnes of oil equivalent energy per year, or 80 times more than in Africa, 40 times more than in South Asia, 15 times more than in East Asia, and 8 times more than in Latin America (1, 2). Similarly, electricity consumption is much higher in industrialized countries, roughly 13,000 kWh per capita in the United States, for example, compared with approximately 600 kWh per capita in developing countries. If, with income growth, people were to consume about one-quarter of the amount of commercial energy consumed by the average citizen in the United States, then, allowing for population growth, total world consumption would eventually rise to around 30 billion tonnes of oil equivalent energy (Btoe) and electricity capacity needs to over 30 million MW, respectively four and ten times today's world levels. Such elementary arithmetic immediately suggests that very large energy demands lie ahead, assuming economic progress.

More specifically, there are three reasons for anticipating large demand increases from developing countries (3–6). First, two billion people are still without electricity, and two billion—nearly half of developing countries' populations—are dependent on biofuels for cooking. Second, barring epidemics or social calamities and assuming some success in family planning programs, the world's population seems set to rise by another four to six billion people over the next few decades before it stabilizes at 10–12 billion (1). These two factors alone, assuming no increase in average per capita consumption of commercial energy in developing countries, point to a five-fold increase in commercial energy consumption over the next few decades.

Third, per capita consumption is expected to grow as a consequence of rising per capita incomes and industrialization. The ratio of the growth of per capita energy consumption to the growth of per capita GNP in developing countries was approximately 2.3 over the past 25 years (1), though somewhat lower (about 1.5) in the high growth economies in East Asia. For electricity the ratios are higher (over 2.0 for most developing countries). Assuming continued economic recovery in Latin America, continued economic growth in Asia, and economic recovery in Africa, long-run economic growth rates of per capita GNP of 3% have been projected by the World Bank (7). Allowing for population growth and for the extension of services to populations still unserved, the long-run growth rate for electricity is set to exceed 7% per year over the next few decades, and for commercial energy in aggregate it will be about 4% per year. Moreover, this scenario makes some provisions for continuous improvements in energy efficiency (8, 19). Scenarios more ambitious with respect to energy efficiency may conceivably point to slower growth, but very large increases in energy demand are nevertheless expected in the coming decades in any scenario involving economic and social success in developing countries (see 9, 10). Figure 1 shows one such scenario for the case of electricity (see 2, 4, and 6 for others).

Such rapid expansions of electricity demand occurred in the industrial countries for the first three-quarters of the present century even as energy efficiency improved. Developing countries are going through the same sort of expansion, but with populations rising to 10 times those of the rich countries in the Organization for Economic Cooperation and Development (OECD). Developing countries may require 5 million MW or more of new capacity in the next

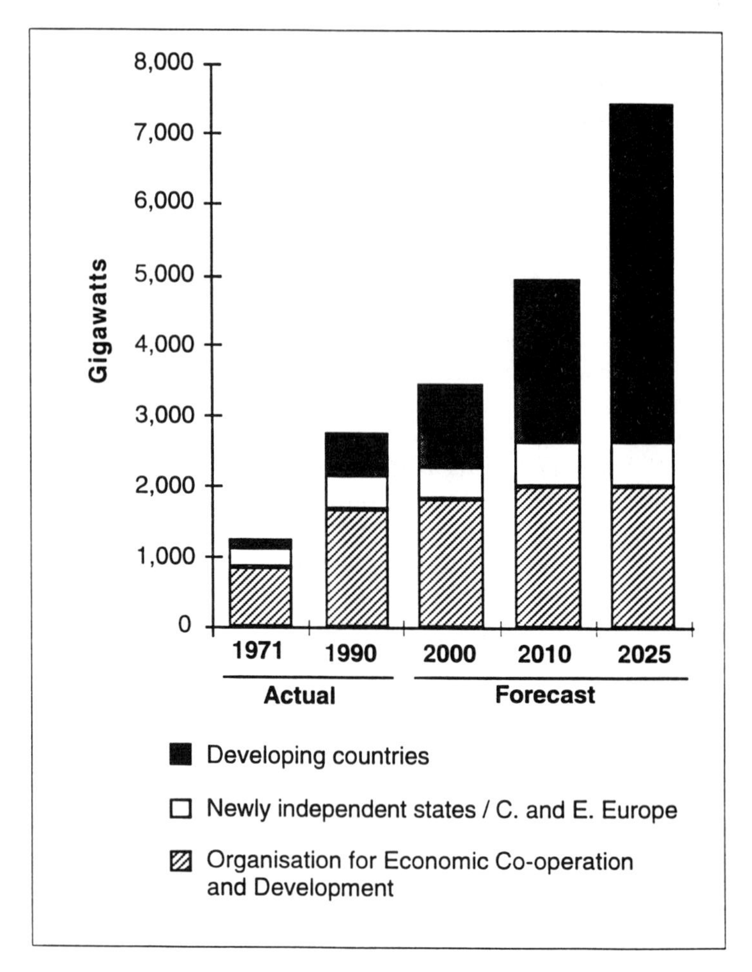

Figure 1 Net installed generating plant capacity, actual and forecast, 1971–2025.

30–40 years (eight times the total current capacity of the United States); even then, per capita electricity consumption would be less than one-quarter of current US consumption.

EIGHT ELEMENTS OF AN ECONOMICALLY EFFICIENT POLICY

Can such enormous growth in energy demand be financed? Can it be reconciled with demands for a cleaner environment? Seen less from the single viewpoint of energy efficiency than from the broader perspective of economic efficiency, there are eight elements in an effective energy policy. Some reflection on the roles of each element indicates why the answer to both questions is yes.

Economic Stability and Growth

A stable and growing economy, free of budgetary distresses and high inflation, is a prerequisite for the success of any microeconomic or social policy. Stability and growth raise savings and improve investment prospects in more efficient energy technologies, as in other areas. These prerequisites create confidence in the future and enable businesses and governments alike to focus less on the short term and begin to address the longer-term environmental problems posed by energy use. Stability and growth also reduce the industry's dependence on public revenue, leaving governments financially able to fulfill their public obligations, such as monitoring pollution, implementing environmental policies, and supporting research and development.

Killick (11) has succinctly outlined the a priori benefits of macroeconomic stability and growth for environmental policy-making. In Mexico in the 1980s, for example, inflation averaged 70% per year, peaking at over 100% per year. Once inflation was reduced, the country became much better positioned to introduce new environmental (and more energy-efficient) policies (12). In a study of the economies of the former Soviet Union (FSU), Dienes, Dobozi, and Radetzki (13) not only document the region's well-known sources of energy inefficiency but conclude "that market reforms can generate significant reductions in energy intensity, given the large reservoir of untapped energy efficiency and the scope for restructuring the economy." These findings assume the closing of energy-intensive industries in the course of restructuring, and improved attitudes toward cost-consciousness brought about by private ownership of industrial and service establishments and residences.

Price Efficiency

A recent survey (14) shows that in 1991 electricity prices averaged less than half of the marginal costs of supply in developing countries, approximately $0.04 (US) per kilowatt hour, as compared with marginal costs of about $0.10 (US) per kWh, a situation that had existed for several years. Preliminary evidence shows that with the commercialization of the electricity industry in many countries and investment by independent power producers prices are moving to more cost-reflecting levels (15, 16). But no matter how cost-reflecting prices are achieved, major gains in energy efficiency would undoubtedly result. The survey of price elasticities in developing countries by Bates (17) showed that short-run own-price elasticities were (numerically) quite low, around –0.1 to –0.2, while long-run price elasticities varied from around –0.2 to 0.8, averaging about –0.5. On this basis, if prices were gradually raised to commercial levels, then for each one US–cent increase in real prices, the following would eventually be achieved:

A 7% reduction of demand, relative to trend levels (assuming a long-run price elasticity of –0.5)
A $25 billion per year improvement in cash flows, and over $120 billion with the roughly $0.06(US)/kWh improvement actually needed. The cash flow improvement arising from a full reform to prices would exceed the currently estimated annual investment requirements of the electricity generation and distribution industry in developing countries, which are installing over 50 GW of new capacity a year.

Good pricing policies would reduce the eventual level of electricity demand in developing countries by about 30%. In absolute terms, this amount is very large. The electricity demand outlook discussed above (Figure 1) shows that the savings in new investment arising from price efficiency would likely be over 1.5 million MW in the period shown, an amount equal to the entire capacity installed in the OECD countries today. At the same time, the greatly improved revenue position would make the (reduced) expansion financially feasible by improving internal cash generation and by making the industry attractive for private finance. Such calculations show that no measure is more important for improving energy efficiency in developing counties than achieving price efficiency, quite apart from the financial rewards to the policy.

Mark Levine and Anthony Churchill have raised a practical question that needs to be addressed [1] about the effects of prices on demand and supply. The abysmally poor supply situation in many developing countries has left much unmet or repressed demand for electricity. The question is: Wouldn't price reforms, by "liberating" the industry from its cash flow and attendant problems, lead to more rapid supply expansion and enable the industry to meet the higher (unrepressed) demands? There are several points to be made here.

First, the improvements in the supply situation would undoubtedly enable the industry to meet repressed demands and, in this sense, level and growth of demand as well as of supplies would be higher than just estimated. When we have attempted to allow for improvement in supply elasticities in simulation studies (see 8, 19) capacity requirements are not reduced in the full reform case by 30%, as just noted, but by 15–20%. These simulations allow for the fact that improving the supply situation would take time, even with radical price reforms.

Second, while savings in demand would be lower than estimated savings when supply response is ignored, the cash flow benefits are far greater. Elec-

[1]Levine raised this question in a US AID seminar on energy efficiency in June 1994. Churchill's analysis has been presented in a conference paper (see 18 and the discussions of it in the conference proceedings).

trical losses in transmission and distribution would be reduced (improving energy efficiency in supply as well as cash flows), as would losses from theft, plant outages, and reserve margins. Electrical losses and losses from theft are as high as 40% in some countries, averaging about 20%, compared with less than 10% in good practice situations. Reserve margins are also very high—in the 20–50% range—compared with 15–20% in the OECD countries, because of poor plant availabilities. For various reasons (including poor maintenance), the thermal efficiencies in fossil-fired power stations were considerably below rated performance.[2] Reductions in losses, and improvements in plant availability and operating practices would enable a significant share of the expected increases in demand to be met without new investment. The improvements in both energy efficiency in supplies and in cash flows arising from price efficiency would therefore be substantial.

Third, repressing demand through blackouts and brownouts, or by simply not connecting people to the networks, is not only an indisputably unsatisfactory way of increasing energy efficiency but probably makes the situation far worse and increases energy inefficiency. When supplies are restricted or unreliable, people use backup generators, which have lower efficiencies and higher pollution levels than grid supplies, kerosene wick lamps (100 times more inefficient than the incandescent lamp), and so forth. The surveys by Lee (22) and Schramm (23) of Nigeria, for example, found that the aggregate capacity of peoples' investments in backup generators exceeded the available capacity from the grid. Aside from huge economic costs, this situation meant that energy efficiency was greatly decreased, not increased, as a consequence of repressed demands.

Price reforms have more complicated effects on energy efficiency than those assumed in simple demand models, but once the benefits of improvements in the efficiency and reliability of supplies are taken into account, the gains in energy efficiency are probably larger than indicated earlier in this section. The financial rewards both from improved revenues and the reductions in the wastes of capacity and energy on the supply side would certainly be larger. In all respects, developing countries would be substantially better off. Examining the problems of supply response and repressed demand therefore strengthens the conclusion as to the importance of price reforms, for both energy and economic efficiency.

Regulatory Reform and Openness to Private Investment

The case for making the public utilities in developing countries self-standing corporations, operating on commercial principles and regulated by bodies independent of special interests, has been made many times over the past several decades. A full discussion of this subject is beyond the scope of the present paper (it was recently reviewed in the World Bank's policy paper on

[2]See 20, 21 for data and references relating to the figures quoted in this sentence.

electric power (20), which has attracted much attention. Suffice it to say that regulatory reforms requiring the public utilities to work on commercial principles, and the growth of private investment in the industry, are forces for both price and cost efficiency, and hence for energy efficiency, for the reasons discussed in the above Price Efficiency section.

In recent years, the public policies of developing countries have undergone a sea change, with dirigism giving way to more commercially-minded policies. Early experiences with privatization in Latin America indicate that private producers are seeking contracts that will push prices to more cost-reflecting levels (24). If this trend continues, private investment may succeed in achieving efficiency where public policies toward state enterprises have often failed. A task of future reviews will be to assess the quantitative achievements of regulatory reform and private investment.

DSM Programs (Nonprice Measures)

Marginal cost pricing, which generally requires prices to vary according to time of day, voltage level and season, is central to DSM. But any satisfactory pricing policy requires support from nonprice measures, of which the following three are now common:

- information and advisory services to raise consumers' awareness of opportunities to save energy and reduce bills;
- efficiency standards and codes and the labeling of equipment and building materials;
- utility-managed rebate schemes to encourage consumers to use high-efficiency devices.

The third measure is the most controversial because it has often depended on large subsidies and intrusive regulation and is at odds with the idea of an industry operating on commercial principles and under independent (arm's-length) regulation. Utility-managed rebate schemes are also not strictly nonprice measures because they depend greatly on the manipulation of prices through the regulatory process to achieve the subsidies, and for this reason too have been open to criticism (24, 25).

The first two measures, on the other hand, are time-honored ways of helping markets work better. They may be particularly important when end-use markets are growing rapidly (as in developing countries) because they can influence choice when new investments are being made; incorporating efficiency in new investments is simpler and cheaper than retrofitting old ones.

It is not an overstatement to say therefore that the controversies (alluded to above) about energy efficiency are disagreements over one-third of merely one of the eight policy elements discussed here, namely utility-managed demand-side management programs. The other seven and two-thirds are not only not controversial, they are mutually complementary and capable of bringing about immense improvements in energy efficiency, economic efficiency, financial returns, and the environment.

Can the controversies over this one particular matter be resolved? Much depends on whether people are willing to acknowledge the existence of alternative approaches to demand management. For example, a good case can be made for rebate schemes for new and innovative products and ideas; private companies themselves in numerous other areas frequently apply such policies; and a good economic case can be made for such schemes based on the analysis of cost curves and transactions costs. (We will return to this theme in the Assessment. Reference should also be made to Baumol's paper in this volume.) But any rebates, when justified, need to be based on the assumption that they will be removed when the market is sufficiently established such that the new products and ideas meet a "market test" and become financially self-sufficient.

We must ask, why should the electricity producers be expected to manage the programs and why should the regulators be expected to approve the rebates? And why should the utilities be expected to "delink profits from the sales of electricity," as many DSM proponents have argued? Such policies raise questions not only about the financial sustainability of the policies, but about commercial propriety (27). For example, what would monopolies and antitrust commissions (or the industry itself) have to say if the oil companies were asked not only to delink their profits from sales, but to cross-subsidize and sell automobiles that have higher mpg ratings than others? Energy efficiency can be encouraged without going to such lengths and undermining the commercial basis of an industry. The utilities and private electricity producers obviously have a role in encouraging consumers to use electricity more efficiently, and this will often support other policies such as those to improve load factors, or peak load or seasonal pricing policies.

But any successful program needs to go beyond the utilities and their customers. The work of energy service companies provides an example, and it is interesting that their client base is increasingly less the utilities and more the suppliers of efficient end-use energy technologies—light bulb manufacturers, refrigerator manufacturers, architects, builders—and the users of the technologies. This change has a commercial logic because the suppliers and the users both stand to gain financially from the changes in practices that the energy service companies can facilitate, and both have more direct interest in what the technologies can deliver in the way of efficiency. This avenue seems more promising than the much-discussed one of unlinking profits from the sales of electricity through utility-managed programs.

Environmental Policies

When the social costs of pollution exceed the costs of abating it, this is another source of economic—and energy—inefficiency. A good environ-

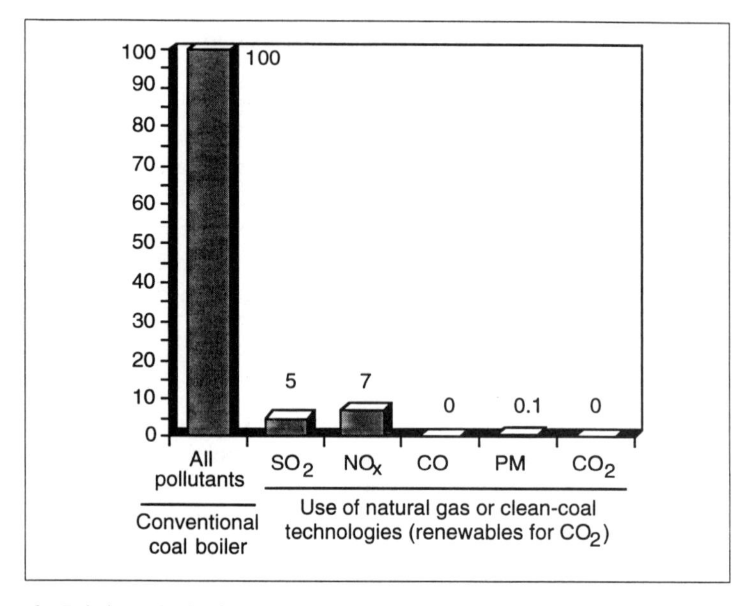

Figure 2 Relative pollution intensities of polluting and low-polluting practices (Electric Power).

mental policy will address the problem by taxing or regulating the pollution (as opposed to the output or sales) of an industry such that the industry will have an incentive to develop and adopt low-polluting practices such as gas-fired power plants, "clean-coal" technologies, and (an emerging prospect) solar energy.

Figure 2 shows how important environmental policies are in abating pollution by comparing the gaseous emissions per unit of output of low-polluting technologies for generating electricity with those of a conventional coal boiler with mechanical controls only (without electrostatic precipitators or bag-house filters). With the partial exception of NO_x, local pollution can be reduced by factors of 20 or more and in some cases eliminated entirely through technologies now available. Incorporating them into the capital stock once an environmental policy has been put in place does of course take time. The process may take as long as 20–30 years, given the long lifetimes of generating stations (unless they are retrofitted), but the evidence is now fairly convincing that local pollution can be reduced substantially even as output expands (see 4, 8, 19, and 37). Solar energy technologies are also showing much promise for addressing global pollution from CO_2 (38). The time required to introduce them on a large scale, should the need arise, would likely be quite long, perhaps half a century or longer (4, chapter 8; 39). But because emissions are the product of output and emissions per unit output, such data indicate that reducing both local and global pollution from energy production and use substantially over the long term should be possible, at much higher energy use levels than today's.

Rural Energy Supplies

If commercial energy supplies are not expanded in developing countries, the number of people dependent on fuelwood for cooking (and also without electricity) will rise from 2 billion today to 5 billion in 30 years. Such an increase would be environmentally and economically untenable: It would be damaging to soils, agriculture, forests, and people's health (28–30). And it would be energy inefficient because cooking with biofuels consumes roughly five or more times the amount of energy as cooking with modern fuels (Figure 3). If economic development is to proceed, modern energy forms will surely need to become available, affordable, and widely used throughout the developing world. These changes will require income growth in rural and urban areas, and all that such growth in turn requires of development policies, supported by investments in "modern" energy forms—oil and gas fuels, renewable energy, and electricity. In addition, because biofuels will be used extensively for a

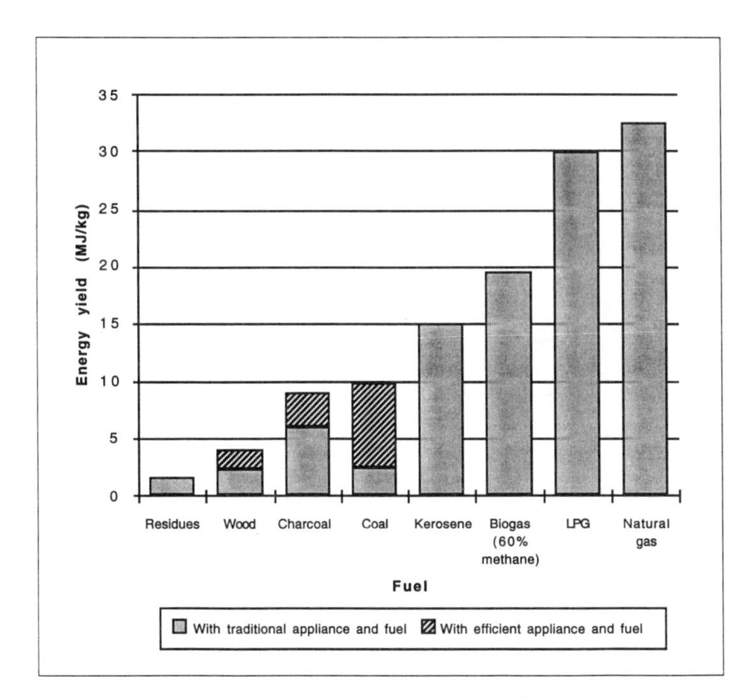

Figure 3 Delivered energy for cooking using various fuels (megajoules per kilogram).

long time, investments in afforestation, supported by extension and education, will be needed to encourage people to use them sustainably with less damage to soils, forests, and health. Much research in recent years has shown that such investments, besides being socially and environmentally desirable, generate good economic returns (40–42).

Research and Development

Recent developments in environmentally-improved practices and energy-efficient technologies have depended on a successful mix of public and private policies. Developments in oil and gas technologies have stemmed mostly from private investment, whereas clean-coal and solar-energy development have depended on a mix of private and public investments in research and development and will continue to do so. Research and development in solar-energy technologies in particular holds much promise, and the inadequacy of present policies in this area raises economic and environmental concerns (31). A proper discussion of research and development is beyond the scope of this paper. It is noted here in order to avoid what would be an omission from any listing of the elements of an efficient policy.[3]

ENERGY EFFICIENCY OR ECONOMIC EFFICIENCY?

A common theme in the above analysis is that, for the purposes of policy-making, economic efficiency is a more appropriate criterion to use than energy efficiency. The two often go together, but when the costs of achieving energy efficiency are high, they may not. Energy efficiency is a technical term for the amount of heat, light, motive power, or refrigeration delivered for a particular service per unit of energy consumed, and of course it is an indispensable concept for engineering and scientific purposes. But for policy-making, economic efficiency is the better concept. As Lee Schipper has said,[4] we need policies that leave countries better off, not worse off, and it would be a pity if the search for energy efficiency and environmental improvements were to lead to the latter. Does the economic efficiency criterion meet Schipper's criterion? The principles of economics on which the economic efficiency criterion is based are precisely intended to give guidance on ways of improving the economic welfare of nations.

Specifically, economic efficiency is measured by the economic rate of return to investment or, alternatively, the net present value of investment. It requires:

[3]See Professor Baumol's paper in this volume for a fuller discussion, and (31a).
[4]At a World Bank Roundtable on Energy Efficiency, September 1994.

1. prices that reflect the marginal costs of supply (price efficiency);
2. marginal costs calculated to include the costs of compliance with environmental laws and regulations;
3. public investments chosen according to their prospective economic rates of return;
4. private investments chosen according to their prospective private (net of tax) financial rates of return;
5. information services and codes to bring clarity into the making of economic transactions, and laws and regulations to avoid dishonest practices.

The first and fourth conditions are generally met when supplies are provided under competitive arrangements, and departures of prices from marginal costs are a good indication that competition is lacking or that an industry is being poorly regulated. If natural monopolies exist, regulation according to the principles of marginal cost pricing and a requirement that investments are chosen according to their prospective rates of return (the third condition) are intended to ensure that monopolistic rents and practices are avoided and that economic efficiency is achieved both in the provision and use of a service. In the case of electricity, for instance, marginal cost pricing points to the importance of time-of-day and (depending on the system) seasonal pricing, as well as pricing according to voltage level. It has recently been accepted by conservationists as central for DSM programs; but the development of both the principles and the practice owe much to the work of Boiteux (33) and Turvey (34–36) 30 years and more ago.

The second and fifth conditions (compliance with environmental policies and with the laws and regulations required for open and honest business) show that an approach based on economic efficiency is not and cannot be that of an economic purist. All economic activities are necessarily undertaken within a framework of laws and regulations set by public authorities in response to a public need or concern, and for this reason the maximization of the returns to investment is very much a constrained optimum. The opposite is also often true, especially with respect to the environment: Laws and regulations themselves will be influenced by economic factors, principally costs and practicability. But once the laws and regulations are established, the other four conditions need to be met if private and public interest in the implementation of a policy are to converge.

Seven and two-thirds of the eight elements discussed above are fully consistent with the aims of improving economic efficiency, and the remaining one-third of the fifth element can be made consistent if (*a*) more weight is given to nonutility demand management programs, and (*b*) the idea of delinking sales from profits is abandoned in favor of a more commercial approach to providing incentives for innovation.

PRIORITIES

What weights should be given to the eight elements discussed above? None has zero weight, but some are far more important than others, depending on the economic circumstances of a country. For instance, to say that reducing the amount of firewood and animal dung used for cooking is a priority in the United States would be silly; but for the billions of people dependent on these fuels in Africa, Asia, and Latin America, increasing the supply of modern energy forms (including renewable energy) will be a social, economic, and environmental priority for several decades. Similarly, rich countries have had nearly 50 years of economic stability and growth, aside from minor recessions, which were insignificant compared with the "lost decade" of the 1980s in Latin America or with the current economic upheavals in East Europe and the countries of the FSU, or with the problems in much of Africa. For most of the developing world, the achievement and maintenance of economic stability and growth—the first of the eight elements—must be the main priority, because success in all seven other elements turns on this one.

Second in importance for developing countries are institutional reforms to promote more commercial pricing and regulatory policies (elements 2 and 3). Some would add the emergence of private or independent producers in electricity generation (element 4), and a good case has been made that they are a force for price efficiency and regulatory reform (20). Good policies in this area would not only enormously improve industry cash flow, as discussed above, leaving it better placed financially to expand supplies to people still not served (element 7) and to respond to environmental policies by investing in low-polluting technologies (element 6), but would also have major benefits for the public revenue, and (under wise administrations) for public policy.

An example is the $120 billion shortfall in electricity sales revenues (discussed above) created by subsidized prices. (This figure excludes the massive, possibly larger, subsidies that the countries of the FSU are trying to eliminate.) This shortfall amounts to 3% of developing countries' GNP, is 50 times the annual energy lending of the World Bank to these regions and is made up directly or indirectly of public revenue. Moreover, it translates into an economic loss of comparable value once the "deadweight" and other losses arising from price and cost inefficiencies are taken into account (19); that is, removal of the subsidies is not merely a matter of transfer payments (lower taxes/higher prices), but first and foremost of reducing the real economic costs of severe fiscal deficits, dilapidated supplies, and of people being deprived of electricity service.

The realization of how much could be financed with price reforms in the electricity sector alone while leaving the countries economically better off is sobering. In the *World Development Report 1992,* for instance (4), we estimated that the annual incremental costs of putting developing countries onto an environmentally-sustainable growth path would amount to roughly 2–3% of GNP, or less than the economic and fiscal gains that would result from electricity price reforms. These costs included significant public investments in soil conservation; investments to control a wide range of emissions and effluents from industry and transport; the requirements of greatly expanded water and sanitation programs (among the main priorities of development); additional resources for agricultural research; greatly expanded primary and secondary education for girls (a much-neglected area); and a $7 billion per year increase in family planning programs (present funding is $5 billion per year), which many consider crucial for lowering world population growth. This is not to say that the resources liberated by electricity or any other energy price reforms should or would be used for these purposes, but such calculations do show how much is to be gained from good regulatory and pricing policies toward the industry. If such policies come second to achieving and maintaining economic stability and growth, therefore, they are a close second.

From a purely environmental perspective, elements 6, 7, and 8 deserve the highest ranking, once the enabling conditions (elements 1–3) are in place or are being put in place. They deserve this ranking because the amounts by which local and, ultimately, global pollution can be reduced through the development and use of the low-polluting technologies are one, two, and sometimes (depending on the pollutant) several orders of magnitude greater than any reductions possible through energy efficiency (Figure 2), whether such efficiency is achieved through the "win-win" macroeconomic polices and price reforms discussed above or through nonprice DSM measures.

Where does this leave us with nonprice measures? Giving nonprice measures the same or even comparable weight to the first three elements discussed above in relation to economic policies, or to the last three discussed in relation to environmental policy, is difficult to do in good conscience. Nonprice measures are important, but pursuing them should not be allowed to detract from the prime importance of the other policy elements. DSM activism has succeeded in putting nonprice measures on the map; but too frequently it has overstepped the bounds of reason and evidence, and often the principles of good policy-making. By detracting from the overarching importance of the other seven elements of policy, DSM activism runs the danger of damaging the environment and the economy. Policies need to be approached from a broader perspective.

CONCLUSIONS

People interested in energy policy, whether in business, finance, government, or the environmental movement, should welcome and support an approach based on economic principles for three reasons. By solving the financing problem and encouraging innovation and cost-efficiency, the economic-efficiency approach will enable all countries to meet energy demands. By giving proper weight to the development and use of low-polluting technologies this approach will enable reduction of local and, over the long-term, global pollution as energy demands grow. And in developing countries especially, an economic approach will enable the industry to play its part in raising living standards for the population at large. Given good policies, there is no reason at all why developing countries, like the industrial countries before them, should not enjoy the benefits of much higher levels of energy consumption than they do today.

ACKNOWLEDGMENTS

This paper is a development of a shorter note on the subject written for the World Bank (32). I would like to thank my colleagues Richard Stern, Karl Jechoutek, Trevor Byer, Amarquaye Armar, Ranjit Lamech, Ernesto Terrado, Anthony Churchill, Robert Saunders, and several others for the many discussions we have had on this subject. I am also grateful to Keith Openshaw and Douglas Barnes for the information contained in Figure 3 and to Paul Wolman for his help with various parts of the paper. Responsibility for the views expressed is of course my own.

Literature Cited

1. World Bank. 1994. *World Dev. Rep. 1994: Infrastructure for Development* (statistical annexes). Washington, DC: World Bank
2. International Energy Agency. 1994. *World Energy Outlook 1994.* Paris: OECD/IEA
3. Anderson D. 1991. Energy and the environment: An economic perspective on recent technological developments and policies. *Wealth Nations Found. Spec. Briefing Pap. 1,* Edinburgh
4. World Bank. 1992. *World Dev. Rep. 1992: Econ. Dev. Environ.,* Washington, DC
5. World Energy Council. 1994. *Energy for Tomorrow's World.* New York: St. Martin's
6. Shell Group of Companies. 1993. *Global Scenarios: 1992–2020.* London: Group Planning, Shell Centre
7. World Bank. 1994. *Global Economic Prospects and the Developing Countries.* Washington, DC: World Bank
8. Anderson D, Cavendish W. 1992. Efficiency and substitution in pollution abatement: three case studies. *World Bank Discuss. Pap. 186.* Washington, DC: World Bank
9. Schipper L, Meyers S. 1992. *Energy Efficiency and Human Activity: Past Trends, Future Prospects.* Cambridge: Cambridge Univ. Press
10. Schipper L. 1993. Energy efficiency: lessons from the past and strategies for the future. *Proc. World Bank Annu. Conf. Dev. Econ.,* pp. 397–427. Washington, DC: World Bank
11. Killick T. 1991. Notes on macro-economic adjustment and the environment. In *Development Research: The Environmental Challenge,* by JT Winpenny, pp. 38–42. London: Overseas Dev. Inst.
12. Eskeland G. 1992. Demand management in environmental protection: fuel taxes and air pollution in Mexico City. *World Bank Country Econ. Dep. Pap.* Washington, DC: World Bank
13. Dienes L, Dobozi I, Radetzki M. 1994. *Energy and Economic Reform in the Former Soviet Union.* New York: St. Martin's

14. Heidarian J, Wu G. 1994. Power sector statistics for developing countries. *Draft Rep., Industry and Energy Dep.*, World Bank, Washington, DC
15. Moscote R. A note on private investments in electric power in Latin America. *Finance Private Dev. Short Note Ser.*, Washington, DC: World Bank. In press
16. World Bank. 1994. Power and energy efficiency status report on the bank's policy and IFCÕs activities. *Joint World Bank/IFC Seminar,* Washington, DC
17. Bates R. 1993. The impact of economic policy on energy and the environment in developing countries. *Annu. Rev. Energy Environ.* 18:479–506
18. Churchill A. 1993. Energy demand and supply in the developing world, 1990–2020: three decades of explosive growth. *Proc. Annu. World Bank Conf. Dev. Econ.*, pp. 441–58. Washington, DC: World Bank
19. Cavendish W, Anderson D. 1994. Efficiency and substitution in pollution abatement. *Oxford Econ. Pap.* 46:774–99
20. World Bank. 1993. The World Bank's role in the electric power sector: a policy paper. Washington, DC: World Bank
21. World Bank. 1993. Energy efficiency and conservation in the developing world: a World Bank policy paper. Washington, DC: World Bank
22. Lee K, Anas A. 1992. The costs of deficient infrastructure: the case of Nigerian manufacturing. *Urban Stud.* 29 (7):1071–92
23. Schramm G. 1993. Nigeria: issues and options in the energy sector. *ESMAP Rep. 11672-UNI,* Ind. Energy Dep. Washington, DC: World Bank
24. Joskow P. 1990. Understanding the unbundled utility: conservation bidding proposals. *Public Util. Fortnightly* 125 (1):18–28
25. Lamech R. 1994. When energy conservation doesn't work: critique of a DSM program. *FPD Note 3.* Washington, DC: World Bank
26. Deleted in proof
27. Anderson D. 1993. Energy efficiency and the economics of pollution abatement. *Annu. Rev. Energy Environ.* 18: 291–318
28. Smith K. 1988. Air pollution: assessing total exposure in developing countries. *Environment* 30(10):16–35
29. Doolette J, Magrath W, eds. 1990. Watershed development in Asia: strategies and technologies. *World Bank Tech. Pap. 127.* Washington, DC: World Bank
30. Newcombe K. 1984. An economic justification for rural afforestation: the case of Ethiopia. *Energy Dep. Pap. 16.* Washington, DC: World Bank
31. Kozloff K, Dower R. 1993. *A New Power Base: Renewable Energy Policies for the Nineties and Beyond.* Washington, DC: World Resour. Inst.
31a. National Research Council (US)/World Bank. 1994. *Symp. Marshaling Tech. Dev., Nov. 28–30. Irvine, Calif.*
32. Anderson D. 1994. Energy production and use: eight elements of an *economically* efficient policy. *ESMAP Connection (Newsl. Joint UNDP/World Bank Energy Sect. Manage. Programme)* 2 (1):5\D8
33. Boiteux M. 1964. Marginal cost pricing and the "tarif vert" of Eléctricité de France. In *Marginal Cost Pricing in Practice*, ed. J Nelson. Englewood Cliffs, NJ: Prentice-Hall
34. Turvey R. 1968. *Optimal Pricing and Investment in Electricity Supply.* London: Allen & Unwin
35. Turvey R. 1970. *Economic Analysis and Public Enterprises.* London: Allen & Unwin
36. Turvey R, Anderson D. 1977. *Electricity Economics: Essays and Case Studies.* Baltimore: Johns Hopkins Univ. Press
37. Organization for Economic Cooperation and Development. 1991. *The State of the Environment.* Paris: OECD
38. Ahmed K. 1994. Renewable energy technologies: A review of the status and costs of selected technologies. *World Bank Tech. Pap. 240, Energy Ser.*, Washington DC: World Bank
39. Anderson D, Bird CD. 1992. Carbon Accumulations and Technical Progress—a simulation study of costs. *Oxford Bull. Econ. Stat.* 54(1):1–29
40. Gregerson H, Draper S, Elz D, eds. 1989. *People and Trees: The Role of Social Forestry in Sustainable Development. Econ. Dev. Inst. Semin. Ser.*, Washington, DC: World Bank
41. Barnes D, Openshaw K, Smith K, van der Plas R. 1994. What makes people cook with improved biomass stoves? A comparative international review of stove programs. *World Bank Tech. Pap. 242, Energy Ser.*, Washington, DC: World Bank
42. Anderson D. 1988. The *Economics of Afforestation: A Case Study in Africa.* Baltimore: Johns Hopkins Univ. Press

FOUR MISCONCEPTIONS ABOUT DEMAND-SIDE MANAGEMENT

Charles J. Cicchetti[1]

Department of Economics, University of Southern California, Los Angeles, California 90089; Arthur Andersen Economic Consulting, 633 West Fifth Street, 26th Floor, Los Angeles, California 90071

KEY WORDS: demand-side management, economists, externalities, marginal-cost pricing, energy services

---

CONTENTS

Introduction: The Public Policy Debate

Some economists aver that a utility should only be permitted to sponsor demand-side management (DSM)[2] programs as a stand-alone business and that the market should be left to determine the appropriate amount of end-user energy conservation. Regulators should work to get the prices right, and then step aside to let the market work. Prices should be set equal to, or closer to, true marginal cost. Electricity consumption has declined significantly during the past decade because of the general increase in utility prices, suggesting that people do indeed respond to price signals. In short, electricity prices and

[1]Many individuals have provided helpful comments, especially Armond Cohen, Amory Lovins, and Kristina M. Sepetys. The views presented here are not necessarily attributable to any of those mentioned, and any remaining errors are solely the responsibility of the author.

[2]By DSM, I mean any utility-financed and -sponsored program designed to reduce electricity use (kWh), reduce capacity or power demand (kW), or shift demand (kW) or energy (kWh) to off-peak periods.

markets, not regulators or regulatory mandates, should inform consumers' decisions to invest or not invest in energy efficiency.

Other economists generally do not dispute the logic just described. They counter, however, by emphasizing that various externalities and market failures exist that work to distort consumer demand and prohibit the markets from functioning properly. Setting prices to reflect direct utility costs might be a good first step toward correcting the problems associated with energy-efficiency programs. But it is only the first step, since direct utility costs do not begin to include the vast range of "costs" not captured in electricity prices. Examples of such costs include air pollution and national security problems associated with dependence on foreign oil supplies. In addition, it is important to note that after more than 20 years of economic testimony, few would conclude that utility tariffs, in fact, reflect the economists' notion of marginal cost.

Despite the potential for considerable savings, many individuals and businesses fail to act on their opportunities to invest in energy efficiency. For many customers, especially residential, electricity bills do not provide sufficient information about the end uses that determine their energy consumption. For example, while most telephone bills provide some indication that long-distance telephone calls are cheaper at certain times, an electricity bill (particularly for residential consumers) provides little or no indication of when electricity consumption is most costly or which appliances consume the most energy. Since much of the success of DSM programs is dependent upon the market penetration of energy-efficient equipment, these programs will continue to struggle until consumers become better informed about the costs associated with their energy consumption. Given environmental externalities and consumer information problems, a pro-active public policy response in the form of utility-sponsored DSM programs seems necessary to redress economic losses.

Nevertheless, many economists remain opposed to utility-sponsored DSM programs. The following sections respond to four common criticisms of utility-sponsored DSM programs:

1. DSM programs impede market reform and competition.
2. Regulators favor conservation even when consumers reject it.
3. The benefits of DSM are grossly exaggerated.
4. The costs of DSM programs are significantly underestimated.

DSM Programs Impede Market Reform and Competition

Even before the recent efforts to reform and restructure the electricity industry, critics claimed that DSM, a regulatory mandate, was a significant impediment to encouraging more competition in electricity markets. It is in fact, however, still an open question how regulators intend to manage the proposed mix of competition and regulation aimed toward—among other things—continued encouragement of DSM. The 1992 Energy Policy Act[3] specifically denied the Federal Energy Regulatory Commission, or FERC, the authority to require retail wheeling.[4] State regulators have generally blocked natural gas and electricity bypass[5] to protect residential customers from paying for stranded investments made by utilities. Coincidentally, prohibiting bypass also protects the market for DSM.

Critics have also argued that utilities should not monopolize the DSM market. In fact, the typical utility role has been to help to create a market, not to monopolize it. Utilities have generally acted as brokers and facilitators, seeking payment for closing deals, but not attempting to take out monopoly profits. Moreover, utility activity will continue to be constrained by market forces, customers, competitors, and entrants, as well as by antitrust restrictions.

DSM supporters deserve credit for their part in encouraging the current movement to restructure and introduce greater competition into the electric industry. Electric utilities are quickly becoming distribution companies that buy electricity and do not produce it. To get the performance incentives right, electric utilities need to be paid based upon how effectively they purchase electricity and how efficiently they purchase the fuels they choose to burn in existing generators. The DSM performance-based incentive schemes developed in the 1980s are being applied to the non-utility generators. Rather than thwarting competition, DSM is indirectly encouraging more of it.

Increased competition in generation is important for economic efficiency. DSM's role in gaining better incentives for this effort is not, however, the only reason for rejecting the notion that DSM programs are inherently anticompetitive. There is an increasing perception that the relevant market for assessing economic efficiency is not simply kWs and kWhs. Rather, energy services—such as lighting, heating, cooling, and pumping—make up the relevant market. Utilities have a major role to play in this emerging market.

To keep electric utilities out of the market developing on the customers' side of the meter makes about as much sense as keeping telephone companies out of the more advanced customer-premise communication, computing, and entertainment services. Energy services can be sold to customers, as they consume hot water, cooling, heating, appliance use, and lighting. Utilities need

[3]1992 Energy Policy Act, Report 102-1018 (5 October 1992).

[4]The term "retail wheeling" refers to the purchase of electricity by an industrial customer from a source other than its local utility.

[5]The term "bypass" refers generally to a customer obtaining service from a supplier without utilizing the facility of the former supplier, i.e. bypassing the former supplier.

to find creative means to define products, package them, and market new services on the other side of the customers' meter. Such means may include sophisticated, computerized load management; acting as brokers for real-time pricing from a competitive generation market; or more efficient home and business energy-using devices.

Regulators Favor Conservation Even When Consumers Reject It

Critics maintain that DSM programs will only succeed if consumers are highly subsidized. In other words, utilities must pay for consumers to consume energy efficiency. Critics also think that regulators will place enormous financial pressure on utilities and that management will go to extremes to force such highly subsidized conservation into their customers' homes and businesses. First, utility conservation programs are not, and should never have been viewed as, give-away programs. Most regulators and utility executives price DSM so that the participants in the program pay their fair share. Utilities are businesses. When they provide products that yield value, they should be able to collect fees and earn income.

Even DSM critics agree that those who do not participate in DSM programs would benefit from DSM by the difference between the marginal cost of electricity and regulated prices, when the latter is less. This suggests some jointness in the production of DSM and electricity, which means that all beneficiaries should pay for DSM. But there are two other reasons why non-participants might be willing to pay to support some additional DSM consumption by other customers. Future energy price levels depend upon today's consumption and how least-cost utility planners respond today to meet that future demand. If DSM is less costly than new generation, then all customers (and economic efficiency) will benefit if DSM is selected over jointly produced new generation.

Nonparticipants also gain or lose based upon the level of externalities associated with electricity generation and DSM. This concept has sometimes been exaggerated by some proponents of environmental adders. DSM critics often reply that the environmental externality issue is overplayed by DSM proponents. Regardless of the difference resulting from exaggeration, reducing environmental damage is not unimportant.

The Benefits of DSM Are Grossly Exaggerated

Critics point out that DSM benefits are based on engineering estimates, not actual DSM performance, and that proponents use exaggerated notions of environmental benefits. DSM is in its infancy, and marketers and promoters may be guilty of exaggeration. Regulation is a process that sifts the wheat from the chaff, however. Regulators are also not prone to overcompensate investor-owned utilities. Increasingly, regulators and utility management are learning how to improve their estimates of expected savings, including adjustments for the so-called "snap-back" effect. True-up tests, and adjustments for actual savings versus planned savings, are increasingly being considered around the nation to establish performance-based DSM utility earnings.

The critics are correct that some people exaggerate the value of environmental benefits. Many DSM proponents fail to recognize that electric utilities have already made significant efforts to internalize environmental costs. They therefore seek too much contribution for DSM from nonparticipants. Neither side is entirely correct. The good news is that regulators are again learning how to price both DSM and kWhs better.

DSM Program Costs Are Significantly Underestimated

Critics also claim that the costs of DSM programs are significantly underestimated. This is simply not so for direct costs. Increasingly, utility-sponsored DSM programs are evaluated by regulators using the full costs (i.e. all attributable costs) as direct reflections of level of effort and actual performance. Two costs given especially noteworthy treatment are customer hassle costs and lost utility revenue. The first is the key for determining how utility and other marketing entities can inform the customers about the value of DSM.

Some critics maintain that the hassle and associated opportunity costs involved in pursuing energy efficiency provide further evidence that conservation is not an economic choice. This conclusion is wrong. Most people do not understand their electricity tariffs or bills. Utility customers pay for electricity some weeks after they consume it, and they do not know how their use of various electricity-using appliances affects their kWh consumption and monthly electric bill. For consumers to develop an understanding of electricity consumption and to evaluate the performance of conservation and efficiency improvement alternatives takes time. It also takes time to evaluate the various alternatives in the market. Utility bills and the price/use communication needs to be fixed. Time-of-use pricing, load management, and two-way communication between the customer and utility need to be implemented. All of these changes are forms of DSM and need to thought of as such. They are also part of the energy purchasing and brokering process that some progressive utilities now recognize is a new service their customers are seeking.

Second, there is the "tyranny of small decisions" phenomenon. It might not be worth it for each of us to make our streets safe, protect ourselves from fire, etc. Yet, we all benefit when such services are provided. If DSM is less costly than generating electricity, except for the fact that customers cannot interpret their electricity bills and comprehend conservation choices, then there is a social gain to be had in centralizing and consolidating these activities within the utility. Nevertheless, consumers of DSM are the beneficiaries and they need to pay for these services.

Policy Conclusion

Advocacy of effective incentives for utility conservation performance is based on two premises. First, an appropriate goal of public policy is to obtain the optimum amount of cost-effective, utility-sponsored conservation. Second, economic incentives are preferable to command-and-control regulation as performance motivators and economic efficiency enhancers.

To encourage DSM activities, an incentive program should make DSM as profitable as other utility activities. DSM savings measurement will become more accurate as utility DSM programs and incentive mechanisms become more prevalent, as more dollars are spent on DSM, and as a greater percentage of utility profits depends on the claimed savings and DSM incentives.

The energy marketplace has heretofore been highly regulated. This is changing. Competition for generation is ubiquitous. Incentive regulation is being used to reform the engineering focus of rate-based cost-of-service regulation. The new utility focus is on the services, not units of energy consumption. Price, not cost, is the main concern. Combining the two, energy services and the price of these services, is where utility management, marketers, and regulators are (or should be) headed. The market is being made more competitive because of the political interest in DSM and because performance-based regulation is becoming more prevalent. The trick is to benefit from these profound changes.

The "carrot and stick" regulatory support for DSM is likely to evolve into a commercial, marketing-oriented service line in which electric utilities and natural gas companies learn a new corporate culture similar to that of their telephone cousins. Many formally bundled utility services will be unbundled and priced separately (again, there are telephone similarities). These will likely include reliability, stand-by, coordination, dispatch, safety, billing, information access, brokering, energy efficiency, and demand-side management.

DSM could be sold successfully (it has already been proven valuable) in a marketing, as opposed to a command-and-control, context. Some electric and natural gas companies are already moving in this direction for their largest customers. This movement will be expanded to all customers. The current impetus for DSM was achieved by utility-mandated programs. What is needed and what will happen is a return to some of DSM marketing ideas of the late 1970s (specifically, building customer relations through free comprehensive energy audits) that were shelved in the 1980s.

Another transitional measure might include spinning off DSM activities into nonregulated divisions for the largest customers that seek utility-provided enhanced energy service contracts. Such spin-offs would help utilities enter into contracts to improve business relations between themselves and their largest commercial and industrial customers. These spin-offs would also help to demonstrate that DSM can succeed in the market without regulatory subsidies.

Innovative marketing and DSM/energy service bundling should be based upon the fact that consumers want heat, light, cooling, motor power, and hot water, as opposed to kWhs, British thermal units (Btus), etc. Using their existing franchise value, utilities should fight to keep competitors from taking away their customers. To do this, utilities need to package security, safety, integrity, reliability, brokering, low-cost energy, DSM, and other services. These actions form the basis for a utility to develop a winning strategy for competing aggressively and successfully in an emerging, highly competitive energy services world.

ENERGY-EFFICIENCY SOLUTIONS: What Commodity Prices Can't Deliver

Ralph Cavanagh
Natural Resources Defense Council, 71 Stevenson Street, Suite 1825, San Francisco, California 94105

KEY WORDS: energy efficiency, environment, competition, global climate, market barriers

---

CONTENTS

The Challenge

Over the past two decades, the brightest feature of US energy policy has been largely successful efforts to accelerate the pace of energy-efficiency improvements. Fortunately, the process is nowhere near complete, as continued progress is essential for meeting compelling environmental and economic objectives, both at home and abroad.

We now confront a debate over how to achieve this progress in a nation—and a world—that seeks more competition in the production of all fuels, including those traditionally controlled by integrated monopolies. That debate typically is not over whether more competition is good for energy users and energy efficiency; rather, dispute centers on the terms on which competition should occur, and the criteria that will determine winners and losers.

The stakes are enormous. In the environmental arena, for example, energy issues now dominate an agenda of unprecedented scope and significance. Forums on global climate change have replaced meetings of the Organization of Petroleum Exporting Countries (OPEC) as priorities for international media coverage. Global treaty commitments to stabilize the atmosphere will require significant, sustained reductions in global consumption of fossil fuels. Urban air pollution has proved more enduring than gasoline lines. Add to this an arsenal of concerns associated with acid rain, other air and water pollutants, and radioactive-waste disposal.

Also, meeting today's US energy needs carries a half trillion dollar annual price tag: about $5000 per household. Every one of those households stands to benefit from new energy-efficient technologies, which can get more work out of less energy at lower cost while reducing our vulnerability to fluctuations in fossil fuel prices.

For developing countries, economic opportunities are even greater. For more than one and one-half billion Chinese citizens, pressing aspirations are hostage to unreliable electrical energy service: "Blackouts are plaguing the largest industrial centers like Shanghai, and shortages are cutting deeply into productivity nationwide" (1). Yet China is falling two-thirds short of annual targets for adding power-plant megawatts. Similar stories can be told from Manila to Port-au-Prince.

An American Success Story

Those who have been suffering abroad from energy shortages can find encouragement in an American success story. From both economic and environmental perspectives, the United States has made remarkable progress over the past 20 years: The US economy has cut its average energy needs per dollar of product by more than one fourth. Over the same period, energy consumption by the average new automobile and refrigerator has been cut in half. Our economy used less oil in 1994 than it did two decades earlier. And regions such as California and the Pacific Northwest have been meeting most of their needs for new generating capacity by systematically improving electric end-use efficiencies.

How did we improve our energy performance? It certainly wasn't solely a response to energy prices, which have been dropping across the board in recent years. We got more energy efficient by mobilizing public and private institutions to seek out energy savings. We set minimum fuel economy standards for vehicles, buildings, and appliances, and pushed most of them up over time. And our electric and gas utilities began investing in energy savings as well as new energy production.

As a result, we were able to raise our energy productivity at times when energy prices were dropping. And by reducing our energy needs, we helped unleash the competitive pressures that have kept those prices down over the past decade.

Electric utilities are a good case in point. In 1993 alone, they invested more than two billion dollars in helping their customers save energy more cheaply than it could be produced. Since the mid-1980s, such programs have saved the equivalent of 80 large power plants, and contributed to a decade-long reduction in inflation-adjusted electricity bills throughout the US economy. In California alone, the net value of verified utility-financed efficiency improvements exceeded two billion dollars between 1990 and 1993 (see 2, 3). Some critics had claimed that efficiency improvements would only yield offsetting increases in consumption, but on-the-ground results from California and other states prove the contrary. "Saving energy means saving money, very little of which is recycled into energy consumption" (4).

And the best should be yet to come; we have only just begun to tap our full energy-efficiency potential. Our automobile fleet now has an average fuel economy of just over 21 miles per gallon—in a world where some vehicles already get 70–100 miles per gallon and performance at even higher levels is within reach. Most of our commercial buildings are still wasting more than half of the electricity they use in lights and computers. Just bringing all US refrigerators up to the efficiency of the best new model would free up enough electricity to power 10 million new households.

The Limits of Price-Based Strategies

Why can't we simply rely on energy price signals to exhaust the national pool of inexpensive savings? Decades of experience demonstrate the futility of such a strategy. Decisions about efficiency levels often are made by people who won't be paying the energy bills, such as landlords or developers of commercial office space. Many buildings are occupied for their entire lives by very temporary owners or renters, each unwilling to make long-term efficiency improvements that would mostly reward subsequent users. And sometimes what looks like apathy about efficiency merely reflects inadequate information or time to evaluate it—ask anyone who has replaced a water heater, furnace, or refrigerator on extremely short notice.

The principal market barrier to energy-efficiency remains what the National Association of Regulatory Commissioners (NARUC) calls the "payback gap." When individual consumers and businesses can be persuaded to invest in energy efficiency, they demand a much higher profit than any generation sponsor earns. Numerous studies demonstrate that conservation typically has to recoup its full costs in three years or less in order to attract investors, even though longer periods may pass before many new power plants begin to earn any income, let alone profits (5–7).

There are many explanations for the widespread reluctance to make long-term energy conservation investments; a comprehensive assessment appears in a recent review by the US Office of Technology Assessment (see 8 (outlining numerous reasons "why energy efficiency has not been implemented to the level that appears economically justified")). The gap between the investment perspectives of conservers and producers invites unnecessary expenditures on energy facilities and fuel that could have been avoided at lower cost by efficiency improvements. Energy price signals alone cannot overcome the problem. As a joint report by two of our national laboratories concluded in March 1994, "a defining characteristic of [energy] market imperfections is that they cannot be overcome by consumers or firms acting independently" (9).

This does not mean that energy price signals do not matter, or that price distortions and subsidies do not inflict economic damage. It does mean that pricing strategies are not by themselves sufficient to realize more than a fraction of a nation's cost-effective energy-efficiency opportunities.

Unleashing Our Energy-Efficiency Opportunities

So what do we need, in addition to unsubsidized energy-commodity prices, in order to sustain energy-efficiency progress? Put aside for the moment whether, in a world generally hostile to broad-based energy taxes, environmental costs ever are likely to figure adequately in such prices. Even if they did—and the goal remains eminently worth pursuing—efficiency opportunities would continue to languish behind market barriers.

Most economists would be hostile, properly, to a response rooted in government efforts to micro-manage energy choices. But more measured approaches have a proven track record. Minimum-performance standards for equipment, buildings, and vehicles are one example. Others draw on the proposition that wherever government intervenes to ensure long-term investment in energy supply, its financial mechanisms should provide access on equal terms to both production and efficiency options.

The classic case is the regulated electric utility, where "integrated resource management" principles call for picking the mix of efficiency and production investments that minimizes life-cycle costs. I would apply those principles assiduously to international financial institutions involved in analogous choices among competing energy investments, such as the World Bank. For governments, including my own, I recommend the same guidelines in evaluating the appropriate balance between oil depletion allowances, ethanol subsidies, renewable energy tax credits, and fuel-economy incentives for automobile manufacturers. If government is going to be in the business of orchestrating energy-resource portfolios, it should take inventory of all plausible supply- and demand-side options and pick the best buys first.

Why not keep government out of such decisions altogether? Whatever the merits in the abstract, no government in modern history has been able to resist. I admit, therefore, to a certain impatience with those who use "what-if" fantasies about aggressive energy taxation and deregulation to justify eliminating energy-efficiency incentives and standards that reliably deliver cost-effective savings. As Northwest power planner Dick Watson likes to put it, successful mountain climbers generally insist on having the next handhold within reach before letting go of the one they are grasping.

Moreover, at least part of the interventionist impulse is founded on principles that most economists are hard-pressed to resist. Already noted has been the persistent failure of most energy prices to incorporate externalities such as climate instability and adventures in oil diplomacy.

For electricity in particular, add a strong policy argument involving the portfolio-management function that utilities discharge under regulatory supervision. Just as multiple decision-makers cannot operate a transmission system reliably, they cannot orchestrate a diversified mix of resources for meeting a healthy economy's electrical service needs at the lowest possible life-cycle costs.[1]

A central theme of modern regulation has been utilities' portfolio management function: choosing and buying the combination of generating resources, purchased power, and demand-side efficiency improvements that will minimize the life-cycle cost of reliable energy services for customers collectively. This is the crucial "integrated resource planning" activity that some critics wrongly label "government-sponsored central planning." Such efforts might include, for example:

1. Removing barriers to cost-effective energy savings
2. Efficiently conducting competitive procurements to meet new energy-resource needs
3. Targeting demand-side investments at strategic locations in order to defer costly new transmission and distribution infrastructure
4. Securing a balanced mix of dispersed, intermittent resources and baseload units
5. Negotiating portfolios of long- and short-term fuel-supply and maintenance contracts for utility-owned generation
6. Integrating, for mutual advantage, those who need additional electricity and those best positioned to achieve improved end-use efficiency (often different and widely separated entities).

In short, I refer here to the decisions that collectively will determine almost everything consequential about our long-term electricity future. None are

[1]In other words, like grid control, resource-portfolio management is a classic "natural monopoly" that cannot be broken up without imposing significant costs on customers and society generally.

For example, those who need additional energy services and those best positioned to achieve improved efficiencies often will be different and widely separated entities; no institution can rival utilities in their capacity to bring both groups together for mutual advantage. Intermittent wind and solar resources that would not fit the demand profile of any small group of customers can be invaluable supplements to an integrated, diversified utility system. Quantitative analysis by the Northwest Power Planning Council confirms that valuable resource diversification can be achieved, and investment in reserves and duplicative generation can be minimized, under effective system-wide portfolio management. See Reference 10 (identifying potential benefits of $2.5 billion from "improved coordination of resource development" over the Northwest power system).

obvious candidates for resolution through contracts between individual buyers and sellers, based on responses to energy-commodity prices. And those who offer government efficiency standards as an alternative to some utilities' portfolio investments need to pay more attention to the way these strategies merge in practice. Utilities' efficiency investments frequently have been crucial in laying the technical and political foundations for improved standards. As rebates work to transform markets and spur innovation in appliance and building design, standards can be applied to lock in the gains with minimal opposition (11). Without the utilities' complementary investments (not to mention their formidable political clout), the standards almost certainly would founder.

Wherever electric power systems retain portfolio managers—which today is virtually everywhere—their regulators should concentrate on giving those managers the right financial incentives to build portfolios that minimize life-cycle costs. Here the US record is decidedly mixed but improving fast; incentives to maximize concrete poured and kilowatt-hours sold have been giving way to much more balanced structures, including opportunities for utilities to share the financial benefits of cost-effective investments in efficiency improvements. In March 1995, a formal ruling by the US Treasury Department ensured favorable tax treatment for energy-efficiency programs operated under such incentive systems (12). All of this is perfectly consistent with the widely anticipated dissolution of vertically integrated utilities; portfolio managers need not own anything in their portfolios.

Nor should electric-service monopolies be rewarded for increases in retail electricity consumption. I find puzzling Anderson's contention that "commercial propriety" is somehow undermined when regulators decline to link utilities' profits with retail electricity sales; arguably it is more consistent with the commercial underpinnings of electricity distribution to tie profits to quality and cost of service, as opposed to commodity volumes. Certainly no argument to the contrary emerges from analogies to commodity businesses that operate without price regulation. As long as significant elements of the electricity business retain natural monopoly characteristics, the monopoly regulators will be entitled to weigh the public interest associated with alternative incentive systems—and to pick those best calculated to minimize the life-cycle cost of reliable service.

Here and in other energy sectors, decisions about industry structure and regulatory intervention should be driven by a healthy respect for market principles and an equally healthy awareness of market barriers. Whether the thinking and acting are global or local, we cannot afford to waste opportunities to save energy more cheaply than it can be produced. Truthful energy prices are part of the answer—but only part.

Literature Cited

1. Tyler P. 1994. China's power needs exceed investor tolerance. *New York Times,* Nov. 7
2. California Public Utilities Commission. 1993. *Decision 93-09-078,* Sept. 17
3. Brown M, Mihlmester P. 1994. *Summary of California DSM Impact Evaluation Studies,* pp. ES-3–ES-4. Aug. 11
4. Schipper L. 1994. Energy efficiency works, and it saves money. Letter to the editor. *New York Times,* Nov. 26
5. Governor Pete Wilson. 1991. *The 1992–1993 California Energy Plan,* Oct:39–41
6. National Association of Regulatory Utilities Commission. 1988. *Least-Cost Utility Planning: A Handbook for Public Utility Commissioners,* Vol. 2, Chapter II, Dec.
7. Alliance to Save Energy. 1987. *Industrial Decision-Making Interviews: Findings and Recommendations,* Jan:20
8. US Congress, Office of Technological Assessment. 1992. *Building Energy Efficiency,* May:73–85
9. Levine MD, Hirst E, Koomey JG, McMahon JE, Sanstad AH. 1994. *Energy Efficiency, Market Failures and Government Policy. LBL-35376, ORNL/CON-383, UC-350.* Oak Ridge TN:Oak Ridge Natl. Lab.
10. Northwest Power Planning Council. 1991. *1991 Northwest Conservation and Electric Power Plan,* Vol. II, Part II, pp. 793–95
11. Goldstein D. 1994. Promoting energy efficiency in the utility sector through coordinated regulations and incentives. *Phys. Soc.* 23(2) (originally presented at a meeting of the American Physical Society in Washington DC, April 1993).
12. US Treasury Department. 1995. *Internal Revenue Bulletin 1995-16, Revenue Ruling 95-32,* Mar. 24

UTILITY-SUBSIDIZED ENERGY-EFFICIENCY PROGRAMS

Paul L. Joskow
Department of Economics, E52-373, MIT, Cambridge, Massachusetts 02139

KEY WORDS: demand-side management, electric utilities, energy efficiency, market imperfections

CONTENTS

Introduction

It is hard to take issue with almost anything that Dennis Anderson writes in his paper, especially as it relates to developing countries. The analytical perspective that he adopts is familiar to most economists. And the public policy prescriptions that he outlines, at least as they apply to developed countries, make good sense and are familiar to most economists who have been involved in the US debate about utility-subsidized energy-efficiency programs. Anderson's paper does not, however, address adequately the arguments that proponents of utility-sponsored energy-efficiency programs in the United States have relied upon to support their case. Nor does it address the most important criticisms that have been made of these programs. Moreover, the problems faced by electricity sectors in developed countries are significantly different from those faced by these sectors in many developing countries. Perhaps for this reason, Anderson's paper and the arguments that have gone on in the past few years in the United States regarding utility-sponsored energy-efficiency programs are like ships passing one another unseen in the night.

Developing-Country Problems Are Different

It is important to understand at the outset that the most serious problems that plague the electricity sectors in developing countries are not major problems for the electricity sectors in the United States or in many other developed countries. The average and marginal prices of electricity are not significantly below the marginal cost of supplying electricity in most regions of the United States. Indeed, in those areas of the United States where utility-sponsored energy-efficiency programs have been promoted most heavily (e.g. California and the Northeast), just the opposite is the case. Regulated electricity prices exceed marginal costs, in both the short run and the long run, in these areas of the country.[1] While average prices are below long-run marginal cost in some regions (e.g. the Northwest), the gap is not large, and rate design is frequently used to close the gap on the margin. There is certainly more that can be done, especially in the area of time-of-use pricing. Significant progress has been made on the pricing front in the United States over the past 20 years, however, and the underpricing of electricity is no longer the primary rationale for utility-sponsored (and subsidized) energy-efficiency programs.

Electric utilities in the United States are financially healthy by and large and have adequate access to the capital funds that are required to make the investments necessary to balance supply and demand. Perhaps more importantly, the gradual movement from an industry organized around utilities that are vertically integrated among generation, transmission, and distribution to an industry that relies heavily on power by contract with independent suppliers has shifted much of the direct financing burden for new generating capacity on to independent power producers. Cheap natural gas and the associated movement to combined-cycle generating technology have also reduced the capital intensity and lead time for new generating facilities. Virtually the entire US population already has access to electricity, demand is projected to grow relatively slowly over the next 20 years, and the electric power networks in the United States are highly reliable and by world standards reasonably efficient. Electricity shortages are not a problem, electricity theft is not pervasive, and people pay their utility bills with at least as much regularity as for other commercial services. Mercifully, we have largely been spared the corruption and inefficiencies that often plague government-owned electric power sectors, especially in developing countries.

On the environmental front, the US electric power sector has been a primary target of regulations controlling air and water pollution and hazardous wastes. Strict regulations to control emissions of oxides of sulfur and nitrogen, carbon

[1]Prices based on average total cost exceed marginal costs as a result of investments in nuclear facilities built on the assumption that fossil fuel prices would continue to escalate at rapid real rates, government policies that forced utilities to contract for too much power at too high a price from cogeneration and small power producers under regulations implementing Title II of the Public Utility Regulatory Policy Act (PURPA), and a general surplus of generating capacity resulting from unanticipated slow demand growth.

monoxide, and particulates have been phased in over the past two decades, and very significant emissions reductions have been accomplished. Although inefficient "command and control" schemes for controlling emissions have been the norm in the United States, the Clean Air Act Amendments of 1990 represent a major change from this tradition. The 1990 law creates a tradeable allowance system for controlling emissions of SO_2 from utility boilers and gives utilities substantial freedom to meet an overall national cap on SO_2 emissions in a least-cost manner.[2] The ethos of emissions caps, tradeable rights to pollute, and flexibility to respond in a least-cost manner to emissions constraints is spreading to other pollutants at the local and regional levels (e.g. NO_x in the Northeastern United States and smog-related air pollutants in the Los Angeles basin.)

On the research and development front, the electric power sector organized the Electric Power Research Institute (EPRI) more than 20 years ago. Roughly $300 million dollars per year is spent by EPRI on a wide range of research projects, increasingly focused on end-use efficiency, pollution abatement, and renewable energy. Additional research funds in these areas are expended by the US government, by individual utilities, and by equipment vendors.

In summary, the main problems discussed in Anderson's paper that motivate electricity policy reform in developing countries are not viewed as major problems in the United States and other developed countries. Moreover, these problems are not the primary rationales used to justify recent expansions of utility-sponsored energy-efficiency programs in the United States.

Rationales For Utility Energy-Efficiency Programs

Proponents of post-1985 utility-sponsored energy-efficiency programs in the United States have not relied (primarily) on arguments that electricity is underpriced or that the objective of public policy should be energy efficiency rather than economic efficiency. To the contrary, proponents of programs through which a utility heavily subsidizes specific energy-efficiency investments made by consumers[3] have wrapped themselves in the economic efficiency banner. They advertise these proposals as promoting overall economic efficiency and as reducing the average customer's electricity bill. Their argument goes something along the following lines:

1. There are large "untapped" economical opportunities to conserve energy: The most important rationale used to justify utility energy-efficiency programs is the assertion that there are huge opportunities to conserve energy at a

[2]Whether or not regulated utilities will in fact make full use of this flexibility is an interesting question that I will not attempt to answer here.

[3]The costs of these programs and any lost net revenues resulting from reduced demand are then recovered in the rates charged to all consumers.

life-cycle cost for the associated equipment that is significantly less than the cost of the electricity supplies that would be avoided because of the improvements in energy efficiency. This view is supported by a number of engineering studies that have identified numerous ways in which residential, commercial, and industrial consumers supposedly can use electricity more (energy) efficiently to provide the same end-use services (light, heating, drive power) and reduce the life-cycle cost of these services very significantly. The largest and cheapest untapped savings are purported to lie in the commercial and industrial sectors. Amory Lovins is the most quoted proponent of this view (1). He argues that US consumers can reduce their electricity consumption by 70%, maintain or improve the level of end-use services provided, and substantially reduce the life-cycle costs of end-use services if consumers would only invest in the most economical energy-efficient equipment. Thus, Lovins argues, there is a huge free lunch out there if consumers would only take advantage of these cheap energy-efficiency options that can both save them money in the long run and significantly reduce the impact of electricity production on the environment by reducing how much must be produced to meet the demand for end-use services.

2. Numerous market barriers and market imperfections keep consumers from taking advantage of these energy-efficiency opportunities: Economists are always suspicious when they are confronted with public policy initiatives predicated on a huge free lunch that consumers refuse to eat on their own. Why, after all, would consumers routinely spend 8 cents per kWh to buy electricity to run standard, inefficient refrigerators when they can buy a more efficient model that allows them to cut electricity consumption by 40% at an incremental cost (for the more efficient refrigerator) of only 3 cents per kWh saved compared to the most likely alternative investment? That is, if they can save 5 cents per kWh times 40% of the kWh used by their current refrigerators by buying an efficient model rather than replacing their refrigerator with a standard, inefficient model, then why aren't equipment vendors, energy service companies, and others out there selling these opportunities to consumers eager to save a buck? Proponents of utility- (and government-) sponsored energy-efficiency programs argue that consumers face enormous market barriers to making "the right" decisions about investments in more energy-efficient equipment (2) even if the price of electricity reflects the appropriate marginal cost of supplying it. These alleged barriers include information market imperfections, information processing imperfections, imperfect capital markets and associated financing constraints, and imperfections in the way rental and real estate markets value long-lived energy-efficiency investments. It is argued that these imperfections do and will exist whether or not electricity is priced at its true marginal cost and that they cannot be remedied merely by giving consumers better information.

3. Utilities are in a unique position to overcome these market barriers and imperfections: If there are serious imperfections in the markets in which energy-efficient equipment is bought and sold, then finding ways to ameliorate the relevant imperfections in a cost-effective manner is an appropriate objective of public policy.[4] Utilities, it is argued, are in a unique position to correct these market imperfections because they (*a*) have access to the funds needed to help to finance investments in energy efficiency, (*b*) deal with all consumers within their franchise areas and have the trust of these customers, (*c*) are generally knowledgeable about energy-efficiency opportunities, and (*d*) are in a unique position as electricity suppliers to "balance investments in supply-side and demand-side resources on a level playing field."

4. Utilities are not "investing in end-use energy efficiency" because the regulatory process gives them incentives to sell more electricity and to focus on minimizing the commodity price of electricity rather than the average customer's bill for end-use services. Once one assumes that utilities should be acting as surrogates for their customers in making end-use energy-efficiency decisions, it is only natural to ask why they aren't already doing it. The answer proposed by advocates of utility programs is that traditional regulation provides utilities with no incentives to do so, especially when the price of electricity exceeds it marginal cost. The phrase "investing in end-use energy efficiency," of course, is nothing more than a euphemism for providing subsidies to customers who participate in utility energy-efficiency programs. The cost of these subsidies is passed on to all customers in higher rates.[5] Utilities have little if any financial incentive to implement such programs. Utilities can increase their profits by selling more electricity when prices exceed marginal cost as long as there is some regulatory lag or, in the long run, the expected regulated return on new investments exceeds the associated cost of capital. Regulatory lag works in the opposite direction in the short run when it comes to expenditures on energy efficiency, and there are no countervailing long-run opportunities to earn a return in excess of the cost of capital given the way conservation expenditures have generally been treated for ratemaking purposes by state public utility commissions. As a result, utility energy-efficiency program proponents argue that regulation must be fixed in two ways. First, utilities should be given incentives to spend money on customer energy efficiency when such investments are cost justified based on an overall "least cost to society criterion." That is, utilities should be able to profit from investments in energy efficiency in much the same way as they can profit from increasing the

[4]Obviously, if the costs of fixing the imperfections or their consequences is greater than the cost of the initial imperfections, then they are not worth trying to fix.

[5]Rates must increase whenever the price of electricity exceeds the marginal cost of supplying it, as has been demonstrated ad nauseum (e.g. 3).

quantities of electricity that they supply. Second, "integrated resource planning" (IRP) procedures must be adopted that force utilities to evaluate "supply-side and demand-side investment opportunities," again based on an overall least-cost-to-society criterion, and to make investment decisions accordingly.

The Economist's Response

The responses of most serious economists who understand market economics have been fairly uniform. They have questioned each of the above four propositions upon which the case for utility-subsidized energy-efficiency programs—as well as the integrated resource planning programs that have accompanied them—have been based.

First, there has been enormous skepticism about the magnitude of the "untapped economical energy-efficiency opportunities" that are actually out there to be had by real consumers in real markets. These estimates routinely ignore important costs that are relevant to consumers and society and rely on optimistic assumptions about the actual savings that will be achieved by these measures. Experience with utility programs, which themselves underestimate costs and overestimate savings, makes it clear that many of these engineering calculations of costs and savings are nothing more than fantasy (4–7). To the extent that the evidence supports a case for "untapped" savings, they are much more modest than most energy-efficiency activists would lead us to believe. Moreover, the important role that prices do play in promoting cost-effective energy conservation decisions is underestimated or ignored completely.

Second, there is fairly compelling evidence that consumers use what appear to be very high implicit discount rates when they evaluate energy-efficiency investments (3). Whether these implicit discount rates reflect poor information about energy-efficiency opportunities, biased forecasts of future energy prices, information processing problems, capital/financial constraints, imperfections in rental and real estate markets, or other causes is unknown, however. Nor is it known whether these problems are unique to energy-efficiency investments or are a more pervasive phenomenon characterizing how consumers make investment decisions more generally in an uncertain world. Indeed, one of the most disappointing things about all of the discussion about market barriers is that it has generated little serious research to give us insight into the nature and sources of these market barriers and how they respond to different policy instruments. It is hard to know how to fix something if you don't know why it is broken.

Third, even if such market barriers exist, it is not at all clear that utilities are in a unique position to overcome the problems. The instruments that a utility has at its disposal are rather limited. It can subsidize energy-efficiency investments by consumers, relying on its regulated monopoly position to recover the costs in the rates charged to all consumers. But why should a utility be in a good position to figure out how millions of diverse customers should be using electricity and making supporting conservation investments? Utilities may be experts regarding the production of electricity, but why should they be better informed about how electricity can be used most efficiently than are appliance and equipment vendors, industrial engineers, and large energy consumers themselves? Moreover, how do we know that the subsidies are actually yielding the efficiency improvements that utilities think that they are going to yield? Many consumers may eagerly accept the subsidies even if the associated investment would be inefficient when evaluated at the full cost rather than the subsidized cost. These issues have gradually led public utility commissions to order utilities to implement large monitoring and evaluation programs to better evaluate the costs and benefits of these programs (6).

Utilities can also offer to provide financing for qualified energy-efficiency investments and require the customers who take advantage of such financing to pay back the loan over time out of the savings that are achieved. Arrangements like this are referred to as "shared savings" contracts. Many economists, including myself, have accepted offering such contracts as an appropriate role for utilities (8). If the problems arise from information market imperfections, financing constraints, and credibility of the suppliers of energy-efficiency services, then "shared savings" contracts of this type offered by utilities appear to provide a solution to several of the assumed market imperfections while retaining appropriate decentralized incentives for customers to choose to participate only when equipment costs, transactions costs, expected future utilization patterns, and energy prices actually make the investments economical. Furthermore, if a utility gives a consumer the relevant information, offers to finance the investment for him or her, and agrees to be compensated for the investment over time out of the projected savings and the consumer still won't make the investment, then it is quite likely that the assumptions upon which the projected savings have been based are wrong. Proponents of utility-subsidized energy conservation programs have generally resisted this kind of decentralized shared savings approach in favor of large direct subsidies to participants.

Utilities can also engage in information dissemination programs of various kinds, but it is far from obvious that having hundreds of utilities promoting diverse energy-efficiency programs and engaging in uncoordinated advertising and promotion is the best way to provide the relevant information to the public, to architects and building engineers, and to wholesalers and retailers of equipment. Utilities cannot cost-effectively fix basic imperfections in capital markets or rental and housing markets, or implement building and appliance efficiency standards themselves. If these imperfections are really important, a broader set of public policy initiatives implemented by government is likely to be required to correct them effectively.

In practice, it is not so much that utilities are in a unique position to overcome these real or imagined market barriers, or that electricity prices are too low to provide appropriate price incentives,[6] but rather that the regulated status of utilities has created a unique opportunity for energy-efficiency proponents to find a mechanism to finance the programs that they think are good for society. The fact of the matter is that environmental activists in the United States captured the utility regulatory process in many states during the late 1980s and induced the regulators to engage in a pervasive "taxation by regulation" program (9) to use the public utility ratemaking process to tax electricity customers as a group to pay for energy-efficiency (and renewable energy) programs that the environmentalists favored. As long as these taxes could be hidden from the public and rationalized as being "least cost" from some perspective, utilities represented a very important political opportunity to advance a particular agenda.

Conclusions

In the end, the rise of large utility-subsidized energy-efficiency programs in the United States in the late 1980s and early 1990s can best be understood from the perspective of political economy rather than from the perspective of neoclassical market economics. These programs represent a partially successful effort to capture the regulatory process and to use the institution of regulated monopoly to raise funds to pursue certain social ends. In the process, many of the interesting issues raised about the imperfections associated with the markets through which energy-efficiency decisions are made, the nature and causes of these imperfections, and the effectiveness of alternative public policies to ameliorate them have largely been ignored. The ability to use utilities in this way depends critically on the continuation of the institution of regulated monopoly electricity suppliers insulated from competition at the retail level. As competition spreads in the electric power sector, it is creating major conflicts between increasing competitive opportunities for customers vs continued reliance on utilities to pursue energy-efficiency programs that raise prices within the context of highly politicized IRP programs that embody a centralized planning philosophy (10). As competition intensifies, new ways will have to be found to achieve the energy-efficiency and environmental goals that motivate utility-subsidized energy-efficiency programs.

Ironically, there may be a better case for energy-efficiency programs—government or utility sponsored—designed to respond to the market imperfections discussed above in developing countries than there is in developed countries. Although it would be desirable from an economic efficiency perspective for

[6]There is agreement, however, that utility subsidies for energy efficiency may very well be justified when the price of electricity is below the marginal cost of electricity (2).

developing countries to raise electricity prices to reflect the marginal cost of supply and to move the electricity sectors to a point of financial self-sufficiency, it will be extremely difficult to jump to a full-cost pricing situation overnight. Especially in developing countries that have democratic governments, very rapid increases in electricity prices can be political suicide for incumbent governments. Full price adjustment will necessarily take time, and there will continue to be political demands for subsidies by some sectors, just as has been the historical experience in developed countries. Indeed, it is not surprising that responsible officials in developing countries become enraged when institutions such as the World Bank require them to adopt economic policies that developed economies have themselves adopted only imperfectly. Moreover, given the evolutionary state of other market institutions in developing countries, many of the market barriers and market imperfections that proponents of energy-efficiency subsidies have argued exist in the United States are likely to be orders of magnitude more severe in developing countries. As a result, it appears to me that programs specifically designed to overcome these barriers are likely to be much more important in developing countries than in developed countries such as the United States, where the relevant market imperfections are likely to be much less severe.

Literature Cited

1. Fickett AP, Gellings CW, Lovins AB. 1990. Efficient use of electricity. *Sci. Am.* 262:64–74
2. Levine MD, Hirst E, Koomey JG, McMahon JE, Sanstad AH. 1994. *Energy Efficiency, Market Failures, and Government Policy. LBL-35376, ORNL/CON-383, UC-350.* Oak Ridge, TN: Oak Ridge Natl. Lab.
3. Joskow PL. 1990. Should conservation proposals be included in competitive bidding programs? In *Competition in Electricity: New Markets and New Structures,* ed. J Plummer, S Troppmann. Arlington, VA: Public Util. Rep.
4. Joskow PL, Marron D. 1992. What does a negawatt really cost? Evidence from electric utility conservation programs. *The Energy J.* 13(4):41–74
5. Joskow PL, Marron D. 1993. What does utility-subsidized energy conservation really cost? *Science* 260:281,370
6. Joskow PL, Marron D. 1993. What does a negawatt really cost? Further thoughts and additional evidence. *The Electr. J.* 6(6):14–26
7. Joskow PL. 1994. More from the guru of energy conservation: There must be a pony! *The Electr. J.* 7(4):50–61
8. Joskow PL. 1990. Understanding the unbundled utility conservation bidding proposals. *Publ. Util. Fortn.* 125(1):18–28
9. Posner R. 1971. Taxation by regulation. *Bell J. Econ. Manage. Sci.* 2:22–50
10. Joskow PL. 1993. *Conflicts Between Competition, Conservation, and Environmental Policies Affecting the Electric Power Industry.* Presented at Public Utility Res. Cent., Univ. Fl., April 29 (Mimeo)

ENERGY EFFICIENCY POLICY AND MARKET FAILURES

Mark D. Levine, Jonathan G. Koomey, James E. McMahon, Alan H. Sanstad
Energy Analysis Program, Energy and Environment Division, Lawrence Berkeley Laboratory, Berkeley, CA 94720

Eric Hirst
Energy Division, Oak Ridge National Laboratory, Oak Ridge, TN 37831

KEY WORDS: economic efficiency, appliance standards, market transformation, demand-side management

---

CONTENTS

INTRODUCTION

The difference between the actual energy used to provide energy services such as lighting, heating, cooling, and refrigeration and the level of energy efficiency that can be provided in a cost-effective way for the same services is defined as the "energy efficiency gap." Understanding the reasons for this gap requires careful quantitative analysis of the engineering and economic characteristics of specific technologies, as well as an assessment of data on adoption of the technologies in the market and of policies to promote energy efficiency.

Our primary objective in this paper is to focus attention on the empirical basis for skepticism about the effectiveness of the market mechanism in yielding cost-effective energy efficiency improvements. We present a series of examples that, in our view, provide evidence for market failures related to energy efficiency.

We discuss the role of energy efficiency policies such as standards, utility demand-side management programs, and other government programs that have been used to overcome these market failures. If they are carefully designed and successfully implemented, such energy efficiency policies can contribute to the reduction of the energy efficiency gap.

We believe that energy efficiency policies aimed at improving energy efficiency at a lower cost than society currently pays for energy services represent good public policy. Programs that lead to increased economic efficiency as well as energy efficiency should continue to be pursued.

IDENTIFICATION OF MARKET FAILURES IN ENERGY SERVICE MARKETS

Consumers and firms require fuels and electricity not for their own sake but rather for the services they can provide, such as lighting, heating and cooling, and refrigeration. At any given time a range of technological options is commercially available to deliver these services, distinguished by energy-efficiency and thus by the cost of the services delivered.

Dozens of studies have indicated the potential for achieving substantial cost-effective energy savings in buildings and equipment, generally in the range of 20–40% compared with current choices for new equipment and buildings. The measures that can be applied to new and existing residential buildings, including costs and energy savings, were discussed extensively in 1991 (1).[1]

Typically, energy-efficient devices have higher initial purchase prices, but lower operating costs than their less efficient counterparts. When two devices provide equivalent energy services, an internal rate-of-return for the incremental investment in the more efficient device can be calculated using purchase prices and operating costs. This rate-of-return can then be compared with social discount rates and with the borrowing or savings rates of the purchaser. When the rate-of-return for the incremental investment in the more efficient device exceeds the social discount rate, but the less efficient device is purchased, we

[1]For a summary of the results of a number of estimates of national energy savings that could be obtained through cost-effective investments in energy efficiency, see References 2 and 3.

infer that energy services are not being obtained at minimum cost from a *societal* perspective. When the rate-of-return also exceeds market interest rates, but the less efficient device is purchased, we infer that energy services are not being obtained at minimum cost from a *private* perspective.[2] Analyses of purchase decisions for many commercial energy-using products have found failure to obtain energy services at minimum cost, implying the existence of market failures for energy efficiency.

The assertion is often made that seemingly cost-effective energy efficiency investments must have irreducible hidden costs that are ignored by purchasers. Several categories of potentially hidden costs may affect purchase decisions. One hidden cost is a reduced level of energy service (e.g. quality of lighting or temperature, or comfort levels with heating systems). Another is irreducible private costs (such as the inconvenience of installing efficient equipment). Other costs include sales, income, and property taxes, and maintenance fees associated with the efficiency measure.

The time and effort needed to learn about, search for, or develop confidence in the performance of new technologies are also hidden costs. These costs are likely to be greatest with relatively untried technologies. Often a new technology will take years to saturate the market even if it is clearly superior to the technology it replaces. Such technology-diffusion time-lags raise analytical and empirical problems that are difficult to address in the standard engineering-economic framework. Thus, in the examples described below, we concentrate on readily available technologies with characteristics that minimize diffusion barriers.

If consumers ignore seemingly cost-effective energy efficiency investments with no hidden costs or time lags, this is a market failure (4). A market failure is "a condition in any market that results in an inefficient allocation of resources." When market failures occur, government policies can be designed to overcome these economic inefficiencies (5).

We provide four examples of cases in which a product (*a*) is commercially available (i.e. can be easily obtained by consumers), (*b*) provides identical amenity or utility to the customer, (*c*) has a much higher return on investment for energy efficiency than either market or social rate of return, (*d*) results in no increased risk to the consumer, and (*e*) causes the consumer no known hidden cost, such as inconvenience in installation, or higher maintenance costs. Our examples are presented primarily to make the *prima facie* case for market failure. Later we will provide evidence that such market imperfections in the investment in energy efficiency are likely to be widespread.

[2]In this paper, we consider an energy efficiency measure cost-effective from a private perspective if it yields an internal return greater than 6–8% real; from a social perspective, a return greater than 3–5% real is considered cost-effective.

Table 1 Characteristics of standard core-coil and efficient core-coil fluorescent ballasts

	Approximate adjusted market share 1987	Capital cost ca. 1989$	Power savings W	Energy savings kWh/year	Marginal CCE ¢/kWh (1989$)	Implie Margin real IR
Ballast type						
Standard core-coil	90%	11.0	0	0	—	—
Efficient core-coil	9%	15.4	11	29	1.4	60.3%

Assumptions: Operation time for offices = 2600 hrs/yr, ballast lifetime = 45,000 hrs = 17.3 yrs, discount rate = real, capital recovery factor (CRF) = 0.0917, and 1988 U.S. average commercial sector electricity price of 7.4¢/kW Capital costs are from (8) and have been adjusted from 1987-$ to 1989-$ using the consumer price index.

Efficient Core-Coil Versus Standard Core-Coil Commercial Fluorescent Ballasts

Efficient core-coil fluorescent ballasts providing equivalent amenity and longer lifetimes than standard core-coil ballasts were introduced in 1976. Table 1 shows that efficient core-coil ballasts offer energy savings at a cost of conserved energy (CCE) of $0.014 kWh^{-1}.[3] This CCE implies a real market discount rate of about 60% for those who purchase standard core-coil ballasts.[4]

Table 1 also shows that standard core-coil ballasts would have accounted for about 90% of all fluorescent ballast sales in 1987, if one excludes the five states in which standards prohibited their sale.[5]

The fact that consumers purchased the less efficient model does not mean

[3]At a 15% real discount rate, the CCE would be 74% higher, or $0.025 kWh^{-1}, which is still significantly cheaper than the price of electricity.

[4]This example assumes 2600 operating hours per year, lower than the number in almost all commercial buildings (6). Efficient core-coil ballasts are even cost effective when operated just 600 hours year^{-1}, and the lights in all types of commercial buildings typically operate for thousands of hours every year. The lowest plausible number of annual operating hours for commercial buildings is around 1300, which would yield a cost of conserved energy of $0.028 kWh^{-1}, still half of the electricity price. These calculations assume a real discount rate of 6% and other parameters as specified in Table 1.

[5]Market shares in Table 1 were adjusted to represent market shares if state standards did not exist in 1987. By the end of 1987, standards prohibiting sale of inefficient core-coil ballasts existed in five states representing about one quarter of the US population (California, New York, Massachusetts, Connecticut, and Florida). Installation and maintenance costs are equivalent for all types of ballasts. Even more efficient electronic ballasts would have made up the remaining 1% of sales in 1987 (they have a capital cost of 33.4, power savings of 33 W, energy savings of 86 kWh/yr, a marginal CCE of 2.8, and an implied marginal real IRR of 26.3%).

that they actually performed a life-cycle cost calculation using a high discount rate; rather, it strongly suggests that cost-effective efficiency measures are ignored by many purchasers.

This example describes a case in which there are no hidden costs because the more efficient device provides equivalent amenity and is based on proven technology that is essentially identical to its inefficient counterpart, but has higher energy savings and a longer lifetime. This technology was introduced in 1976 and was widely available and easily obtained well before 1987. Even though this ballast saves energy and costs significantly less than electricity, we found that a large majority of consumers still purchased the less efficient standard core-coil ballast.[6]

High Efficiency Versus Low Efficiency Residential Refrigerator/Freezers

Between 1977 and 1979, two refrigerators of differing efficiency and price were displayed next to each other and sold by the same national retailer. Both models were top-freezer, auto-defrost, roughly 17 cubic feet in volume, and had identical features. The price of the higher efficiency model, which used 410 kWh $year^{-1}$ less electricity, was $60 more than the less efficient one. The absolute difference in prices for the refrigerators was constant over the analysis period, even though the purchase price for both models was reduced by rebates at various times. The high efficiency model was advertised widely and a prominent consumer magazine recommended it. The efficient and inefficient models together accounted for significant fractions of total unit sales of all models of that particular brand (roughly 35–75%, depending on the region).

Using a 6% real discount rate and a lifetime of 20 years, the more efficient refrigerator saved energy at a cost of 1.3¢ kWh^{-1} (1979$), lower than the electricity prices in every state at that time. Prices in 1979 were lowest in Washington ($0.015 kWh^{-1}), while the national average was about $0.054 kWh^{-1}.

Because the more efficient refrigerator provided service identical to the less efficient refrigerator at a lower life-cycle cost, one might expect purchasers to choose the more efficient refrigerator. However, from 1977–1979, the inefficient model was still purchased by around 45% of purchasers of either refrigerator in the Midwest, 35–40% in the East, 54–69% in the South, and 57–67%

[6]In testimony before the US House of Representatives in 1988, Ernest Freegard, Vice President of Advanced Transformer Company described this consumer behavior as follows: "The ballast industry vigorously promoted energy efficient ballasts and found the market acceptance to be slow...It continues to surprise us that it has been so difficult to sell such a bargain" (7).

in the Pacific region.[7] Thus, 35–70% of the purchasers of these two models chose the inefficient model, in spite of the low cost of conserved energy.

Over 60% of buyers in the Pacific region used implied real discount rates exceeding 34%; 59% of buyers in the South used discount rates exceeding 41%; 45% of buyers in the Midwest used discount rates exceeding 56%; and 40% of buyers in the East used discount rates exceeding 58% (9). These implicit rates of return are significantly higher than those prevailing in the capital markets (typically 4–12% real).

The only potential hidden cost in this example is consumer uncertainty about the savings of the more efficient refrigerator. However, consumers trusted other information provided by the retailer about the refrigerator's performance enough to purchase it, so distrust of the energy savings claim seems an unlikely reason for consumers' preference for the inefficient model. Two possible explanations are that consumers do not understand how to assess an investment in energy efficiency (or even the idea of treating the $60 as an investment) and that the salesperson provided poor information, perhaps because of inadequate knowledge on the part of the salesperson.

Whatever the cause, this consumer behavior clearly indicates that market failures affected the market for more efficient refrigerators. The energy efficient refrigerator saves energy at a cost below the price of electricity in all usage situations and in all US locations, yet about half of all consumers who had a choice between otherwise identical refrigerators purchased the inefficient model.

Low-Power Office Equipment

For years, the highly competitive and technologically advanced computer industry failed to incorporate energy-saving features that could be added at no cost to the purchaser to computers, monitors, and printers. Specifically, manufacturers ignored a feature that switches these products to a low-power state (which is often a power reduction of 70%) after a specified period of inactivity. Consumer utility is little affected because the computer springs back to life very rapidly when the user starts working again. Energy use can be substantially reduced because even heavily-used personal computers sit idle for significant parts of the day.

The rapid proliferation of personal computers and associated printers has resulted in significant increases in electricity use in commercial buildings. Office-equipment electricity use is projected to be about 70 TWh in 1995 (10),

[7]These purchasing patterns reflect the influence of electricity price, at least qualitatively: In 1979, the Pacific and Southern regions had prices of \$0.034 kWh^{-1} and \$0.04 kWh^{-1}, respectively, and the Midwest and East had prices of \$0.055 kWh^{-1}.

accounting for approximately 7% of anticipated total commercial sector electricity use (11).

In 1992, the US Environmental Protection Agency (EPA) launched the Energy Star Computer program. This voluntary program calls for manufacturers to incorporate the low-power option in their computer products. Manufacturers that meet the Energy Star criteria are eligible to display the Energy Star Logo on their products, product literature, and advertisements (12) to indicate that their products are energy efficient. As of June 1994, the EPA had signed partnership agreements with manufacturers of more than 75% of all desktop computers and 90% of all laser printers sold in the United States (13).

We cite this example to illustrate that electricity use was essentially ignored by this highly competitive and technologically advanced industry prior to the EPA's program. A product development specialist for a major computer manufacturer explained to the EPA that they once considered incorporating lower-power states into their mainstream product line, but that the marketing department advised them that it would not be worth the effort.

This example focuses on features that can be added to computers, monitors, and printers to make them more energy efficient. The resulting technologies have no hidden costs and have been available to manufacturers for many years. Adding the features saves energy at a cost below the price of electricity. Even in the competitive computer industry, this highly cost-effective energy savings option was universally ignored until the EPA increased its visibility and created an atmosphere in which adopting the strategy was beneficial to the industry (and ignoring it was disadvantageous).

Standby Power in Color Televisions

A small amount of power is needed at all times to allow the remote control to turn televisions on and off. Even when the television is operating, this power allows the remote control to adjust volume and change channels. In 1987, DOE found that out of twenty-five 19" and 20" color television models, more than 70% had standby power of 2 or more W. The average standby power for the entire sample of television models was 4.4 W (14).

Televisions with standby losses greater than 2 W typically feed power to the tuner using a resistor. By replacing this resistor with a transformer, the standby power can be reduced to 2 W. The additional manufacturing cost associated with the transformer is $2.15 (in 1987$). The cost of the resistor is subtracted from that of the transformer, and markups are applied to calculate the consumer price. The consumer price of reducing this standby power was estimated to be $4.30. This investment would save 21 kWh $year^{-1}$.

Using a real discount rate of 6% and a lifetime of 11.5 years, these costs and energy savings translate into a cost of conserved energy of $0.025 kWh^{-1}. This CCE is lower than the average price of electricity by a factor of about three, yet more than 70% of televisions did not have this simple technology as of 1987. At a 15% real discount rate, the CCE becomes \$0.038 kWh^{-1}, which is about half the average price of electricity in the United States.

Approximately 23 million color televisions were sold in the United States in 1993. Changes of a few watts in standby power may sound small for individual televisions, but with hundreds of millions of televisions consuming standby power every hour of the day, 2.4 W savings per television adds up to approximately 450 megawatts and 3.2 TWh of annual savings. Such savings would be worth about \$240 million $year^{-1}$ if evaluated at national average electricity prices.

This example describes a cost-effective improvement in television efficiency that has no hidden costs, because the consumer would see no difference in performance and reliability would not be affected. Standby power use is constant, so there are no usage-related variations. The cost of reducing standby losses is based on well-known, simple technology. The technology to reduce standby losses to 2W or less is well within the capabilities of all television manufacturers. Reducing standby power saves energy at a cost below the price of electricity, yet most manufacturers choose not to add this option. This is a case in which consumers and most manufacturers have ignored an energy-efficient device because it is a minor feature of the product and consumer appeal rests on many other product attributes. Yet, aggregated nationally, foregone energy and dollar savings are not trivial.

A CASE STUDY SHOWING THE IMPACTS OF A POLICY TO OVERCOME MARKET FAILURES

California was the first state to promulgate energy-efficiency regulations for residential appliances, which limited the sale of refrigerators to those with energy consumption lower than a specified maximum. The range of other appliances that could be legally sold was similarly restricted on the basis of energy efficiency or annual energy consumption. In 1986, after several states had adopted their own energy efficiency regulations, manufacturers and environmental groups negotiated a set of national efficiency standards (16). These national appliance standards were encoded in the *National Appliance Energy Conservation Act of 1987*.[8] This Act established national efficiency standards for refrigerators and several other appliances and required the US Department of Energy to issue other standards and to update all standards at defined intervals. Since then, the US Department of Energy has updated those standards

[8]An amendment to this Act went into effect in April 1991 prohibiting the use of standard core-coil ballasts (15). Stricter fluorescent ballast standards are currently being considered for adoption.

for refrigerators, freezers, and other products and continues to study updates for possible future rules (17).

Cumulative expenditures by the federal government for the appliance efficiency program from 1979–1993 total about $50 million (C Adams, personal communication). These expenses included the development of test procedures for measuring efficiencies, technical analyses to provide an engineering and economic basis for the standard levels selected, administrative costs associated with public hearings, publication of laws and supporting technical documents, and management of the program.

The benefits of the program have been determined by using end-use forecasting models, which account for the primary effect (increased efficiency of new appliances) and important secondary effects (more expensive appliances, fuel switching, and changes in operating practices). The results suggest a cumulative net benefit of $46 billion to the nation for appliances sold from 1990–2015. This consists of a net present cost of $32 billion for higher-priced appliances and a net present savings of $78 billion. These benefits, as well as the cumulative energy savings of the standards to date (those already promulgated), are shown in Figure 1.

Figure 2, which shows the effects of standards that have been promulgated, illustrates the substantial reductions in residential energy demand growth owing to the standards. Proposed standards for eight products will cause significant additional declines in residential energy demand if adopted. The lowest curve in Figure 2 shows the impact of the standards proposed by the US DOE in March 1994 for eight products, assuming they are implemented in 1998.[9]

Included in this accounting are the projected increased costs to consumers and the value of energy savings (calculated at average energy prices). The costs to manufacturers are included and (mostly) passed on to consumers. However, the benefits do not include the value of decreased emissions of air pollutants and carbon dioxide. The quantities saved represent 1.5–2% of total national emissions of SO_x, NO_x, and CO_2 (17).

It has been suggested that consumers' choice of appliance models may be restricted by the standards, thereby creating a hidden cost. Considering refrigerators as an example, however, more models were available after the regulations became effective in 1990 than were available in 1986 (19). Moreover, real prices of refrigerators have declined from 1986–1991 (20). The average refrigerator purchased since the introduction of efficiency standards is larger and most likely to be a side-by-side model (21). The engineering cost data with manufacturer and dealer markups added suggest that the more-efficient

[9]In response to public comments, the US Department of Energy is conducting additional analyses before making a final ruling.

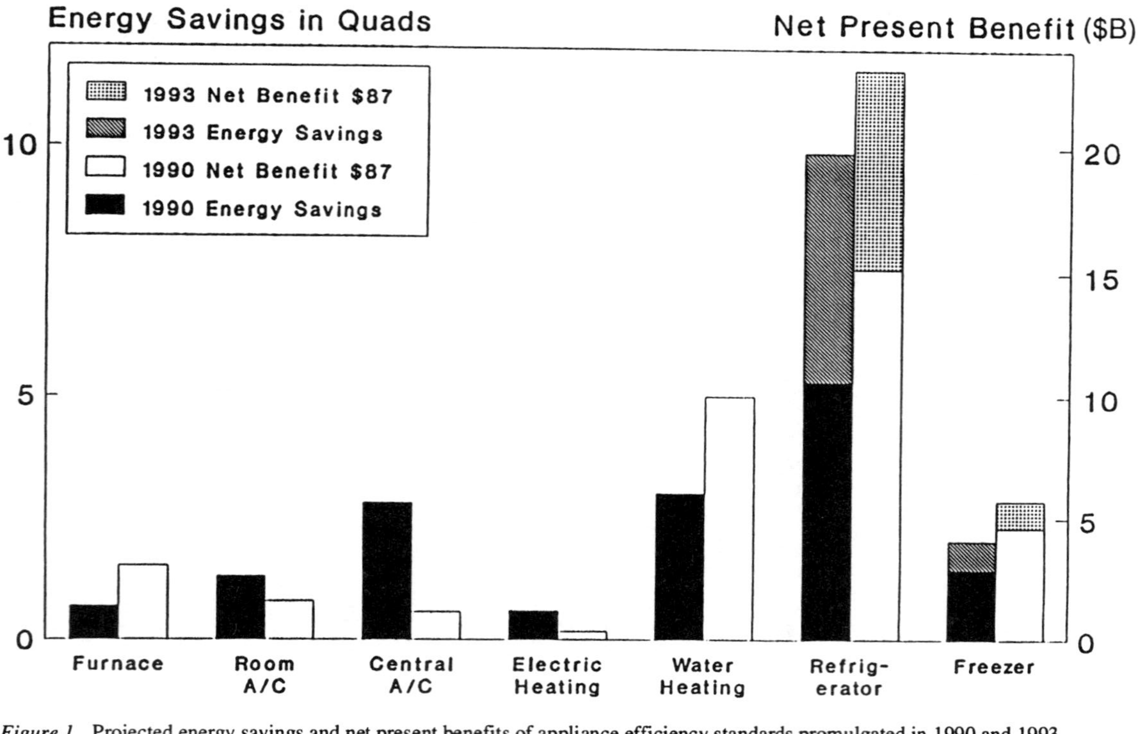

Figure 1 Projected energy savings and net present benefits of appliance efficiency standards promulgated in 1990 and 1993.

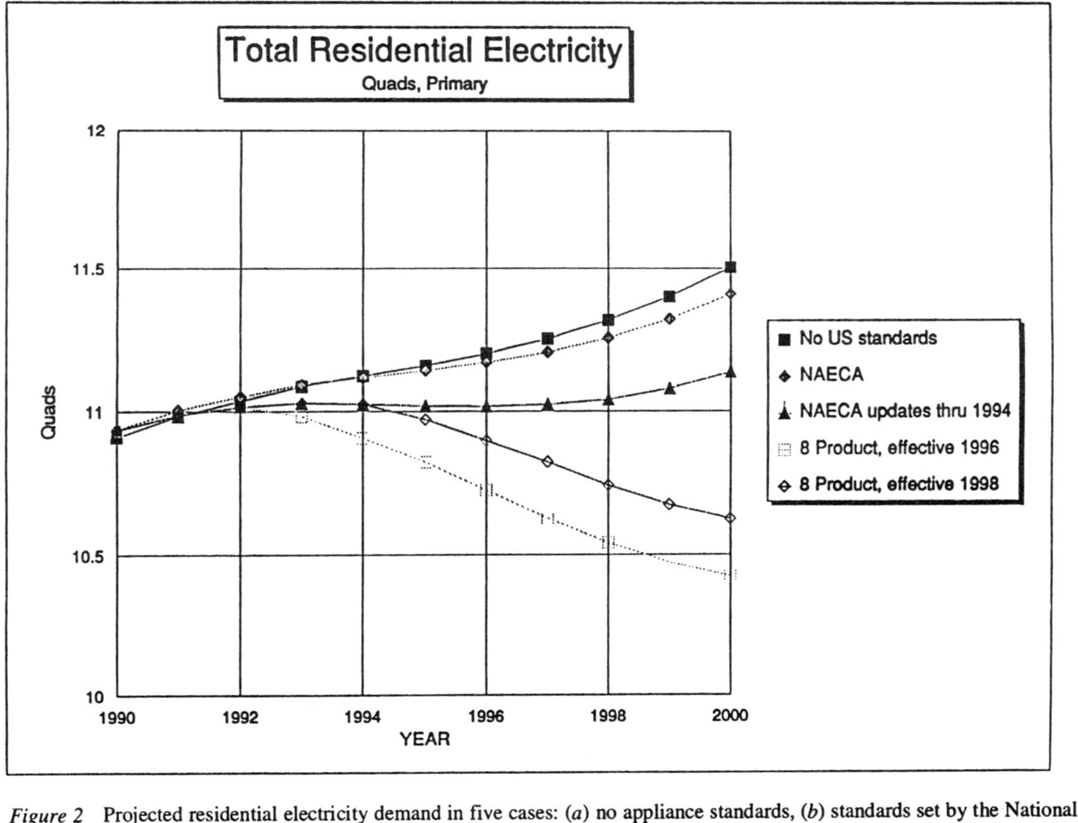

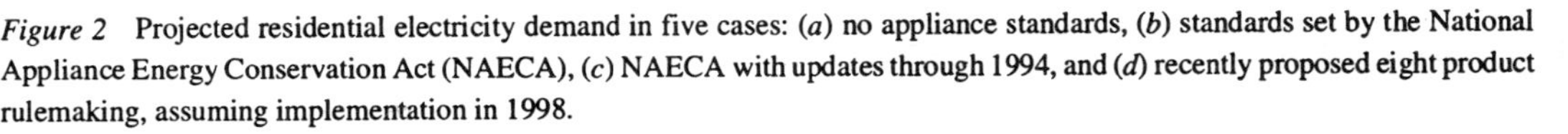

Figure 2 Projected residential electricity demand in five cases: (*a*) no appliance standards, (*b*) standards set by the National Appliance Energy Conservation Act (NAECA), (*c*) NAECA with updates through 1994, and (*d*) recently proposed eight product rulemaking, assuming implementation in 1998.

appliances required by the federal standards pay back their incremental investment in three years or less (22); poststandard price increases are difficult to detect using actual pre- and post-standard price data on refrigerators.

Another potential criticism of the standards is that they may cause the consumer to sacrifice a valued amenity. As noted above, such a criticism is largely ruled out by the standards that have been separately developed for various classes of appliances with differing features. For refrigerators, for example, some of the different classes include freezing compartments that are top-mounted, side-mounted, and bottom-mounted. For purposes of rulemaking, each class of refrigerator, as well as each size, must meet a different standard. (In practice, an equation specifies the size-dependence of the standard level within each refrigerator class.)

We believe that the large cost-effective energy savings resulting from appliance standards—0.15 quadrillion Btu ($1 billion) per year at present, eight times this amount in 15–20 years—provides evidence of market failure related to energy efficiency. No evidence exists that these energy-efficiency gains would have come about in the absence of standards. Although no definitive research on what constitutes the market imperfections has been done, we speculate that a variety of factors indicate imperfections.

We believe that consumers are interested in various appliance features, but not primarily in energy efficiency. The transaction costs of finding energy-efficient models with all the other desired features are likely to be high. The manufacturers know that consumers value attributes other than energy efficiency and undoubtedly also believe that consumers may not be able to evaluate the true energy cost-savings of greater efficiency. In a competitive market, manufacturers are motivated to minimize costs. Thus, given a choice between many models (an expensive proposition) and one efficiency level for each set of features, manufacturers chose the latter. This choice does not mean that consumers do not value or benefit from cost-effective levels of efficiency, but suggests that, before efficiency standards were established, most consumers were poorly informed about these benefits and could only avoid lower-efficiency purchases by investing substantial time and effort.

This market failure is important because the benefits of high-efficiency appliances are substantial. Could more economically-efficient methods have been used to address this failure? Prior to the adoption of efficiency standards, some suggested that information and labeling programs could achieve significant energy efficiency gains. However, such programs were in place for many years before the NAECA standards went into effect and had only limited impact (19).

Each year, the 100 million households in the United States spend about $67 billion for residential appliances (including heating and cooling equipment) to purchase about 55 million major appliances and 12 million heating and air conditioning systems. If the $2–4 million that the government spent annually (over 15 years) on appliance standards had been spent instead on consumer education, this would amount to about $0.02–0.04 per household per year. It is hard to believe that such expenditures would affect many purchase decisions and it is inconceivable that the impact would be in any way comparable to that of the standards.

We believe that appliance standards have had a large beneficial impact on residential energy expenditures. The standards yield a benefit–to–cost ratio to consumers of more than 2.5 (at a real discount rate of 6%), not including the value of environmental externalities. The standards will reduce power plant construction by more than 40 500-MW units over the next 20 years, have not reduced the choice of appliances or the services they provide, and have drawn almost no objection from consumers. Although some may argue that such a policy theoretically ought to have major flaws (or hidden costs), the evidence indicates otherwise. The positive impact of appliance standards is additional evidence for substantial imperfections in the market that have made such large savings possible.

DO ENERGY EFFICIENCY POLICIES LEAD TO INCREASES IN ECONOMIC EFFICIENCY?

The second and third sections of this paper indicate the empirical bases for our belief that systematic underinvestment in energy efficiency occurs in many markets. The literature on this topic is considerable and consists in part of speculation about likely causes for this underinvestment. The literature also includes theoretical work on market factors that would account for this consumer behavior and assessments of factors that would make such behavior economically desirable or undesirable (23–26). However, little empirical data exists as yet to provide quantitative information or predictive value to help answer the question of why consumers underinvest in energy efficiency.[10]

Without detailed knowledge of why consumers fail to adequately invest in energy efficiency, designing policies to overcome the barriers is more difficult. However, as empiricists, we believe that a trial-and-error approach, in which policies are designed on best-available information, implemented (at times on a trial basis), evaluated, modified, and implemented (often on a very extensive basis), can be highly effective. Investing funds in monitoring and assessing the policies is critical, especially for energy-efficiency policies whose impacts

[10]We strongly support the goal of initiating work to gain knowledge about reasons that consumers do not choose energy efficiency measures that are apparently cost-effective for differing products in various markets. Joe Eto (LBL) and Eric Hirst (ORNL) are organizing a DOE-funded workshop among researchers to investigate research projects that can increase understanding of this topic.

are not easily measured. Such monitoring and assessment is not usually done, with the unfortunate result that we know much less than we need to about the effects of policies and programs for promoting energy efficiency.

Despite the inadequate data, experience with policies and programs in the United States is sufficient for drawing some conclusions about whether programs that have increased energy efficiency have also increased economic efficiency. We will use this historical experience as the basis for criteria for public policy on energy efficiency.

Appliance Energy-Efficiency Standards

We have seen above that the appliance standards have had very positive economic impacts: on the US energy economy; on almost all consumers of appliances, heating, and cooling equipment; and on the environment. Impacts on appliance manufacturers are difficult to quantify, but there is no indication of adverse effects. Thus, this policy, pursued very actively in the United States since 1987, has resulted in substantial gains in both energy and economic efficiency.

Utility Demand-Side Management (DSM) Programs

Viewpoints on utility DSM have undergone an interesting evolution in the United States. Until the mid to late 1980s, DSM was generally advocated by environmental and energy-efficiency proponents and opposed by most utilities. Starting in 1989, when the rules for return on investment were changed by state public-utility commissioners in certain key states (mostly on the two coasts), which allowed utilities to gain additional profits through large-scale DSM programs, most utilities in these states became DSM advocates. The main opposition to the utility/environmental alliance came from industry groups (especially in the Northeast) that were concerned that the DSM programs might increase electricity prices and/or create undesirable cross-subsidies. During the period 1989–1993, utility DSM investment grew from about $870 million to about $2.4 billion per year (27). Although the investment in DSM is still growing, the powerful forces for restructuring the industry could also lead to reduced DSM investments in the coming years.

Utility DSM programs in which a large proportion of total costs are paid by customers are unlikely to be compatible with many of the forces that are driving utility restructuring and with some of the proposed approaches to restructuring. Many observers believe that a restructured and competitive utility industry will produce major economic savings for the nation and that such a restructured industry will have little room for utility DSM (28, 29). Individuals with this view are opposed to requiring utilities to implement DSM programs in order to achieve the increased savings they expect from a restructured industry.

We do not share these views because we believe that utility restructuring and the continuation of DSM programs need not be incompatible (30). No matter how the industry is restructured, a regulated entity for transmitting and distributing electricity to consumers will almost certainly be needed. A wires charge, levied against the regulated elements of a restructured industry, can be used to continue utility programs that are in the public interest.[11] But a crucial question is whether these utility DSM programs are in fact cost-effective. If they are not, then we believe they should not be pursued. However, if they increase economic efficiency (and reduce externalities), then considerable effort to determine how to maintain such programs in a restructured industry is justified.

We now turn to the empirical data critical to our discussion. A widely publicized series of papers by Joskow and Marron has led many observers to conclude that utility DSM programs are not likely to be cost-effective in many cases (31, 32). Joskow and Marron evaluated DSM programs in ten utilities, obtaining partial estimates of costs of about \$0.04 kWh^{-1} of saved electricity. They observed that the data were incomplete, and by enumerating all costs that appeared to be omitted, rendered a guess that the actual costs could be two to three times higher (i.e. \$0.08 to \$0.12 per kWh saved). The latter estimates are higher than the average cost of electricity in most regions, leading the authors to conclude that many of the programs encountered "...were either only marginally cost-effective or clearly wasteful."

Joskow and Marron's conceptual framework is very reasonable. Empirically, however, they are on less solid ground. The data were not reported on a program-by-program basis; rather, they were from a group of very varied DSM programs within a utility. Some of the programs were pilot efforts or low-income DSM, which are not intended to be justified solely by economics. In addition, few of the energy savings estimates were based on carefully measured data, and the cost estimates were grossly incomplete; the most significant omission was the portion of the efficiency investment paid by the consumer. They counted only the costs borne by the utility.

Because of these data problems, the data from these ten utilities are difficult, if not impossible, for us to interpret. We find little support for the assumption that omitted costs and/or overestimated savings mean many utility DSM programs are not cost-effective. Unfortunately, this conclusion was widely publicized.

An attempt to overcome the data inadequacies of the Joskow and Marron effort was made in a recent research project by Eto, Vine, Shown, Sonnenblick, and Payne (33). Data related to increasing the energy efficiency of commercial lighting were assembled from 20 utility DSM programs. These programs

[11]How and where such a wires charge could or should be levied, as well as how one could best employ funds obtained in this way, are important issues beyond the scope of this paper.

represent more than $170 million of utility DSM programs. The Eto, et al assessment differed from that of Joskow and Marron in key ways: (*a*) Half the programs provided estimates of energy savings based on measured consumption data (vs none in Joskow and Marron) and the remaining ones provided sufficient information to obtain meaningful estimates; (*b*) All programs were of a similar nature (commercial lighting), although they clearly differed in the specific measures offered or details of implementation; and (*c*) Critically, few unreported costs were associated with the programs.[12]

Eto, et al developed conclusions about a group of DSM programs based on empirical data that are significantly better than those used by Joskow and Marron. Although the data are not perfect, they show no evidence of any systematic bias toward over- or underestimation of energy savings and program costs. The major limitation is that the work has, thus far, addressed only one type of utility DSM activity (although one of the most popular and most widely implemented).

The results of the analysis showed that the weighted average cost of electricity saved was $0.039 kWh^{-1}. These costs varied among the 20 utilities, depending mainly on the avoided cost of electricity for the utility, which the authors believe determined how much the utility wished to invest in lighting efficiency. (High avoided costs led to programs that sought high energy savings; lower avoided costs led to more modest efforts.)

We stress the Eto, et al results because we are impressed with empirical results based on careful analysis of meaningful data. We do not conclude from this study that all or even most utility DSM programs are cost-effective, because that analysis has not yet been done.

We also emphasize these results for another reason. Although one can use anecdotal data to illustrate that programs do not work or that they are successful, a careful analysis across similar programs—in which no effort is made to choose only good or bad ones—is much more useful for deriving an overall sense of whether utility DSM programs are or are not cost-effective and why. We could have discussed countless individual studies or a few of the surveys of utility costs and savings based on estimated data. Such a discussion is in evidence in the debate between Amory Lovins and Paul Joskow (34). We believe that an analytical approach such as Joskow's, but with much better data than Joskow used, is the best way to obtain a balanced judgment of the costs and benefits of utility DSM.

The data so far suggest that utility DSM programs, when evaluated using measured data, can be cost-effective. Our knowledge of energy efficiency technology other than lighting efficiency (performance and costs) and of the

[12]Customer costs were included in all programs. Measurement and evaluation costs, amounting to 3% of the utility costs, were not included.

reluctance of the market to adopt many technologies (e.g. the major energy-efficiency improvements in household appliances and heating and cooling equipment came to the market only through the mandatory appliance standards program) suggests to us that many other DSM programs can be cost-effective, if they are properly designed and implemented.

Although the arguments above are compelling, our support for selected utility DSM is based on a rationale that is more complex. The most valuable utility DSM programs are those that aim to transform markets. Evidence for this argument is provided by an assessment of the costs and benefits of utility DSM programs that was carried out by New England Electric Service (35). In this case, the costs were well-characterized and the energy savings were measured. The programs, well-known to be highly successful, were clearly cost-effective. However, the interesting feature of the analysis was that the majority of customers, who had previously been skeptical, indicated that the utility program convinced them of the value of energy-efficient lighting. The customers indicated that in the future they would install such measures without utility funds. Furthermore, these participants likely talked with nonparticipants, modifying their views of the technology as well. A simple calculation suggested that the program's benefits would be much higher than the costs if skepticism could be overcome in the process of implementing the program. However, these results are based on reported intentions, not actual behavior. We are unaware of any efforts to assess changes in behavior as a result of participation in or exposure to a utility DSM program or its technologies.

Utility DSM programs are desirable public policy instruments for the following reasons. Utility DSM is a means of opening up large markets for better energy technology, which will benefit energy users. Most energy users (utility customers) have limited expertise in evaluating the energy efficiency of devices and (especially) systems. Energy efficiency in devices is often most effectively increased through standards (e.g. the appliance standards). But systems—such as the lighting systems involved in the commercial lighting programs, or the energy-efficient design, construction, and operation of buildings—need to be brought to market in much more subtle ways. Utility programs, often combined with state building energy codes, but designed to encourage energy efficiency beyond code levels, appear to be most effective in bringing certain types of energy systems into broad market segments.

If our analysis is correct, then the best of the DSM programs will fade away long before they have achieved a large market penetration. Once they have proved the benefit of the system, the market can and should do the rest. In this sense, many utility DSM programs can be seen as large-scale demonstration programs.

However, utility DSM should not fade away, but rather evolve. When efficient lighting systems are accepted by the market, programs can focus on new technologies and systems that are at the same stage lighting was in a few years ago. For example, performing measurements and tests of commercial buildings prior to or just following occupancy to assure the proper operation of energy systems, including the development and implementation of the technology to do this more cheaply, has the potential to save large amounts of energy in commercial buildings in a highly cost-effective manner. Extending lighting programs to daylighted zones (i.e. areas within about 15 unobstructed feet of windows) in new and existing buildings can actually lower first costs, as well as operating costs, if done when (downsized) air-conditioning systems are installed. Other next generation utility programs are already under active consideration.

Other Market Transformation Programs

Our discussion of the rationale for utility DSM programs has led us to recognize their value as agents of market transformation. Geller and Nadel (36) provided a comprehensive review of programs that have, in the past decade or so, begun to achieve this goal. They describe not only appliance standards and utility DSM, but also research and development (to assure that new technologies and systems are continuously available), demonstration programs, incentives to commercialize new technologies, education programs, bulk purchases, and building codes.

Two recent innovative types of programs for promoting energy efficiency are particularly notable. The first type of program offers manufacturers a multi-million dollar reward for producing a product that is significantly more efficient than products on the market. The Consortium for Energy Efficiency (CEE) has successfully carried out such projects, the most notable being the Golden Carrot program for efficient refrigerators. A second type of program provides other kinds of incentives for industries to implement energy efficiency. For example, the Energy Star programs of the Environmental Protection Agency in which major companies are effectively persuaded into agreeing to implement cost-effective energy efficiency measures in their own plants and/or operations have proved effective. Some of these programs (e.g. Green Lights) have been funded through utility DSM programs.

Geller and Nadel have described how these programs have resulted in major changes in the energy efficiency of many energy-consuming products: refrigerators, fluorescent ballasts, adjustable-speed drives, personal computers, windows, residential construction practices in the Pacific Northwest, and automobiles and light trucks. The data presented in their paper suggest that most of these programs have produced highly cost-effective energy efficiency measures in the marketplace and that they have had large rather than localized impacts. We believe that the evidence from these programs strongly supports the view that energy efficiency programs, by and large, yield substantial gains in economic efficiency.

SUMMARY AND CONCLUSIONS

The beginning of this paper devoted considerable effort to demonstrating that the market for energy efficiency is far from perfect. For those who do not believe this proposition, understanding why energy efficiency policies have been so successful in cost-effectively reducing energy use is difficult.

One can also turn the argument around: The success of policies designed to promote energy efficiency in bringing about substantial, cost-effective energy savings provides a strong argument for market failures. The available empirical data are compelling.

Although we have many theories about what market factors may retard the acceptance of cost-effective energy efficiency, we have only educated guesses about the cause of the market failures because empirical evidence is lacking. This is a key area for fruitful research.

We reject arguments about whether markets for energy efficiency are effective that are based only on theory and not data. Theory can help guide the gathering and analysis of data, but the questions of greatest interest depend on the analysis of meaningful data. We are generally not impressed by anecdotal "data" or by studies that rely on data of low quality or data that cannot be generalized.

Two essential conditions must be met for energy efficiency policies to be good public policy: (*a*) evidence shows that the market is rejecting energy-efficiency technologies or systems that are cost-effective and; (*b*) careful evaluation shows that the benefits of the policy outweigh its costs. We believe these two conditions are met for a wide variety of policies that have been adopted to promote energy efficiency.

The fulfillment of these conditions, combined with a recognition that careful design, political feasibility, and measurement of the impacts of the policy are essential ingredients for success, leads to our view that many programs and policies to promote energy efficiency are beneficial to society.

ACKNOWLEDGMENTS

We thank the following people for helpful comments on an earlier version of this paper: Albert Nichols (National Economic Research Association); Daniel Kirschner (Environmental Defense Fund); Henry Lee (Harvard University); Howard Gruenspecht (Department of Energy); Howard Geller (American Council for an Energy Efficient Economy); Marilyn Brown (Oak Ridge National Laboratory); and Paul Meagher (Electric Power Research Institute). We also thank Lynn Price (Lawrence Berkeley Laboratory) for her comments and writing assistance. This paper is partially based on *Energy Efficiency, Market Failures, and Government Policy*, Lawrence Berkeley Laboratory and Oak Ridge National Laboratory, March 1994. This work was supported by the Office of Energy Efficiency and Renewable Energy, US Department of Energy under Contract Number DE-AC03-76SF00098.

Literature Cited

1. Koomey JG, Atkinson C, Meier A, McMahon JE, Boghosian S, et al. 1991. *The Potential for Electricity Efficiency Improvements in the U.S. Residential Sector*. Berkeley, CA: Lawrence Berkeley Lab. 238 pp.
2. Fickett AP, Gellings CW, Lovins AB. 1990. Efficient use of electricity. *Sci. Am.* 263(3):64–74
3. Rosenfeld AH, Atkinson C, Koomey J, Meier A, Mowris RJ, Price L. 1993. Conserved energy supply curves for U.S. buildings. *Contemp. Policy Issues* 11(1): 45–68
4. Koomey JG. 1990. *Energy efficiency choices in new office buildings: an investigation of market failures and corrective policies*. PhD thesis. Energy Resour. Group, Univ. Calif., Berkeley. 310 pp.
5. Sutherland R. 1991. Market barriers to energy-efficiency investments. *Energy J.* 12(3):15–34
6. Piette MA, Krause F, Verderber R. 1989. *Technology assessment: energy-efficient commercial lighting. LBL Rep. 27032*, Lawrence Berkeley Lab., Berkeley, Calif.
7. US House of Representatives. 1988. *Hearing before the Subcommittee on Energy and Power of the Committee on Energy and Commerce*, 2nd sess. on H.R. 4158, March 23
8. Geller HS, Miller PM. 1988. *Lighting Ballast Efficiency Standards: Analysis of Electricity and Economic Savings*. Washington, DC: Am. Counc. Energy-Efficient Econ.
9. Meier A, Whittier J. 1983. Consumer discount rates implied by purchases of energy-efficient refrigerators. *Energy* 8 (12):957
10. Harris J, Roturier J, Norford LK, Rabl A. 1988. Technology assessment: electronic office equipment. *LBL Rep. 25558*, Lawrence Berkeley Lab., Berkeley, Calif.
11. US Department of Energy. 1994. *Annual Energy Outlook with Projections to 2010. DOE/EIA-0383(94)*. Washington, DC: Energy Info. Admin.
12. Johnson BJ, Zoi CR. 1992. EPA Energy Star computers: the next generation of office equipment. *Proc. ACEEE Summer Study Energy Effic. Bldgs*. Washington, DC: Am. Counc. Energy-Effic. Econ.
13. US Environmental Protection Agency. 1994. *Purchasing an Energy Star Computer, EPA 430-K-94-006*. Washington, DC: US EPA
14. US Department of Energy. 1988. *Technical Support Document: Energy Conservation Standards for Consumer Products: Refrigerators, Furnaces, and Television Sets. DOE/CE-0239*. Washington, DC: DOE
15. California Energy Commission. 1990. *Advanced Lighting Technologies Application Guidelines: Energy Efficient and Electronic Ballasts*. Sacramento, CA: CEC
16. California Energy Commission. 1983. California's appliance standards: an historical review, analysis, and recommendations. *Staff Rep. P400-83-020*, Calif. Energy Comm., Sacramento, Calif.
17. McMahon JE, Berman D, Chan P, Koomey J, Levine MD, Stoft S. 1990. Impacts of U.S. appliance energy performance standards on consumers, manufacturers, electric utilities, and the environment. *Proc. ACEEE Summer Study Energy Effic. Bldgs*., pp. 7.107–7.116. Washington, DC: Am. Counc. Energy-Effic. Econ.
18. Deleted in proof
19. McMahon JE. 1991. Appliance energy labeling in the U.S.A. *Consum. Policy Rev.* 1(2):87–92
20. Association of Home Appliance Manufacturers (AHAM). 1993. *Major Home Appliance Industry Fact Book*. Chicago: AHAM
21. Association of Home Appliance Manufacturers (AHAM). 1993. *Energy Efficiency and Consumption Trends*. Chicago: AHAM. 13 pp.
22. US Department of Energy. 1989. *Tech-*

nical Support Document: Energy Conservation Standards for Consumer Products: Refrigerators and Furnaces. DOE/CE-0277. Washington, DC: DOE
23. Howarth RB, Andersson B. 1993. Market barriers to energy efficiency. *Energy Econ.* 15(4):262–72
24. Levine MD, Koomey JG, Price L, Geller H, Nadel S. 1995. Electricity end-use efficiency: experience with technologies, markets, and policies throughout the world. *Energy: Int. J.* 20(1):37–61
25. Sanstad A, Howarth R. 1994. 'Normal' markets, market imperfections, and energy efficiency. *Energy Policy* 22(10): 811–18
26. Sutherland R. 1994. Energy efficiency or the efficient use of energy resources *Energy Sourc.* 16:261–72
27. Hirst E. 1994. *Costs and Effects of Electric-Utility DSM Programs: 1989 through 1997. ORNL/CON-392.* Oak Ridge, TN: Oak Ridge Natl. Lab.
28. Blumstein C, Bushnell J. 1994. A guide to the Blue Book: issues in California's electric industry restructuring and reform. *Electr. J.* 7(7):18–29
29. Anderson JA. 1994. *Comments of the Electricity Consumers Resource Council ("ELCON") re: Order Instituting Rulemaking on the Commission's Proposed Policies Governing Restructuring California's Electric Services Industry and Reforming Regulation and Order Instituting Investigation on the Commission's Proposed Policies Governing Restructuring California's Electric Services Industry and Reforming Regulation,* Before the Public Utilities Commission of the State of California, June 23
30. Wiel S. 1994. The impact of power sector restructuring on building energy efficiency: the roles of IRP and DSM. *Proc. ACEEE Summer Study Energy Effic. Bldgs.* Washington, DC: Am. Counc. Energy-Effic. Econ.
31. Joskow PL, Marron DB. 1993. What does a negawatt really cost? Evidence from utility conservation programs. *Energy J.* 13(4):41–74
32. Joskow PL, Marron DB. 1993. What does utility-subsidized energy efficiency really cost? *Science* 260:281
33. Eto J, Vine E, Shown L, Sonnenblick R, Payne C. 1994. *The Cost and Performance of Utility Commercial Lighting Programs.* Berkeley, CA: Lawrence Berkeley Lab. 34967
34. Lovins AB. 1993. The cost of energy efficiency. Letters to the Editor. *Science* 261:969–70
35. Levine MD, Sonnenblick R. 1994. On the assessment of utility demand-side management programs. *Energy Policy* 22(10):848–56
36. Geller H, Nadel S. 1994. Market transformation programs: past results, future directions. *Proc. ACEEE Summer Study Energy Effic. Bldgs,* 10:187–97. Washington, DC/Berkeley, CA: Am. Counc. Energy-Effic. Econ.

DEMAND-SIDE MANAGEMENT: AN NTH-BEST SOLUTION?

Albert L. Nichols
National Economic Research Associates, One Main Street, Cambridge, Massachusetts 02142

CONTENTS

Introduction

Supporters of demand-side management (DSM) programs in the United States have hailed them as costless ways to achieve environmental goals, in particular reductions in emissions of CO_2. Evaluations of programs routinely suggest that they are highly cost-effective, with savings in generation and distribution costs exceeding the costs of the programs; the environmental gains, according to this view, are bonuses. The most enthusiastic supporters have argued that many conservation measures are even better than the proverbial free lunch: "It is a lunch you are paid to eat" (1). In contrast, many economists (including myself) have criticized DSM programs—particularly those relying upon rebates (subsidies) for the purchase of energy-efficient equipment—as economically inefficient (e.g. see 2–4). Some of the key criticisms include:

1. Many studies of the effectiveness of DSM programs fail to measure the energy savings appropriately, relying instead on engineering calculations that overstate the real savings by substantial margins.
2. The analyses used to "prove" that DSM programs are cost-beneficial typically omit some potentially important costs (e.g. amenities such as the quality of light or the disruption caused by retrofits to buildings) and often use far lower discount rates than those employed by households and firms in making a wide array of decisions, not just those associated with energy conservation. DSM rebates do not eliminate these real economic costs, only

obscure them by making a large enough payment that customers are willing to bear them.

3. Even without DSM, utilities already "pay" their customers to conserve in the form of the price charged for electricity. Electricity prices in most of the United States are as high as, if not higher than, marginal costs. Thus, end users already have an adequate (if not excessive) incentive to choose energy-efficient equipment.
4. Where the price of electricity is equal to or greater than its marginal cost, DSM rebate programs inevitably involve a significant transfer from rate-payers as a whole (who fund the programs) to program participants (who collect the rebates).
5. The "market barriers" routinely cited by DSM supporters are far from unique to the electricity market, and virtually none (with the exception of environmental effects) qualify as "market failures." Information is imperfect in energy markets, but the problems do not appear to be more significant than in most other markets.
6. Environmental benefits may justify some otherwise cost-ineffective programs, but those benefits should be evaluated explicitly. Moreover, policies aimed directly at the environmental issues of concern—e.g. emission fees or tradeable emission allowances—almost certainly will provide more cost-effective solutions than will generalized energy-efficiency programs.

Most of these criticisms apply to the use of rebate programs to promote energy efficiency in developing countries, as Anderson's paper suggests. There also are some significant differences, however, that make me less confident than Anderson in rejecting the use of DSM rebates in developing countries. The most important of these differences is that the price of electricity in many developing countries is far less than its marginal cost of production. Although raising the price to cover costs would be the most economically efficient solution, it may not be possible, in which case there may be a legitimate role for the kinds of subsidies typically offered by DSM programs in the United States. Another important difference between the United States and many developing countries lies in the stringency of environmental regulations; to the extent that such regulations are relatively weak in most developing countries, the externalities associated with each unit of energy used may be quite large, so that conservation may yield significant environmental improvements (and net benefits).

The Gap Between Price and Marginal Cost

The figures cited by Anderson suggest that, on average, the marginal cost of electricity is about 2.5 times higher than its price in developing countries. As a result, customers have inadequate incentives to use electricity efficiently;

although conserving one kWh of electricity on average will reduce generation and distribution costs by $0.10, customers will not find it in their interest to do so unless the cost of conservation is less than $0.04. The first-best solution is obvious, at least to economists, and is the one that Anderson lists under his second principle: Raise the price so that it equals (or is at least closer to) marginal cost. Such a policy has much to recommend it. Prices closer to marginal costs will provide broad incentives to use energy more efficiently, in economic as well as physical terms, thus increasing welfare. They also will reduce, if not eliminate, the large public subsidies needed to fund systems currently, thus freeing funds for other purposes or allowing reductions in taxes (and their accompanying deadweight losses) on other portions of the economy.

Unfortunately, it is far from obvious that most developing countries will adopt this recommendation. Whatever social and political forces have led them to hold down prices are not likely to go away, at least not quickly. If electricity prices in a developing country are going to remain well below marginal costs for a significant period of time, the economic case against DSM is much less clear. Standard economic theory tells us that in a second- (nth-)best world, if a good is underpriced and its price cannot be raised, then it may be efficient to subsidize the prices of close substitutes. That is essentially what DSM rebate programs do; they lower the price of energy-efficient equipment. So long as the implicit subsidy for conservation is not greater than the gap between the price of electricity and its marginal cost, it can be justified in terms of economic efficiency, at least in theory. Indeed, if the gap between price and marginal cost is great enough, the savings to the supplier of electricity can be large enough to more than offset the cost of the DSM program itself (including the rebate payments and the loss in revenues from reduced sales); i.e. the DSM program can actually reduce the net subsidy needed from the government.[1]

One important alternative to DSM rebates under these circumstances is the use of efficiency standards and building codes, which Anderson recommends. Certainly the most basic improvements in efficiency—those that make sense in virtually all circumstances when evaluated using the real marginal cost of electricity rather than its price—probably should be achieved through standards and codes rather than equipment rebates. For less basic improvements in energy efficiency, however, there may be enough heterogeneity among end users that increased efficiency only makes sense (even from a broader social perspective) for some users. For example, a given increment in the energy efficiency of a particular type of equipment may yield net benefits (from the perspective of social cost) in applications in which the equipment is heavily utilized, but not

[1]If the overall subsidy to the electricity sector is to be reduced, the free-rider rate (the fraction of those collecting the rebate who would have purchased the efficient equipment in any event) also must be relatively modest.

in those in which use is intermittent. Under such circumstances, a subsidy may be more cost-beneficial than a uniform standard, because it better accommodates differences across applications; with a subsidy, intensive users will be more likely than others to find it in their own interest to choose the efficient equipment. In contrast, a standard imposes the same minimum level of efficiency on all users.

Environmental Regulation

As Anderson suggests, the most cost-effective approach to environmental protection is the use of economic instruments that target the pollution itself, rather than generalized incentives or rules promoting energy conservation. Emission charges or tradeable allowances internalize environmental damages and achieve economic efficiency at several different levels:

1. In choosing levels of emission control for particular technologies, firms control the level at which the marginal cost of control is equal to the emission charge or the value of the emission allowance. As a result, marginal control costs are equalized across sources.
2. Because the damages that remain after control are reflected in the costs of alternative technologies, firms have the appropriate incentive to choose technologies with the lowest social costs.
3. Because the damages that remain after control and the selection of technologies are reflected in the firm's production costs and hence in its selling prices, consumers face the appropriate incentives in choosing among competing goods.

Thus, if appropriate emission charges or tradeable allowances are in place, there is no need for separate conservation policies aimed at environmental considerations.

The situation with emission standards (which are far more widely used than charges or allowances) is less clear. Emission standards typically focus on emissions per unit of input (e.g. grams per joule of heat input) or output (e.g. grams per kWh). The implicit allocation of pollution thus rises with output levels. As a result, firms do not have fully appropriate incentives to consider reduced output as a "control" method; the value of residual damages (those remaining after the application of required control technologies) is not reflected in the price of the product (electricity), and end users do not have fully appropriate incentives to conserve. Standards also may fail to provide fully appropriate incentives to firms in choosing among alternative production technologies, because often those technologies are subject to different standards. For example, standards often are more stringent for new plants than for existing ones. Similarly, standards often allow higher emissions from some fuels (e.g. coal) than from others (e.g. gas).

In the United States and most other developed countries, stringent emission standards for most pollutants—particularly for new plants—mean that the residual damages associated with production are small enough that the inefficiency that results from not fully reflecting those residual damages in product prices generally is quite modest.[2] The situation appears to be very different in many developing countries, however; environmental regulations often are minimal, so that the externalities associated with the production of electricity and other goods may be a large fraction of the private cost (especially if prices are below private marginal costs of production). Thus, the rationale for conservation based on environmental externalities may be substantially stronger than in the United States and other developed countries.

As with gaps between the price of electricity and its marginal cost of production, DSM clearly is not the first-best solution to environmental problems associated with the production of electricity or other types of energy. The first-best solutions would be emission charges or tradeable allowances that efficiently internalized environmental effects. Such solutions may not be politically feasible in many circumstances, however; governments unwilling to raise prices even enough to cover out-of-pocket costs and avoid large subsidies are not likely to find it easy to raise prices even higher to deal with environmental fees. Administrative problems also may be severe; governments that find it difficult to stop extensive theft of electricity are likely to have an even harder time monitoring emissions and collecting fees or enforcing allowance requirements. Similar problems may make it hard to enforce environmental standards as well. In contrast, rebates or efficiency standards for new equipment may be much easier to implement, both politically and administratively. Under such circumstances, DSM rebates or efficiency standards may be substantially more cost-effective methods of environmental protection than they typically are in the United States and other developed countries.

Concluding Remarks

Anderson lays out a series of generally sensible economic principles for guiding energy policy in developing countries. Along with most economists who read the paper, I suspect, I applaud his emphasis on privatizing the generation and distribution of electricity and on increasing its price to cover the costs of production. I also share his conviction that environmental considerations generally can be addressed most cost-effectively through emission charges, tradeable allowances, or environmental standards, rather than through programs to

[2]As numerous studies have shown, the far more serious inefficiencies from standards result from the fact that they do not account for differences across sources in the marginal costs of control. Emission charges or tradeable allowances can yield significant savings by giving firms the flexibility to reallocate control efforts to sources with the lowest control costs.

promote lower energy consumption. The vast bulk of environmental gains in the United States and other developed countries clearly have come from such measures—which reduce environmental damage per unit of output—rather than from reductions in energy use.

Relative to these first-best policies, DSM programs that offer rebates for the purchase of energy-efficient equipment are clearly inferior. In an nth-best world, however, where for some time to come the price of electricity in many developing countries is likely to remain far below its marginal cost and environmental regulations are minimal, DSM programs using rebates or mandatory efficiency standards may play useful, albeit limited, roles.

Literature Cited

1. Fickett AP, Gellings CW, Lovins AB. 1990. Efficient use of electricity. *Sci. Am.* 263(3):65–74
2. Nichols AL. 1994. Demand-side management: overcoming market barriers or obscuring real costs? *Energy Policy* 22: 657–65
3. Joskow PL, Marron DB. 1992. What does a negawatt really cost? *Energy J.* 13:1–34
4. Ruff LE. 1988. Least-cost planning and demand-side management: six common fallacies and one simple truth. *Public Util. Fortnightly.* 121(9):19–26

ROUNDTABLE ON ENERGY EFFICIENCY AND THE ECONOMISTS—AN ASSESSMENT[1]

Dennis Anderson
The World Bank, 181 H Street NW, Washington, DC 20433

CONTENTS

INTRODUCTION

Anyone trying to follow the debate between conservationists and economists (and many others in the industry) on energy efficiency might understandably decide there is little possibility for reconciling their views on policies. Views range from high levels of government intervention, frequently inspired by conservationists, to essentially noninterference with market development other than to tax or regulate pollution. The six papers in this roundtable are not this extreme but clearly reflect a broad range of views, and differences are bound to remain. However, areas of agreement can also be found. If we use the eight elements discussed in my paper as a checklist, we find that the main disagreements concern the respective roles of prices and nonprice measures in energy policy. The main possibilities for progress lie with conservationists, on the one hand, recognizing the importance of prices in achieving allocative efficiency, and not overreacting to known sources of market inefficiency; and with econo-

[1]Based on the papers received from Joskow; Nichols; Cicchetti; Cavanagh; and Levine, Koomey, McMahon, Sanstad, and Hirst

mists, on the other, seeking to translate work on the frontier of their subject—on risk and information theory, on regulation, on the roles of law and government, and on property rights—into operational policies. If this can be done, we shall improve economic efficiency measured from the starting points of either the conservationist or the neoclassical economist.

ROLE OF PRICES

Cavanagh notes an "American success story...[of] the past 20 years; the US economy has cut its average energy consumption needs per dollar of product by more than one fourth," which he attributes mostly to the setting of efficiency standards and demand-side activism. But prices also played a large role in cutting consumption. Eighteen years ago, Socolow (1) remarked that "the most significant pressure toward lower energy intensities is [being] exerted by rising energy prices," and the evidence has surely borne out this view. In a masterly study of how the US economy responded to the oil price shocks of 1972–1987, Jorgenson and Wilcoxen (2) concluded that "price-induced energy conservation" acted to stabilize CO_2 emissions over the period. In a recent study, Dargay and Gately (3) showed that the price increases in the first half of the period have had a lasting effect, mainly because the changes in technologies and practices they brought about were eventually incorporated into the capital stock and its management.

Oil-importing countries certainly do not want to repeat the experience of the oil price shocks, and the aims of the studies cited were not to suggest that such shocks are desirable, but to see what could be learned about price effects on economic activity. We also know from econometric studies of short- and long-run price elasticities that gradual changes in prices over time will bring about changes in energy demand, and that differences in energy prices between countries help to explain much of the observed differences in their energy demand and supply patterns, such as the large differences between North America and Europe in vehicle fuel efficiency in the 1970s and before. The evidence is conclusive and consistently points to the importance of prices from the perspectives of finance, economic efficiency, energy efficiency and the environment [4].

Nevertheless, few economists would dispute that nonprice measures have an important role in improving economic efficiency in energy production and use, especially if the net benefits of pollution abatement are included in the definition of economic efficiency. All markets in open societies and in all sectors of economic activity—finance, manufacturing, food, health, education and services—function within a framework of laws, legislation, standards and public and private information services designed to improve the clarity and integrity of economic transactions and in so doing improve economic efficiency. Energy is neither an exception nor, certainly, the most important case; but we do know that when consumers make decisions about energy production and use, an elaborate framework of regulations, laws, and institutions stands behind them, as in most other economic activities. In this sense, pure market transactions, in which prices, along with other economic variables such as income and wealth are the sole influences, are rare. But if such frameworks are "the ghost in the machine," it is equally true that good pricing policies are fundamental for efficiency.

Making these un-novel points is necessary because DSM activists often leap from the legitimate conclusion that market inefficiencies exist to the corollary that government interventions are needed to achieve efficiency. Few economists would take a laissez-faire position in this area and would be hard put to defend it if they did. However, economic policy-making over the past 50 years, especially in developing regions, has shown that interventions to address market failures may often decrease market—and, by implication, energy—efficiency too and, to improve it, they need to be thoroughly thought through and based on evidence. In the present case, the economist's concern is that interventions, however well-meant, may introduce yet greater sources of both economic and energy inefficiency; for example, by politicizing the regulatory process and introducing arbitrary subsidies (Joskow). The debate is not therefore over the desirability of achieving economic efficiency in energy production and use but how best to achieve it. The task ahead will be to find a mix of price and nonprice measures that follow the best traditions of public policy-making. The papers presented in this roundtable offer constructive suggestions in this regard.

IMPROVING MARKET EFFICIENCY

Economists will agree with Nichols and Joskow that we first need to understand better the sources and nature of market deficiencies and how they respond to different policy instruments. Recent articles edited by Huntington, Schipper, and Sanstad (5), as well as Schipper and Meyers' book (20), provide the basis of a useful synthesis, as do papers by Jaffe and Stavins (6,7). A well-known principle of policy making is to focus the policy directly on the problem. Thus, if pollution is the problem, taxing or regulating it directly is less costly and more effective than an indirect approach such as restricting the output of the polluting activity. For example, lead emissions from vehicles and particulate matter emissions from coal-fired power stations have been successfully addressed in the industrial countries less by energy conservation programs (their contributions were not negligible but nevertheless small) than by policies respectively targeted on the use of lead as an octane enhancer and on the emissions of particulates from power stations. In fact, these types of pollution were reduced nearly 100% even as vehicle fuel consumption and coal use for power generation increased.

The six authors in this roundtable present six targeted policies on which there is some agreement. There is, however, one area in which views differ substantially, which is the role of the electric utilities in delivering energy efficiency services in the consumer-markets.

Better Metering and Itemized Billing Systems

Cicchetti reminds us of the crudeness of today's electricity metering and billing systems, which have remained substantially unchanged for decades. He argues, as have others, that energy services need to be sold to consumers as they use hot water, cooling systems, various other appliances, and lighting. Expenditures on the equipment required to provide the services, such as lamps, light bulbs, refrigerators, and heaters are identifiable. But expenditures on the electricity the equipment uses are not identifiable: "...[C]ustomers pay for electricity some weeks after they consume it, and they do not know how their use of various electricity-using appliances affects their kWh consumption and monthly electricity bill. ...Utility bills and the price/use communication need to be fixed. Time-of-use pricing, load management, and two-way communication between the customer and utility need to be implemented. All of these are forms of DSM and need to be thought of as such." Developments in metering and communications technologies that permit itemized billing, including not only billing by time of use, but by the main appliance categories, would indeed represent significant progress toward the achievement of price and energy efficiency. This area of research and development holds much promise (see 23–26).[2]

External Costs

Cavanagh is right to note that economists favor interventions to address the problem of external cost. The literature on the subject, which begins with Pigou's early papers on externalities and spans almost a century, is founded on the premise that external costs, if unaddressed, are a source of economic inefficiency. Moreover, economists are not wedded to the economic instruments of policy, such as taxing pollution, though price or "market-based" incentives are generally preferred, and the case is well summed up by Nichols. In a seminal paper, Coase showed 35 years ago that when property rights are well-defined, negotiated solutions between the polluting and polluted parties are feasible without public intervention (8,9). Is an example of this to be found

[2]I am grateful to David Eskanazi of the Electric Power Research Institute for drawing these materials to my notice and also to an excellent series of technical papers from several countries presented at an IERE Workshop on "New Issues in Metering and Communication"-Clamart (France), September 19–20.

in today's discussions of participatory approaches to development and the environment, such as natural resource management and the resettlement of people displaced by hydro schemes (10, 11)? In a broad sense, these approaches are about establishing property rights and creating a negotiating position for economically disadvantaged parties. Economists must agree that environmentalists deserve credit for these approaches, which improve economic efficiency by resolving property rights disputes (this is rarely a question of social niceties, but an unavoidably disquieting political process in situations where social injustices would otherwise happen) and by drawing on the considerable knowledge of local people about the management of local natural resources.

Over the past two decades, economists have also come to recognize the role of the regulatory approach to pollution control and, as Joskow notes, see yet another promising new approach: establishing tradable emissions permits within an overall emissions cap for some pollutants. Bohm and Russell (12) have summed up the economist's position well: "All the alternatives [for reducing pollution directly] are promising in some situations....If the classic case for the absolute superiority of effluent charges is flawed by the simplicity of the necessary assumptions, the arguments for the superiority of rigid forms of regulation suffer equally from unstated assumptions and static views of the world. There is no substitute for careful analysis of the available alternatives in the specific policy context at issue....That said, we are still tempted to stress the advantage of economic incentive systems,...at least as a complement to the regulatory approach." Noneconomists have profoundly influenced the development of the economic theory and practice of environmental policy.

Removing Unnecessary Subsidies

All of this roundtable's authors agree on the economic and environmental desirability of eliminating unnecessary subsidies. As several have noted, electric power subsidies pose the greatest problem in the developing countries where a removal of subsidies would likely reduce waste in electricity consumption by as much as 30%, relative to trend levels, and greatly improve their fiscal situations. In addition, Nichols and Joskow believe as I do that nonprice measures would lead to further economic gains in the form of greater managerial efficiency and better maintenance of equipment because the management of developing countries' electricity utilities is often seriously undermined not only by cash flow problems but also by high levels of government control.[3]

[3]Rob Socolow has rightly drawn to my attention a common source of misunderstanding on the importance of removing subsidies, commenting that the historical "impregnability of...subsidies is one of the reasons environmentalists are unwilling to spend their energies attacking them, even though there would be a more durable alliance...[between environmentalists and such institutions as national Treasuries, the IMF and the World Bank] if they were to do so. Their silence is usually viewed as disagreement, when it is often dismay." Is there a way forward here?

Nichols makes the valid point that price reforms may need to be undertaken gradually, if only to avoid the imposition of shocks. Price elasticities are numerically much lower in the short run than in the long run (3,4), meaning that the real economic gains in allocative efficiency are long-term not short-term, and a good case can be made for phasing in reforms gradually (13). An exception may be when an economy is in a fiscal and monetary crisis partly as a result of deficit financing of state industries, as happened in much of Latin America in the 1980s, and is happening in much of the former Soviet Union (FSU) today. Shock treatments are sometimes unavoidable, and when prices in the energy sector are severely deformed and the economy is in fiscal distress, such treatments are a good candidate for rapid reform.

Dismantling Unnecessary Government Interventions

Although none of us developed the point, introducing nonprice measures in developing countries also requires care. Many countries are emerging from an era of fiscal crises, inflation, and slow economic growth brought about by extraordinarily high levels of government control.

For this reason, attempts to introduce integrated resource planning and exert new controls through the regulatory process are open to question. In the 1970s, attempts were made to introduce integrated resource planning to rural development in what became known as "integrated rural development" (14).[4] The idea was that, instead of concentrating on one or two inputs to the rural economy, such as research, extension, and rural credit services to improve agricultural productivity, an optimal mix of all relevant inputs would be provided simultaneously: rural roads, marketing facilities for crops and livestock, health clinics, electricity, telecommunications, water and sanitation, primary and secondary education, vocational training, and other inputs. The programs focused on small farmers, who achieve higher crop yields than large land-holders. The landless were given direct support through, for example, vocational training and credit and extension services for small enterprises, which are capable of improving employment and earnings opportunities. Financing was provided as needed in support of land reform and land settlement. The economic costs and benefits were estimated using methods developed by leading economists in which shadow prices reflected actual resource costs (15, 16, 17), and a "reality" check on financial returns was provided by estimating the programs' impact on farm and business budgets using actual prices. The logic was impeccable, the programs were well thought out, and were well supported by the research of hundreds of scholars. Synergies between all such inputs were and still are demonstrable: The programs were morally defensible in that

[4]Recalled in remarks by my colleague Richard Stern at a World Bank–sponsored conference on energy efficiency, in September 1994.

they would reduce poverty, and the *ex ante* estimates of the economic benefits to the rural population and to the national economies were consistently high.

Yet the programs were generally not a success, for reasons which are worth noting by all those interested in improving energy efficiency and the environment in developing countries using methods of integrated resource planning. First, the enabling conditions worked against the programs, principally in the form of agricultural pricing, marketing, and exchange rate policies that undermined the financial rewards to the farmers both of their own exertions and of the programs providing supporting services. Poor enabling conditions also inhibited growth and the development of markets for the farm and nonfarm products of rural areas, further reducing the rewards to rural investments. Second, fiscal crises led many governments to starve the supporting services of cash, and many field agents of their wages, and the services often deteriorated badly. Third, an inheritance of central planning and bureaucratic management had greatly weakened the capacity of local institutions to implement the programs.

In most developing countries, by comparison with the industrial countries, the electricity sector is also emerging from a legacy of state control and a weakness of institutions capable of providing independent, "arms'-length" regulation. Both the industry and its regulators are especially vulnerable to the (generally unstated) costs of new attempts to exert control over investment decisions and, for that matter, over the decisions of consumers. The fiscal situations of governments too have often been undermined by subsidies to the electricity industry. Therefore, an integrated or holistic approach to resource allocation in the industry is not wrong in principle, but an approach cannot be described as integrated if it neglects the economic setting, the enabling conditions for investment, the role of prices, and the principles of independent regulation—a mistake Turvey never made a generation ago (21, 22).

Subsidies for Innovation

The case for public financial support for research and development (R&D) and the demonstration of new technologies is generally accepted. This issue is revisited in a paper in this volume by Baumol (18) who concludes that any subsidies need to be closely related to an analysis of technological developments and cost curves. When cost curves are steeply declining, and risks, indivisibilities, and other factors represent genuine barriers to entry, a good case for subsidy can be made on grounds of efficiency, less of a case when cost curves are flat. But subsidies for a particular technology should not become *permanent features* of policy; they should meet a market test and taper off as costs decline and markets develop. (My emphasis.)

Is this another area where economists and noneconomists are in fact on common ground, but using different languages? Baumol has presented us with a distinguished and carefully considered economic essay on the implications for policy of indivisibilities, high start-up costs, and declining cost curves. In a separate comment on the papers in this roundtable, Socolow notes:[5]

> ...the first generation of...energy efficiency technologies (including some social technologies like labeling) are actually R&D. There is a learning curve for energy efficiency investments, and few of the first generation efforts yield the expected savings, even though savings will be achieved by later generations....From this point of view, the recipients of DSM investments are guinea pigs, and for *that* reason should be subsidized. R&D has at least two dimensions. New components may simply not work as well as expected....And the performance of new components, provided it is carefully monitored after deployment, may reveal opportunities for further system enhancement...

Levine and his coauthors similarly conclude that "the best of the DSM programs [and the case for subsidy] will fade away long before they have achieved a large market penetration. *Once they have proved the benefit of the system*, the market can and should do the rest. In this sense, many utility DSM programs can be seen as large-scale demonstration programs." (My emphasis.)

Standards and Codes for Equipment and Buildings

The argument for setting standards and codes for the energy efficiency of appliances and buildings varies with the circumstances of each case. Mark Levine and colleagues present interesting examples of how standards have improved energy efficiency in the United States. Standards can be used in two ways. They can function as a labeling device to inform consumers about the performance of equipment or materials, allowing them to choose among alternatives which may have different standards and costs. Standards can also be used to "force" the pace of technical change.

Standards used as a labeling device are not controversial. This practice provides clarity to market transactions, and also integrity, because the suppliers can be held to account if the declared standards are not met. In both respects, economic efficiency increases. Furthermore, if itemized billing is implemented as discussed above, consumers will be more readily able to check the performance of equipment against its rating, another example of price and nonprice measures serving complementary functions.

However, when standards and codes are used to force the pace of technical change, more basic questions arise. First, what is the evidence of market inefficiency or that economically beneficial results are possible from the device or practice in question, but are somehow obstructed by market failures? Second, what is the precise nature of the market failure, and is a standard or code necessary to support the development and application of a technology or

[5]Personal communication.

management practice? Or is a market-based incentive better? And third, which institutions are best placed to implement the policy?

The evidence used in support of mandatory standards for energy efficiency rightly continues to be questioned, in the present roundtable and elsewhere. For electricity use, the argument for standards is based mainly on engineering estimates of efficiency gains. Such estimates are an important source of information, but in a thoughtful review of methods and evidence, Fels and Keating (19) found that "engineering-based estimates of energy savings are often inaccurate, and in general overstate the savings achieved," by factors of 2 or more in several cases. Lee Schipper commented (personal communication) that this may partly be due to the difficulties of identifying the effects of efficiency improvements in particular appliances on the overall energy consumption of households (many of the studies reviewed by Fels and Keating used metered household kWh consumption). Another reason for the overestimation, however, may be that the new materials and appliances, by reducing the unit costs of energy use, lead people to use more of the services provided by the energy, such as heat, light, and air-conditioning. In other words, energy efficiency may lead to benefits in the form of *additional* consumption; it is not merely about energy savings. The principles of economic cost-benefit analysis allow for such possibilities by identifying two components in the benefit stream: the cost-savings, relative to the costs of using the old appliance, and the benefits of additional consumption brought about by cost reductions.[6]

The answer to the second question is that economists will continue to favor market-based incentives over standards as a way to encourage technical change on the grounds that incentives are more cost-effective and less likely to stifle innovation through overregulation. Economists will also favor remedies tailored to address specific inefficiencies over remedies that appeal vaguely to market imperfections and consumer irrationalities. Joskow, Levine and colleagues, and Nichols agree that even when energy inefficiencies are identified, we are generally not well-informed about *why* they exist. For this reason, tailoring the policies to the problem is difficult, and, without critical inquiry, creating arbitrary policies may be counterproductive at worst and futile at best.

Institutions

The third question, regarding which institutions are best placed to deliver energy-efficiency policies, creates controversy mainly in the case of electricity,

[6]Cost-benefit methods thus provide the better guide to policy because they recognize the benefits as well as the costs of energy use. They also prompt the question, which Levine and colleagues raised at the beginning of their paper, of whether it is more appropriate to focus on the economic efficiency gap (when it exists) as opposed to the energy efficiency gap. See also Jaffe and Stavins (6).

with respect to utility-managed DSM programs. As Cicchetti's paper shows, thinking about DSM is evolving, and DSM clearly means different things to different people, a source of much confusion. In the past, DSM programs have required the utilities to market electrical equipment and appliances, with support from subsidies or give-away programs, financed out of general electricity revenues. Cicchetti comments that such programs "are not, and should never have been viewed as, give-away programs." However, the arbitrariness of some subsidies, as well as the ferocity of the lobbying to introduce them, has nevertheless raised questions about the propriety and efficiency of the regulatory process.

The controversies are clear in all the papers presented. Can they be resolved? Three points are worth recalling. The first is to avoid arbitrary subsidies. Many subsidies have been rationalized because they lower energy use and, at the same time, lower costs (27). This argument by itself is not a good reason for subsidy; no one, for example (at least to my knowledge), argues for subsidizing winter woollies on the grounds that they enable people in northern climes to turn down the heating and save energy. As discussed earlier, a good case for subsidies or tax credits can be made in support of innovation. However, the subsidies need to be justified by the start-up costs and economic merits of the new approach, the real impediments to market entry (as opposed to vague blame on market failure), and the shapes of the cost curves.

The second point worth recalling is that other industries are often better placed than the electric utilities to deliver innovative products and services. Appliance manufacturers, for example, can better deliver appliances, builders and architects efficient lighting and air-conditioning. A good case can be made for applying tax credits and subsidies to the products and services of these industries rather than the electricity-supply industry.

The third point, made by Cichetti, is that few developments would do more to help people's decisions on electricity use than itemized billing using new metering and communications technologies. We need to review technical options and costs, but the hope is that developments in this area will be encouraged by the electricity industry and by public policy. In all manner of goods and services, people are bombarded with information, sometimes accurate, sometimes slanted, and sometimes false, and have learned to be skeptical of, or to discount, the claims of the purveyors, whether public, private, or other. And isn't such skepticism a perfectly rational response to a long-known source of economic inefficiency? Gresham's law of four centuries—bad money drives out good—remains intact it seems and, as Akerlof's paper on the market for "lemons" showed 25 years ago, is not confined to financial markets (28); there is no reason for thinking that markets in energy efficiency will be an historical exception . What we need is information consumers can trust, conveyed in an uncomplicated way. If consumers were able to see in their electricity bills the costs of the kWh they use for various purposes during times of peak and off-peak demand—for heating, air conditioning, lighting and appliance circuits, and so forth—they would have credible and convincing evidence upon which to act.

Literature Cited

1. Socolow RH. 1977. The coming age of conservation. *Annu. Rev. Energy* 2:239–89
2. Jorgenson DW, Wilcoxen PJ. 1993. Energy prices, productivity and economic growth. *Annu. Rev. Energy Environ.* 18: 343–95
3. Dargay J, Gately D. 1995. The response of world energy and oil demand to income growth and changes in oil prices. *Annu. Rev. Energy Environ.* 20: In press
4. Bates RW. 1993. The impact of energy policy on energy and the environment in developing countries. *Annu. Rev. Energy Environ.* 18:479–506
5. Huntington, Schipper L, Sanstad AH. 1994. Editors introduction. *Energy Policy: Special Issue on Markets for Energy Efficiency* 22(10):657–65
6. Jaffe AB, Stavins RN. 1994. The energy efficiency gap: What does it mean? *Energy Policy* 22(10):804–11
7. Jaffe AB, Stavins RN. 1994. The energy paradox and the diffusion of conservation technology. *Resour. Energy Econ.* 16(2):91–122
8. Coase RH. 1960. The problem of social cost. *J. Law Econ.* III, Oct:1–44
9. Turvey R. 1963. On divergencies between social cost and private cost. *Economica XXX*:309–13
10. Gutman PS. 1994. Involuntary resettlement in hydropower projects. *Annu. Rev. Energy Environ.* 19:189–210
11. Cernea MM. 1988. Involuntary resettlement in development projects. *Tech. Pap 80.* Washington, DC: World Bank
12. Bohm P, Russell CS. 1985. Comparative analysis of alternative policy investments. In *Handbook of Natural Resource and Energy Economics,* ed. AV Kneese, JL Sweeney, 1:395–460. Amsterdam: North Holland
13. Cavendish W, Anderson D. 1994. Efficiency and substitution in pollution abatement. *Oxford Econ. Papers* 46: 774–99
14. World Bank. 1975. *Rural development: Sector policy paper.* Washington, DC: World Bank
15. Little I, Mirrlees J. 1974. *Project Appraisal and Planning for Developing Countries.* New York: Basic
16. Squire L, van der Tak HG. 1975. *Economic Analysis of Projects.* Baltimore: Johns Hopkins Univ. Press
17. Ray A. 1984. *Cost-Benefit Analysis: Issues and Methodologies.* Baltimore/London: Johns Hopkins Univ. Press
18. Baumol WJ. 1995. Environmental industries with substantial startup costs as contributors to trade competitiveness. *Annu. Rev. Energy Environ.* 20: In press
19. Fels MF, Keating, KM. 1993. Measurement of energy savings from demand-side management programs in US electric utilities. *Annu. Rev. Energy Environ.* 18:57–88
20. Schipper L, Meyers S. 1992. *Energy Efficiency and Human Activity: Past Trends, Future Prospects.* Cambridge, UK: Cambridge Univ. Press
21. Turvey R. 1968. *Optimal Pricing and Investment in Electricity Supply.* London: Allen & Unwin
22. Turvey R, Anderson D. 1977. *Electricity Economics: Essays and Case Studies.* Baltimore: Johns Hopkins Univ. Press
23. Timberman M. 1992. Advanced metering: benefits on both sides. *Electr. Power Res. Inst. J.* Apr./May:18–25
24. Douglas J. 1990 Reaching out with two-way communications. *Electr. Power Res. Inst. J.* Sept:5–13
25. Moore, Taylor. 1993 Framework for utility data highways. *Electr. Power Res. Inst. J.* June:34–42

26. Lamesh, Ranjit. 1994. The bad side of demand-side management. *Finance/Private Sector Dev. Ser. Note 3.* Washington, DC: World Bank. See also the responses and counter responses to this note (in the same series) by Watson, Seamans, Lamesh (forthcoming)
27. Akerlof G. 1970. The market for "lemons": quality uncertainty and the market mechanism. *Q. J. Econ.* 84:448–500

SUBJECT INDEX

CUMULATIVE INDEXES

CONTRIBUTING AUTHORS, VOLUMES 1–20

G

H

J

K

L

M

CHAPTER TITLES, VOLUMES 1–20

ECONOMICS

POLICY

ANNUAL REVIEWS

a nonprofit scientific publisher
4139 El Camino Way
P.O. Box 10139
Palo Alto, CA 94303-0139 • USA

ORDER FORM

ORDER TOLL FREE
1.800.523.8635
from USA and Canada

Fax: 1.415.855.9815

nnual Reviews publications may be ordered directly from our office; through stockists, ooksellers and subscription agents, worldwide; and through participating professional ocieties. **Prices are subject to change without notice. We do not ship on approval.**

Individuals: Prepayment required on new accounts. in US dollars, checks drawn on a US bank.

Institutional Buyers: Include purchase order. Calif. Corp. #161041 • ARI Fed. I.D. #94-1156476

Students / Recent Graduates: $10.00 discount from retail price, per volume. ***Requirements:*** **1.** be a degree candidate at, or a graduate within the past three years from, an accredited institution; **2.** present proof of status (photocopy of your student I.D. or proof of date of graduation); **3.** Order direct from Annual Reviews; **4.** prepay. This discount **does not** apply to standing orders, *Index on Diskette,* Special Publications, ARPR, or institutional buyers.

Professional Society Members: Many Societies offer *Annual Reviews* to members at reduced rates. Check with your society or contact our office for a list of participating societies.

California orders add applicable sales tax. • **Canadian orders** add 7% GST. Registration #R 121 449-029.

- **Postage paid** by Annual Reviews (4th class bookrate/surface mail). UPS ground service is available at $2.00 extra per book within the contiguous 48 states only. UPS air service or US airmail is available to any location at actual cost. UPS requires a street address. P.O. Box, APO, FPO, not acceptable.
- **Standing Orders:** Set up a standing order and the new volume in series is sent automatically each year upon publication. Each year you can save 10% by prepayment of prerelease invoices sent 90 days prior to the publication date. Cancellation may be made at any time.
- **Prepublication Orders:** Advance orders may be placed for any volume and will be charged to your account upon receipt. Volumes not yet published will be shipped during month of publication indicated.

NOTE: For copies of individual articles from any *Annual Review,* or copies of any article cited in an *Annual Review,* call **Annual Reviews Preprints and Reprints (ARPR)** toll free 1-800-347-8007 (fax toll free 1-800-347-8008) from the USA or Canada. From elsewhere call 1-415-259-5017.

ANNUAL REVIEWS SERIES *Volumes not listed are no longer in print*		**Prices, postpaid, per volume. USA/other countries**	Regular Order Please send Volume(s):	Standing Order Begin with Volume:
❑ *Annual Review of*	**ANTHROPOLOGY**			
Vols. 1-20	(1972-91)	$41 / $46		
Vols. 21-22	(1992-93)	$44 / $49		
Vol. 23-24	(1994 and Oct. 1995)	$47 / $52	Vol(s). ______	Vol. ______
❑ *Annual Review of*	**ASTRONOMY AND ASTROPHYSICS**			
Vols. 1, 5-14, 16-29	(1963, 67-76, 78-91)	$53 / $58		
Vols. 30-31	(1992-93)	$57 / $62		
Vol. 32-33	(1994 and Sept. 1995)	$60 / $65	Vol(s). ______	Vol. ______
❑ *Annual Review of*	**BIOCHEMISTRY**			
Vols. 31-34, 36-60	(1962-65,67-91)	$41 / $47		
Vols. 61-62	(1992-93)	$46 / $52		
Vol. 63-64	(1994 and July 1995)	$49 / $55	Vol(s). ______	Vol. ______
❑ *Annual Review of*	**BIOPHYSICS AND BIOMOLECULAR STRUCTURE**			
Vols. 1-20	(1972-91)	$55 / $60		
Vols. 21-22	(1992-93)	$59 / $64		
Vol. 23-24	(1994 and June 1995)	$62 / $67	Vol(s). ______	Vol. ______